A TREATISE ON
ELECTRICITY AND MAGNETISM

VOL. 1 - SECOND EDITION

James Clerk Maxwell
(June 13, 1831 – November 5, 1879)

JAMES CLERK MAXWELL

A Treatise on Electricity and Magnetism
- James Clerk Maxwell

IN TWO VOLUMES AND IN THREE EDITIONS

VOLUME 1
- First Edition (1873)
- Second Edition (1881)
- Third Edition (1892)

VOLUME 2
- First Edition (1873)
- Second Edition (1881)
- Third Edition (1892)

A TREATISE ON
ELECTRICITY AND MAGNETISM

VOL. 1 - SECOND EDITION

BY

JAMES CLERK MAXWELL

(June 13, 1831 – November 5, 1879)

James Clerk Maxwell FRSE FRS) was a Scottish scientist in the field of mathematical physics. His most notable achievement was to formulate the classical theory of electromagnetic radiation, bringing together for the first time electricity, magnetism, and light as different manifestations of the same phenomenon.

DIAMOND™
BOOKS
www.diamondbooks .ca

TORONTO, CANADA – 2020

DIAMOND BOOKS - CANADA

DIAMOND BOOKS

Toronto, ON, CANADA
http://www.**diamondbooks** .ca

BIBLIOGRAPHIC INFORMATION

A TREATISE ON ELECTRICITY AND MAGNETISM

VOL. 1 - SECOND EDITION

FIRST PUBLISHED IN 1881
OXFORD
AT THE CLARENDON PRESS
LONDON

PUBLISHED IN CANADA

Published in Canada by DIAMOND BOOKS ™, an imprint
of
DIAMOND PUBLISHERS CANADA
http://www.diamondpublishers .com

REPUBLISHED EDITION : AUGUST, 2020.

PAPERBACK EDITION : ISBN: 978-1-77375-143-6

PRINTED IN CANADA

Clarendon Press Series

A TREATISE

ON

ELECTRICITY AND MAGNETISM

BY

JAMES CLERK MAXWELL, M.A.

LL.D. EDIN., D.C.L., F.R.SS. LONDON AND EDINBURG

HONORARY FELLOW OF TRINITY COLLEGE

AND LATE PROFESSOR OF EXPERIMENTAL PHYSICS IN THE UNIVERSITY OF CAMBRIDGE

VOL. I

SECOND EDITION

Oxford

AT THE CLARENDON PRESS

1881

Clarendon Press Series

A TREATISE

ON

ELECTRICITY AND MAGNETISM

BY

JAMES CLERK MAXWELL, M.A.

LL.D. EDIN., D.C.L., F.R.SS. LONDON AND EDINBURGH
HONORARY FELLOW OF TRINITY COLLEGE,
AND PROFESSOR OF EXPERIMENTAL PHYSICS IN THE UNIVERSITY OF CAMBRIDGE

VOL. I

SECOND EDITION

Oxford

AT THE CLARENDON PRESS

1881

PREFACE TO THE FIRST EDITION.

THE fact that certain bodies, after being rubbed, appear to attract other bodies, was known to the ancients. In modern times, a great variety of other phenomena have been observed, and have been found to be related to these phenomena of attraction. They have been classed under the name of *Electric* phenomena, amber, ἤλεκτρον, having been the substance in which they were first described.

Other bodies, particularly the loadstone, and pieces of iron and steel which have been subjected to certain processes, have also been long known to exhibit phenomena of action at a distance. These phenomena, with others related to them, were found to differ from the electric phenomena, and have been classed under the name of *Magnetic* phenomena, the loadstone, μάγνης, being found in the Thessalian Magnesia.

These two classes of phenomena have since been found to be related to each other, and the relations between the various phenomena of both classes, so far as they are known, constitute the science of Electromagnetism.

In the following Treatise I propose to describe the

most important of these phenomena, to shew how they may be subjected to measurement, and to trace the mathematical connexions of the quantities measured. Having thus obtained the data for a mathematical theory of electromagnetism, and having shewn how this theory may be applied to the calculation of phenomena, I shall endeavour to place in as clear a light as I can the relations between the mathematical form of this theory and that of the fundamental science of Dynamics, in order that we may be in some degree prepared to determine the kind of dynamical phenomena among which we are to look for illustrations or explanations of the electromagnetic phenomena.

In describing the phenomena, I shall select those which most clearly illustrate the fundamental ideas of the theory, omitting others, or reserving them till the reader is more advanced.

The most important aspect of any phenomenon from a mathematical point of view is that of a measurable quantity. I shall therefore consider electrical phenomena chiefly with a view to their measurement, describing the methods of measurement, and defining the standards on which they depend.

In the application of mathematics to the calculation of electrical quantities, I shall endeavour in the first place to deduce the most general conclusions from the data at our disposal, and in the next place to apply the results to the simplest cases that can be chosen. I shall avoid, as much as I can, those questions which, though they have elicited the skill of mathematicians, have not enlarged our knowledge of science.

The internal relations of the different branches of the science which we have to study are more numerous and complex than those of any other science hitherto developed. Its external relations, on the one hand to dynamics, and on the other to heat, light, chemical action, and the constitution of bodies, seem to indicate the special importance of electrical science as an aid to the interpretation of nature.

It appears to me, therefore, that the study of electromagnetism in all its extent has now become of the first importance as a means of promoting the progress of science.

The mathematical laws of the different classes of phenomena have been to a great extent satisfactorily made out.

The connexions between the different classes of phenomena have also been investigated, and the probability of the rigorous exactness of the experimental laws has been greatly strengthened by a more extended knowledge of their relations to each other.

Finally, some progress has been made in the reduction of electromagnetism to a dynamical science, by shewing that no electromagnetic phenomenon is contradictory to the supposition that it depends on purely dynamical action.

What has been hitherto done, however, has by no means exhausted the field of electrical research. It has rather opened up that field, by pointing out subjects of enquiry, and furnishing us with means of investigation.

It is hardly necessary to enlarge upon the beneficial

results of magnetic research on navigation, and the importance of a knowledge of the true direction of the compass, and of the effect of the iron in a ship. But the labours of those who have endeavoured to render navigation more secure by means of magnetic observations have at the same time greatly advanced the progress of pure science.

Gauss, as a member of the German Magnetic Union, brought his powerful intellect to bear on the theory of magnetism, and on the methods of observing it, and he not only added greatly to our knowledge of the theory of attractions, but reconstructed the whole of magnetic science as regards the instruments used, the methods of observation, and the calculation of the results, so that his memoirs on Terrestrial Magnetism may be taken as models of physical research by all those who are engaged in the measurement of any of the forces in nature.

The important applications of electromagnetism to telegraphy have also reacted on pure science by giving a commercial value to accurate electrical measurements, and by affording to electricians the use of apparatus on a scale which greatly transcends that of any ordinary laboratory. The consequences of this demand for electrical knowledge, and of these experimental opportunities for acquiring it, have been already very great, both in stimulating the energies of advanced electricians, and in diffusing among practical men a degree of accurate knowledge which is likely to conduce to the general scientific progress of the whole engineering profession.

There are several treatises in which electrical and magnetic phenomena are described in a popular way. These, however, are not what is wanted by those who have been brought face to face with quantities to be measured, and whose minds do not rest satisfied with lecture-room experiments.

There is also a considerable mass of mathematical memoirs which are of great importance in electrical science, but they lie concealed in the bulky Transactions of learned societies; they do not form a connected system; they are of very unequal merit, and they are for the most part beyond the comprehension of any but professed mathematicians.

I have therefore thought that a treatise would be useful which should have for its principal object to take up the whole subject in a methodical manner, and which should also indicate how each part of the subject is brought within the reach of methods of verification by actual measurement.

The general complexion of the treatise differs considerably from that of several excellent electrical works, published, most of them, in Germany, and it may appear that scant justice is done to the speculations of several eminent electricians and mathematicians. One reason of this is that before I began the study of electricity I resolved to read no mathematics on the subject till I had first read through Faraday's *Experimental Researches on Electricity*. I was aware that there was supposed to be a difference between Faraday's way of conceiving phenomena and that of the mathematicians, so that neither he nor they were satisfied with each other's language. I had also the conviction that this discrepancy did not arise from either party being wrong. I was first convinced of this by Sir William Thomson *, to whose advice and assistance, as well as to his published papers, I owe most of what I have learned on the subject.

As I proceeded with the study of Faraday, I perceived that his method of conceiving the phenomena was also a mathematical one, though not exhibited in the conventional form of mathematical symbols. I also found that these methods were capable of being expressed in the ordinary mathematical forms, and thus compared with those of the professed mathematicians.

For instance, Faraday, in his mind's eye, saw lines of force traversing all space where the mathematicians saw centres of force attracting at a distance: Faraday saw a medium where they saw nothing but distance: Faraday sought the seat of the phenomena in real actions going on in the medium, they were satisfied that they had found it in a power of action at a distance impressed on the electric fluids.

When I had translated what I considered to be Faraday's ideas into a mathematical form, I found that in general the results of the two methods coincided, so that the same phenomena were accounted for, and the same laws of action deduced by both methods, but that Faraday's methods resembled those

* I take this opportunity of acknowledging my obligations to Sir W. Thomson and to Professor Tait for many valuable suggestions made during the printing of this work.

in which we begin with the whole and arrive at the parts by anlaysis, while the ordinary mathematical methods were founded on the principle of beginning with the parts and building up the whole by synthesis.

I also found that several of the most fertile methods of research discovered by the mathematicians could be expressed much better in terms of ideas derived from Faraday than in their original form.

The whole theory, for instance, of the potential, considered as a quantity which satisfies a certain partial differential equation, belongs essentially to the method which I have called that of Faraday. According to the other method, the potential, if it is to be considered at all, must be regarded as the result of a summation of the electrified particles divided each by its distance from a given point. Hence many of the mathematical discoveries of Laplace, Poisson, Green and Gauss find their proper place in this treatise, and their appropriate expression in terms of conceptions mainly derived from Faraday.

Great progress has been made in electrical science, chiefly in Germany, by cultivators of the theory of action at a distance. The valuable electrical measurements of W. Weber are interpreted by him according to this theory, and the electromagnetic speculation which was originated by Gauss, and carried on by Weber, Riemann, J. and C. Neumann, Lorenz, &c. is founded on the theory of action at a distance, but depending either directly on the relative velocity of the particles, or on the gradual propagation of something, whether potential or force, from the one particle to the other. The great success which these eminent men have attained in the application of mathematics to electrical phenomena, gives, as is natural, additional weight to their theoretical speculations, so that those who, as students of electricity, turn to them as the greatest authorities in mathematical electricity, would probably imbibe, along with their mathematical methods, their physical hypotheses.

These physical hypotheses, however, are entirely alien from the way of looking at things which I adopt, and one object which I have in view is that some of those who wish to study electricity may, by reading this treatise, come to see that there is another way of treating the subject, which is no less fitted to explain the phenomena, and which, though in some parts it may appear less definite, corresponds, as I think, more faithfully with our actual knowledge, both in what it affirms and in what it leaves undecided.

In a philosophical point of view, moreover, it is exceedingly important that two methods should be compared, both of which have succeeded in explaining the principal electromagnetic phenomena, and both of which have attempted to explain the propagation of light as an electromagnetic phenomenon, and have actually calculated its velocity, while at the same time the fundamental conceptions of what actually takes place, as well as most of the secondary conceptions of the quantities concerned, are radically different.

I have therefore taken the part of an advocate rather than that of a judge, and have rather exemplified one method than attempted to give an impartial description of both. I have no doubt that the method which I have called the German one will also find its supporters, and will be expounded with a skill worthy of its ingenuity.

I have not attempted an exhaustive account of electrical phenomena, experiments, and apparatus. The student who desires to read all that is known on these subjects will find great assistance from the *Traité d'Electricité* of Professor A. de la Rive, and from several German treatises, such as Wiedemann's *Galvanismus*, Riess' *Reibungselektricität*, Beer's *Einleitung in die Elektrostatik*, &c.

I have confined myself almost entirely to the mathematical treatment of the subject, but I would recommend the student, after he has learned, experimentally if possible, what are the phenomena to be observed, to read carefully Faraday's *Experimental Researches in Electricity*. He will there find a strictly contemporary historical account of some of the greatest electrical discoveries and investigations, carried on in an order and succession which could hardly have been improved if the results had been known from the first, and expressed in the language of a man who devoted much of his attention to the methods of accurately describing scientific operations and their results *.

It is of great advantage to the student of any subject to read the original memoirs on that subject, for science is always most completely assimilated when

* *Life and Letters of Faraday*, vol. i. p. 395.

it is in the nascent state, and in the case of Faraday's *Researches* this is comparatively easy, as they are published in a separate form, and may be read consecutively. If by anything I have here written I may assist any student in understanding Faraday's modes of thought and expression, I shall regard it as the accomplishment of one of my principal aims—to communicate to others the same delight which I have found myself in reading Faraday's *Researches*.

The description of the phenomena, and the elementary parts of the theory of each subject, will be found in the earlier chapters of each of the four Parts into which this treatise is divided. The student will find in these chapters enough to give him an elementary acquaintance with the whole science.

The remaining chapters of each Part are occupied with the higher parts of the theory, the processes of numerical calculation, and the instruments and methods of experimental research.

The relations between electromagnetic phenomena and those of radiation, the theory of molecular electric currents, and the results of speculation on the nature of action at a distance, are treated of in the last four chapters of the second volume.

Feb. 1, 1873.

PREFACE TO THE SECOND EDITION.

WHEN I was asked to read the proof-sheets of the second edition of the *Electricity and Magnetism* the work of printing had already reached the ninth chapter, the greater part of which had been revised by the author.

Those who are familiar with the first edition will see from a comparison with the present how extensive were the changes intended by Professor Maxwell both in the substance and in the treatment of the subject, and how much this edition has suffered from his premature death. The first nine chapters were in some cases entirely re-written, much new matter being added and the former contents rearranged and simplified.

From the ninth chapter onwards the present edition is little more than a reprint. The only liberties I have taken have been in the insertion here and there of a step in the mathematical reasoning where it seemed to be an advantage to the reader, and of a few foot-notes on parts of the subject which my own experience or that of pupils attending my classes shewed to require further elucidation. These footnotes are in square brackets.

There were two parts of the subject in the treatment

of which it was known to me that the Professor contemplated considerable changes: viz. the mathematical theory of the conduction of electricity in a network of wires, and the determination of coefficients of induction in coils of wire. In these subjects I have not found myself in a position to add, from the Professor's notes, anything substantial to the work as it stood in the former edition, with the exception of a numerical table, printed in vol. ii, pp. 317–319. This table will be found very useful in calculating coefficients of induction in circular coils of wire.

In a work so original, and containing so many details of new results, it was impossible but that there should be a few errors in the first edition. I trust that in the present edition most of these will be found to have been corrected. I have the greater confidence in expressing this hope as, in reading some of the proofs, I have had the assistance of various friends conversant with the work, among whom I may mention particularly my brother Professor Charles Niven, and Mr. J. J. Thomson, Fellow of Trinity College, Cambridge.

W. D. NIVEN.

TRINITY COLLEGE, CAMBRIDGE,
Oct. 1, 1881.

CONTENTS.

PRELIMINARY.

ON THE MEASUREMENT OF QUANTITIES.

Art.		Page
1.	The expression of a quantity consists of two factors, the numerical value, and the name of the concrete unit	1
2.	Dimensions of derived units	1
3–5.	The three fundamental units—Length, Time and Mass	2, 3
6.	Derived units	5
7.	Physical continuity and discontinuity	6
8.	Discontinuity of a function of more than one variable	7
9.	Periodic and multiple functions	8
10.	Relation of physical quantities to directions in space	8
11.	Meaning of the words Scalar and Vector	9
12.	Division of physical vectors into two classes, Forces and Fluxes	10
13.	Relation between corresponding vectors of the two classes	11
14.	Line-integration appropriate to forces, surface-integration to fluxes	12
15.	Longitudinal and rotational vectors	13
16.	Line-integrals and potentials	13
17.	Hamilton's expression for the relation between a force and its potential	15
18.	Cyclic regions and geometry of position	16
19.	The potential in an acyclic region is single valued	17
20.	System of values of the potential in a cyclic region	18
21.	Surface-integrals	19
22.	Surfaces, tubes, and lines of flow	21
23.	Right-handed and left-handed relations in space	24
24.	Transformation of a line-integral into a surface-integral	25
25.	Effect of Hamilton's operation ∇ on a vector function	28
26.	Nature of the operation ∇^2	29

PART I.

ELECTROSTATICS.

CHAPTER I.

DESCRIPTION OF PHENOMENA.

Art. Page

27. Electrification by friction. Electrification is of two kinds, to which the names of Vitreous and Resinous, or Positive and Negative, have been given 31
28. Electrification by induction 32
29. Electrification by conduction. Conductors and insulators .. 33
30. In electrification by friction the quantity of the positive electrification is equal to that of the negative electrification .. 34
31. To charge a vessel with a quantity of electricity equal and opposite to that of an excited body 34
32. To discharge a conductor completely into a metallic vessel .. 35
33. Test of electrification by gold-leaf electroscope 35
34. Electrification, considered as a measurable quantity, may be called Electricity 36
35. Electricity may be treated as a physical quantity 37
36. Theory of Two fluids 38
37. Theory of One fluid 40
38. Measurement of the force between electrified bodies 41
39. Relation between this force and the quantities of electricity .. 42
40. Variation of the force with the distance 43
41, 42. Definition of the electrostatic unit of electricity.—Its dimensions 43, 44
43. Proof of the law of electric force 44
44. Electric field 45
45. Electromotive force and potential 46
46. Equipotential surfaces. Example of their use in reasoning about electricity 47
47. Lines of force 48
48. Electric tension 49
49. Electromotive force 49
50. Capacity of a conductor. Electric Accumulators 49
51. Properties of bodies.—Resistance 50
52. Specific Inductive capacity of a dielectric 52
53. 'Absorption' of electricity 53

Art. Page
54. Impossibility of an absolute charge 54
55. Disruptive discharge.—Glow 54
56. Brush 57
57. Spark 57
58. Electrical phenomena of Tourmaline 58
59. Plan of the treatise, and sketch of its results 59
60. Electric polarization and displacement 61
61. The motion of electricity analogous to that of an incompressible fluid 64
62. Peculiarities of the theory of this treatise 65

CHAPTER II.

ELEMENTARY MATHEMATICAL THEORY OF ELECTRICITY.

63. Definition of electricity as a mathematical quantity 68
64. Volume-density, surface-density, and line-density 68
65. Definition of the electrostatic unit of electricity 70
66. Law of force between electrified bodies 70
67. Resultant force between two bodies 71
68. Resultant intensity at a point 71
69. Line-integral of electric intensity; electromotive force 72
70. Electric potential 73
71. Resultant intensity in terms of the potential 74
72. The potential of all points of a conductor is the same 75
73. Potential due to an electrified system 76
74 *a*. Proof of the law of the inverse square. Cavendish's experiments 76
74 *b*. Cavendish's experiments repeated in a modified form 77
74 *c*, *d*, *e*. Theory of the experiments 79–81
75. Surface-integral of electric induction 82
76. Induction through a closed surface due to a single centre of force 83
77. Poisson's extension of Laplace's equation 84
78 *a*, *b*, *c*. Conditions to be fulfilled at an electrified surface .. 85–88
79. Resultant force on an electrified surface 88
80. The electrification of a conductor is entirely on the surface .. 90
81. A distribution of electricity on lines or points is physically impossible 91
82. Lines of electric induction 92
83 *a*. Specific inductive capacity 94
83 *b*. Apparent distribution of electricity 94

CHAPTER III.

ON ELECTRICAL WORK AND ENERGY IN A SYSTEM OF CONDUCTORS.

Art. Page

84. On the superposition of electrified systems. Expression for the energy of a system of conductors 96

85 *a*. Change of the energy in passing from one state to another .. 97

85 *b*. Relations between the potentials and the charges 98

86. Theorems of reciprocity 98

87. Theory of a system of conductors. Coefficients of potential. Capacity. Coefficients of induction 100

88. Dimensions of the coefficients 103

89 *a*. Necessary relations among the coefficients of potential 103

89 *b*. Relations derived from physical considerations 104

89 *c*. Relations among coefficients of capacity and induction 105

89 *d*. Approximation to capacity of one conductor 105

89 *e*. The coefficients of potential changed by a second conductor .. 106

90 *a*. Approximate determination of the coefficients of capacity and induction of two conductors 107

90 *b*. Similar determination for two condensers 107

91. Relative magnitudes of coefficients of potential 109

92. And of induction 110

93 *a*. Mechanical force on a conductor expressed in terms of the charges of the different conductors of the system.. 110

93 *b*. Theorem in quadratic functions 111

93 *c*. Work done by the electric forces during the displacement of a system when the potentials are maintained constant 111

94. Comparison of electrified systems 112

CHAPTER IV.

GENERAL THEOREMS.

95 *a*, *b*. Two opposite methods of treating electrical questions 115, 116

96 *a*. Green's Theorem 118

96 *b*. When one of the functions is many valued 120

96 *c*. When the region is multiply connected 120

96 *d*. When one of the functions becomes infinite in the region .. 121

97 *a*, *b*. Applications of Green's method 123, 124

98. Green's Function 125

99 *a*. Energy of a system expressed as a volume integral 126

Art. Page

99 *b*. Proof of unique solution for the potential when its value is given at every point of a closed surface 127

100 *a–e*. Thomson's Theorem 129–132

101 *a–h*. Expression for the energy when the dielectric constants are different in different directions. Extension of Green's Theorem to a heterogeneous medium 133–137

102 *a*. Method of finding limiting values of electrical coefficients .. 138

102 *b*. Approximation to the solution of problems of the distribution of electricity on conductors at given potentials 140

102 *c*. Application to the case of a condenser with slightly curved plates 142

CHAPTER V.

MECHANICAL ACTION BETWEEN TWO ELECTRICAL SYSTEMS.

103. Expression for the force at any point of the medium in terms of the potentials arising from the presence of the two systems 144

104. In terms of the potential arising from both systems 145

105. Nature of the stress in the medium which would produce the same force 146

106. Further determination of the type of stress 148

107. Modification of the expressions at the surface of a conductor.. 149

108. Discussion of the integral of Art. 104 expressing the force when taken over all space 151

109. Statements of Faraday relative to the longitudinal tension and lateral pressure of the lines of force 153

110. Objections to stress in a fluid considered 153

111. Statement of the theory of electric polarization 154

CHAPTER VI.

POINTS AND LINES OF EQUILIBRIUM.

112. Conditions for a point of equilibrium 157

113. Number of points of equilibrium 158

114. At a point or line of equilibrium there is a conical point or a line of self-intersection of the equipotential surface 159

115. Angles at which an equipotential surface intersects itself .. 160

116. The equilibrium of an electrified body cannot be stable 161

CHAPTER VII.

FORMS OF EQUIPOTENTIAL SURFACES AND LINES OF FLOW.

Art. Page
117. Practical importance of a knowledge of these forms in simple cases 164
118. Two electrified points, ratio 4 : 1. (Fig. I) 165
119. Two electrified points, ratio 4 : −1. (Fig. II) 166
120. An electrified point in a uniform field of force (Fig. III) .. 167
121. Three electrified points. Two spherical equipotential surfaces. (Fig. IV) 167
122. Faraday's use of the conception of lines of force 168
123. Method employed in drawing the diagrams 169

CHAPTER VIII.

SIMPLE CASES OF ELECTRIFICATION.

124. Two parallel planes 172
125. Two concentric spherical surfaces 174
126. Two coaxal cylindric surfaces 176
127. Longitudinal force on a cylinder, the ends of which are surrounded by cylinders at different potentials 177

CHAPTER IX.

SPHERICAL HARMONICS.

128. Heine, Todhunter, Ferrers 179
129 *a*. Singular points 179
129 *b*. Definition of an axis 180
129 *c*. Construction of points of different orders 181
129 *d*. Potential of such points. Surface harmonics Y_n 182
130 *a*. Solid harmonics. $H_n = r^n Y_n$ 182
130 *b*. There are $2n+1$ independent constants in a solid harmonic of the nth order 183
131 *a*. Potential due to a spherical shell 184
131 *b*. Expressed in harmonics.. 184
131 *c*. Mutual potential of shell and external system 185
132. Value of $\iint Y_m Y_n\, ds$ 186
133. Trigonometrical expressions for Y_n.. 187
134. Value of $\iint Y_m Y_n ds$, when $m = n$ 189
135 *a*. Special case when Y_m is a zonal harmonic 190
135 *b*. Laplace's expansion of a surface harmonic 190
136. Conjugate harmonics 192

Art. Page
137. Standard harmonics of any order 192
138. Zonal harmonics 193
139. Laplace's coefficient or Biaxal harmonic 194
140 *a*. Tesseral harmonics. Their trigonometrical expansion .. 194
140 *b*. Notations used by various authors 197
140 *c*. Forms of the tesseral and sectorial harmonics 197
141. Surface integral of the square of a tesseral harmonic 198
142 *a*. Determination of a given tesseral harmonic in the expansion of a function 199
142 *b*. The same in terms of differential coefficients of the function.. 199
143. Figures of various harmonics 200
144 *a*. Spherical conductor in a given field of force 201
144 *b*. Spherical conductor in a field for which Green's function is known 201
145 *a*. Distribution of electricity on a nearly spherical conductor .. 204
145 *b*. When acted on by external electrical force 206
145 *c*. When enclosed in a nearly spherical and nearly concentric vessel 207
146. Equilibrium of electricity on two spherical conductors 208

CHAPTER X.

CONFOCAL SURFACES OF THE SECOND DEGREE.

147. The lines of intersection of two systems and their intercepts by the third system 215
148. The characteristic equation of V in terms of ellipsoidal coordinates 216
149. Expression of α, β, γ in terms of elliptic functions 217
150. Particular solutions of electrical distribution on the confocal surfaces and their limiting forms 218
151. Continuous transformation into a figure of revolution about the axis of z 221
152. Transformation into a figure of revolution about the axis of x.. 222
153. Transformation into a system of cones and spheres.. 223
154. Confocal paraboloids.. 223

CHAPTER XI.

THEORY OF ELECTRIC IMAGES.

155. Thomson's method of electric images 226
156. When two points are oppositely and unequally electrified, the surface for which the potential is zero is a sphere 227

Art. Page
157. Electric images.. 228
158. Distribution of electricity on the surface of the sphere 230
159. Image of any given distribution of electricity.. 231
160. Resultant force between an electrified point and sphere 232
161. Images in an infinite plane conducting surface 234
162. Electric inversion 235
163. Geometrical theorems about inversion 236
164. Application of the method to the problem of Art. 158 237
165. Finite systems of successive images.. 238
166. Case of two spherical surfaces intersecting at an angle $\frac{\pi}{n}$ .. 240
167. Enumeration of the cases in which the number of images is finite 241
168. Case of two spheres intersecting orthogonally.. 242
169. Case of three spheres intersecting orthogonally 245
170. Case of four spheres intersecting orthogonally 246
171. Infinite series of images. Case of two concentric spheres .. 247
172. Any two spheres not intersecting each other 249
173. Calculation of the coefficients of capacity and induction 251
174. Calculation of the charges of the spheres, and of the force between them 253
175. Distribution of electricity on two spheres in contact. Proof sphere 255
176. Thomson's investigation of an electrified spherical bowl.. .. 257
177. Distribution on an ellipsoid, and on a circular disk at potential V 257
178. Induction on an uninsulated disk or bowl by an electrified point in the continuation of the plane or spherical surface .. 258
179. The rest of the sphere supposed uniformly electrified 259
180. The bowl maintained at potential V and uninfluenced 259
181. Induction on the bowl due to a point placed anywhere 260

CHAPTER XII.

CONJUGATE FUNCTIONS IN TWO DIMENSIONS.

182. Cases in which the quantities are functions of x and y only .. 262
183. Conjugate functions.. 263
184. Conjugate functions may be added or subtracted 264
185. Conjugate functions of conjugate functions are themselves conjugate 265
186. Transformation of Poisson's equation 267
187. Additional theorems on conjugate functions 268

Art. Page
188. Inversion in two dimensions 268
189. Electric images in two dimensions 269
190. Neumann's transformation of this case 270
191. Distribution of electricity near the edge of a conductor formed by two plane surfaces 272
192. Ellipses and hyperbolas. (Fig. X) 273
193. Transformation of this case. (Fig. XI) 274
194. Application to two cases of the flow of electricity in a conducting sheet 276
195. Application to two cases of electrical induction 276
196. Capacity of a condenser consisting of a circular disk between two infinite planes 277
197. Case of a series of equidistant planes cut off by a plane at right angles to them 279
198. Case of a furrowed surface 280
199. Case of a single straight groove 281
200. Modification of the results when the groove is circular 281
201. Application to Sir W. Thomson's guard-ring 284
202. Case of two parallel plates cut off by a perpendicular plane. (Fig. XII) 285
203. Case of a grating of parallel wires. (Fig. XIII) 286
204. Case of a single electrified wire transformed into that of the grating 287
205. The grating used as a shield to protect a body from electrical influence 288
206. Method of approximation applied to the case of the grating .. 289

CHAPTER XIII.

ELECTROSTATIC INSTRUMENTS.

207. The frictional electrical machine 292
208. The electrophorus of Volta 293
209. Production of electrification by mechanical work.—Nicholson's Revolving Doubler 294
210. Principle of Varley's and Thomson's electrical machines 294
211. Thomson's water-dropping machine 297
212. Holtz's electrical machine 298
213. Theory of regenerators applied to electrical machines 298
214. On electrometers and electroscopes. Indicating instruments and null methods. Difference between registration and measurement 300
215. Coulomb's Torsion Balance for measuring charges 301

Art. Page

216. Electrometers for measuring potentials. Snow-Harris's and Thomson's 304
217. Principle of the guard-ring. Thomson's Absolute Electrometer 305
218. Heterostatic method.. 308
219. Self-acting electrometers.—Thomson's Quadrant Electrometer 309
220. Measurement of the electric potential of a small body 312
221. Measurement of the potential at a point in the air.. 313
222. Measurement of the potential of a conductor without touching it 314
223. Measurement of the superficial density of electrification. The proof plane 315
224. A hemisphere used as a test 316
225. A circular disk.. 317
226. On electric accumulators. The Leyden jar 319
227. Accumulators of measurable capacity 320
228. The guard-ring accumulator 321
229. Comparison of the capacities of accumulators.. 323

PART II.

ELECTROKINEMATICS.

CHAPTER I.

THE ELECTRIC CURRENT.

230. Current produced when conductors are discharged.. 326
231. Transference of electrification 326
232. Description of the voltaic battery 327
233. Electromotive force 328
234. Production of a steady current 328
235. Properties of the current 329
236. Electrolytic action 329
237. Explanation of terms connected with electrolysis 330
238. Different modes of passage of the current 330
239. Magnetic action of the current.. 331
240. The Galvanometer 332

CHAPTER II.

CONDUCTION AND RESISTANCE.

Art. Page

241. Ohm's Law 333
242. Generation of heat by the current. Joule's Law 334
243. Analogy between the conduction of electricity and that of heat 335
244. Differences between the two classes of phenomena 335
245. Faraday's doctrine of the impossibility of an absolute charge .. 336

CHAPTER III.

ELECTROMOTIVE FORCE BETWEEN BODIES IN CONTACT.

246. Volta's law of the contact force between different metals at the same temperature 337
247. Effect of electrolytes 338
248. Thomson's voltaic current in which gravity performs the part of chemical action 338
249. Peltier's phenomenon. Deduction of the thermoelectric electromotive force at a junction 338
250. Seebeck's discovery of thermoelectric currents 340
251. Magnus's law of a circuit of one metal 340
252. Cumming's discovery of thermoelectric inversions 342
253. Thomson's deductions from these facts, and discovery of the reversible thermal effects of electric currents in copper and in iron 342
254. Tait's law of the electromotive force of a thermoelectric pair .. 343

CHAPTER IV.

ELECTROLYSIS.

255. Faraday's law of electrochemical equivalents 345
256. Clausius's theory of molecular agitation 347
257. Electrolytic polarization 347
258. Test of an electrolyte by polarization 348
259. Difficulties in the theory of electrolysis 348
260. Molecular charges 349
261. Secondary actions observed at the electrodes 351
262. Conservation of energy in electrolysis 353
263. Measurement of chemical affinity as an electromotive force .. 354

CHAPTER V.

ELECTROLYTIC POLARIZATION.

Art. Page
264. Difficulties of applying Ohm's law to electrolytes 356
265. Ohm's law nevertheless applicable 356
266. The effect of polarization distinguished from that of resistance 356
267. Polarization due to the presence of the ions at the electrodes. The ions not in a free state 357
268. Relation between the electromotive force of polarization and the state of the ions at the electrodes 358
269. Dissipation of the ions and loss of polarization 359
270. Limit of polarization 359
271. Ritter's secondary pile compared with the Leyden jar 360
272. Constant voltaic elements.—Daniell's cell 363

CHAPTER VI.

MATHEMATICAL THEORY OF THE DISTRIBUTION OF ELECTRIC CURRENTS.

273. Linear conductors 367
274. Ohm's Law 367
275. Linear conductors in series 367
276. Linear conductors in multiple arc 368
277. Resistance of conductors of uniform section 369
278. Dimensions of the quantities involved in Ohm's law 370
279. Specific resistance and conductivity in electromagnetic measure 371
280. Linear systems of conductors in general 371
281. Reciprocal property of any two conductors of the system .. 373
282 *a*, *b*. Conjugate conductors 373, 374
283. Heat generated in the system 374
284. The heat is a minimum when the current is distributed according to Ohm's law 375

CHAPTER VII.

CONDUCTION IN THREE DIMENSIONS.

285. Notation 376
286. Composition and resolution of electric currents 376
287. Determination of the quantity which flows through any surface 377
288. Equation of a surface of flow 378

Art. Page

289. Relation between any three systems of surfaces of flow 378
290. Tubes of flow 378
291. Expression for the components of the flow in terms of surfaces of flow 379
292. Simplification of this expression by a proper choice of parameters 379
293. Unit tubes of flow used as a complete method of determining the current 379
294. Current-sheets and current-functions 380
295. Equation of 'continuity' 380
296. Quantity of electricity which flows through a given surface .. 382

CHAPTER VIII.

RESISTANCE AND CONDUCTIVITY IN THREE DIMENSIONS.

297. Equations of resistance 383
298. Equations of conduction 384
299. Rate of generation of heat 384
300. Conditions of stability 385
301. Equation of continuity in a homogeneous medium 386
302. Solution of the equation 386
303. Theory of the coefficient *T*. It probably does not exist .. 387
304. Generalized form of Thomson's theorem 388
305. Proof without symbols 389
306. Strutt's method applied to a wire of variable section.—Lower limit of the value of the resistance 390
307. Higher limit 393
308. Lower limit for the correction for the ends of the wire 395
309. Higher limit 396

CHAPTER IX.

CONDUCTION THROUGH HETEROGENEOUS MEDIA.

310. Surface-conditions 398
311. Spherical surface 400
312. Spherical shell 401
313. Spherical shell placed in a field of uniform flow 402
314. Medium in which small spheres are uniformly disseminated .. 403
315. Images in a plane surface 404
316. Method of inversion not applicable in three dimensions 405
317. Case of conduction through a stratum bounded by parallel planes 405

Art. Page
318. Infinite series of images. Application to magnetic induction .. 406
319. On stratified conductors. Coefficients of conductivity of a conductor consisting of alternate strata of two different substances 407
320. If neither of the substances has the rotatory property denoted by T the compound conductor is free from it 408
321. If the substances are isotropic the direction of greatest resistance is normal to the strata 408
322. Medium containing parallelepipeds of another medium 409
323. The rotatory property cannot be introduced by means of conducting channels 410
324. Construction of an artificial solid having given coefficients of longitudinal and transverse conductivity 411

CHAPTER X.

CONDUCTION IN DIELECTRICS.

325. In a strictly homogeneous medium there can be no internal charge 412
326. Theory of a condenser in which the dielectric is not a perfect insulator 413
327. No residual charge due to simple conduction 414
328. Theory of a composite accumulator 414
329. Residual charge and electrical absorption 416
330. Total discharge 418
331. Comparison with the conduction of heat 419
332. Theory of telegraph cables and comparison of the equations with those of the conduction of heat 421
333. Opinion of Ohm on this subject 422
334. Mechanical illustration of the properties of a dielectric 423

CHAPTER XI.

MEASUREMENT OF THE ELECTRIC RESISTANCE OF CONDUCTORS.

335. Advantage of using material standards of resistance in electrical measurements 426
336. Different standards which have been used and different systems which have been proposed 426
337. The electromagnetic system of units 427
338. Weber's unit, and the British Association unit or Ohm 427
339. Professed value of the Ohm 10,000,000 metres per second .. 427

Art. Page
340. Reproduction of standards 428
341. Forms of resistance coils 429
342. Coils of great resistance 430
343. Arrangement of coils in series 430
344. Arrangement in multiple arc 431
345. On the comparison of resistances. (1) Ohm's method 432
346. (2) By the differential galvanometer 432
347. (3) By Wheatstone's Bridge 436
348. Estimation of limits of error in the determination 437
349. Best arrangement of the conductors to be compared 438
350. On the use of Wheatstone's Bridge 440
351. Thomson's method for small resistances 442
352. Matthiessen and Hockin's method for small resistances 444
353. Comparison of great resistances by the electrometer 446
354. By accumulation in a condenser 447
355. Direct electrostatic method 447
356. Thomson's method for the resistance of a galvanometer 448
357. Mance's method of determining the resistance of a battery .. 449
358. Comparison of electromotive forces 452

CHAPTER XII.

ELECTRIC RESISTANCE OF SUBSTANCES.

359. Metals, electrolytes, and dielectrics 454
360. Resistance of metals 455
361. Resistance of mercury 456
362. Table of resistance of metals 457
363. Resistance of electrolytes 458
364. Experiments of Paalzow 458
365. Experiments of Kohlrausch and Nippoldt 459
366. Resistance of dielectrics 460
367. Gutta-percha 462
368. Glass 462
369. Gases 463
370. Experiments of Wiedemann and Rühlmann 463

ELECTRICITY AND MAGNETISM.

PRELIMINARY.

ON THE MEASUREMENT OF QUANTITIES.

1.] EVERY expression of a Quantity consists of two factors or components. One of these is the name of a certain known quantity of the same kind as the quantity to be expressed, which is taken as a standard of reference. The other component is the number of times the standard is to be taken in order to make up the required quantity. The standard quantity is technically called the Unit, and the number is called the Numerical Value of the quantity.

There must be as many different units as there are different kinds of quantities to be measured, but in all dynamical sciences it is possible to define these units in terms of the three fundamental units of Length, Time, and Mass. Thus the units of area and of volume are defined respectively as the square and the cube whose sides are the unit of length.

Sometimes, however, we find several units of the same kind founded on independent considerations. Thus the gallon, or the volume of ten pounds of water, is used as a unit of capacity as well as the cubic foot. The gallon may be a convenient measure in some cases, but it is not a systematic one, since its numerical relation to the cubic foot is not a round integral number.

2.] In framing a mathematical system we suppose the fundamental units of length, time, and mass to be given, and deduce all the derivative units from these by the simplest attainable definitions.

The formulae at which we arrive must be such that a person

of any nation, by substituting for the different symbols the numerical values of the quantities as measured by his own national units, would arrive at a true result.

Hence, in all scientific studies it is of the greatest importance to employ units belonging to a properly defined system, and to know the relations of these units to the fundamental units, so that we may be able at once to transform our results from one system to another.

This is most conveniently done by ascertaining the *dimensions* of every unit in terms of the three fundamental units. When a given unit varies as the nth power of one of these units, it is said to be of *n dimensions* as regards that unit.

For instance, the scientific unit of volume is always the cube whose side is the unit of length. If the unit of length varies, the unit of volume will vary as its third power, and the unit of volume is said to be of three dimensions with respect to the unit of length.

A knowledge of the dimensions of units furnishes a test which ought to be applied to the equations resulting from any lengthened investigation. The dimensions of every term of such an equation, with respect to each of the three fundamental units, must be the same. If not, the equation is absurd, and contains some error, as its interpretation would be different according to the arbitrary system of units which we adopt *.

The Three Fundamental Units.

3.] (1) *Length.* The standard of length for scientific purposes in this country is one foot, which is the third part of the standard yard preserved in the Exchequer Chambers.

In France, and other countries which have adopted the metric system, it is the mètre. The mètre is theoretically the ten millionth part of the length of a meridian of the earth measured from the pole to the equator; but practically it is the length of a standard preserved in Paris, which was constructed by Borda to correspond, when at the temperature of melting ice, with the value of the preceding length as measured by Delambre. The mètre has not been altered to correspond with new and more accurate measurements of the earth, but the arc of the meridian is estimated in terms of the original mètre.

* The theory of dimensions was first stated by Fourier, *Théorie de Chaleur*, § 160.

In astronomy the mean distance of the earth from the sun is sometimes taken as a unit of length.

In the present state of science the most universal standard of length which we could assume would be the wave length in vacuum of a particular kind of light, emitted by some widely diffused substance such as sodium, which has well-defined lines in its spectrum. Such a standard would be independent of any changes in the dimensions of the earth, and should be adopted by those who expect their writings to be more permanent than that body.

In treating of the dimensions of units we shall call the unit of length $[L]$. If l is the numerical value of a length, it is understood to be expressed in terms of the concrete unit $[L]$, so that the actual length would be fully expressed by $l\,[L]$.

4.] (2) *Time.* The standard unit of time in all civilized countries is deduced from the time of rotation of the earth about its axis. The sidereal day, or the true period of rotation of the earth, can be ascertained with great exactness by the ordinary observations of astronomers; and the mean solar day can be deduced from this by our knowledge of the length of the year.

The unit of time adopted in all physical researches is one second of mean solar time.

In astronomy a year is sometimes used as a unit of time. A more universal unit of time might be found by taking the periodic time of vibration of the particular kind of light whose wave length is the unit of length.

We shall call the concrete unit of time $[T]$, and the numerical measure of time t.

5.] (3) *Mass.* The standard unit of mass is in this country the avoirdupois pound preserved in the Exchequer Chambers. The grain, which is often used as a unit, is defined to be the 7000th part of this pound.

In the metrical system it is the gramme, which is theoretically the mass of a cubic centimètre of distilled water at standard temperature and pressure, but practically it is the thousandth part of the standard kilogramme preserved in Paris.

The accuracy with which the masses of bodies can be compared by weighing is far greater than that hitherto attained in the measurement of lengths, so that all masses ought, if possible, to be compared directly with the standard, and not deduced from experiments on water.

In descriptive astronomy the mass of the sun or that of the

earth is sometimes taken as a unit, but in the dynamical theory of astronomy the unit of mass is deduced from the units of time and length, combined with the fact of universal gravitation. The astronomical unit of mass is that mass which attracts another body placed at the unit of distance so as to produce in that body the unit of acceleration.

In framing a universal system of units we may either deduce the unit of mass in this way from those of length and time already defined, and this we can do to a rough approximation in the present state of science; or, if we expect * soon to be able to determine the mass of a single molecule of a standard substance, we may wait for this determinatien before fixing a universal standard of mass.

We shall denote the concrete unit of mass by the symbol $[M]$ in treating of the dimensions of other units. The unit of mass will be taken as one of the three fundamental units. When, as in the French system, a particular substance, water, is taken as a standard of density, then the unit of mass is no longer independent, but varies as the unit of volume, or as $[L^3]$.

If, as in the astronomical system, the unit of mass is defined with respect to its attractive power, the dimensions of $[M]$ are $[L^3 T^{-2}]$.

For the acceleration due to the attraction of a mass m at a distance r is by the Newtonian Law $\frac{m}{r^2}$. Suppose this attraction to act for a very small time t on a body originally at rest, and to cause it to describe a space s, then by the formula of Galileo,

$$s = \tfrac{1}{2} f t^2 = \tfrac{1}{2} \frac{m}{r^2} t^2;$$

whence $m = 2\frac{r^2 s}{t^2}$. Since r and s are both lengths, and t is a time, this equation cannot be true unless the dimensions of m are $[L^3 T^{-2}]$. The same can be shewn from any astronomical equation in which the mass of a body appears in some but not in all of the terms †.

* See Prof. J. Loschmidt, 'Zur Grösse der Luftmolecule,' *Academy of Vienna*, Oct. 12, 1865; G. J. Stoney on 'The Internal Motions of Gases,' *Phil. Mag.*, Aug. 1868; and Sir W. Thomson on 'The Size of Atoms,' *Nature*, March 31, 1870.

† If a centimetre and a second are taken as units, the astronomical unit of mass would be about $1{\cdot}537 \times 10^7$ grammes, or 15·37 tonnes according to Baily's repetition of Cavendish's experiment. Baily adopts 5·6604 as the result of all his experiments as the mean density of the earth, and this, with the values used by Baily for the dimensions of the earth and the intensity of gravitation at its surface, gives the above value as the direct result of his experiments.

Derived Units.

6.] The unit of Velocity is that velocity in which unit of length is described in unit of time. Its dimensions are $[LT^{-1}]$.

If we adopt the units of length and time derived from the vibrations of light, then the unit of velocity is the velocity of light.

The unit of Acceleration is that acceleration in which the velocity increases by unity in unit of time. Its dimensions are $[LT^{-2}]$.

The unit of Density is the density of a substance which contains unit of mass in unit of volume. Its dimensions are $[ML^{-3}]$.

The unit of Momentum is the momentum of unit of mass moving with unit of velocity. Its dimensions are $[MLT^{-1}]$.

The unit of Force is the force which produces unit of momentum in unit of time. Its dimensions are $[MLT^{-2}]$.

This is the absolute unit of force, and this definition of it is implied in every equation in Dynamics. Nevertheless, in many text books in which these equations are given, a different unit of force is adopted, namely, the weight of the national unit of mass; and then, in order to satisfy the equations, the national unit of mass is itself abandoned, and an artificial unit is adopted as the dynamical unit, equal to the national unit divided by the numerical value of the intensity of gravity at the place. In this way both the unit of force and the unit of mass are made to depend on the value of the intensity of gravity, which varies from place to place, so that statements involving these quantities are not complete without a knowledge of the intensity of gravity in the places where these statements were found to be true.

The abolition, for all scientific purposes, of this method of measuring forces is mainly due to the introduction by Gauss of a general system of making observations of magnetic force in countries in which the intensity of gravity is different. All such forces are now measured according to the strictly dynamical method deduced from our definitions, and the numerical results are the same in whatever country the experiments are made.

The unit of Work is the work done by the unit of force acting through the unit of length measured in its own direction. Its dimensions are $[ML^2T^{-2}]$.

The Energy of a system, being its capacity of performing work, is measured by the work which the system is capable of performing by the expenditure of its whole energy.

The definitions of other quantities, and of the units to which they are referred, will be given when we require them.

In transforming the values of physical quantities determined in terms of one unit, so as to express them in terms of any other unit of the same kind, we have only to remember that every expression for the quantity consists of two factors, the unit and the numerical part which expresses how often the unit is to be taken. Hence the numerical part of the expression varies inversely as the magnitude of the unit, that is, inversely as the various powers of the fundamental units which are indicated by the dimensions of the derived unit.

On Physical Continuity and Discontinuity.

7.] A quantity is said to vary continuously if, when it passes from one value to another, it assumes all the intermediate values.

We may obtain the conception of continuity from a consideration of the continuous existence of a particle of matter in time and space. Such a particle cannot pass from one position to another without describing a continuous line in space, and the coordinates of its position must be continuous functions of the time.

In the so-called 'equation of continuity,' as given in treatises on Hydrodynamics, the fact expressed is that matter cannot appear in or disappear from an element of volume without passing in or out through the sides of that element.

A quantity is said to be a continuous function of its variables if, when the variables alter continuously, the quantity itself alters continuously.

Thus, if u is a function of x, and if, while x passes continuously from x_0 to x_1, u passes continuously from u_0 to u_1, but when x passes from x_1 to x_2, u passes from u_1' to u_2, u_1' being different from u_1, then u is said to have a discontinuity in its variation with respect to x for the value $x = x_1$, because it passes abruptly from u_1 to u_1' while x passes continuously through x_1.

If we consider the differential coefficient of u with respect to x for the value $x = x_1$ as the limit of the fraction

$$\frac{u_2 - u_0}{x_2 - x_0},$$

when x_2 and x_0 are both made to approach x_1 without limit, then, if x_0 and x_2 are always on opposite sides of x_1, the ultimate value of the numerator will be $u_1' - u_1$, and that of the denominator will be zero. If u is a quantity physically continuous, the discontinuity can exist only with respect to the particular variable x. We must in this case admit that it has an infinite differential coefficient when $x = x_1$. If u is not physically continuous, it cannot be differentiated at all.

It is possible in physical questions to get rid of the idea of discontinuity without sensibly altering the conditions of the case. If x_0 is a very little less than x_1, and x_2 a very little greater than x_1, then u_0 will be very nearly equal to u_1 and u_2 to u_1'. We may now suppose u to vary in any arbitrary but continuous manner from u_0 to u_2 between the limits x_0 and x_2. In many physical questions we may begin with a hypothesis of this kind, and then investigate the result when the values of x_0 and x_2 are made to approach that of x_1 and ultimately to reach it. If the result is independent of the arbitrary manner in which we have supposed u to vary between the limits, we may assume that it is true when u is discontinuous.

Discontinuity of a Function of more than One Variable.

8.] If we suppose the values of all the variables except x to be constant, the discontinuity of the function will occur for particular values of x, and these will be connected with the values of the other variables by an equation which we may write

$$\phi = \phi\,(x, y, z, \&c.) = 0.$$

The discontinuity will occur when $\phi = 0$. When ϕ is positive the function will have the form $F_2\,(x, y, z, \&c.)$. When ϕ is negative it will have the form $F_1\,(x, y, z, \&c.)$. There need be no necessary relation between the forms F_1 and F_2.

To express this discontinuity in a mathematical form, let one of the variables, say x, be expressed as a function of ϕ and the other variables, and let F_1 and F_2 be expressed as functions of ϕ, y, z, &c. We may now express the general form of the function by any formula which is sensibly equal to F_2 when ϕ is positive, and to F_1 when ϕ is negative. Such a formula is the following—

$$F = \frac{F_1 + e^{n\phi} F_2}{1 + e^{n\phi}}.$$

As long as n is a finite quantity, however great, F will be a continuous function, but if we make n infinite F will be equal to F_2 when ϕ is positive, and equal to F_1 when ϕ is negative.

Discontinuity of the Derivatives of a Continuous Function.

The first derivatives of a continuous function may be discontinuous. Let the values of the variables for which the discontinuity of the derivatives occurs be connected by the equation

$$\phi = \phi(x, y, z \ldots) = 0,$$

and let F_1 and F_2 be expressed in terms of ϕ and $n-1$ other variables, say $(y, z \ldots)$.

Then, when ϕ is negative, F_1 is to be taken, and when ϕ is positive F_2 is to be taken, and, since F is itself continuous, when ϕ is zero, $F_1 = F_2$.

Hence, when ϕ is zero, the derivatives $\frac{dF_1}{d\phi}$ and $\frac{dF_2}{d\phi}$ may be different, but the derivatives with respect to any of the other variables, such as $\frac{dF_1}{dy}$ and $\frac{dF_2}{dy}$, must be the same. The discontinuity is therefore confined to the derivative with respect to ϕ, all the other derivatives being continuous.

Periodic and Multiple Functions.

9.] If u is a function of x such that its value is the same for x, $x+a$, $x+na$, and all values of x differing by a, u is called a periodic function of x, and a is called its period.

If x is considered as a function of u, then, for a given value of u, there must be an infinite series of values of x differing by multiples of a. In this case x is called a multiple function of u, and a is called its cyclic constant.

The differential coefficient $\frac{dx}{du}$ has only a finite number of values corresponding to a given value of u.

On the Relation of Physical Quantities to Directions in Space.

10.] In distinguishing the kinds of physical quantities, it is of great importance to know how they are related to the directions of those coordinate axes which we usually employ in defining the positions of things. The introduction of coordinate axes into geometry by Des Cartes was one of the greatest steps in mathematical progress, for it reduced the methods of geometry to calculations performed on numerical quantities. The position of a point is made to depend on the length of three lines which are always drawn in determinate directions, and the line joining two points is in like manner considered as the resultant of three lines.

But for many purposes of physical reasoning, as distinguished from calculation, it is desirable to avoid explicitly introducing the Cartesian coordinates, and to fix the mind at once on a point of space instead of its three coordinates, and on the magnitude and direction of a force instead of its three components. This mode of contemplating geometrical and physical quantities is more primitive and more natural than the other, although the ideas connected with it did not receive their full development till Hamilton made the next great step in dealing with space, by the invention of his Calculus of Quaternions.

As the methods of Des Cartes are still the most familiar to students of science, and as they are really the most useful for purposes of calculation, we shall express all our results in the Cartesian form. I am convinced, however, that the introduction of the ideas, as distinguished from the operations and methods of Quaternions, will be of great use to us in the study of all parts of our subject, and especially in electrodynamics, where we have to deal with a number of physical quantities, the relations of which to each other can be expressed far more simply by a few expressions of Hamilton's, than by the ordinary equations.

11.] One of the most important features of Hamilton's method is the division of quantities into Scalars and Vectors.

A Scalar quantity is capable of being completely defined by a single numerical specification. Its numerical value does not in any way depend on the directions we assume for the coordinate axes.

A Vector, or Directed quantity, requires for its definition three numerical specifications, and these may most simply be understood as having reference to the directions of the coordinate axes.

Scalar quantities do not involve direction. The volume of a geometrical figure, the mass and the energy of a material body, the hydrostatical pressure at a point in a fluid, and the potential at a point in space, are examples of scalar quantities.

A vector quantity has direction as well as magnitude, and is such that a reversal of its direction reverses its sign. The displacement of a point, represented by a straight line drawn from its original to its final position, may be taken as the typical vector quantity, from which indeed the name of Vector is derived.

The velocity of a body, its momentum, the force acting on it, an electric current, the magnetization of a particle of iron, are instances of vector quantities.

There are physical quantities of another kind which are related to directions in space, but which are not vectors. Stresses and strains in solid bodies are examples of these, and so are some of the properties of bodies considered in the theory of elasticity and in the theory of double refraction. Quantities of this class require for their definition *nine* numerical specifications. They are expressed in the language of Quaternions by linear and vector functions of a vector.

The addition of one vector quantity to another of the same kind is performed according to the rule given in Statics for the composition of forces. In fact, the proof which Poisson gives of the 'parallelogram of forces' is applicable to the composition of any quantities such that turning them end for end is equivalent to a reversal of their sign.

When we wish to denote a vector quantity by a single symbol, and to call attention to the fact that it is a vector, so that we must consider its direction as well as its magnitude, we shall denote it by a German capital letter, as $\mathfrak{A}$, $\mathfrak{B}$, &c.

In the calculus of Quaternions, the position of a point in space is defined by the vector drawn from a fixed point, called the origin, to that point. If we have to consider any physical quantity whose value depends on the position of the point, that quantity is treated as a function of the vector drawn from the origin. The function may be itself either scalar or vector. The density of a body, its temperature, its hydrostatic pressure, the potential at a point, are examples of scalar functions. The resultant force at a point, the velocity of a fluid at a point, the velocity of rotation of an element of the fluid, and the couple producing rotation, are examples of vector functions.

12.] Physical vector quantities may be divided into two classes, in one of which the quantity is defined with reference to a line, while in the other the quantity is defined with reference to an area.

For instance, the resultant of an attractive force in any direction may be measured by finding the work which it would do on a body if the body were moved a short distance in that direction and dividing it by that short distance. Here the attractive force is defined with reference to a line.

On the other hand, the flux of heat in any direction at any point of a solid body may be defined as the quantity of heat which crosses a small area drawn perpendicular to that direction divided by that area and by the time. Here the flux is defined with reference to an area.

There are certain cases in which a quantity may be measured with reference to a line as well as with reference to an area.

Thus, in treating of the displacements of elastic solids, we may direct our attention either to the original and the actual position of a particle, in which case the displacement of the particle is measured by the line drawn from the first position to the second, or we may consider a small area fixed in space, and determine what quantity of the solid passes across that area during the displacement.

In the same way the velocity of a fluid may be investigated either with respect to the actual velocity of the individual particles, or with respect to the quantity of the fluid which flows through any fixed area.

But in these cases we require to know separately the density of the body as well as the displacement or velocity, in order to apply the first method, and whenever we attempt to form a molecular theory we have to use the second method.

In the case of the flow of electricity we do not know anything of its density or its velocity in the conductor, we only know the value of what, on the fluid theory, would correspond to the product of the density and the velocity. Hence in all such cases we must apply the more general method of measurement of the flux across an area.

In electrical science, electromotive and magnetic intensity belong to the first class, being defined with reference to lines. When we wish to indicate this fact, we may refer to them as Intensities.

On the other hand, electric and magnetic induction, and electric currents, belong to the second class, being defined with reference to areas. When we wish to indicate this fact, we shall refer to them as Fluxes.

Each of these forces may be considered as producing, or tending to produce, its corresponding flux. Thus, electromotive intensity produces electric currents in conductors, and tends to produce them in dielectrics. It produces electric induction in dielectrics, and probably in conductors also. In the same sense, magnetic intensity produces magnetic induction.

13.] In some cases the flux is simply proportional to the force and in the same direction, but in other cases we can only affirm that the direction and magnitude of the flux are functions of the direction and magnitude of the force.

The case in which the components of the flux are *linear* functions of those of the force is discussed in the chapter on the Equations of Conduction, Art. 297. There are in general nine coefficients which determine the relation between the force and the flux. In certain cases we have reason to believe that six of these coefficients form three pairs of equal quantities. In such cases the relation between the line of direction of the force and the normal plane of the flux is of the same kind as that between a diameter of an ellipsoid and its conjugate diametral plane. In Quaternion language, the one vector is said to be a linear and vector function of the other, and when there are three pairs of equal coefficients the function is said to be self-conjugate.

In the case of magnetic induction in iron, the flux, (the magnetization of the iron,) is not a linear function of the magnetizing force. In all cases, however, the product of the force and the flux resolved in its direction, give a result of scientific importance, and this is always a scalar quantity.

14.] There are two mathematical operations of frequent occurrence which are appropriate to these two classes of vectors, or directed quantities.

In the case of forces, we have to take the integral along a line of the product of an element of the line, and the resolved part of the force along that element. The result of this operation is called the Line-integral of the force. It represents the work done on a body carried along the line. In certain cases in which the line-integral does not depend on the form of the line, but only on the positions of its extremities, the line-integral is called the Potential.

In the case of fluxes, we have to take the integral, over a surface, of the flux through every element of the surface. The result of this operation is called the Surface-integral of the flux. It represents the quantity which passes through the surface.

There are certain surfaces across which there is no flux. If two of these surfaces intersect, their line of intersection is a line of flux. In those cases in which the flux is in the same direction as the force, lines of this kind are often called Lines of Force. It would be more correct, however, to speak of them in electrostatics and magnetics as Lines of Induction, and in electrokinematics as Lines of Flow.

15.] There is another distinction between different kinds of directed quantities, which, though very important in a physical point of view, is not so necessary to be observed for the sake of the mathematical methods. This is the distinction between longitudinal and rotational properties.

The direction and magnitude of a quantity may depend upon some action or effect which takes place entirely along a certain line, or it may depend upon something of the nature of rotation about that line as an axis. The laws of combination of directed quantities are the same whether they are longitudinal or rotational, so that there is no difference in the mathematical treatment of the two classes, but there may be physical circumstances which indicate to which class we must refer a particular phenomenon. Thus, electrolysis consists of the transfer of certain substances along a line in one direction, and of certain other substances in the opposite direction, which is evidently a longitudinal phenomenon, and there is no evidence of any rotational effect about the direction of the force. Hence we infer that the electric current which causes or accompanies electrolysis is a longitudinal, and not a rotational phenomenon.

On the other hand, the north and south poles of a magnet do not differ as oxygen and hydrogen do, which appear at opposite places during electrolysis, so that we have no evidence that magnetism is a longitudinal phenomenon, while the effect of magnetism in rotating the plane of polarized light distinctly shews that magnetism is a rotational phenomenon.

On Line-integrals.

16.] The operation of integration of the resolved part of a vector quantity along a line is important in physical science generally, and should be clearly understood.

Let x, y, z be the coordinates of a point P on a line whose length, measured from a certain point A, is s. These coordinates will be functions of a single variable s.

Let R be the numerical value of the vector quantity at P, and let the tangent to the curve at P make with the direction of R the angle ϵ, then $R\cos\epsilon$ is the resolved part of R along the line, and the integral

$$L = \int_0^s R\cos\epsilon\, ds$$

is called the line-integral of R along the line s.

We may write this expression

$$L = \int_0^s \left(X\frac{dx}{ds} + Y\frac{dy}{ds} + Z\frac{dz}{ds}\right) ds,$$

where X, Y, Z are the components of R parallel to x, y, z respectively.

This quantity is, in general, different for different lines drawn between A and P. When, however, within a certain region, the quantity
$$X\,dx + Y\,dy + Z\,dz = -D\Psi,$$
that is, when it is an exact differential within that region, the value of L becomes
$$L = \Psi_A - \Psi_P,$$
and is the same for any two forms of the path between A and P, provided the one form can be changed into the other by continuous motion without passing out of this region.

On Potentials.

The quantity Ψ is a scalar function of the position of the point, and is therefore independent of the directions of reference. It is called the Potential Function, and the vector quantity whose components are X, Y, Z is said to have a potential Ψ, if

$$X = -\left(\frac{d\Psi}{dx}\right), \qquad Y = -\left(\frac{d\Psi}{dy}\right), \qquad Z = -\left(\frac{d\Psi}{dz}\right).$$

When a potential function exists, surfaces for which the potential is constant are called Equipotential surfaces. The direction of R at any point of such a surface coincides with the normal to the surface, and if n be a normal at the point P, then $R = -\dfrac{d\Psi}{dn}$.

The method of considering the components of a vector as the first derivatives of a certain function of the coordinates with respect to these coordinates was invented by Laplace * in his treatment of the theory of attractions. The name of Potential was first given to this function by Green †, who made it the basis of his treatment of electricity. Green's essay was neglected by mathematicians till 1846, and before that time most of its important theorems had been rediscovered by Gauss, Chasles, Sturm, and Thomson ‡.

* Méc. Céleste, liv. iii.

† Essay on the Application of Mathematical Analysis to the Theories of Electricity and Magnetism, Nottingham, 1828. Reprinted in *Crelle's Journal*, and in Mr. Ferrers' edition of Green's Works.

‡ Thomson and Tait, *Natural Philosophy*, § 483.

In the theory of gravitation the potential is taken with the opposite sign to that which is here used, and the resultant force in any direction is then measured by the rate of *increase* of the potential function in that direction. In electrical and magnetic investigations the potential is defined so that the resultant force in any direction is measured by the *decrease* of the potential in that direction. This method of using the expression makes it correspond in sign with potential energy, which always decreases when the bodies are moved in the direction of the forces acting on them.

17.] The geometrical nature of the relation between the potential and the vector thus derived from it receives great light from Hamilton's discovery of the form of the operator by which the vector is derived from the potential.

The resolved part of the vector in any direction is, as we have seen, the first derivative of the potential with respect to a coordinate drawn in that direction, the sign being reversed.

Now if i, j, k are three unit vectors at right angles to each other, and if X, Y, Z are the components of the vector $\mathfrak{F}$ resolved parallel to these vectors, then

$$\mathfrak{F} = iX + jY + kZ; \tag{1}$$

and by what we have said above, if Ψ is the potential,

$$\mathfrak{F} = -\left(i\frac{d\Psi}{dx} + j\frac{d\Psi}{dy} + k\frac{d\Psi}{dz}\right). \tag{2}$$

If we now write ∇ for the operator,

$$i\frac{d}{dx} + j\frac{d}{dy} + k\frac{d}{dz}, \tag{3}$$

$$\mathfrak{F} = -\nabla\Psi. \tag{4}$$

The symbol of operation ∇ may be interpreted as directing us to measure, in each of three rectangular directions, the rate of increase of Ψ, and then, considering the quantities thus found as vectors, to compound them into one. This is what we are directed to do by the expression (3). But we may also consider it as directing us first to find out in what direction Ψ increases fastest, and then to lay off in that direction a vector representing this rate of increase.

M. Lamé, in his *Traité des Fonctions Inverses*, uses the term Differential Parameter to express the magnitude of this greatest rate of increase, but neither the term itself, nor the mode in which

Lamé uses it, indicates that the quantity referred to has direction as well as magnitude. On those rare occasions in which I shall have to refer to this relation as a purely geometrical one, I shall call the vector 𝔉 the space-variation of the scalar function Ψ, using the phrase to indicate the direction, as well as the magnitude, of the most rapid decrease of Ψ.

18.] There are cases, however, in which the conditions

$$\frac{dZ}{dy}-\frac{dY}{dz}=0,\quad \frac{dX}{dz}-\frac{dZ}{dx}=0,\quad \text{and}\quad \frac{dY}{dx}-\frac{dX}{dy}=0,$$

which are those of $X\,dx+Y\,dy+Z\,dz$ being a complete differential, are satisfied throughout a certain region of space, and yet the line-integral from A to P may be different for two lines, each of which lies wholly within that region. This may be the case if the region is in the form of a ring, and if the two lines from A to P pass through opposite segments of the ring. In this case, the one path cannot be transformed into the other by continuous motion without passing out of the region.

We are here led to considerations belonging to the Geometry of Position, a subject which, though its importance was pointed out by Leibnitz and illustrated by Gauss, has been little studied. The most complete treatment of this subject has been given by J. B. Listing *.

Let there be p points in space, and let l lines of any form be drawn joining these points so that no two lines intersect each other, and no point is left isolated. We shall call a figure composed of lines in this way a Diagram. Of these lines, $p-1$ are sufficient to join the p points so as to form a connected system. Every new line completes a loop or closed path, or, as we shall call it, a Cycle. The number of independent cycles in the diagram is therefore $\kappa = l-p+1$.

Any closed path drawn along the lines of the diagram is composed of these independent cycles, each being taken any number of times and in either direction.

The existence of cycles is called Cyclosis, and the number of cycles in a diagram is called its Cyclomatic number.

Cyclosis in Surfaces and Regions.

Surfaces are either complete or bounded. Complete surfaces are either infinite or closed. Bounded surfaces are limited by one or

* *Der Census Raümlicher Complexe*, Gött. Abh., Bd. x. S. 97 (1861).

more closed lines, which may in the limiting cases become *double* finite lines or points.

A finite region of space is bounded by one or more closed surfaces. Of these one is the external surface, the others are included in it and exclude each other, and are called internal surfaces.

If the region has one bounding surface, we may suppose that surface to contract inwards without breaking its continuity or cutting itself. If the region is one of simple continuity, such as a sphere, this process may be continued till it is reduced to a point; but if the region is like a ring, the result will be a closed curve; and if the region has multiple connexions, the result will be a diagram of lines, and the cyclomatic number of the diagram will be that of the region. The space outside the region has the same cyclomatic number as the region itself. Hence, if the region is bounded by internal as well as external surfaces, its cyclomatic number is the sum of those due to all the surfaces.

When a region encloses within itself other regions, it is called a Periphractic region.

The number of internal bounding surfaces of a region is called its periphractic number. A closed surface is also periphractic, its periphractic number being unity.

The cyclomatic number of a closed surface is twice that of either of the regions which it bounds. To find the cyclomatic number of a bounded surface, suppose all the boundaries to contract inwards, without breaking continuity, till they meet. The surface will then be reduced to a point in the case of an acyclic surface, or to a linear diagram in the case of cyclic surfaces. The cyclomatic number of the diagram is that of the surface.

19.] THEOREM I. *If throughout any acyclic region*

$$X\,dx + Y\,dy + Z\,dz = -D\Psi,$$

the value of the line-integral from a point A to a point P taken along any path within the region will be the same.

We shall first shew that the line-integral taken round any closed path within the region is zero.

Suppose the equipotential surfaces drawn. They are all either closed surfaces or are bounded entirely by the surface of the region, so that a closed line within the region, if it cuts any of the surfaces at one part of its path, must cut the same surface in the opposite direction at some other part of its path, and the corresponding portions of the line-integral being equal and opposite, the total value is zero.

Hence if AQP and $AQ'P$ are two paths from A to P, the line-integral for $AQ'P$ is the sum of that for AQP and the closed path $AQ'PQA$. But the line-integral of the closed path is zero, therefore those of the two paths are equal.

Hence if the potential is given at any one point of such a region, that at any other point is determinate.

20.] THEOREM II. *In a cyclic region in which the equation*

$$X\,dx + Y\,dy + Z\,dz = -D\Psi$$

is everywhere satisfied, the line-integral from A to P, along a line drawn within the region, will not in general be determinate unless the channel of communication between A and P be specified.

Let K be the cyclomatic number of the region, then K sections of the region may be made by surfaces which we may call Diaphragms, so as to close up K of the channels of communication, and reduce the region to an acyclic condition without destroying its continuity.

The line-integral from A to any point P taken along a line which does not cut any of these diaphragms will be, by the last theorem, determinate in value.

Now let A and P be taken indefinitely near to each other, but on opposite sides of a diaphragm, and let K be the line-integral from A to P.

Let A' and P' be two other points on opposite sides of the same diaphragm and indefinitely near to each other, and let K' be the line-integral from A' to P'. Then $K' = K$.

For if we draw AA' and PP', nearly coincident, but on opposite sides of the diaphragm, the line-integrals along these lines will be equal. Suppose each equal to L, then K', the line-integral of $A'P'$, is equal to that of $A'A + AP + PP' = -L + K + L = K =$ that of AP.

Hence the line-integral round a closed curve which passes through one diaphragm of the system in a given direction is a constant quantity K. This quantity is called the Cyclic constant corresponding to the given cycle.

Let any closed curve be drawn within the region, and let it cut the diaphragm of the first cycle p times in the positive direction and p' times in the negative direction, and let $p - p' = n_1$. Then the line-integral of the closed curve will be $n_1 K_1$.

Similarly the line-integral of any closed curve will be

$$n_1 K_1 + n_2 K_2 + \dots + n_K K_K;$$

where n_K represents the excess of the number of positive passages of the curve through the diaphragm of the cycle K over the number of negative passages.

If two curves are such that one of them may be transformed into the other by continuous motion without at any time passing through any part of space for which the condition of having a potential is not fulfilled, these two curves are called Reconcileable curves. Curves for which this transformation cannot be effected are called Irreconcileable curves *.

The condition that $X\,dx + Y\,dy + Z\,dz$ is a complete differential of some function Ψ for all points within a certain region, occurs in several physical investigations in which the directed quantity and the potential have different physical interpretations.

In pure kinematics we may suppose X, Y, Z to be the components of the displacement of a point of a continuous body whose original coordinates are x, y, z; the condition then expresses that these displacements constitute a *non-rotational strain* †.

If X, Y, Z represent the components of the velocity of a fluid at the point x, y, z, then the condition expresses that the motion of the fluid is irrotational.

If X, Y, Z represent the components of the force at the point x, y, z, then the condition expresses that the work done on a particle passing from one point to another is the difference of the potentials at these points, and the value of this difference is the same for all reconcileable paths between the two points.

On Surface-Integrals.

21.] Let dS be the element of a surface, and ϵ the angle which a normal to the surface drawn towards the positive side of the surface makes with the direction of the vector quantity R, then $\iint R \cos\epsilon\, dS$ is called the *surface-integral of R over the surface S.*

THEOREM III. *The surface-integral of the flux inwards through a closed surface may be expressed as the volume-integral of its convergence taken within the surface.* (See Art. 25.)

Let X, Y, Z be the components of R, and let l, m, n be the

* See Sir W. Thomson 'On Vortex Motion,' *Trans. R. S. Edin.*, 1867–8.

† See Thomson and Tait's *Natural Philosophy*, § 190 (*i*).

direction-cosines of the normal to S measured inwards. Then the surface-integral of R over S is

$$\iint R\cos\epsilon\, dS = \iint Xl\, dS + \iint Ym\, dS + \iint Zn\, dS; \tag{1}$$

the values of X, Y, Z being those at a point in the surface, and the integrations being extended over the whole surface.

If the surface is a closed one, then, when y and z are given, the coordinate x must have an even number of values, since a line parallel to x must enter and leave the enclosed space an equal number of times provided it meets the surface at all.

At each entrance

$$l\, dS = dy\, dz,$$

and at each exit

$$l\, dS = -dy\, dz.$$

Let a point travelling from $x = -\infty$ to $x = +\infty$ first enter the space when $x = x_1$, then leave it when $x = x_2$, and so on; and let the values of X at these points be X_1, X_2, &c., then

$$\iint X l\, dS = \iint \{(X_1 - X_2) + (X_3 - X_4) + \text{\&c.} + (X_{2n-1} - X_{2n})\}\, dy\, dz. \tag{2}$$

If X is a quantity which is continuous, and has no infinite values between x_1 and x_2, then

$$X_2 - X_1 = \int_{x_1}^{x_2} \frac{dX}{dx}\, dx; \tag{3}$$

where the integration is extended from the first to the second intersection, that is, along the first segment of x which is within the closed surface. Taking into account all the segments which lie within the closed surface, we find

$$\iint X l\, dS = -\iiint \frac{dX}{dx}\, dx\, dy\, dz, \tag{4}$$

the double integration being confined to the closed surface, but the triple integration being extended to the whole enclosed space. Hence, if X, Y, Z are continuous and finite within a closed surface S, the total surface-integral of R over that surface will be

$$\iint R\cos\epsilon\, dS = -\iiint \left(\frac{dX}{dx} + \frac{dY}{dy} + \frac{dZ}{dz}\right) dx\, dy\, dz, \tag{5}$$

the triple integration being extended over the whole space within S.

Let us next suppose that X, Y, Z are not continuous within the closed surface, but that at a certain surface $F(x, y, z) = 0$ the values of X, Y, Z alter abruptly from X, Y, Z on the negative side of the surface to X', Y', Z' on the positive side.

If this discontinuity occurs, say, between x_1 and x_2, the value of X_2-X_1 will be

$$\int_{x_1}^{x_2} \frac{dX}{dx}\,dx+(X'-X), \tag{6}$$

where in the expression under the integral sign only the finite values of the derivative of X are to be considered.

In this case therefore the total surface-integral of R over the closed surface will be expressed by

$$\iint R\cos\epsilon\, dS = -\iiint\left(\frac{dX}{dx}+\frac{dY}{dy}+\frac{dZ}{dz}\right)dx\,dy\,dz+\iint(X'-X)\,dy\,dz$$
$$+\iint(Y'-Y)\,dz\,dx+\iint(Z'-Z)\,dx\,dy; \tag{7}$$

or, if l', m', n' are the direction-cosines of the normal to the surface of discontinuity, and dS' an element of that surface,

$$\iint R\cos\epsilon\, dS = -\iiint\left(\frac{dX}{dx}+\frac{dY}{dy}+\frac{dZ}{dz}\right)dx\,dy\,dz$$
$$+\iint\{l'(X'-X)+m'(Y'-Y)+n'(Z'-Z)\}\,dS', \tag{8}$$

where the integration of the last term is to be extended over the surface of discontinuity.

If at every point where X, Y, Z are continuous

$$\frac{dX}{dx}+\frac{dY}{dy}+\frac{dZ}{dz}=0, \tag{9}$$

and at every surface where they are discontinuous

$$l'X'+m'Y'+n'Z' = l'X+m'Y+n'Z, \tag{10}$$

then the surface-integral over every closed surface is zero, and the distribution of the vector quantity is said to be Solenoidal.

We shall refer to equation (9) as the General solenoidal condition, and to equation (10) as the Superficial solenoidal condition.

22.] Let us now consider the case in which at every point within the surface S the equation

$$\frac{dX}{dx}+\frac{dY}{dy}+\frac{dZ}{dz}=0 \tag{11}$$

is satisfied. We have as a consequence of this the surface-integral over the closed surface equal to zero.

Now let the closed surface S consist of three parts S_1, S_0, and S_2. Let S_1 be a surface of any form bounded by a closed line L_1. Let S_0 be formed by drawing lines from every point of L_1 always coinciding with the direction of R. If l, m, n are the direction-cosines of the normal at any point of the surface S_0, we have

$$R\cos\epsilon = Xl + Ym + Zn = 0. \tag{12}$$

Hence this part of the surface contributes nothing towards the value of the surface-integral.

Let S_2 be another surface of any form bounded by the closed curve L_2 in which it meets the surface S_0.

Let Q_1, Q_0, Q_2 be the surface-integrals of the surfaces S_1, S_0, S_2, and let Q be the surface-integral of the closed surface S. Then

$$Q = Q_1 + Q_0 + Q_2 = 0; \tag{13}$$

and we know that

$$Q_0 = 0; \tag{14}$$

therefore

$$Q_2 = -Q_1; \tag{15}$$

or, in other words, the surface-integral over the surface S_2 is equal and opposite to that over S_1 whatever be the form and position of S_2, provided that the intermediate surface S_0 is one for which R is always tangential.

If we suppose L_1 a closed curve of small area, S_0 will be a tubular surface having the property that the surface-integral over every complete section of the tube is the same.

Since the whole space can be divided into tubes of this kind provided

$$\frac{dX}{dx} + \frac{dY}{dy} + \frac{dZ}{dz} = 0, \tag{16}$$

a distribution of a vector quantity consistent with this equation is called a Solenoidal Distribution.

On Tubes and Lines of Flow.

If the space is so divided into tubes that the surface-integral for every tube is unity, the tubes are called Unit tubes, and the surface-integral over any finite surface S bounded by a closed curve L is equal to the *number* of such tubes which pass through S in the positive direction, or, what is the same thing, the number which pass through the closed curve L.

Hence the surface-integral of S depends only on the form of its boundary L, and not on the form of the surface within its boundary.

On Periphractic Regions.

If, throughout the whole region bounded externally by the single closed surface S, the solenoidal condition

$$\frac{dX}{dx} + \frac{dY}{dy} + \frac{dZ}{dz} = 0$$

is satisfied, then the surface-integral taken over any closed surface drawn within this region will be zero, and the surface-integral taken over a bounded surface within the region will depend only on the form of the closed curve which forms its boundary.

It is not, however, generally true that the same results follow if the region within which the solenoidal condition is satisfied is bounded otherwise than by a single surface.

For if it is bounded by more than one continuous surface, one of these is the external surface and the others are internal surfaces, and the region S is a periphractic region, having within it other regions which it completely encloses.

If within one of these enclosed regions, say, that bounded by the closed surface S_1, the solenoidal condition is not satisfied, let

$$Q_1 = \iint R \cos \epsilon \, dS_1$$

be the surface-integral for the surface enclosing this region, and let Q_2, Q_3, &c. be the corresponding quantities for the other enclosed regions S_2, S_3, &c.

Then, if a closed surface S' is drawn within the region S, the value of its surface-integral will be zero only when this surface S' does not include any of the enclosed regions S_1, S_2, &c. If it includes any of these, the surface-integral is the sum of the surface-integrals of the different enclosed regions which lie within it.

For the same reason, the surface-integral taken over a surface bounded by a closed curve is the same for such surfaces only, bounded by the closed curve, as are reconcileable with the given surface by continuous motion of the surface within the region S.

When we have to deal with a periphractic region, the first thing to be done is to reduce it to an aperiphractic region by drawing lines L_1, L_2, &c. joining the internal surfaces S_1, S_2, &c. to the external surface S. Each of these lines, provided it joins surfaces which were not already in continuous connexion, reduces the periphractic number by unity, so that the whole number of lines to be drawn to remove the periphraxy is equal to the periphractic number, or the number of internal surfaces. In drawing these lines we must remember that any line joining surfaces which are already connected does not diminish the periphraxy, but introduces cyclosis. When these lines have been drawn we may assert that if the solenoidal condition is satisfied in the region S, any closed surface drawn entirely within S, and not cutting any of the lines, has its surface-integral zero. If it cuts any line, say L_1, once or any odd number of times, it encloses the surface S_1 and the surface-integral is Q_1.

The most familiar example of a periphractic region within which the solenoidal condition is satisfied is the region surrounding a mass attracting or repelling inversely as the square of the distance.

In this case we have

$$X = m\frac{x}{r^3}, \quad Y = m\frac{y}{r^3}, \quad Z = m\frac{z}{r^3};$$

where m is the mass, supposed to be at the origin of coordinates.

At any point where r is finite

$$\frac{dX}{dx} + \frac{dY}{dy} + \frac{dZ}{dz} = 0,$$

but at the origin these quantities become infinite. For any closed surface not including the origin, the surface-integral is zero. If a closed surface includes the origin, its surface-integral is $4\pi m$.

If, for any reason, we wish to treat the region round m as if it were not periphractic, we must draw a line from m to an infinite distance, and in taking surface-integrals we must remember to add $4\pi m$ whenever this line crosses from the negative to the positive side of the surface.

On Right-handed and Left-handed Relations in Space.

23.] In this treatise the motions of translation along any axis and of rotation about that axis will be assumed to be of the same sign when their directions correspond to those of the translation and rotation of an ordinary or right-handed screw *.

For instance, if the actual rotation of the earth from west to east is taken positive, the direction of the earth's axis from south to north will be taken positive, and if a man walks forward in the positive direction, the positive rotation is in the order, head, right-hand, feet, left-hand.

* The combined action of the muscles of the arm when we turn the upper side of the right-hand outwards, and at the same time thrust the hand forwards, will impress the right-handed screw motion on the memory more firmly than any verbal definition. A common corkscrew may be used as a material symbol of the same relation.

Professor W. H. Miller has suggested to me that as the tendrils of the vine are right-handed screws and those of the hop left-handed, the two systems of relations in space might be called those of the vine and the hop respectively.

The system of the vine, which we adopt, is that of Linnæus, and of screw-makers in all civilized countries except Japan. De Candolle was the first who called the hop-tendril right-handed, and in this he is followed by Listing, and by most writers on the circular polarization of light. Screws like the hop-tendril are made for the couplings of railway-carriages, and for the fittings of wheels on the left side of ordinary carriages, but they are always called left-handed screws by those who use them.

If we place ourselves on the positive side of a surface, the positive direction along its bounding curve will be opposite to the motion of the hands of a watch with its face towards us.

This is the right-handed system which is adopted in Thomson and Tait's *Natural Philosophy*, § 243, and in Tait's *Quaternions*. The opposite, or left-handed system, is adopted in Hamilton's *Quaternions* (*Lectures*, p. 76, and *Elements*, p. 108, and p. 117 note). The operation of passing from the one system to the other is called, by Listing, *Perversion*.

The reflexion of an object in a mirror is a perverted image of the object.

When we use the Cartesian axes of x, y, z, we shall draw them so that the ordinary conventions about the cyclic order of the symbols lead to a right-handed system of directions in space. Thus, if x is drawn eastward and y northward, z must be drawn upward.

The areas of surfaces will be taken positive when the order of integration coincides with the cyclic order of the symbols. Thus, the area of a closed curve in the plane of xy may be written either

$$\int x\,dy \quad \text{or} \quad -\int y\,dx;$$

the order of integration being x, y in the first expression, and y, x in the second.

This relation between the two products $dx\,dy$ and $dy\,dx$ may be compared with the rule for the product of two perpendicular vectors in the method of Quaternions, the sign of which depends on the order of multiplication; and with the reversal of the sign of a determinant when the adjoining rows or columns are exchanged.

For similar reasons a volume-integral is to be taken positive when the order of integration is in the cyclic order of the variables x, y, z, and negative when the cyclic order is reversed.

We now proceed to prove a theorem which is useful as establishing a connexion between the surface-integral taken over a finite surface and a line-integral taken round its boundary.

24.] THEOREM IV. *A line-integral taken round a closed curve may be expressed in terms of a surface-integral taken over a surface bounded by the curve.*

Let X, Y, Z be the components of a vector quantity $\mathfrak{A}$ whose line-integral is to be taken round a closed curve s.

Let S be any continuous finite surface bounded entirely by the closed curve s, and let ξ, η, ζ be the components of another vector quantity 𝔅, related to X, Y, Z by the equations

$$\xi = \frac{dZ}{dy} - \frac{dY}{dz}, \quad \eta = \frac{dX}{dz} - \frac{dZ}{dx}, \quad \zeta = \frac{dY}{dx} - \frac{dX}{dy}. \tag{1}$$

Then the surface-integral of 𝔅 taken over the surface S is equal to the line-integral of 𝔄 taken round the curve s. It is manifest that ξ, η, ζ satisfy of themselves the solenoidal condition

$$\frac{d\xi}{dx} + \frac{d\eta}{dy} + \frac{d\zeta}{dz} = 0.$$

Let l, m, n be the direction-cosines of the normal to an element of the surface dS, reckoned in the positive direction. Then the value of the surface-integral of 𝔅 may be written

$$\iint (l\xi + m\eta + n\zeta)\, dS. \tag{2}$$

In order to form a definite idea of the meaning of the element dS, we shall suppose that the values of the coordinates x, y, z for every point of the surface are given as functions of two independent variables α and β. If β is constant and α varies, the point (x, y, z) will describe a curve on the surface, and if a series of values is given to β, a series of such curves will be traced, all lying on the surface S. In the same way, by giving a series of constant values to α, a second series of curves may be traced, cutting the first series, and dividing the whole surface into elementary portions, any one of which may be taken as the element dS.

The projection of this element on the plane of $y\,z$ is, by the ordinary formula,

$$l\,dS = \left(\frac{dy}{d\alpha}\frac{dz}{d\beta} - \frac{dy}{d\beta}\frac{dz}{d\alpha}\right) d\beta\, d\alpha. \tag{3}$$

The expressions for $m\,dS$ and $n\,dS$ are obtained from this by substituting x, y, z in cyclic order.

The surface-integral which we have to find is

$$\iint (l\xi + m\eta + n\zeta)\, dS; \tag{4}$$

or, substituting the values of ξ, η, ζ in terms of X, Y, Z,

$$\iint \left(m\frac{dX}{dz} - n\frac{dX}{dy} + n\frac{dY}{dx} - l\frac{dY}{dz} + l\frac{dZ}{dy} - m\frac{dZ}{dx}\right) dS. \tag{5}$$

The part of this which depends on X may be written

$$\iint \left\{ \frac{dX}{dz}\left(\frac{dz}{d\alpha}\frac{dx}{d\beta} - \frac{dz}{d\beta}\frac{dx}{d\alpha}\right) - \frac{dX}{dy}\left(\frac{dx}{d\alpha}\frac{dy}{d\beta} - \frac{dx}{d\beta}\frac{dy}{d\alpha}\right)\right\} d\beta\, d\alpha; \tag{6}$$

adding and subtracting $\frac{dX}{dx}\frac{dx}{da}\frac{dx}{d\beta}$, this becomes

$$\iint\left\{\frac{dx}{d\beta}\left(\frac{dX}{dx}\frac{dx}{da}+\frac{dX}{dy}\frac{dy}{da}+\frac{dX}{dz}\frac{dz}{da}\right)\right.$$
$$\left.-\frac{dx}{da}\left(\frac{dX}{dx}\frac{dx}{d\beta}+\frac{dX}{dy}\frac{dy}{d\beta}+\frac{dX}{dz}\frac{dz}{d\beta}\right)\right\}d\beta\, da; \quad (7)$$

$$=\iint\left(\frac{dX}{da}\frac{dx}{d\beta}-\frac{dX}{d\beta}\frac{dx}{da}\right)d\beta\, da. \quad (8)$$

Let us now suppose that the curves for which a is constant form a series of closed curves surrounding a point on the surface for which a has its minimum value, a_0, and let the last curve of the series, for which $a = a_1$, coincide with the closed curve s.

Let us also suppose that the curves for which β is constant form a series of lines drawn from the point at which $a = a_0$ to the closed curve s, the first, β_0, and the last, β_1, being identical.

Integrating (8) by parts, the first term with respect to a and the second with respect to β, the double integrals destroy each other and the expression becomes

$$\int_{\beta_0}^{\beta_1}\left(X\frac{dx}{d\beta}\right)_{a=a_1}d\beta-\int_{\beta_0}^{\beta_1}\left(X\frac{dx}{d\beta}\right)_{a=a_0}d\beta-\int_{a_0}^{a_1}\left(X\frac{dx}{da}\right)_{\beta=\beta_1}da$$
$$+\int_{a_0}^{a_1}\left(X\frac{dx}{da}\right)_{\beta=\beta_0}da. \quad (9)$$

Since the point (a, β_1) is identical with the point (a, β_0), the third and fourth terms destroy each other; and since there is but one value of x at the point where $a = a_0$, the second term is zero, and the expression is reduced to the first term:

Since the curve $a = a_1$ is identical with the closed curve s, we may write the expression in the form

$$\int X\frac{dx}{ds}ds, \quad (10)$$

where the integration is to be performed round the curve s. We may treat in the same way the parts of the surface-integral which depend upon Y and Z, so that we get finally,

$$\iint(l\xi+m\eta+n\zeta)\,dS=\int\left(X\frac{dx}{ds}+Y\frac{dy}{ds}+Z\frac{dz}{ds}\right)ds; \quad (11)$$

where the first integral is extended over the surface S, and the second round the bounding curve s *.

* This theorem was given by Professor Stokes, *Smith's Prize Examination*, 1854, question 8. It is proved in Thomson and Tait's *Natural Philosophy*, § 190 (*j*).

On the effect of the operator ∇ on a vector function.

25.] We have seen that the operation denoted by ∇ is that by which a vector quantity is deduced from its potential. The same operation, however, when applied to a vector function, produces results which enter into the two theorems we have just proved (III and IV). The extension of this operator to vector displacements, and most of its further development, is due to Professor Tait *.

Let σ be a vector function of ρ, the vector of a variable point. Let us suppose, as usual, that

$$\rho = ix + jy + kz,$$

and $$\sigma = iX + jY + kZ;$$

where X, Y, Z are the components of σ in the directions of the axes.

We have to perform on σ the operation

$$\nabla = i\frac{d}{dx} + j\frac{d}{dy} + k\frac{d}{dz}.$$

Performing this operation, and remembering the rules for the multiplication of i, j, k, we find that $\nabla\sigma$ consists of two parts, one scalar and the other vector.

The scalar part is

$$S\nabla\sigma = -\left(\frac{dX}{dx} + \frac{dY}{dy} + \frac{dZ}{dz}\right), \text{ see Theorem III,}$$

and the vector part is

$$V\nabla\sigma = i\left(\frac{dZ}{dy} - \frac{dY}{dz}\right) + j\left(\frac{dX}{dz} - \frac{dZ}{dx}\right) + k\left(\frac{dY}{dx} - \frac{dX}{dy}\right).$$

If the relation between X, Y, Z and ξ, η, ζ is that given by equation (1) of the last theorem, we may write

$$V\nabla\sigma = i\xi + j\eta + k\zeta. \text{ See Theorem IV.}$$

It appears therefore that the functions of X, Y, Z which occur in the two theorems are both obtained by the operation ∇ on the vector whose components are X, Y, Z. The theorems themselves may be written

$$\iiint S\nabla\sigma\, ds = \iint S.\sigma Uv\, ds, \quad \text{(III)}$$

and $$\int S\sigma\, d\rho = \iint S.\nabla\sigma\, Uv\, ds; \quad \text{(IV)}$$

* See *Proc. R. S. Edin.*, April 28, 1862. 'On Green's and other allied Theorems,' *Trans. R. S. Edin.*, 1869-70, a very valuable paper; and 'On some Quaternion Integrals,' *Proc. R. S. Edin.*, 1870-71.

where $d\varsigma$ is an element of a volume, ds of a surface, $d\rho$ of a curve, and $U\nu$ a unit-vector in the direction of the normal.

To understand the meaning of these functions of a vector, let us suppose that σ_0 is the value of σ at a point P, and let us examine the value of $\sigma-\sigma_0$ in the neighbourhood of P. If we draw a closed surface round P, then, if the surface-integral of σ over this surface is directed inwards, $S\nabla\sigma$ will be positive, and the vector $\sigma-\sigma_0$ near the point P will be on the whole directed towards P, as in the figure (1).

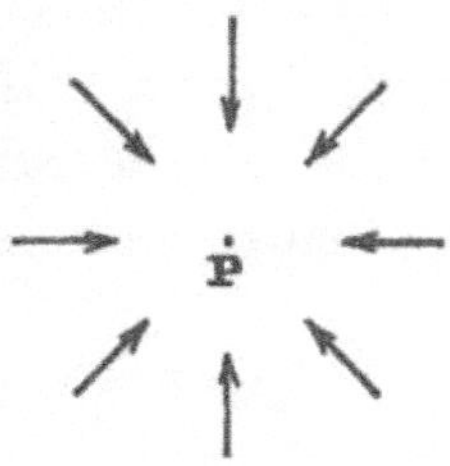

Fig. 1.

I propose therefore to call the scalar part of $\nabla\sigma$ the *convergence* of σ at the point P.

To interpret the vector part of $\nabla\sigma$, let us suppose ourselves to be looking in the direction of the vector whose components are ξ, η, ζ, and let us examine the vector $\sigma-\sigma_0$ near the point P. It will appear as in the figure (2), this vector being arranged on the whole tangentially in the direction opposite to the hands of a watch.

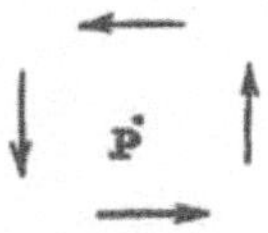

Fig. 2.

I propose (with great diffidence) to call the vector part of $\nabla\sigma$ the *rotation* of σ at the point P.

In Fig. 3 we have an illustration of rotation combined with convergence.

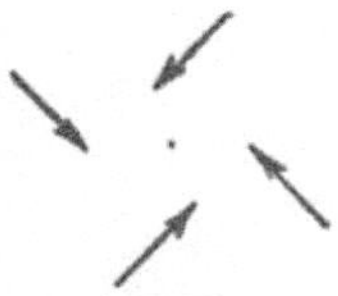
Fig. 3.

Let us now consider the meaning of the equation

$$V\nabla\sigma = 0.$$

This implies that $\nabla\sigma$ is a scalar, or that the vector σ is the space-variation of some scalar function Ψ.

26.] One of the most remarkable properties of the operator ∇ is that when repeated it becomes

$$\nabla^2 = -\left(\frac{d^2}{dx^2} + \frac{d^2}{dy^2} + \frac{d^2}{dz^2}\right),$$

an operator occurring in all parts of Physics, which we may refer to as Laplace's Operator.

This operator is itself essentially scalar. When it acts on a scalar function the result is scalar, when it acts on a vector function the result is a vector.

If, with any point P as centre, we draw a small sphere whose radius is r, then if q_0 is the value of q at the centre, and $\bar{q}$ the mean value of q for all points within the sphere,

$$q_0 - \bar{q} = \tfrac{1}{10} r^2 \nabla^2 q;$$

so that the value at the centre exceeds or falls short of the mean value according as $\nabla^2 q$ is positive or negative.

I propose therefore to call $\nabla^2 q$ the *concentration* of q at the point P, because it indicates the excess of the value of q at that point over its mean value in the neighbourhood of the point.

If q is a scalar function, the method of finding its mean value is well known. If it is a vector function, we must find its mean value by the rules for integrating vector functions. The result of course is a vector.

PART I.

ELECTROSTATICS.

CHAPTER I.

DESCRIPTION OF PHENOMENA.

Electrification by Friction.

27.] Experiment I*. Let a piece of glass and a piece of resin, neither of which exhibits any electrical properties, be rubbed together and left with the rubbed surfaces in contact. They will still exhibit no electrical properties. Let them be separated. They will now attract each other.

If a second piece of glass be rubbed with a second piece of resin, and if the pieces be then separated and suspended in the neighbourhood of the former pieces of glass and resin, it may be observed—

(1) That the two pieces of glass repel each other.

(2) That each piece of glass attracts each piece of resin.

(3) That the two pieces of resin repel each other.

These phenomena of attraction and repulsion are called Electrical phenomena, and the bodies which exhibit them are said to be *electrified*, or to be *charged with electricity*.

Bodies may be electrified in many other ways, as well as by friction.

The electrical properties of the two pieces of glass are similar to each other but opposite to those of the two pieces of resin: the glass attracts what the resin repels and repels what the resin attracts.

* See Sir W. Thomson 'On the Mathematical Theory of Electricity,' *Cambridge and Dublin Mathematical Journal*, March, 1848.

If a body electrified in any manner whatever behaves as the glass does, that is, if it repels the glass and attracts the resin, the body is said to be *vitreously* electrified, and if it attracts the glass and repels the resin it is said to be *resinously* electrified. All electrified bodies are found to be either vitreously or resinously electrified.

It is the established practice of men of science to call the vitreous electrification positive, and the resinous electrification negative. The exactly opposite properties of the two kinds of electrification justify us in indicating them by opposite signs, but the application of the positive sign to one rather than to the other kind must be considered as a matter of arbitrary convention, just as it is a matter of convention in mathematical diagrams to reckon positive distances towards the right hand.

No force, either of attraction or of repulsion, can be observed between an electrified body and a body not electrified. When, in any case, bodies not previously electrified are observed to be acted on by an electrified body, it is because they have become *electrified by induction*.

Electrification by Induction.

28.] EXPERIMENT II *. Let a hollow vessel of metal be hung up by white silk threads, and let a similar thread be attached to the lid of the vessel so that the vessel may be opened or closed without touching it.

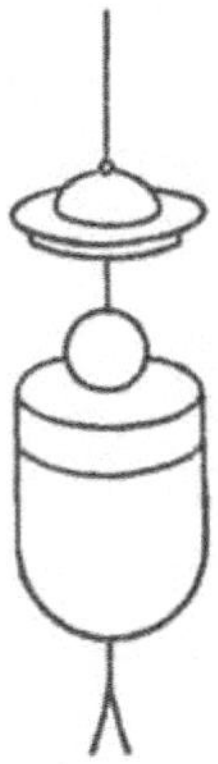

Fig. 4.

Let the pieces of glass and resin be similarly suspended and electrified as before.

Let the vessel be originally unelectrified, then if an electrified piece of glass is hung up within it by its thread without touching the vessel, and the lid closed, the outside of the vessel will be found to be vitreously electrified, and it may be shewn that the electrification outside of the vessel is exactly the same in whatever part of the interior space the glass is suspended.

If the glass is now taken out of the vessel without touching it, the electrification of the glass will be the same as before it was put in, and that of the vessel will have disappeared.

This electrification of the vessel, which depends on the glass

* This, and several experiments which follow, are due to Faraday, 'On Static Electrical Inductive Action,' *Phil. Mag.*, 1843, or *Exp. Res.*, vol. ii. p. 279.

being within it, and which vanishes when the glass is removed, is called electrification by Induction.

Similar effects would be produced if the glass were suspended near the vessel on the outside, but in that case we should find an electrification, vitreous in one part of the outside of the vessel and resinous in another. When the glass is inside the vessel the whole of the outside is vitreously and the whole of the inside resinously electrified.

Electrification by Conduction.

29.] EXPERIMENT III. Let the metal vessel be electrified by induction, as in the last experiment, let a second metallic body be suspended by white silk threads near it, and let a metal wire, similarly suspended, be brought so as to touch simultaneously the electrified vessel and the second body.

The second body will now be found to be vitreously electrified, and the vitreous electrification of the vessel will have diminished.

The electrical condition has been transferred from the vessel to the second body by means of the wire. The wire is called a *conductor* of electricity, and the second body is said to be *electrified by conduction.*

Conductors and Insulators.

EXPERIMENT IV. If a glass rod, a stick of resin or gutta-percha, or a white silk thread, had been used instead of the metal wire, no transfer of electricity would have taken place. Hence these latter substances are called Non-conductors of electricity. Non-conductors are used in electrical experiments to support electrified bodies without carrying off their electricity. They are then called Insulators.

The metals are good conductors; air, glass, resins, gutta-percha, vulcanite, paraffin, &c. are good insulators; but, as we shall see afterwards, all substances resist the passage of electricity, and all substances allow it to pass, though in exceedingly different degrees. This subject will be considered when we come to treat of the motion of electricity. For the present we shall consider only two classes of bodies, good conductors, and good insulators.

In Experiment II an electrified body produced electrification in the metal vessel while separated from it by air, a non-conducting medium. Such a medium, considered as transmitting these electrical effects without conduction, has been called by Faraday a Dielectric medium, and the action which takes place through it is called Induction.

In Experiment III the electrified vessel produced electrification in the second metallic body through the medium of the wire. Let us suppose the wire removed, and the electrified piece of glass taken out of the vessel without touching it, and removed to a sufficient distance. The second body will still exhibit vitreous electrification, but the vessel, when the glass is removed, will have resinous electrification. If we now bring the wire into contact with both bodies, conduction will take place along the wire, and all electrification will disappear from both bodies, shewing that the electrification of the two bodies was equal and opposite.

30.] Experiment V. In Experiment II it was shewn that if a piece of glass, electrified by rubbing it with resin, is hung up in an insulated metal vessel, the electrification observed outside does not depend on the position of the glass. If we now introduce the piece of resin with which the glass was rubbed into the same vessel, without touching it or the vessel, it will be found that there is no electrification outside the vessel. From this we conclude that the electrification of the resin is exactly equal and opposite to that of the glass. By putting in any number of bodies, electrified in any way, it may be shewn that the electrification of the outside of the vessel is that due to the algebraic sum of all the electrifications, those being reckoned negative which are resinous. We have thus a practical method of adding the electrical effects of several bodies without altering the electrification of each.

31.] Experiment VI. Let a second insulated metallic vessel, B, be provided, and let the electrified piece of glass be put into the first vessel A, and the electrified piece of resin into the second vessel B. Let the two vessels be then put in communication by the metal wire, as in Experiment III. All signs of electrification will disappear.

Next, let the wire be removed, and let the pieces of glass and of resin be taken out of the vessels without touching them. It will be found that A is electrified resinously and B vitreously.

If now the glass and the vessel A be introduced together into a larger insulated vessel C, it will be found that there is no electrification outside C. This shews that the electrification of A is exactly equal and opposite to that of the piece of glass, and that of B may be shewn in the same way to be equal and opposite to that of the piece of resin.

We have thus obtained a method of charging a vessel with a quantity of electricity exactly equal and opposite to that of an electrified body without altering the electrification of the latter, and we may in this way charge any number of vessels with exactly equal quantities of electricity of either kind, which we may take for provisional units.

32.] EXPERIMENT VII. Let the vessel *B*, charged with a quantity of positive electricity, which we shall call, for the present, unity, be introduced into the larger insulated vessel *C* without touching it. It will produce a positive electrification on the outside of *C*. Now let *B* be made to touch the inside of *C*. No change of the external electrification will be observed. If *B* is now taken out of *C* without touching it, and removed to a sufficient distance, it will be found that *B* is completely discharged, and that *C* has become charged with a unit of positive electricity.

We have thus a method of transferring the charge of *B* to *C*.

Let *B* be now recharged with a unit of electricity, introduced into *C* already charged, made to touch the inside of *C*, and removed. It will be found that *B* is again completely discharged, so that the charge of *C* is doubled.

If this process is repeated, it will be found that however highly *C* is previously charged, and in whatever way *B* is charged, when *B* is first entirely enclosed in *C*, then made to touch *C*, and finally removed without touching *C*, the charge of *B* is completely transferred to *C*, and *B* is entirely free from electrification.

This experiment indicates a method of charging a body with any number of units of electricity. We shall find, when we come to the mathematical theory of electricity, that the result of this experiment affords an accurate test of the truth of the theory.

33.] Before we proceed to the investigation of the law of electrical force, let us enumerate the facts we have already established.

By placing any electrified system inside an insulated hollow conducting vessel, and examining the resultant effect on the outside of the vessel, we ascertain the character of the total electrification of the system placed inside, without any communication of electricity between the different bodies of the system.

The electrification of the outside of the vessel may be tested with great delicacy by putting it in communication with an electroscope.

We may suppose the electroscope to consist of a strip of gold leaf hanging between two bodies charged, one positively, and the other negatively. If the gold leaf becomes electrified it will incline towards the body whose electrification is opposite to its own. By increasing the electrification of the two bodies and the delicacy of the suspension, an exceedingly small electrification of the gold leaf may be detected.

When we come to describe electrometers and multipliers we shall find that there are still more delicate methods of detecting electrification and of testing the accuracy of our theories, but at present we shall suppose the testing to be made by connecting the hollow vessel with a gold leaf electroscope.

This method was used by Faraday in his very admirable demonstration of the laws of electrical phenomena *.

34.] I. The total electrification of a body, or system of bodies, remains always the same, except in so far as it receives electrification from or gives electrification to other bodies.

In all electrical experiments the electrification of bodies is found to change, but it is always found that this change is due to want of perfect insulation, and that as the means of insulation are improved, the loss of electrification becomes less. We may therefore assert that the electrification of a body placed in a perfectly insulating medium would remain perfectly constant.

II. When one body electrifies another by conduction, the total electrification of the two bodies remains the same, that is, the one loses as much positive or gains as much negative electrification as the other gains of positive or loses of negative electrification.

For if the two bodies are enclosed in the hollow vessel, no change of the total electrification is observed.

III. When electrification is produced by friction, or by any other known method, equal quantities of positive and negative electrification are produced.

For the electrification of the whole system may be tested in the hollow vessel, or the process of electrification may be carried on within the vessel itself, and however intense the electrification of the parts of the system may be, the electrification of the whole, as indicated by the gold leaf electroscope, is invariably zero.

The electrification of a body is therefore a physical quantity capable of measurement, and two or more electrifications can be combined experimentally with a result of the same kind as when

* 'On Static Electrical Inductive Action,' *Phil. Mag.*, 1843, or *Exp. Res.*, vol. ii. p. 249.

two quantities are added algebraically. We therefore are entitled to use language fitted to deal with electrification as a quantity as well as a quality, and to speak of any electrified body as 'charged with a certain quantity of positive or negative electricity.'

35.] While admitting electricity, as we have now done, to the rank of a physical quantity, we must not too hastily assume that it is, or is not, a substance, or that it is, or is not, a form of energy, or that it belongs to any known category of physical quantities. All that we have hitherto proved is that it cannot be created or annihilated, so that if the total quantity of electricity within a closed surface is increased or diminished, the increase or diminution must have passed in or out through the closed surface.

This is true of matter, and is expressed by the equation known as the Equation of Continuity in Hydrodynamics.

It is not true of heat, for heat may be increased or diminished within a closed surface, without passing in or out through the surface, by the transformation of some other form of energy into heat, or of heat into some other form of energy.

It is not true even of energy in general if we admit the immediate action of bodies at a distance. For a body outside the closed surface may make an exchange of energy with a body within the surface. But if all apparent action at a distance is the result of the action between the parts of an intervening medium, it is conceivable that in all cases of the increase or diminution of the energy within a closed surface we may be able, when the nature of this action of the parts of the medium is clearly understood, to trace the passage of the energy in or out through that surface.

There is, however, another reason which warrants us in asserting that electricity, as a physical quantity, synonymous with the total electrification of a body, is not, like heat, a form of energy. An electrified system has a certain amount of energy, and this energy can be calculated by multiplying the quantity of electricity in each of its parts by another physical quantity, called the Potential of that part, and taking half the sum of the products. The quantities 'Electricity' and 'Potential,' when multiplied together, produce the quantity 'Energy.' It is impossible, therefore, that electricity and energy should be quantities of the same category, for electricity is only one of the factors of energy, the other factor being 'Potential.'

Energy, which is the product of these factors, may also be considered as the product of several other pairs of factors, such as

A Force	× A distance through which the force is to act.
A Mass	× Gravitation acting through a certain height.
A Mass	× Half the square of its velocity.
A Pressure	× A volume of fluid introduced into a vessel at that pressure.
A Chemical Affinity	× A chemical change, measured by the number of electro-chemical equivalents which enter into combination.

If we ever should obtain distinct mechanical ideas of the nature of electric potential, we may combine these with the idea of energy to determine the physical category in which 'Electricity' is to be placed.

36.] In most theories on the subject, Electricity is treated as a substance, but inasmuch as there are two kinds of electrification which, being combined, annul each other, and since we cannot conceive of two substances annulling each other, a distinction has been drawn between Free Electricity and Combined Electricity.

Theory of Two Fluids.

In what is called the Theory of Two Fluids, all bodies, in their unelectrified state, are supposed to be charged with equal quantities of positive and negative electricity. These quantities are supposed to be so great that no process of electrification has ever yet deprived a body of all the electricity of either kind. The process of electrification, according to this theory, consists in taking a certain quantity P of positive electricity from the body A and communicating it to B, or in taking a quantity N of negative electricity from B and communicating it to A, or in some combination of these processes.

The result will be that A will have $P+N$ units of negative electricity over and above its remaining positive electricity, which is supposed to be in a state of combination with an equal quantity of negative electricity. This quantity $P+N$ is called the Free electricity, the rest is called the Combined, Latent, or Fixed electricity.

In most expositions of this theory the two electricities are called 'Fluids,' because they are capable of being transferred from one body to another, and are, within conducting bodies, extremely mobile. The other properties of fluids, such as their inertia, weight, and elasticity, are not attributed to them by those who have used the theory for merely mathematical purposes; but the use of the word Fluid has been apt to mislead the vulgar, including many men of science who are not natural philosophers, and who have seized on the word Fluid as the only term in the statement of the theory which seemed intelligible to them.

We shall see that the mathematical treatment of the subject has been greatly developed by writers who express themselves in terms of the 'Two Fluids' theory. Their results, however, have been deduced entirely from data which can be proved by experiment, and which must therefore be true, whether we adopt the theory of two fluids or not. The experimental verification of the mathematical results therefore is no evidence for or against the peculiar doctrines of this theory.

The introduction of two fluids permits us to consider the negative electrification of A and the positive electrification of B as the effect of *any one* of three different processes which would lead to the same result. We have already supposed it produced by the transfer of P units of positive electricity from A to B, together with the transfer of N units of negative electricity from B to A. But if $P+N$ units of positive electricity had been transferred from A to B, or if $P+N$ units of negative electricity had been transferred from B to A, the resulting 'free electricity' on A and on B would have been the same as before, but the quantity of 'combined electricity' in A would have been less in the second case and greater in the third than it was in the first.

It would appear therefore, according to this theory, that it is possible to alter not only the amount of free electricity in a body, but the amount of combined electricity. But no phenomena have ever been observed in electrified bodies which can be traced to the varying amount of their combined electricities. Hence either the combined electricities have no observable properties, or the amount of the combined electricities is incapable of variation. The first of these alternatives presents no difficulty to the mere mathematician, who attributes no properties to the fluids except those of attraction and repulsion, for he conceives the two fluids simply to annul one another, like $+e$ and $-e$, and their combination to be a true mathematical zero. But to those who cannot use the word Fluid without thinking of a substance it is difficult to conceive how the combination of the two fluids can have no properties at all, so that the addition of more or less of the combination to a body shall not in any way affect it, either by increasing its mass or its weight, or altering some of its other properties. Hence it has been supposed by some, that in every process of electrification exactly equal quantities of the two fluids are transferred in opposite directions, so that the total quantity of the two fluids in any body taken together remains always the same. By this new law they 'contrive to save appearances,' forgetting that there would have been no need of the law except to reconcile the 'two fluids' theory with facts, and to prevent it from predicting non-existent phenomena.

Theory of One Fluid.

37.] In the theory of One Fluid everything is the same as in the theory of Two Fluids except that, instead of supposing the two substances equal and opposite in all respects, one of them, generally the negative one, has been endowed with the properties and name of Ordinary Matter, while the other retains the name of The Electric Fluid. The particles of the fluid are supposed to repel one another according to the law of the inverse square of the distance, and to attract those of matter according to the same law. Those of matter are supposed to repel each other and attract those of electricity.

If the quantity of the electric fluid in a body is such that a particle of the electric fluid outside the body is as much repelled by the electric fluid in the body as it is attracted by the matter of the body, the body is said to be Saturated. If the quantity of fluid in the body is greater than that required for saturation, the excess is called the Redundant fluid, and the body is said to be Overcharged. If it is less, the body is said to be Undercharged, and the quantity of fluid which would be required to saturate it is sometimes called the Deficient fluid. The number of units of electricity required to saturate one gramme of ordinary matter must be very great, because a gramme of gold may be beaten out to an area of a square metre, and when in this form may have a negative charge of at least 60,000 units of electricity. In order to saturate the gold leaf, this quantity of electric fluid must be communicated to it, so that the whole quantity required to saturate it must be greater than this. The attraction between the matter and the fluid in two saturated bodies is supposed to be a very little greater than the repulsion between the two portions of matter and that between the two portions of fluid. This residual force is supposed to account for the attraction of gravitation.

This theory does not, like the Two-Fluid theory, explain too much. It requires us, however, to suppose the mass of the electric fluid so small that no attainable positive or negative electrification has yet perceptibly increased or diminished either the mass or the weight of a body, and it has not yet been able to assign sufficient reasons why the vitreous rather than the resinous electrification should be supposed due to an *excess* of electricity.

One objection has sometimes been urged against this theory by men who ought to have reasoned better. It has been said that the doctrine that the particles of matter uncombined with electricity *repel* one another, is in direct antagonism with the well-established fact that every particle of matter *attracts* every other particle throughout the universe. If the theory of One Fluid were true we should have the heavenly bodies repelling one another.

But it is manifest that the heavenly bodies, according to this theory, if they consisted of matter uncombined with electricity, would be in the highest state of negative electrification, and would repel each other. We have no reason to believe that they are in such a highly electrified state, or could be maintained in that state. The earth and all the bodies whose attraction has been observed are rather in an unelectrified state, that is, they contain the normal charge of electricity, and the only action between them is the residual force lately mentioned. The artificial manner, however, in which this residual force is introduced is a much more valid objection to the theory.

In the present treatise I propose, at different stages of the investigation, to test the different theories in the light of additional classes of phenomena. For my own part, I look for additional light on the nature of electricity from a study of what takes place in the space intervening between the electrified bodies. Such is the essential character of the mode of investigation pursued by Faraday in his *Experimental Researches*, and as we go on I intend to exhibit the results, as developed by Faraday, W. Thomson, &c., in a connected and mathematical form, so that we may perceive what phenomena are explained equally well by all the theories, and what phenomena indicate the peculiar difficulties of each theory.

Measurement of the Force between Electrified Bodies.

38.] Forces may be measured in various ways. For instance, one of the bodies may be suspended from one arm of a delicate balance, and weights suspended from the other arm, till the body, when unelectrified, is in equilibrium. The other body may then be placed at a known distance beneath the first, so that the attraction or repulsion of the bodies when electrified may increase or diminish the apparent weight of the first. The weight which must be added to or taken from the other arm, when expressed in dynamical measure, will measure the force between the bodies. This arrangement was used by Sir W. Snow Harris, and is that adopted in Sir W. Thomson's absolute electrometers. See Art. 217.

It is sometimes more convenient to use a torsion-balance, in which a horizontal arm is suspended by a fine wire or fibre, so as to be capable of vibrating about the vertical wire as an axis, and the body is attached to one end of the arm and acted on by the force in the tangential direction, so as to turn the arm round the vertical axis, and so twist the suspension wire through a certain angle. The torsional rigidity of the wire is found by observing the time of oscillation of the arm, the moment of inertia of the arm being otherwise known, and from the angle of torsion and the torsional rigidity the force of attraction or repulsion can be deduced. The torsion-balance was devised by Michell for the determination of the force of gravitation between small bodies, and was used by Cavendish for this purpose. Coulomb, working independently of these philosophers, reinvented it, thoroughly studied its action, and successfully applied it to discover the laws of electric and magnetic forces; and the torsion-balance has ever since been used in all researches where small forces have to be measured. See Art. 215.

39.] Let us suppose that by either of these methods we can measure the force between two electrified bodies. We shall suppose the dimensions of the bodies small compared with the distance between them, so that the result may not be much altered by any inequality of distribution of the electrification on either body, and we shall suppose that both bodies are so suspended in air as to be at a considerable distance from other bodies on which they might induce electrification.

It is then found that if the bodies are placed at a fixed distance and charged respectively with e and e' of our provisional units of electricity, they will repel each other with a force proportional to the product of e and e'. If either e or e' is negative, that is, if one of the charges is vitreous and the other resinous, the force will be attractive, but if both e and e' are negative the force is again repulsive.

We may suppose the first body, *A*, charged with *m* units of vitreous and *n* units of resinous electricity, which may be conceived separately placed within the body, as in Experiment V.

Let the second body, *B*, be charged with m' units of positive and n' units of negative electricity.

Then each of the *m* positive units in *A* will repel each of the m' positive units in *B* with a certain force, say *f*, making a total effect equal to $mm'f$.

Since the effect of negative electricity is exactly equal and opposite to that of positive electricity, each of the *m* positive units in *A* will attract each of the n' negative units in *B* with the same force *f*, making a total effect equal to $mn'f$.

Similarly the *n* negative units in *A* will attract the m' positive units in *B* with a force $nm'f$, and will repel the n' negative units in *B* with a force $nn'f$.

The total repulsion will therefore be $(mm' + nn')f$; and the total attraction will be $(mn' + m'n)f$.

The resultant repulsion will be

$$(mm' + nn' - mn' - nm')f \quad \text{or} \quad (m-n)(m'-n')f.$$

Now $m-n=e$ is the algebraical value of the charge on *A*, and $m'-n'=e'$ is that of the charge on *B*, so that the resultant repulsion may be written $ee'f$, the quantities *e* and e' being always understood to be taken with their proper signs.

Variation of the Force with the Distance.

40.] Having established the law of force at a fixed distance, we may measure the force between bodies charged in a constant manner and placed at different distances. It is found by direct measurement that the force, whether of attraction or repulsion, varies inversely as the square of the distance, so that if *f* is the repulsion between two units at unit distance, the repulsion at distance *r* will be fr^{-2}, and the general expression for the repulsion between *e* units and e' units at distance *r* will be

$$fee'r^{-2}.$$

Definition of the Electrostatic Unit of Electricity.

41.] We have hitherto used a wholly arbitrary standard for our unit of electricity, namely, the electrification of a certain piece of glass as it happened to be electrified at the commencement of our experiments. We are now able to select a unit on a definite principle, and in order that this unit may belong to a general system we define it so that f may be unity, or in other words—

The electrostatic unit of electricity is that quantity of positive electricity which, when placed at unit of distance from an equal quantity, repels it with unit of force.

This unit is called the Electrostatic unit to distinguish it from the Electromagnetic unit, to be afterwards defined.

We may now write the general law of electrical action in the simple form $F = ee' r^{-2}$; or,

The repulsion between two small bodies charged respectively with e and e' units of electricity is numerically equal to the product of the charges divided by the square of the distance.

Dimensions of the Electrostatic Unit of Quantity.

42.] If $[Q]$ is the concrete electrostatic unit of quantity itself, and e, e' the numerical values of particular quantities; if $[L]$ is the unit of length, and r the numerical value of the distance; and if $[F]$ is the unit of force, and F the numerical value of the force, then the equation becomes

$$F[F] = ee'r^{-2}[Q^2][L^{-2}];$$

whence

$$[Q] = [LF^{\frac{1}{2}}]$$
$$= [L^{\frac{3}{2}}T^{-1}M^{\frac{1}{2}}].$$

This unit is called the Electrostatic Unit of electricity. Other units may be employed for practical purposes, and in other departments of electrical science, but in the equations of electrostatics quantities of electricity are understood to be estimated in electrostatic units, just as in physical astronomy we employ a unit of mass which is founded on the phenomena of gravitation, and which differs from the units of mass in common use.

Proof of the Law of Electrical Force.

43.] The experiments of Coulomb with the torsion-balance may be considered to have established the law of force with a certain approximation to accuracy. Experiments of this kind, however, are rendered difficult, and in some degree uncertain, by several disturbing causes, which must be carefully traced and corrected for.

In the first place, the two electrified bodies must be of sensible dimensions relative to the distance between them, in order to be capable of carrying charges sufficient to produce measurable forces.

The action of each body will then produce an effect on the distribution of electricity on the other, so that the charge cannot be considered as evenly distributed over the surface, or collected at the centre of gravity; but its effect must be calculated by an intricate investigation. This, however, has been done as regards two spheres by Poisson in an extremely able manner, and the investigation has been greatly simplified by Sir W. Thomson in his *Theory of Electrical Images*. See Arts. 172–175.

Another difficulty arises from the action of the electricity induced on the sides of the case containing the instrument. By making the inner surface of the instrument of metal, this effect can be rendered definite and measurable.

An independent difficulty arises from the imperfect insulation of the bodies, on account of which the charge continually decreases. Coulomb investigated the law of dissipation, and made corrections for it in his experiments.

The methods of insulating charged conductors, and of measuring electrical effects, have been greatly improved since the time of Coulomb, particularly by Sir W. Thomson; but the perfect accuracy of Coulomb's law of force is established, not by any direct experiments and measurements (which may be used as illustrations of the law), but by a mathematical consideration of the phenomenon described as Experiment VII, namely, that an electrified conductor B, if made to touch the inside of a hollow closed conductor C and then withdrawn without touching C, is perfectly discharged, in whatever manner the outside of C may be electrified. By means of delicate electroscopes it is easy to shew that no electricity remains on B after the operation, and by the mathematical theory given at Art. 74, this can only be the case if the force varies inversely as the square of the distance, for if the law were of any different form B would be electrified.

The Electric Field.

44.] The Electric Field is the portion of space in the neighbourhood of electrified bodies, considered with reference to electric phenomena. It may be occupied by air or other bodies, or it may be a so-called vacuum, from which we have withdrawn every substance which we can act upon with the means at our disposal.

If an electrified body be placed at any part of the electric field it will, in general, produce a sensible disturbance in the electrification of the other bodies.

But if the body is very small, and its charge also very small, the electrification of the other bodies will not be sensibly disturbed, and we may consider the position of the body as determined by its centre of mass. The force acting on the body will then be proportional to its charge, and will be reversed when the charge is reversed.

Let e be the charge of the body, and F the force acting on the body in a certain direction, then when e is very small F is proportional to e, or

$$F = Re,$$

where R depends on the distribution of electricity on the other bodies in the field. If the charge e could be made equal to unity without disturbing the electrification of other bodies we should have $F = R$.

We shall call R the Resultant Electromotive Intensity at the given point of the field. When we wish to express the fact that this quantity is a vector we shall denote it by the German letter $\mathfrak{E}$.

Electromotive Force and Potential.

45.] If the small body carrying the small charge e be moved from one given point, A, to another B, along a given path, it will experience at each point of its course a force Re, where R varies from point to point of the course. Let the whole work done on the body by the electrical force be Ee, then E is called the Total Electromotive Force along the path AB. If the path forms a complete circuit, and if the total electromotive force round the circuit does not vanish, the electricity cannot be in equilibrium but a current will be produced. Hence in Electrostatics the electromotive force round any closed circuit must be zero, so that if A and B are two points on the circuit, the electromotive force from A to B is the same along either of the two paths into which the circuit is broken, and since either of these can be altered independently of the other, the electromotive force from A to B is the same for all paths from A to B.

If B is taken as a point of reference for all other points, then the electromotive force from A to B is called the Potential of A. It depends only on the position of A. In mathematical investigations, B is generally taken at an infinite distance from the electrified bodies.

A body charged positively tends to move from places of greater positive potential to places of smaller positive, or of negative, potential, and a body charged negatively tends to move in the opposite direction.

In a conductor the electrification is free to move relatively to the conductor. If therefore two parts of a conductor have different potentials, positive electricity will move from the part having greater potential to the part having less potential as long as that difference continues. A conductor therefore cannot be in electrical equilibrium unless every point in it has the same potential. This potential is called the Potential of the Conductor.

Equipotential Surfaces.

46.] If a surface described or supposed to be described in the electric field is such that the electric potential is the same at every point of the surface it is called an Equipotential surface.

An electrified particle constrained to rest upon such a surface will have no tendency to move from one part of the surface to another, because the potential is the same at every point. An equipotential surface is therefore a surface of equilibrium or a level surface.

The resultant force at any point of the surface is in the direction of the normal to the surface, and the magnitude of the force is such that the work done on an electrical unit in passing from the surface V to the surface V' is $V-V'$.

No two equipotential surfaces having different potentials can meet one another, because the same point cannot have more than one potential, but one equipotential surface may meet itself, and this takes place at all points and along all lines of equilibrium.

The surface of a conductor in electrical equilibrium is necessarily an equipotential surface. If the electrification of the conductor is positive over the whole surface, then the potential will diminish as we move away from the surface on every side, and the conductor will be surrounded by a series of surfaces of lower potential.

But if (owing to the action of external electrified bodies) some regions of the conductor are charged positively and others negatively, the complete equipotential surface will consist of the surface of the conductor itself together with a system of other surfaces, meeting the surface of the conductor in the lines which divide the positive from the negative regions. These lines will be lines of equilibrium, and an electrified particle placed on one of these lines will experience no force in any direction.

When the surface of a conductor is charged positively in some parts and negatively in others, there must be some other electrified body in the field besides itself. For if we allow a positively electrified particle, starting from a positively charged part of the surface, to move always in the direction of the resultant force upon it, the potential at the point will continually diminish till the point reaches either a negatively charged surface at a potential less than that of the first conductor, or moves off to an infinite distance. Since the potential at an infinite distance is zero, the latter case can only occur when the potential of the conductor is positive.

In the same way a negatively electrified particle, moving off from a negatively charged part of the surface, must either reach a positively charged surface, or pass off to infinity, and the latter case can only happen when the potential of the conductor is negative.

Therefore, if both positive and negative charge exist on a conductor, there must be some other body in the field whose potential has the same sign as that of the conductor but a greater numerical value, and if a conductor of any form is alone in the field the charge of every part is of the same sign as the potential of the conductor.

The interior surface of a hollow conducting vessel containing no charged bodies is entirely free from charge. For if any part of the surface were charged positively, a positively electrified particle moving in the direction of the force upon it, must reach a negatively charged surface at a lower potential. But the whole interior surface has the same potential. Hence it can have no charge.

A conductor placed inside the vessel and communicating with it, may be considered as bounded by the interior surface. Hence such a conductor has no charge.

Lines of Force.

47.] The line described by a point moving always in the direction of the resultant intensity is called a Line of force. It cuts the equipotential surfaces at right angles. The properties of lines of force will be more fully explained afterwards, because Faraday has expressed many of the laws of electrical action in terms of his conception of lines of force drawn in the electric field, and indicating both the direction and the intensity at every point.

Electric Tension.

48.] Since the surface of a conductor is an equipotential surface, the resultant force is normal to the surface, and it will be shewn in Art. 78 that it is proportional to the superficial density of the electrification. Hence the electricity on any small area of the surface will be acted on by a force tending *from* the conductor and proportional to the product of the resultant force and the density, that is, proportional to the square of the resultant force.

This force, which acts outwards as a tension on every part of the conductor, will be called electric Tension. It is measured like ordinary mechanical tension, by the force exerted on unit of area.

The word Tension has been used by electricians in several vague senses, and it has been attempted to adopt it in mathematical language as a synonym for Potential; but on examining the cases in which the word has been used, I think it will be more consistent with usage and with mechanical analogy to understand by tension a pulling force of so many pounds weight per square inch exerted on the surface of a conductor or elsewhere. We shall find that the conception of Faraday, that this electric tension exists not only at the electrified surface but all along the lines of force, leads to a theory of electric action as a phenomenon of stress in a medium.

Electromotive Force.

49.] When two conductors at different potentials are connected by a thin conducting wire, the tendency of electricity to flow along the wire is measured by the difference of the potentials of the two bodies. The difference of potentials between two conductors or two points is therefore called the Electromotive force between them.

Electromotive force cannot in all cases be expressed in the form of a difference of potentials. These cases, however, are not treated of in Electrostatics. We shall consider them when we come to heterogeneous circuits, chemical actions, motions of magnets, inequalities of temperature, &c.

Capacity of a Conductor.

50.] If one conductor is insulated while all the surrounding conductors are kept at the zero potential by being put in communication with the earth, and if the conductor, when charged with a quantity E of electricity, has a potential V, the ratio of E to V is called the Capacity of the conductor. If the conductor is completely enclosed within a conducting vessel without touching it, then the charge on the inner conductor will be equal and opposite to the charge on the inner surface of the outer conductor, and will be equal to the capacity of the inner conductor multiplied by the difference of the potentials of the two conductors.

Electric Accumulators.

A system consisting of two conductors whose opposed surfaces are separated from each other by a thin stratum of an insulating medium is called an electric Accumulator. The two conductors are called the Electrodes and the insulating medium is called the Dielectric. The capacity of the accumulator is directly proportional to the area of the opposed surfaces and inversely proportional to the thickness of the stratum between them. A Leyden jar is an accumulator in which glass is the insulating medium. Accumulators are sometimes called Condensers, but I prefer to restrict the term 'condenser' to an instrument which is used not to hold electricity but to increase its superficial density.

PROPERTIES OF BODIES IN RELATION TO STATICAL ELECTRICITY.

Resistance to the Passage of Electricity through a Body.

51.] When a charge of electricity is communicated to any part of a mass of metal the electricity is rapidly transferred from places of high to places of low potential till the potential of the whole mass becomes the same. In the case of pieces of metal used in ordinary experiments this process is completed in a time too short to be observed, but in the case of very long and thin wires, such as those used in telegraphs, the potential does not become uniform till after a sensible time, on account of the resistance of the wire to the passage of electricity through it.

The resistance to the passage of electricity is exceedingly different in different substances, as may be seen from the tables at Arts. 362, 366, and 369, which will be explained in treating of Electric Currents.

All the metals are good conductors, though the resistance of lead is 12 times that of copper or silver, that of iron 6 times, and that of mercury 60 times that of copper. The resistance of all metals increases as their temperature rises.

Many liquids conduct electricity by electrolysis. This mode of conduction will be considered in Part II. For the present, we may regard all liquids containing water and all damp bodies as conductors, far inferior to the metals, but incapable of insulating a charge of electricity for a sufficient time to be observed. The resistance of electrolytes diminishes as the temperature rises.

On the other hand, the gases at the atmospheric pressure, whether dry or moist, are insulators so nearly perfect when the electric tension is small that we have as yet obtained no evidence of electricity passing through them by ordinary conduction. The gradual loss of charge by electrified bodies may in every case be traced to imperfect insulation in the supports, the electricity either passing through the substance of the support or creeping over its surface. Hence, when two charged bodies are hung up near each other, they will preserve their charges longer if they are electrified in opposite ways, than if they are electrified in the same way. For though the electromotive force tending to make the electricity pass through the air between them is much greater when they are oppositely electrified, no perceptible loss occurs in this way. The actual loss takes place through the supports, and the electromotive force through the supports is greatest when the bodies are electrified in the same way. The result appears anomalous only when we expect the loss to occur by the passage of electricity through the air between the bodies. The passage of electricity through gases takes place, in general, by disruptive discharge, and does not begin till the electromotive force has reached a certain value. The value of the electromotive force which can exist in a dielectric without a discharge taking place is called the Electric Strength of the dielectric. The electric strength of air diminishes as the pressure is reduced from the atmospheric pressure to that of about three millimetres of mercury. When the pressure is still further reduced, the electric strength rapidly increases; and when the exhaustion is carried to the highest degree hitherto attained, the electromotive force required to produce a spark of a quarter of an inch is greater than that which will give a spark of eight inches in air at the ordinary pressure.

A vacuum, that is to say, that which remains in a vessel after we have removed everything which we can remove from it, is therefore an insulator of very great electric strength.

The electric strength of hydrogen is much less than that of air.

Certain kinds of glass when cold are marvellously perfect insulators, and Sir W. Thomson has preserved charges of electricity for years in bulbs hermetically sealed. The same glass, however, becomes a conductor at a temperature below that of boiling water.

Gutta-percha, caoutchouc, vulcanite, paraffin, and resins are good insulators, the resistance of gutta-percha at 75° F. being about 6×10^{19} times that of copper.

Ice, crystals, and solidified electrolytes, are also insulators.

Certain liquids, such as naphtha, turpentine, and some oils, are insulators, but inferior to the best solid insulators.

DIELECTRICS.

Specific Inductive Capacity.

52.] All bodies whose insulating power is such that when they are placed between two conductors at different potentials the electromotive force acting on them does not immediately distribute their electricity so as to reduce the potential to a constant value, are called by Faraday Dielectrics.

It appears from the hitherto unpublished researches of Cavendish that he had, before 1773, measured the capacity of plates of glass, rosin, beeswax, and shellac, and had determined the ratio in which their capacity exceeded that of plates of air of the same dimensions.

Faraday, to whom these researches were unknown, discovered that the capacity of an accumulator depends on the nature of the insulating medium between the two conductors, as well as on the dimensions and relative position of the conductors themselves. By substituting other insulating media for air as the dielectric of the accumulator, without altering it in any other respect, he found that when air and other gases were employed as the insulating medium the capacity of the accumulator remained sensibly the same, but that when shellac, sulphur, glass, &c. were substituted for air, the capacity was increased in a ratio which was different for each substance.

By a more delicate method of measurement Boltzmann succeeded in observing the variation of the inductive capacity of gases at different pressures.

This property of dielectrics, which Faraday called Specific Inductive Capacity, is also called the Dielectric Constant of the substance. It is defined as the ratio of the capacity of an accumulator when its dielectric is the given substance, to its capacity when the dielectric is a vacuum.

If the dielectric is not a good insulator, it is difficult to measure its inductive capacity, because the accumulator will not hold a charge for a sufficient time to allow it to be measured; but it is certain that inductive capacity is a property not confined to good insulators, and it is probable that it exists in all bodies.

Absorption of Electricity.

53.] It is found that when an accumulator is formed of certain dielectrics, the following phenomena occur.

When the accumulator has been for some time electrified and is then suddenly discharged and again insulated, it becomes recharged in the same sense as at first, but to a smaller degree, so that it may be discharged again several times in succession, these discharges always diminishing. This phenomenon is called that of the Residual Discharge.

The instantaneous discharge appears always to be proportional to the difference of potentials at the instant of discharge, and the ratio of these quantities is the true capacity of the accumulator; but if the contact of the discharger is prolonged so as to include some of the residual discharge, the apparent capacity of the accumulator, calculated from such a discharge, will be too great.

The accumulator if charged and left insulated appears to lose its charge by conduction, but it is found that the proportionate rate of loss is much greater at first than it is afterwards, so that the measure of conductivity, if deduced from what takes place at first, would be too great. Thus, when the insulation of a submarine cable is tested, the insulation appears to improve as the electrification continues.

Thermal phenomena of a kind at first sight analogous take place in the case of the conduction of heat when the opposite sides of a body are kept at different temperatures. In the case of heat we know that they depend on the heat taken in and given out by the body itself. Hence, in the case of the electrical phenomena, it has been supposed that electricity is absorbed and emitted by the parts of the body. We shall see, however, in Art. 329, that the phenomena can be explained without the hypothesis of absorption of electricity, by supposing the dielectric in some degree heterogeneous.

That the phenomenon called Electric Absorption is not an actual absorption of electricity by the substance may be shewn by charging the substance in any manner with electricity while it is surrounded by a closed metallic insulated vessel. If, when the substance is charged and insulated, the vessel be instantaneously discharged and then left insulated, no charge is ever communicated to the vessel by the gradual dissipation of the electrification of the charged substance within it.

54.] This fact is expressed by the statement of Faraday that it is impossible to charge matter with an absolute and independent charge of one kind of electricity *.

In fact it appears from the result of every experiment which has been tried that in whatever way electrical actions may take place among a system of bodies surrounded by a metallic vessel, the charge on the outside of that vessel is not altered.

Now if any portion of electricity could be forced into a body so as to be absorbed in it, or to become latent, or in any way to exist in it, without being connected with an equal portion of the opposite electricity by lines of induction, or if, after having being absorbed, it could gradually emerge and return to its ordinary mode of action, we should find some change of electrification in the surrounding vessel.

As this is never found to be the case, Faraday concluded that it is impossible to communicate an absolute charge to matter, and that no portion of matter can by any change of state evolve or render latent one kind of electricity or the other. He therefore regarded induction as 'the essential function both in the first development and the consequent phenomena of electricity.' His 'induction' is (1298) a polarized state of the particles of the dielectric, each particle being positive on one side and negative on the other, the positive and the negative electrification of each particle being always exactly equal.

Disruptive Discharge †.

55.] If the electromotive intensity at any point of a dielectric is gradually increased, a limit is at length reached at which there is a sudden electrical discharge through the dielectric, generally accompanied with light and sound, and with a temporary or permanent rupture of the dielectric.

The intensity of the electromotive force when this takes place is a measure of what we may call the electric strength of the dielectric. It depends on the nature of the dielectric, and is greater in dense air than in rare air, and greater in glass than in air, but

* *Exp. Res.*, vol. i. series xi. ¶ ii. 'On the Absolute Charge of Matter,' and (1244).
† See Faraday, *Exp. Res.*, vol. i., series xii. and xiii.

in every case, if the electromotive force be made great enough, the dielectric gives way and its insulating power is destroyed, so that a current of electricity takes place through it. It is for this reason that distributions of electricity for which the electromotive intensity becomes anywhere infinite cannot exist.

The Electric Glow.

Thus, when a conductor having a sharp point is electrified, the theory, based on the hypothesis that it retains its charge, leads to the conclusion that as we approach the point the superficial density of the electricity increases without limit, so that at the point itself the surface-density, and therefore the resultant electrical force, would be infinite. If the air, or other surrounding dielectric, had an invincible insulating power, this result would actually occur; but the fact is, that as soon as the resultant force in the neighbourhood of the point has reached a certain limit, the insulating power of the air gives way, so that the air close to the point becomes a conductor. At a certain distance from the point the resultant force is not sufficient to break through the insulation of the air, so that the electric current is checked, and the electricity accumulates in the air round the point.

The point is thus surrounded by particles of air charged with electricity of the same kind with its own. The effect of this charged air round the point is to relieve the air at the point itself from part of the enormous electromotive force which it would have experienced if the conductor alone had been electrified. In fact the surface of the electrified body is no longer pointed, because the point is enveloped by a rounded mass of charged air, the surface of which, rather than that of the solid conductor, may be regarded as the outer electrified surface.

If this portion of charged air could be kept still, the electrified body would retain its charge, if not on itself at least in its neighbourhood, but the charged particles of air being free to move under the action of electrical force, tend to move away from the electrified body because it is charged with the same kind of electricity. The charged particles of air therefore tend to move off in the direction of the lines of force and to approach those surrounding bodies which are oppositely electrified. When they are gone, other uncharged particles take their place round the point, and since these cannot shield those next the point itself from the excessive electric tension, a new discharge takes place, after which the newly charged particles move off, and so on as long as the body remains electrified.

In this way the following phenomena are produced:—At and close to the point there is a steady glow, arising from the constant discharges which are taking place between the point and the air very near it.

The charged particles of air tend to move off in the same general direction, and thus produce a current of air from the point, consisting of the charged particles, and probably of others carried along by them. By artificially aiding this current we may increase the glow, and by checking the formation of the current we may prevent the continuance of the glow *.

The electric wind in the neighbourhood of the point is sometimes very rapid, but it soon loses its velocity, and the air with its charged particles is carried about with the general motions of the atmosphere, and constitutes an invisible electric cloud. When the charged particles come near to any conducting surface, such as a wall, they induce on that surface a charge opposite to their own, and are then attracted towards the wall, but since the electromotive force is small they may remain for a long time near the wall without being drawn up to the surface and discharged. They thus form an electrified atmosphere clinging to conductors, the presence of which may sometimes be detected by the electrometer. The electrical forces, however, acting between large masses of charged air and other bodies are exceedingly feeble compared with the ordinary forces which produce winds, and which depend on inequalities of density due to differences of temperature, so that it is very improbable that any observable part of the motion of ordinary thunder clouds arises from electrical causes.

The passage of electricity from one place to another by the motion of charged particles is called Electrical Convection or Convective Discharge.

The electrical glow is therefore produced by the constant passage of electricity through a small portion of air in which the tension is very high, so as to charge the surrounding particles of air which are continually swept off by the electric wind, which is an essential part of the phenomenon.

The glow is more easily formed in rare air than in dense air, and more easily when the point is positive than when it is negative.

* See Priestley's *History of Electricity*, pp. 117 and 591; and Cavendish's 'Electrical Researches,' *Phil. Trans.*, 1771, § 4, or Art. 125 of Reprint of Cavendish.

This and many other differences between positive and negative electrification must be studied by those who desire to discover something about the nature of electricity. They have not, however, been satisfactorily brought to bear upon any existing theory.

The Electric Brush.

56.] The electric brush is a phenomenon which may be produced by electrifying a blunt point or small ball so as to produce an electric field in which the tension diminishes as the distance increases, but in a less rapid manner than when a sharp point is used. It consists of a succession of discharges, ramifying as they diverge from the ball into the air, and terminating either by charging portions of air or by reaching some other conductor. It is accompanied by a sound, the pitch of which depends on the interval between the successive discharges, and there is no current of air as in the case of the glow.

The Electric Spark.

57.] When the tension in the space between two conductors is considerable all the way between them, as in the case of two balls whose distance is not great compared with their radii, the discharge, when it occurs, usually takes the form of a spark, by which nearly the whole electrification is discharged at once.

In this case, when any part of the dielectric has given way, the parts on either side of it in the direction of the electric force are put into a state of greater tension so that they also give way, and so the discharge proceeds right through the dielectric, just as when a little rent is made in the edge of a piece of paper a tension applied to the paper in the direction of the edge causes the paper to be torn through, beginning at the rent, but diverging occasionally where there are weak places in the paper. The electric spark in the same way begins at the point where the electric tension first overcomes the insulation of the dielectric, and proceeds from that point, in an apparently irregular path, so as to take in other weak points, such as particles of dust floating in air.

All these phenomena differ considerably in different gases, and in the same gas at different densities. Some of the forms of electrical discharge through rare gases are exceedingly remarkable. In some cases there is a regular alternation of luminous and dark strata, so that if the electricity, for example, is passing along a tube containing a very small quantity of gas, a number of luminous disks will be seen arranged transversely at nearly equal intervals along the axis of the tube and separated by dark strata. If the strength of the current be increased a new disk will start into existence, and it and the old disks will arrange themselves in closer order. In a tube described by Mr. Gassiot * the light of each of the disks is bluish on the negative and reddish on the positive side, and bright red in the central stratum.

These, and many other phenomena of electrical discharge, are exceedingly important, and when they are better understood they will probably throw great light on the nature of electricity as well as on the nature of gases and of the medium pervading space. At present, however, they must be considered as outside the domain of the mathematical theory of electricity.

Electric Phenomena of Tourmaline.

58.] Certain crystals of tourmaline, and of other minerals, possess what may be called Electric Polarity. Suppose a crystal of tourmaline to be at a uniform temperature, and apparently free from electrification on its surface. Let its temperature be now raised, the crystal remaining insulated. One end will be found positively and the other end negatively electrified. Let the surface be deprived of this apparent electrification by means of a flame or otherwise, then if the crystal be made still hotter, electrification of the same kind as before will appear, but if the crystal be cooled the end which was positive when the crystal was heated will become negative.

These electrifications are observed at the extremities of the crystallographic axis. Some crystals are terminated by a six-sided pyramid at one end and by a three-sided pyramid at the other. In these the end having the six-sided pyramid becomes positive when the crystal is heated.

Sir W. Thomson supposes every portion of these and other hemihedral crystals to have a definite electric polarity, the intensity of which depends on the temperature. When the surface is passed through a flame, every part of the surface becomes electrified to such an extent as to exactly neutralize, for all external points, the effect of the internal polarity. The crystal then has no external electrical action, nor any tendency to change its mode of electrification. But if it be heated or cooled the interior polariza-

* *Intellectual Observer*, March, 1866.

tion of each particle of the crystal is altered, and can no longer be balanced by the superficial electrification, so that there is a resultant external action.

Plan of this Treatise.

59.] In the following treatise I propose first to explain the ordinary theory of electrical action, which considers it as depending only on the electrified bodies and on their relative position, without taking account of any phenomena which may take place in the intervening media. In this way we shall establish the law of the inverse square, the theory of the potential, and the equations of Laplace and Poisson. We shall next consider the charges and potentials of a system of electrified conductors as connected by a system of equations, the coefficients of which may be supposed to be determined by experiment in those cases in which our present mathematical methods are not applicable, and from these we shall determine the mechanical forces acting between the different electrified bodies.

We shall then investigate certain general theorems by which Green, Gauss, and Thomson have indicated the conditions of solution of problems in the distribution of electricity. One result of these theorems is, that if Poisson's equation is satisfied by any function, and if at the surface of every conductor the function has the value of the potential of that conductor, then the function expresses the actual potential of the system at every point. We also deduce a method of finding problems capable of exact solution.

In Thomson's theorem, the total energy of the system is expressed in the form of the integral of a certain quantity extended over the whole space between the electrified bodies, and also in the form of an integral extended over the electrified surfaces only. The equality of these two expressions may be thus interpreted physically. We may conceive the physical relation between the electrified bodies, either as the result of the state of the intervening medium, or as the result of a direct action between the electrified bodies at a distance. If we adopt the latter conception, we may determine the law of the action, but we can go no further in speculating on its cause. If, on the other hand, we adopt the conception of action through a medium, we are led to enquire into the nature of that action in each part of the medium.

It appears from the theorem, that if we are to look for the seat of the electric energy in the different parts of the dielectric medium, the amount of energy in any small part must depend on the square of the resultant electromotive intensity at that place multiplied by a coefficient called the specific inductive capacity of the medium.

It is better, however, in considering the theory of dielectrics from the most general point of view, to distinguish between the electromotive intensity at any point and the electric polarization of the medium at that point, since these directed quantities, though related to one another, are not, in some solid substances, in the same direction. The most general expression for the electric energy of the medium per unit of volume is half the product of the electromotive intensity and the electric polarization multiplied by the cosine of the angle between their directions. In all fluid dielectrics the electromotive intensity and the electric polarization are in the same direction and in a constant ratio.

If we calculate on this hypothesis the total energy residing in the medium, we shall find it equal to the energy due to the electrification of the conductors on the hypothesis of direct action at a distance. Hence the two hypotheses are mathematically equivalent.

If we now proceed to investigate the mechanical state of the medium on the hypothesis that the mechanical action observed between electrified bodies is exerted through and by means of the medium, as in the familiar instances of the action of one body on another by means of the tension of a rope or the pressure of a rod, we find that the medium must be in a state of mechanical stress.

The nature of this stress is, as Faraday pointed out*, a tension along the lines of force combined with an equal pressure in all directions at right angles to these lines. The magnitude of these stresses is proportional to the energy of the electrification per unit of volume, or, in other words, to the square of the resultant electromotive intensity multiplied by the specific inductive capacity of the medium.

This distribution of stress is the only one consistent with the observed mechanical action on the electrified bodies, and also with the observed equilibrium of the fluid dielectric which surrounds them. I have therefore thought it a warrantable step in scientific

* *Exp. Res.*, series xi. 1297.

procedure to assume the actual existence of this state of stress, and to follow the assumption into its consequences. Finding the phrase *electric tension* used in several vague senses, I have attempted to confine it to what I conceive to have been in the mind of some of those who have used it, namely, the state of stress in the dielectric medium which causes motion of the electrified bodies, and leads, when continually augmented, to disruptive discharge. Electric tension, in this sense, is a tension of exactly the same kind, and measured in the same way, as the tension of a rope, and the dielectric medium, which can support a certain tension and no more, may be said to have a certain strength in exactly the same sense as the rope is said to have a certain strength. Thus, for example, Thomson has found that air at the ordinary pressure and temperature can support an electric tension of 9600 grains weight per square foot before a spark passes.

60.] From the hypothesis that electric action is not a direct action between bodies at a distance, but is exerted by means of the medium between the bodies, we have deduced that this medium must be in a state of stress. We have also ascertained the character of the stress, and compared it with the stresses which may occur in solid bodies. Along the lines of force there is tension, and perpendicular to them there is pressure, the numerical magnitude of these forces being equal, and each proportional to the square of the resultant intensity at the point. Having established these results, we are prepared to take another step, and to form an idea of the nature of the electric polarization of the dielectric medium.

An elementary portion of a body may be said to be polarized when it acquires equal and opposite properties on two opposite sides. The idea of internal polarity may be studied to the greatest advantage as exemplified in permanent magnets, and it will be explained at greater length when we come to treat of magnetism.

The electric polarization of an elementary portion of a dielectric is a forced state into which the medium is thrown by the action of electromotive force, and which disappears when that force is removed. We may conceive it to consist in what we may call an electrical displacement, produced by the electromotive intensity. When the electromotive force acts on a conducting medium it produces a current through it, but if the medium is a non-conductor or dielectric, the current cannot flow through the medium, but the electricity is displaced within the medium in the direction of the electromotive intensity, the extent of this displacement depending on the magnitude of the electromotive intensity, so that if the electromotive intensity increases or diminishes, the electric displacement increases and diminishes in the same ratio.

The amount of the displacement is measured by the quantity of electricity which crosses unit of area, while the displacement increases from zero to its actual amount. This, therefore, is the measure of the electric polarization.

The analogy between the action of electromotive force in producing electric displacement and of ordinary mechanical force in producing the displacement of an elastic body is so obvious that I have ventured to call the ratio of the electromotive intensity to the corresponding electric displacement the *coefficient of electric elasticity* of the medium. This coefficient is different in different media, and varies inversely as the specific inductive capacity of each medium.

The variations of electric displacement evidently constitute electric currents. These currents, however, can only exist during the variation of the displacement, and therefore, since the displacement cannot exceed a certain value without causing disruptive discharge, they cannot be continued indefinitely in the same direction, like the currents through conductors.

In tourmaline, and other pyro-electric crystals, it is probable that a state of electric polarization exists, which depends upon temperature, and does not require an external electromotive force to produce it. If the interior of a body were in a state of permanent electric polarization, the outside would gradually become charged in such a manner as to neutralize the action of the internal polarization for all points outside the body. This external superficial charge could not be detected by any of the ordinary tests, and could not be removed by any of the ordinary methods for discharging superficial electrification. The internal polarization of the substance would therefore never be discovered unless by some means, such as change of temperature, the amount of the internal polarization could be increased or diminished. The external electrification would then be no longer capable of neutralizing the external effect of the internal polarization, and an apparent electrification would be observed, as in the case of tourmaline.

If a charge e is uniformly distributed over the surface of a sphere, the resultant force at any point of the medium surrounding the sphere is numerically equal to the charge e divided by the square of the distance from the centre of the sphere. This resultant force, according to our theory, is accompanied by a displacement of electricity in a direction outwards from the sphere.

If we now draw a concentric spherical surface of radius r, the whole displacement, E, through this surface will be proportional to the resultant force multiplied by the area of the spherical surface. But the resultant force is directly as the charge e and inversely as the square of the radius, while the area of the surface is directly as the square of the radius.

Hence the whole displacement, E, is proportional to the charge e, and is independent of the radius.

To determine the ratio between the charge e, and the quantity of electricity, E, displaced outwards through any one of the spherical surfaces, let us consider the work done upon the medium in the region between two concentric spherical surfaces, while the displacement is increased from E to $E+\delta E$. If V_1 and V_2 denote the potentials at the inner and the outer of these surfaces respectively, the electromotive force by which the additional displacement is produced is $V_1 - V_2$, so that the work spent in augmenting the displacement is $(V_1 - V_2)\delta E$.

If we now make the inner surface coincide with that of the electrified sphere, and make the radius of the other infinite, V_1 becomes V, the potential of the sphere, and V_2 becomes zero, so that the whole work done in the surrounding medium is $V\delta E$.

But by the ordinary theory, the work done in augmenting the charge is $V\delta e$, and if this is spent, as we suppose, in augmenting the displacement, $\delta E = \delta e$, and since E and e vanish together, $E = e$, or—

The displacement outwards through any spherical surface concentric with the sphere is equal to the charge on the sphere.

To fix our ideas of electric displacement, let us consider an accumulator formed of two conducting plates A and B, separated by a stratum of a dielectric C. Let W be a conducting wire joining A and B, and let us suppose that by the action of an electromotive force a quantity Q of positive electricity is transferred along the wire from B to A. The positive electrification of A and the negative electrification of B will produce a certain electromotive force acting from A towards B in the dielectric stratum, and this will produce an electric displacement from A towards B within the dielectric. The amount of this displacement, as measured by the quantity of electricity forced across an imaginary section of the dielectric dividing it into two strata, will be, according to our theory, exactly Q. See Arts. 75, 76, 111.

It appears, therefore, that at the same time that a quantity Q of electricity is being transferred along the wire by the electromotive force from B towards A, so as to cross every section of the wire, the same quantity of electricity crosses every section of the dielectric from A towards B by reason of the electric displacement.

The displacements of electricity during the discharge of the accumulator will be the reverse of these. In the wire the discharge will be Q from A to B, and in the dielectric the displacement will subside, and a quantity of electricity Q will cross every section from B towards A.

Every case of charge or discharge may therefore be considered as a motion in a closed circuit, such that at every section of the circuit the same quantity of electricity crosses in the same time, and this is the case, not only in the voltaic circuit where it has always been recognised, but in those cases in which electricity has been generally supposed to be accumulated in certain places.

61.] We are thus led to a very remarkable consequence of the theory which we are examining, namely, that the motions of electricity are like those of an *incompressible* fluid, so that the total quantity within an imaginary fixed closed surface remains always the same. This result appears at first sight in direct contradiction to the fact that we can charge a conductor and then introduce it into the closed space, and so alter the quantity of electricity within that space. But we must remember that the ordinary theory takes no account of the electric displacement in the substance of dielectrics which we have been investigating, but confines its attention to the electrification at the bounding surfaces of the conductors and dielectrics. In the case of the charged conductor let us suppose the charge to be positive, then if the surrounding dielectric extends on all sides beyond the closed surface there will be electric polarization, accompanied with displacement from within outwards all over the closed surface, and the surface-integral of the displacement taken over the surface will be equal to the charge on the conductor within.

Thus when the charged conductor is introduced into the closed space there is immediately a displacement of a quantity of electricity equal to the charge through the surface from within outwards, and the whole quantity within the surface remains the same.

The theory of electric polarization will be discussed at greater length in Chapter V, and a mechanical illustration of it will be given in Art. 334, but its importance cannot be fully understood till we arrive at the study of electromagnetic phenomena.

62.] The peculiar features of the theory are:—

That the energy of electrification resides in the dielectric medium, whether that medium be solid, liquid, or gaseous, dense or rare, or even what is called a vacuum, provided it be still capable of transmitting electrical action.

That the energy in any part of the medium is stored up in the form of a state of constraint called electric polarization, the amount of which depends on the resultant electromotive intensity at the place.

That electromotive force acting on a dielectric produces what we have called electric displacement, the relation between the intensity and the displacement being in the most general case of a kind to be afterwards investigated in treating of conduction, but in the most important cases the displacement is in the same direction as the force, and is numerically equal to the intensity multiplied by $\frac{1}{4\pi}K$, where K is the specific inductive capacity of the dielectric.

That the energy per unit of volume of the dielectric arising from the electric polarization is half the product of the electromotive intensity and the electric displacement, multiplied, if necessary, by the cosine of the angle between their directions.

That in fluid dielectrics the electric polarization is accompanied by a tension in the direction of the lines of induction, combined with an equal pressure in all directions at right angles to the lines of induction, the tension or pressure per unit of area being numerically equal to the energy per unit of volume at the same place.

That the surface of any elementary portion into which we may conceive the volume of the dielectric divided must be conceived to be charged so that the surface-density at any point of the surface is equal in magnitude to the displacement through that point of the surface *reckoned inwards*. If the displacement is in the positive direction, the surface of the element will be charged negatively on the positive side of the element, and positively on the negative side. These superficial charges will in general destroy one another when consecutive elements are considered, except where the dielectric has an internal charge, or at the surface of the dielectric.

That whatever electricity may be, and whatever we may understand by the movement of electricity, the phenomenon which we have called electric displacement is a movement of electricity in the same sense as the transference of a definite quantity of electricity through a wire is a movement of electricity, the only difference being that in the dielectric there is a force which we have called electric elasticity which acts against the electric displacement, and forces the electricity back when the electromotive force is removed; whereas in the conducting wire the electric elasticity is continually giving way, so that a current of true conduction is set up, and the resistance depends, not on the total quantity of electricity displaced from its position of equilibrium, but on the quantity which crosses a section of the conductor in a given time.

That in every case the motion of electricity is subject to the same condition as that of an incompressible fluid, namely, that at every instant as much must flow out of any given closed surface as flows into it.

It follows from this that every electric current must form a closed circuit. The importance of this result will be seen when we investigate the laws of electro-magnetism.

Since, as we have seen, the theory of direct action at a distance is mathematically identical with that of action by means of a medium, the actual phenomena may be explained by the one theory as well as by the other, provided suitable hypotheses be introduced when any difficulty occurs. Thus, Mossotti has deduced the mathematical theory of dielectrics from the ordinary theory of attraction merely by giving an electric instead of a magnetic interpretation to the symbols in the investigation by which Poisson has deduced the theory of magnetic induction from the theory of magnetic fluids. He assumes the existence within the dielectric of small conducting elements, capable of having their opposite surfaces oppositely electrified by induction, but not capable of losing or gaining electricity on the whole, owing to their being insulated from each other by a non-conducting medium. This theory of dielectrics is consistent with the laws of electricity, and may be actually true. If it is true, the specific inductive capacity of a dielectric may be greater, but cannot be less, than that of a vacuum. No instance has yet been found of a dielectric having an inductive capacity less than that of a vacuum, but if such should be discovered, Mossotti's physical theory must be abandoned, although his formulae would all remain exact, and would only require us to alter the sign of a coefficient.

In many parts of physical science, equations of the same form are found applicable to phenomena which are certainly of quite different natures, as, for instance, electric induction through dielectrics, conduction through conductors, and magnetic induction. In all these cases the relation between the force and the effect produced is expressed by a set of equations of the same kind, so that when a problem in one of these subjects is solved, the problem and its solution may be translated into the language of the other subjects and the results in their new form will still be true.

CHAPTER II.

ELEMENTARY MATHEMATICAL THEORY OF STATICAL ELECTRICITY.

Definition of Electricity as a Mathematical Quantity.

63.] We have seen that the properties of charged bodies are such that the charge of one body may be equal to that of another, or to the sum of the charges of two bodies, and that when two bodies are equally and oppositely charged they have no electrical effect on external bodies when placed together within a closed insulated conducting vessel. We may express all these results in a concise and consistent manner by describing an electrified body as *charged* with a certain *quantity of electricity*, which we may denote by e. When the charge is positive, that is, according to the usual convention, vitreous, e will be a positive quantity. When the charge is negative or resinous, e will be negative, and the quantity $-e$ may be interpreted either as a negative quantity of vitreous electricity or as a positive quantity of resinous electricity.

The effect of adding together two equal and opposite charges of electricity, $+e$ and $-e$, is to produce a state of no charge expressed by zero. We may therefore regard a body not charged as virtually charged with equal and opposite charges of indefinite magnitude, and a charged body as virtually charged with unequal quantities of positive and negative electricity, the algebraic sum of these charges constituting the observed electrification. It is manifest, however, that this way of regarding an electrified body is entirely artificial, and may be compared to the conception of the velocity of a body as compounded of two or more different velocities, no one of which is the actual velocity of the body.

ON ELECTRIC DENSITY.

Distribution in Three Dimensions.

64.] *Definition.* The electric volume-density at a given point in space is the limiting ratio of the quantity of electricity within a sphere whose centre is the given point to the volume of the sphere, when its radius is diminished without limit.

We shall denote this ratio by the symbol ρ, which may be positive or negative.

Distribution over a Surface.

It is a result alike of theory and of experiment, that, in certain cases, the charge of a body is entirely on the surface. The density at a point on the surface, if defined according to the method given above, would be infinite. We therefore adopt a different method for the measurement of surface-density.

Definition. The electric density at a given point on a surface is the limiting ratio of the quantity of electricity within a sphere whose centre is the given point to the area of the surface contained within the sphere, when its radius is diminished without limit.

We shall denote the surface-density by the symbol σ.

Those writers who supposed electricity to be a material fluid or a collection of particles, were obliged in this case to suppose the electricity distributed on the surface in the form of a stratum of a certain thickness θ, its density being ρ_0, or that value of ρ which would result from the particles having the closest contact of which they are capable. It is manifest that on this theory

$$\rho_0 \theta = \sigma.$$

When σ is negative, according to this theory, a certain stratum of thickness θ is left entirely devoid of positive electricity, and filled entirely with negative electricity, or, on the theory of one fluid, with matter.

There is, however, no experimental evidence either of the electric stratum having any thickness, or of electricity being a fluid or a collection of particles. We therefore prefer to do without the symbol for the thickness of the stratum, and to use a special symbol for surface-density.

Distribution on a Line.

It is sometimes convenient to suppose electricity distributed on a line, that is, a long narrow body of which we neglect the thickness. In this case we may define the line-density at any point to be the limiting ratio of the charge on an element of the line to the length of that element when the element is diminished without limit.

If λ denotes the line-density, then the whole quantity of electricity on a curve is $e = \int \lambda\, ds$, where ds is the element of the curve. Similarly, if σ is the surface-density, the whole quantity of electricity on the surface is

$$e = \iint \sigma\, dS,$$

where dS is the element of surface.

If ρ is the volume-density at any point of space, then the whole electricity within a certain volume is

$$e = \iiint \rho\, dx\, dy\, dz.$$

where $dx\, dy\, dz$ is the element of volume. The limits of integration in each case are those of the curve, the surface, or the portion of space considered.

It is manifest that e, λ, σ and ρ are quantities differing in kind, each being one dimension in space lower than the preceding, so that if l be a line, the quantities e, $l\lambda$, $l^2\sigma$, and $l^3\rho$ will be all of the same kind, and if $[L]$ be the unit of length, and $[\lambda]$, $[\sigma]$, $[\rho]$ the units of the different kinds of density, $[e]$, $[L\lambda]$, $[L^2\sigma]$, and $[L^3\rho]$ will each denote one unit of electricity.

Definition of the Unit of Electricity.

65.] Let A and B be two points the distance between which is the unit of length. Let two bodies, whose dimensions are small compared with the distance AB, be charged with equal quantities of positive electricity and placed at A and B respectively, and let the charges be such that the force with which they repel each other is the unit of force, measured as in Art. 6. Then the charge of either body is said to be the unit of electricity.

If the charge of the body at B were a unit of negative electricity, then, since the action between the bodies would be reversed, we should have an attraction equal to the unit of force. If the charge of A were also negative, and equal to unity, the force would be repulsive, and equal to unity.

Since the action between any two portions of electricity is not affected by the presence of other portions, the repulsion between e units of electricity at A and e' units at B is ee', the distance AB being unity. See Art. 39.

Law of Force between Charged Bodies.

66.] Coulomb shewed by experiment that the force between

charged bodies whose dimensions are small compared with the distance between them, varies inversely as the square of the distance. Hence the repulsion between two such bodies charged with quantities e and e' and placed at a distance r is

$$\frac{ee'}{r^2}.$$

We shall prove in Art. 74 that this law is the only one consistent with the observed fact that a conductor, placed in the inside of a closed hollow conductor and in contact with it, is deprived of all electrical charge. Our conviction of the accuracy of the law of the inverse square of the distance may be considered to rest on experiments of this kind, rather than on the direct measurements of Coulomb.

Resultant Force between Two Bodies.

67.] In order to calculate the resultant force between two bodies we might divide each of them into its elements of volume, and consider the repulsion between the electricity in each of the elements of the first body and the electricity in each of the elements of the second body. We should thus get a system of forces equal in number to the product of the numbers of the elements into which we have divided each body, and we should have to combine the effects of these forces by the rules of Statics. Thus, to find the component in the direction of x we should have to find the value of the sextuple integral

$$\iiiiiint \frac{\rho\rho'(x-x')\,dx\,dy\,dz\,dx'\,dy'\,dz'}{\{(x-x')^2+(y-y')^2+(z-z')^2\}^{\frac{3}{2}}},$$

where x, y, z are the coordinates of a point in the first body at which the electrical density is ρ, and x', y', z', and ρ' are the corresponding quantities for the second body, and the integration is extended first over the one body and then over the other.

Resultant Intensity at a Point.

68.] In order to simplify the mathematical process, it is convenient to consider the action of an electrified body, not on another body of any form, but on an indefinitely small body, charged with an indefinitely small amount of electricity, and placed at any point of the space to which the electrical action extends. By making the charge of this body indefinitely small we render insensible its disturbing action on the charge of the first body.

Let e be the charge of the small body, and let the force acting on it when placed at the point (x, y, z) be Re, and let the direction-cosines of the force be l, m, n, then we may call R the resultant electrical Intensity at the point (x, y, z).

If X, Y, Z denote the components of R, then

$$X = Rl, \qquad Y = Rm, \qquad Z = Rn.$$

In speaking of the resultant electrical intensity at a point, we do not necessarily imply that any force is actually exerted there, but only that if an electrified body were placed there it would be acted on by a force Re, where e is the charge of the body *.

Definition. The Resultant electric Intensity at any point is the force which would be exerted on a small body charged with the unit of positive electricity, if it were placed there without disturbing the actual distribution of electricity.

This force not only tends to move a body charged with electricity, but to move the electricity within the body, so that the positive electricity tends to move in the direction of R and the negative electricity in the opposite direction. Hence the quantity R is also called the Electromotive Intensity at the point (x, y, z).

When we wish to express the fact that the resultant intensity is a vector, we shall denote it by the German letter $\mathfrak{E}$. If the body is a dielectric, then, according to the theory adopted in this treatise, the electricity is displaced within it, so that the quantity of electricity which is forced in the direction of $\mathfrak{E}$ across unit of area fixed perpendicular to $\mathfrak{E}$ is

$$\mathfrak{D} = \frac{1}{4\pi} K \mathfrak{E};$$

where $\mathfrak{D}$ is the displacement, $\mathfrak{E}$ the resultant intensity, and K the specific inductive capacity of the dielectric.

If the body is a conductor, the state of constraint is continually giving way, so that a current of conduction is produced and maintained as long as $\mathfrak{E}$ acts on the medium.

Line-Integral of Electric Intensity, or Electromotive Force along an Arc of a Curve.

69.] The Electromotive force along a given arc AP of a curve is numerically measured by the work which would be done by the

* The Electric and Magnetic Intensity correspond, in electricity and magnetism, to the intensity of gravity, commonly denoted by g, in the theory of heavy bodies.

electric force on a unit of positive electricity carried along the curve from A, the beginning, to P, the end of the arc.

If s is the length of the arc, measured from A, and if the resultant intensity R at any point of the curve makes an angle ϵ with the tangent drawn in the positive direction, then the work done on unit of electricity in moving along the element of the curve ds will be

$$R \cos \epsilon \, ds,$$

and the total electromotive force E will be

$$E = \int_0^s R \cos \epsilon \, ds,$$

the integration being extended from the beginning to the end of the arc.

If we make use of the components of the intensity, the expression becomes

$$E = \int_0^s \left(X \frac{dx}{ds} + Y \frac{dy}{ds} + Z \frac{dz}{ds} \right) ds.$$

If X, Y, and Z are such that $X dx + Y dy + Z dz$ is the complete differential of $-V$, a function of x, y, z, then

$$E = \int_A^P (X dx + Y dy + Z dz) = -\int_A^P dV = V_A - V_P;$$

where the integration is performed in any way from the point A to the point P, whether along the given curve or along any other line between A and P.

In this case V is a scalar function of the position of a point in space, that is, when we know the coordinates of the point, the value of V is determinate, and this value is independent of the position and direction of the axes of reference. See Art. 16.

On Functions of the Position of a Point.

In what follows, when we describe a quantity as a function of the position of a point, we mean that for every position of the point the function has a determinate value. We do not imply that this value can always be expressed by the same formula for all points of space, for it may be expressed by one formula on one side of a given surface and by another formula on the other side.

On Potential Functions.

70.] The quantity $X dx + Y dy + Z dz$ is an exact differential whenever the force arises from attractions or repulsions whose intensity is a function of the distances from any number of points.

For if r_1 be the distance of one of the points from the point (x, y, z), and if R_1 be the repulsion, then

$$X_1 = R_1 \frac{x - x_1}{r_1} = R_1 \frac{dr_1}{dx},$$

with similar expressions for Y_1 and Z_1, so that

$$X_1\, dx + Y_1\, dy + Z_1\, dz = R_1\, dr_1;$$

and since R_1 is a function of r_1 only, $R_1\, dr_1$ is an exact differential of some function of r_1, say $-V_1$.

Similarly for any other force R_2, acting from a centre at distance r_2,
$$X_2 dx + Y_2 dy + Z_2 dz = R_2 dr_2 = -dV_2.$$

But $X = X_1 + X_2 +$ &c. and Y and Z are compounded in the same way, therefore

$$X\, dx + Y dy + Z\, dz = -dV_1 - dV_2 + \text{\&c.} = -dV.$$

The integral of this quantity, under the condition that it vanishes at an infinite distance, is called the Potential Function.

The use of this function in the theory of attractions was introduced by Laplace in the calculation of the attraction of the earth. Green, in his essay 'On the Application of Mathematical Analysis to Electricity,' gave it the name of the Potential Function. Gauss, working independently of Green, also used the word Potential. Clausius and others have applied the term Potential to the work which would be done if two bodies or systems were removed to an infinite distance from one another. We shall follow the use of the word in recent English works, and avoid ambiguity by adopting the following definition due to Sir W. Thomson.

Definition of Potential. The Potential at a Point is the work which would be done on a unit of positive electricity by the electric forces if it were placed at that point without disturbing the electric distribution, and carried from that point to an infinite distance: or, what comes to the same thing, the work which must be done by an external agent in order to bring the unit of positive electricity from an infinite distance (or from any place where the potential is zero) to the given point.

71.] *Expressions for the Resultant Intensity and its components in terms of the Potential.*

Since the total electromotive force along any arc AB is

$$E_{AB} = V_A - V_B,$$

if we put ds for the arc AB we shall have for the force resolved in the direction of ds,

$$R\cos\epsilon = -\frac{dV}{ds};$$

whence, by assuming ds parallel to each of the axes in succession, we get

$$X = -\frac{dV}{dx}, \qquad Y = -\frac{dV}{dy}, \qquad Z = -\frac{dV}{dz};$$

$$R = \left\{\left|\frac{dV}{dx}\right|^2 + \left|\frac{dV}{dy}\right|^2 + \left|\frac{dV}{dz}\right|^2\right\}^{\frac{1}{2}}.$$

We shall denote the intensity itself, whose magnitude, or tensor, is R and whose components are X, Y, Z, by the German letter $\mathfrak{E}$, as in Arts. 17 and 68.

The Potential at all Points within a Conductor is the same.

72.] A conductor is a body which allows the electricity within it to move from one part of the body to any other when acted on by electromotive force. When the electricity is in equilibrium there can be no electromotive force acting within the conductor. Hence $R = 0$ throughout the whole space occupied by the conductor. From this it follows that

$$\frac{dV}{dx} = 0, \qquad \frac{dV}{dy} = 0, \qquad \frac{dV}{dz} = 0;$$

and therefore for every point of the conductor

$$V = C,$$

where C is a constant quantity.

Since the potential at all points within the substance of the conductor is C, the quantity C is called the Potential *of* the conductor. C may be defined as the work which must be done by external agency in order to bring a unit of electricity from an infinite distance to the conductor, the distribution of electricity being supposed not to be disturbed by the presence of the unit.

It will be shewn at Art. 246 that in general when two bodies of different kinds are in contact, an electromotive force acts from one to the other through the surface of contact, so that when they are in equilibrium the potential of the latter is higher than that of the former. For the present, therefore, we shall suppose all our conductors made of the same metal, and at the same temperature.

If the potentials of the conductors A and B be V_A and V_B respectively, then the electromotive force along a wire joining A and B will be

$$V_A - V_B$$

in the direction AB, that is, positive electricity will tend to pass from the conductor of higher potential to the other.

Potential, in electrical science, has the same relation to Electricity that Pressure, in Hydrostatics, has to Fluid, or that Temperature, in Thermodynamics, has to Heat. Electricity, Fluids, and Heat all tend to pass from one place to another, if the Potential, Pressure, or Temperature is greater in the first place than in the second. A fluid is certainly a substance, heat is as certainly not a substance, so that though we may find assistance from analogies of this kind in forming clear ideas of formal relations of electrical quantities, we must be careful not to let the one or the other analogy suggest to us that electricity is either a substance like water, or a state of agitation like heat.

Potential due to any Electrical System.

73.] Let there be a single electrified point charged with a quantity e of electricity, and let r be the distance of the point x', y', z' from it, then

$$V = \int_r^\infty R\,dr = \int_r^\infty \frac{e}{r^2}\,dr = \frac{e}{r}.$$

Let there be any number of electrified points whose coordinates are (x_1, y_1, z_1), (x_2, y_2, z_2), &c. and their charges e_1, e_2, &c., and let their distances from the point (x', y', z') be r_1, r_2, &c., then the potential of the system at (x', y', z') will be

$$V = \Sigma\left(\frac{e}{r}\right).$$

Let the electric density at any point (x, y, z) within an electrified body be ρ, then the potential due to the body is

$$V = \iiint \frac{\rho}{r}\,dx\,dy\,dz\,;$$

where $$r = \{(x-x')^2+(y-y')^2+(z-z')^2\}^{\frac{1}{2}},$$

the integration being extended throughout the body.

On the Proof of the Law of the Inverse Square.

74 *a*.] The fact that the force between electrified bodies is inversely as the square of the distance may be considered to be established by Coulomb's direct experiments with the torsion-balance. The results, however, which we derive from such experiments must be regarded as affected by an error depending on the probable error of each experiment, and unless the skill of the operator be very great,

the probable error of an experiment with the torsion-balance is considerable.

A far more accurate verification of the law of force may be deduced from an experiment similar to that described at Art 32 (Exp. VII).

Cavendish, in his hitherto unpublished work on electricity, makes the evidence of the law of force depend on an experiment of this kind.

He fixed a globe on an insulating support, and fastened two hemispheres by glass rods to two wooden frames hinged to an axis so that the hemispheres, when the frames were brought together, formed an insulated spherical shell concentric with the globe.

The globe could then be made to communicate with the hemispheres by means of a short wire, to which a silk string was fastened so that the wire could be removed without discharging the apparatus.

The globe being in communication with the hemispheres, he charged the hemispheres by means of a Leyden jar, the potential of which had been previously measured by an electrometer, and immediately drew out the communicating wire by means of the silk string, removed and discharged the hemispheres, and tested the electrical condition of the globe by means of a pith ball electrometer.

No indication of any charge of the globe could be detected by the pith ball electrometer, which at that time (1773) was considered the most delicate electroscope.

Cavendish next communicated to the globe a known fraction of the charge formerly communicated to the hemispheres, and tested the globe again with his electrometer.

He thus found that the charge of the globe in the original experiment must have been less than $\frac{1}{60}$ of the charge of the whole apparatus, for if it had been greater it would have been detected by the electrometer.

He then calculated the ratio of the charge of the globe to that of the hemispheres on the hypothesis that the repulsion is inversely as a power of the distance differing slightly from 2, and found that if this difference was $\frac{1}{50}$ there would have been a charge on the globe equal to $\frac{1}{57}$ of that of the whole apparatus, and therefore capable of being detected by the electrometer.

74 *b*.] The experiment has recently been repeated at the Cavendish Laboratory in a somewhat different manner.

The hemispheres were fixed on an insulating stand, and the globe fixed in its proper position within them by means of an ebonite ring. By this arrangement the insulating support of the globe was never exposed to the action of any sensible electric force, and therefore never became charged, so that the disturbing effect of electricity creeping along the surface of the insulators was entirely removed.

Instead of removing the hemispheres before testing the potential of the globe, they were left in their position, but discharged to earth. The effect of a given charge of the globe on the electrometer was not so great as if the hemispheres had been removed, but this disadvantage was more than compensated by the perfect security afforded by the conducting vessel against all external electric disturbances.

The short wire which made the connexion between the shell and the globe was fastened to a small metal disk which acted as a lid to a small hole in the shell, so that when the wire and the lid were lifted up by a silk string, the electrode of the electrometer could be made to dip into the hole and rest on the globe within.

The electrometer was Thomson's Quadrant Electrometer described in Art. 219. The case of the electrometer and one of the electrodes were always connected to earth, and the testing electrode was connected to earth till the electricity of the shell had been discharged.

To estimate the original charge of the shell, a small brass ball was placed on an insulating support at a considerable distance from the shell.

The operations were conducted as follows :—

The shell was charged by communication with a Leyden jar.

The small ball was connected to earth so as to give it a negative charge by induction, and was then left insulated.

The communicating wire between the globe and the shell was removed by a silk string.

The shell was then discharged, and kept connected to earth.

The testing electrode was disconnected from earth, and made to touch the globe, passing through the hole in the shell.

Not the slightest effect on the electrometer could be observed.

To test the sensitiveness of the apparatus the shell was disconnected from earth and the small ball was discharged to earth. The electrometer then showed a positive deflection, *D*.

The negative charge of the brass ball was about $\frac{1}{54}$ of the original charge of the shell, and the positive charge induced by the ball when the shell was put to earth was about $\frac{1}{9}$ of that of the ball.

Hence when the ball was put to earth the potential of the shell, as indicated by the electrometer, was about $\frac{1}{158}$ of its original potential.

But if the repulsion had been as r^{q-2}, the potential of the globe would have been $-0{\cdot}1478\,q$ of that of the shell by equation 22, p. 81.

Hence if $\pm d$ be the greatest deflexion of the electrometer which could escape observation, and D the deflexion observed in the second part of the experiment, q cannot exceed

$$\pm \frac{1}{72} \frac{d}{D}.$$

Now even in a rough experiment D was more than $300\,d$, so that q cannot exceed

$$\pm \frac{1}{21600}.$$

Theory of the Experiment.

74 c.] To find the potential at any point due to a uniform spherical shell, the repulsion between two units of matter being any given function of the distance.

Let $\phi(r)$ be the repulsion between two units at distance r, and let $f(r)$ be such that

$$\frac{df(r)}{dr}\left(=f'(r)\right) = r\int_r^\infty \phi(r)\,dr. \tag{1}$$

Let the radius of the shell be a, and its surface density σ, then, if α denotes the whole mass of the shell,

$$\alpha = 4\pi a^2 \sigma. \tag{2}$$

Let b denote the distance of the given point from the centre of the shell, and let r denote its distance from any given point of the shell.

If we refer the point on the shell to spherical coordinates, the pole being the centre of the shell, and the axis the line drawn to the given point, then

$$r^2 = a^2 + b^2 - 2\,ab\cos\theta. \tag{3}$$

The mass of the element of the shell is

$$\sigma a^2 \sin\theta\, d\phi\, d\theta, \tag{4}$$

and the potential due to this element at the given point is

$$\sigma a^2 \sin\theta \frac{f'(r)}{r}\, d\theta\, d\phi; \tag{5}$$

and this has to be integrated with respect to ϕ from $\phi = 0$ to $\phi = 2\pi$, which gives

$$2\pi\sigma a^2 \sin\theta \frac{f' r}{r}\, d\theta, \tag{6}$$

which has to be integrated from $\theta = 0$ to $\theta = \pi$.

Differentiating (3) we find

$$r\,dr = ab \sin\theta\, d\theta. \tag{7}$$

Substituting the value of $d\theta$ in (6) we obtain

$$2\pi\sigma \frac{a}{b} f'(r)\, dr, \tag{8}$$

the integral of which is

$$V = 2\pi\sigma \frac{a}{b} \{f(r_1) - f(r_2)\}, \tag{9}$$

when r_1 is the greatest value of r, which is always $a+b$, and r_1 is the least value of r, which is $b-a$ when the given point is outside the shell and $a-b$ when it is within the shell.

If we write α for the whole charge of the shell, and V for its potential at the given point, then for a point outside the shell

$$V = \frac{\alpha}{2ab} \{f(b+a) - f(b-a)\}. \tag{10}$$

For a point on the shell itself

$$V = \frac{\alpha}{2a^2} f(2a), \tag{11}$$

and for a point inside the shell

$$V = \frac{\alpha}{2ab} \{f(a+b) - f(a-b)\}. \tag{12}$$

We have next to determine the potentials of two concentric spherical shells, the radii of the outer and inner shells being a and b, and their charges α and β.

Calling the potential of the outer shell A, and that of the inner B, we have by what precedes

$$A = \frac{\alpha}{2a^2} f(2a) + \frac{\beta}{2ab} \{f(a+b) - f(a-b)\}, \tag{13}$$

$$B = \frac{\beta}{2b^2} f(2b) + \frac{\alpha}{2ab} \{f(a+b) - f(a-b)\}. \tag{14}$$

In the first part of the experiment the shells communicate by the short wire and are both raised to the same potential, say V.

By putting $A = B = V$, and solving the equations (13) and (14) for β, we find the charge of the inner shell

$$\beta = 2Vb \frac{bf(2a) - a[f(a+b) - f(a-b)]}{f(2a)f(2b) - [f(a+b) - f(a-b)]^2}. \tag{15}$$

In the experiment of Cavendish, the hemispheres forming the outer shell were removed to a distance which we may suppose infinite, and discharged. The potential of the inner shell (or globe) would then become

$$B_1 = \frac{\beta}{2b^2} f(2b). \tag{16}$$

In the form of the experiment as repeated at the Cavendish Laboratory the outer shell was left in its place, but connected to earth, so that $A = 0$. In this case we find for the potential of the inner shell in terms of V

$$B_2 = V\left\{1 - \frac{a}{b}\,\frac{f(a+b)-f(a-b)}{f(2a)}\right\}. \tag{17}$$

74*d*.] Let us now assume, with Cavendish, that the law of force is some inverse power of the distance, not differing much from the inverse square, and let us put

$$\phi(r) = r^{q-2}; \tag{18}$$

then

$$f(r) = \frac{1}{1-q^2}\,r^{q+1}. \tag{19}$$

If we suppose q to be small, we may expand this by the exponential theorem in the form

$$f(r) = \frac{1}{1-q^2}\,r\left\{1 + q\log r + \frac{1}{1.2}(q\log r)^2 + \&c.\right\}; \tag{20}$$

and if we neglect terms involving q^2, equations (16) and (17) become

$$B_1 = \tfrac{1}{2}\,\frac{a}{a-b}\,Vq\left[\log\frac{4a^2}{a^2-b^2} - \frac{a}{b}\log\frac{a+b}{a-b}\right], \tag{21}$$

$$B_2 = \tfrac{1}{2}Vq\left[\log\frac{4a^2}{a^2-b^2} - \frac{a}{b}\log\frac{a+b}{a-b}\right], \tag{22}$$

from which we may determine q in terms of the results of the experiment.

74*e*.] Laplace gave the first demonstration that no function of the distance except the inverse square satisfies the condition that a uniform spherical shell exerts no force on a particle within it*.

If we suppose that β in equation (15) is always zero, we may apply the method of Laplace to determine the form of $f(r)$. We have by (15),

$$bf(2a) - af(a+b) + af(a-b) = 0.$$

Differentiating twice with respect to b, and dividing by a, we find

$$f''(a+b) = f''(a-b).$$

If this equation is generally true

$$f''(r) = C_0, \text{ a constant.}$$

* *Méc. Cél.*, I. 2.

Hence, $$f'(r) = C_0 r + C_1;$$

and by (1) $$\int_r^\infty \phi(r)\,dr = \frac{f'(r)}{r} = C_0 + \frac{C_1}{r},$$

$$\phi(r) = \frac{C_1}{r^2}.$$

We may observe, however, that though the assumption of Cavendish, that the force varies as some power of the distance, may appear less general than that of Laplace, who supposes it to be any function of the distance, it is the only one consistent with the fact that similar figures can be electrified so as to have similar electrical properties.

For if the force were any function of the distance except a power of the distance, the ratio of the force at two different distances would not be a function of the ratio of the distances, but would depend on the absolute value of the distances, and would therefore involve the ratios of these distances to an absolutely fixed length.

Indeed Cavendish himself points out that on his own hypothesis as to the constitution of the electric fluid, it is impossible for the distribution of electricity to be accurately similar in two conductors geometrically similar, unless the charges are proportional to the volumes. For he supposes the particles of the electric fluid to be closely pressed together near the surface of the body, and this is equivalent to supposing that the law of repulsion is no longer the inverse square, but that as soon as the particles come into contact, their repulsion begins to increase at a much greater rate with any further diminution of their distance.

Surface-Integral of Electric Induction, and Electric Displacement through a surface.

75.] Let R be the resultant intensity at any point of the surface, and ϵ the angle which R makes with the normal drawn towards the positive side of the surface, then $R\cos\epsilon$ is the component of the intensity normal to the surface, and if dS is the element of the surface, the electric displacement through dS will be, by Art. 68,

$$\frac{1}{4\pi} KR\cos\epsilon\, dS,$$

since we do not at present consider any dielectric except air, $K=1$.

We may, however, avoid introducing at this stage the theory of electric displacement, by calling $R\cos\epsilon\, dS$ the Induction through the element dS. This quantity is well known in mathematical physics, but the name of induction is borrowed from Faraday. The surface-integral of induction is

$$\iint R \cos \epsilon \, dS,$$

and it appears by Art. 21, that if X, Y, Z are the components of R, and if these quantities are continuous within a region bounded by a closed surface S, the induction reckoned from within outwards is

$$\iint R \cos \epsilon \, dS = \iiint \left(\frac{dX}{dx} + \frac{dY}{dy} + \frac{dZ}{dz}\right) dx\, dy\, dz,$$

the integration being extended through the whole space within the surface.

Induction through a Closed Surface due to a Single Centre of Force.

76.] Let a quantity e of electricity be supposed to be placed at a point O, and let r be the distance of any point P from O, the force at that point is $R = er^{-2}$ in the direction OP.

Let a line be drawn from O in any direction to an infinite distance. If O is without the closed surface this line will either not cut the surface at all, or it will issue from the surface as many times as it enters. If O is within the surface the line must first issue from the surface, and then it may enter and issue any number of times alternately, ending by issuing from it.

Let ϵ be the angle between OP and the normal to the surface drawn outwards where OP cuts it, then where the line issues from the surface, $\cos \epsilon$ will be positive, and where it enters, $\cos \epsilon$ will be negative.

Now let a sphere be described with centre O and radius unity, and let the line OP describe a conical surface of small angular aperture about O as vertex.

This cone will cut off a small element $d\omega$ from the surface of the sphere, and small elements dS_1, dS_2, &c. from the closed surface at the different places where the line OP intersects it.

Then, since any one of these elements dS intersects the cone at a distance r from the vertex and at an obliquity ϵ,

$$dS = r^2 \sec \epsilon \, d\omega\,;$$

and, since $R = er^{-2}$, we shall have

$$R \cos \epsilon \, dS = \pm e \, d\omega\,;$$

the positive sign being taken when r issues from the surface, and the negative where it enters it.

If the point O is without the closed surface, the positive values are equal in number to the negative ones, so that for any direction of r,

$$\Sigma R \cos\epsilon\, dS = 0,$$

and therefore

$$\iint R \cos\epsilon\, dS = 0,$$

the integration being extended over the whole closed surface.

If the point O is within the closed surface the radius vector OP first issues from the closed surface, giving a positive value of $e\,d\omega$, and then has an equal number of entrances and issues, so that in this case

$$\Sigma R \cos\epsilon\, dS = e\,d\omega.$$

Extending the integration over the whole closed surface, we shall include the whole of the spherical surface, the area of which is 4π, so that

$$\iint R \cos\epsilon\, dS = e \iint d\omega = 4\pi e.$$

Hence we conclude that the total induction outwards through a closed surface due to a centre of force e placed at a point O is zero when O is without the surface, and $4\pi e$ when O is within the surface.

Since in air the displacement is equal to the induction divided by 4π, the displacement through a closed surface, reckoned outwards, is equal to the electricity within the surface.

Corollary. It also follows that if the surface is not closed but is bounded by a given closed curve, the total induction through it is ωe, where ω is the solid angle subtended by the closed curve at O. This quantity, therefore, depends only on the closed curve, and the form of the surface of which it is the boundary may be changed in any way, provided it does not pass from one side to the other of the centre of force.

On the Equations of Laplace and Poisson.

77.] Since the value of the total induction of a single centre of force through a closed surface depends only on whether the centre is within the surface or not, and does not depend on its position in any other way, if there are a number of such centres e_1, e_2, &c. within the surface, and e_1', e_2', &c. without the surface, we shall have

$$\iint R \cos\epsilon\, dS = 4\pi e;$$

where e denotes the algebraical sum of the quantities of electricity at all the centres of force within the closed surface, that is, the total electricity within the surface, resinous electricity being reckoned negative.

If the electricity is so distributed within the surface that the density is nowhere infinite, we shall have by Art. 64,

$$4\pi e = 4\pi \iiint \rho\, dx\, dy\, dz,$$

and by Art. 75,

$$\iint R \cos \epsilon\, dS = \iiint \left(\frac{dX}{dx} + \frac{dY}{dy} + \frac{dZ}{dz}\right) dx\, dy\, dz.$$

If we take as the closed surface that of the element of volume $dx\, dy\, dz$, we shall have, by equating these expressions,

$$\frac{dX}{dx} + \frac{dY}{dy} + \frac{dZ}{dz} = 4\pi\rho;$$

and if a potential V exists, we find by Art. 71,

$$\frac{d^2 V}{dx^2} + \frac{d^2 V}{dy^2} + \frac{d^2 V}{dz^2} + 4\pi\rho = 0.$$

This equation, in the case in which the density is zero, is called Laplace's Equation. In its more general form it was first given by Poisson. It enables us, when we know the potential at every point, to determine the distribution of electricity.

We shall denote, as in Art. 26, the quantity

$$\frac{d^2 V}{dx^2} + \frac{d^2 V}{dy^2} + \frac{d^2 V}{dz^2} \text{ by } -\nabla^2 V,$$

and we may express Poisson's equation in words by saying that the electric density multiplied by 4π is the concentration of the potential. Where there is no electrification, the potential has no concentration, and this is the interpretation of Laplace's equation.

By Art. 72, V is constant within a conductor. Hence within a conductor the volume-density is zero, and the whole charge must be on the surface.

If we suppose that in the superficial and linear distributions of electricity the volume-density ρ remains finite, and that the electricity exists in the form of a thin stratum or a narrow fibre, then, by increasing ρ and diminishing the depth of the stratum or the section of the fibre, we may approach the limit of true superficial or linear distribution, and the equation being true throughout the process will remain true at the limit, if interpreted in accordance with the actual circumstances.

Variation of the Potential at a Charged Surface.

78*a*.] The potential function, V, must be physically continuous in the sense defined in Art. 7, except at the bounding surface of two different media, in which case, as we shall see in Art. 246, there may be a difference of potential between the substances, so that when the electricity is in equilibrium, the potential at a point in one substance is higher than the potential at the contiguous point in the other substance by a constant quantity, C, depending on the natures of the two substances and on their temperatures.

But the first derivatives of V with respect to x, y, or z may be discontinuous, and, by Art. 8, the points at which this discontinuity occurs must lie in a surface, the equation of which may be expressed in the form

$$\phi = \phi(x, y, z) = 0. \tag{1}$$

This surface separates the region in which ϕ is negative from the region in which ϕ is positive.

Let V_1 denote the potential at any given point in the negative region, and V_2 that at any given point in the positive region, then at any point in the surface at which $\phi = 0$, and which may be said to belong to both regions,

$$V_1 + C = V_2, \tag{2}$$

where C is the constant excess of potential, if any, in the substance on the positive side of the surface.

Let l, m, n be the direction-cosines of the normal ν_2 drawn from a given point of the surface into the positive region. Those of the normal ν_1 drawn from the same point into the negative region will be $-l$, $-m$, and $-n$.

The rates of variation of V along the normals are

$$\frac{dV_1}{d\nu_1} = -l\frac{dV_1}{dx} - m\frac{dV_1}{dy} - n\frac{dV_1}{dz}, \tag{3}$$

$$\frac{dV_2}{d\nu_2} = l\frac{dV_2}{dx} + m\frac{dV_2}{dy} + n\frac{dV_2}{dz}. \tag{4}$$

Let any line be drawn on the surface, and let its length, measured from a fixed point in it, be s, then at every point of the surface, and therefore at every point of this line, $V_2 - V_1 = C$. Differentiating this equation with respect to s, we get

$$\left(\frac{dV_2}{dx} - \frac{dV_1}{dx}\right)\frac{dx}{ds} + \left(\frac{dV_2}{dy} - \frac{dV_1}{dy}\right)\frac{dy}{ds} + \left(\frac{dV_2}{dz} - \frac{dV_1}{dz}\right)\frac{dz}{ds} = 0; \tag{5}$$

and since the normal is perpendicular to this line

$$l\frac{dx}{ds} + m\frac{dy}{ds} + n\frac{dz}{ds} = 0. \tag{6}$$

From (3), (4), (5), (6) we find

$$\frac{dV_2}{dx} - \frac{dV_1}{dx} = l\left(\frac{dV_1}{dv_1} + \frac{dV_2}{dv_2}\right), \tag{7}$$

$$\frac{dV_2}{dy} - \frac{dV_1}{dy} = m\left(\frac{dV_1}{dv_1} + \frac{dV_2}{dv_2}\right), \tag{8}$$

$$\frac{dV_2}{dz} - \frac{dV_1}{dz} = n\left(\frac{dV_1}{dv_1} + \frac{dV_2}{dv_2}\right). \tag{9}$$

If we consider the variation of the electromotive intensity at a point in passing through the surface, that component of the intensity which is normal to the surface may change abruptly at the surface, but the other two components parallel to the tangent plane remain continuous in passing through the surface.

78 *b*.] To determine the charge of the surface, let us consider a closed surface which is partly in the positive region and partly in the negative region, and which therefore encloses a portion of the surface of discontinuity.

The surface integral,

$$\iint R \cos\epsilon\, dS,$$

extended over this surface, is equal to $4\pi e$, where e is the quantity of electricity within the closed surface.

Proceeding as in Art. 21, we find

$$\iint R\cos\epsilon\, dS = \iiint \left(\frac{dX}{dx} + \frac{dY}{dy} + \frac{dZ}{dz}\right) dx\, dy\, dz$$
$$+ \iint \{l(X_2 - X_1) + m(Y_2 - Y_1) + n(Z_2 - Z_1)\}\, dS, \quad (2)$$

where the triple integral is extended throughout the closed surface, and the double integral over the surface of discontinuity.

Substituting for the terms of this equation their values from (7), (8), (9),

$$4\pi e = \iiint 4\pi\rho\, dx\, dy\, dz - \iint \left(\frac{dV_1}{dv_1} + \frac{dV_2}{dv_2}\right) dS. \tag{11}$$

But by the definition of the volume-density, ρ, and the surface-density, σ,

$$4\pi e = 4\pi \iiint \rho\, dx\, dy\, dz + 4\pi \iint \sigma\, dS. \tag{12}$$

Hence, comparing the last terms of these two equations,

$$\frac{dV_1}{dv_1} + \frac{dV_2}{dv_2} + 4\pi\sigma = 0. \tag{13}$$

This equation is called the characteristic equation of V at an electrified surface of which the surface-density is σ.

78c.] If V is a function of x, y, z which, throughout a given continuous region of space, satisfies Laplace's equation

$$\frac{d^2V}{dx^2} + \frac{d^2V}{dy^2} + \frac{d^2V}{dz^2} = 0,$$

and if throughout a finite portion of this region V is constant and equal to C, then V must be constant and equal to C throughout the whole region in which Laplace's equation is satisfied.

If V is not equal to C throughout the whole region, let S be the surface which bounds the finite portion within which $V = C$.

At the surface S, $V = C$.

Let ν be a normal drawn outwards from the surface S. Since S is the boundary of the continuous region for which $V = C$, the value of V as we travel from the surface along the normal begins to differ from C. Hence $\frac{dV}{d\nu}$ just outside the surface may be positive or negative, but cannot be zero except for normals drawn from the boundary line between a positive and a negative area.

But if ν' is the normal drawn inwards from the surface S, $V' = C$ and $\frac{dV'}{d\nu'} = 0$.

Hence, at every point of the surface except certain boundary lines,

$$\frac{dV}{d\nu} + \frac{dV'}{d\nu'} (= -4\pi\sigma)$$

is a finite quantity, positive or negative, and therefore the surface S has a continuous distribution of electricity over all parts of it except certain boundary lines which separate positively from negatively charged areas.

Laplace's equation is not satisfied at the surface S except at points lying on certain lines on the surface. The surface S therefore, within which $V = C$, includes the whole of the continuous region within which Laplace's equation is satisfied.

Force Acting on a Charged Surface.

79.] The general expression for the components of the force acting on a charged body parallel to the three axes are of the form

$$A = \iiint \rho X \, dx \, dy \, dz, \tag{14}$$

with similar expressions for B and C, the components parallel to y and z.

But at a charged surface ρ is infinite, and X is discontinuous, so that we cannot calculate the force directly from expressions of this form.

We have proved, however, that the discontinuity affects only that component of the intensity which is normal to the charged surface, the other two components being continuous.

Let us therefore assume the axis of x normal to the surface at the given point, and let us also assume, at least in the first part of our investigation, that X is not really discontinuous, but that it changes continuously from X_1 to X_2 while x changes from x_1 to x_2. If the result of our calculation gives a definite limiting value for the force when x_2-x_1 is diminished without limit, we may consider it correct when $x_2 = x_1$, and the charged surface has no thickness.

Substituting for ρ its value as found in Art. 77,

$$A = \frac{1}{4\pi}\iiint\left(\frac{dX}{dx}+\frac{dY}{dy}+\frac{dZ}{dz}\right)X\,dx\,dy\,dz. \tag{15}$$

Integrating this expression with respect to x from $x = x_1$ to $x = x_2$ it becomes

$$A = \frac{1}{4\pi}\iint\left[\tfrac{1}{2}(X_2^2-X_1^2)+\int_{x_1}^{x_2}\left(\frac{dY}{dy}+\frac{dZ}{dz}\right)X\,dx\right]dy\,dz. \tag{16}$$

This is the value of A for a stratum parallel to yz of which the thickness is x_2-x_1.

Since Y and Z are continuous, $\frac{dY}{dy}+\frac{dZ}{dz}$ is finite, and since X is also finite,

$$\int_{x_1}^{x_2}\left(\frac{dY}{dy}+\frac{dZ}{dz}\right)X\,dx < C(x_2-x_1),$$

where C is the greatest value of $\left(\frac{dY}{dy}+\frac{dZ}{dz}\right)X$ between $x = x_1$ and $x = x_2$.

Hence when x_2-x_1 is diminished without limit this term must ultimately vanish, leaving

$$A = \iint\frac{1}{8\pi}(X_2^2-X_1^2)\,dy\,dz, \tag{17}$$

where X_1 is the value of X on the negative and X_2 on the positive side of the surface.

But by Art. 78, $\quad X_2-X_1 = \frac{dV_1}{dx}-\frac{dV_2}{dx} = 4\pi\sigma, \qquad (18)$

so that we may write

$$A = \iint\tfrac{1}{2}(X_2+X_1)\,\sigma\,dy\,dz. \tag{19}$$

Here $dy\,dz$ is the element of the surface, σ is the surface-density, and $\frac{1}{2}(X_2+X_1)$ is the arithmetical mean of the electromotive intensity on the two sides of the surface.

Hence an element of a charged surface is acted on by a force, the component of which normal to the surface is equal to the charge of the element into the arithmetical mean of the normal electromotive intensities on the two sides of the surface.

Since the other two components of the electromagnetic intensity are not discontinuous, there can be no ambiguity in estimating the corresponding components of the force acting on the surface.

We may now suppose the direction of the normal to the surface to be in any direction with respect to the axes, and write the general expressions for the components of the force on the element of surface dS,

$$\left.\begin{aligned} A &= \tfrac{1}{2}(X_1+X_2)\,\sigma\,dS, \\ B &= \tfrac{1}{2}(Y_1+Y_2)\,\sigma\,dS, \\ C &= \tfrac{1}{2}(Z_1+Z_2)\,\sigma\,dS. \end{aligned}\right\} \qquad (20)$$

Charged Surface of a Conductor.

80.] We have already shewn (Art. 72) that throughout the substance of a conductor in electric equilibrium $X = Y = Z = 0$, and therefore V is constant.

Hence $$\frac{dX}{dx} + \frac{dY}{dy} + \frac{dZ}{dz} = 4\pi\rho = 0,$$

and therefore ρ must be zero throughout the substance of the conductor, or there can be no electricity in the interior of the conductor.

Hence a superficial distribution of electricity is the only possible distribution in a conductor in equilibrium.

A distribution throughout the mass of a body can exist only when the body is a non-conductor.

Since the resultant intensity within the conductor is zero, the resultant intensity just outside the conductor must be in the direction of the normal and equal to $4\pi\sigma$, acting outwards from the conductor.

This relation between the surface-density and the resultant intensity close to the surface of a conductor is known as Coulomb's Law, Coulomb having ascertained by experiment that the intensity of the electric force near a given point of the surface of a conductor is normal to the surface and *proportional* to the surface-density at the given point. The numerical relation

$$R = 4\pi\sigma$$

was established by Poisson.

The force acting on an element, dS, of the charged surface of a conductor is, by Art. 79, (since the intensity is zero on the inner side of the surface,)

$$\tfrac{1}{2} R\sigma dS = 2\pi\sigma d^2 S = \frac{1}{8\pi} R^2 dS.$$

This force acts outwards from the conductor, whether the charge of the surface is positive or negative.

Its value in dynes per square centimetre is

$$\tfrac{1}{2} R\sigma = 2\pi\sigma^2 = \frac{1}{8\pi} R^2,$$

acting as a tension outwards from the surface of the conductor.

81.] If we now suppose an elongated body to be electrified, we may, by diminishing its lateral dimensions, arrive at the conception of an electrified line.

Let ds be the length of a small portion of the elongated body, and let c be its circumference, and σ the surface density of the electricity on its surface; then, if λ is the charge per unit of length, $\lambda = c\sigma$, and the resultant electrical intensity close to the surface will be

$$4\pi\sigma = 4\pi\frac{\lambda}{c}.$$

If, while λ remains finite, c be diminished indefinitely, the intensity at the surface will be increased indefinitely. Now in every dielectric there is a limit beyond which the intensity cannot be increased without a disruptive discharge. Hence a distribution of electricity in which a finite quantity is placed on a finite portion of a line is inconsistent with the conditions existing in nature.

Even if an insulator could be found such that no discharge could be driven through it by an infinite force, it would be impossible to charge a linear conductor with a finite quantity of electricity, for an infinite electromotive force would be required to bring the electricity to the linear conductor.

In the same way it may be shewn that a point charged with a finite quantity of electricity cannot exist in nature. It is convenient, however, in certain cases, to speak of electrified lines and points, and we may suppose these represented by electrified wires, and by small bodies of which the dimensions are negligible compared with the principal distances concerned.

Since the quantity of electricity on any given portion of a wire at a given potential diminishes indefinitely when the diameter of the wire is indefinitely diminished, the distribution of electricity on bodies of considerable dimensions will not be sensibly affected by the introduction of very fine metallic wires into the field, such as are used to form electrical connexions between these bodies and the earth, an electrical machine, or an electrometer.

On Lines of Force.

82.] If a line be drawn whose direction at every point of its course coincides with that of the resultant intensity at that point, the line is called a Line of Force.

In every part of the course of a line of force, it is proceeding from a place of higher potential to a place of lower potential.

Hence a line of force cannot return into itself, but must have a beginning and an end. The beginning of a line of force must be in a positively charged surface, and the end of a line of force must be in a negatively charged surface.

The beginning and the end of the line are called corresponding points on the positive and negative surface respectively.

If the line of force moves so that its beginning traces a closed curve on the positive surface, its end will trace a corresponding closed curve on the negative surface, and the line of force itself will generate a tubular surface called a tube of induction. Such a tube is called a Solenoid *.

Since the force at any point of the tubular surface is in the tangent plane, there is no induction across the surface. Hence if the tube does not contain any electrified matter, by Art. 77 the total induction through the closed surface formed by the tubular surface and the two ends is zero, and the values of $\iint R\cos\epsilon\, dS$ for the two ends must be equal in magnitude but opposite in sign.

If these surfaces are the surfaces of conductors

$$\epsilon = 0 \quad \text{and} \quad R = -4\pi\sigma,$$

and $\iint R\cos\epsilon\, dS$ becomes $-4\pi\iint\sigma\, dS$, or the charge of the surface multiplied by 4π.

Hence the positive charge of the surface enclosed within the closed curve at the beginning of the tube is numerically equal to the negative charge enclosed within the corresponding closed curve at the end of the tube.

* From σωλήν, a tube. Faraday uses (3271) the term 'Sphondyloid' in the same sense.

Several important results may be deduced from the properties of lines of force.

The interior surface of a closed conducting vessel is entirely free from charge, and the potential at every point within it is the same as that of the conductor, provided there is no insulated and charged body within the vessel.

For since a line of force must begin at a positively charged surface and end at a negatively charged surface, and since no charged body is within the vessel, a line of force, if it exists within the vessel, must begin and end on the interior surface of the vessel itself.

But the potential must be higher at the beginning of a line of force than at the end of the line, whereas we have proved that the potential at all points of a conductor is the same.

Hence no line of force can exist in the space within a hollow vessel, provided no charged body be placed inside it.

If a conductor within a closed hollow vessel is placed in communication with the vessel, its potential becomes the same as that of the vessel, and its surface becomes continuous with the inner surface of the vessel. The conductor is therefore free from charge.

If we suppose any charged surface divided into elementary portions such that the charge of each element is unity, and if solenoids having these elements for their bases are drawn through the field of force, then the surface-integral for any other surface will be represented by the number of solenoids which it cuts. It is in this sense that Faraday uses his conception of lines of force to indicate not only the direction but the amount of the force at any place in the field.

We have used the phrase Lines of Force because it has been used by Faraday and others. In strictness, however, these lines should be called Lines of Electric Induction.

In the ordinary cases the lines of induction indicate the direction and magnitude of the resultant electromotive intensity at every point, because the intensity and the induction are in the same direction and in a constant ratio. There are other cases, however, in which it is important to remember that these lines indicate primarily the induction, and that the intensity is directly indicated by the equipotential surfaces, being normal to these surfaces and inversely proportional to the distances of consecutive surfaces.

On Specific Inductive Capacity.

83*a*.] In the preceding investigation of surface-integrals we have adopted the ordinary conception of direct action at a distance, and have not taken into consideration any effects depending on the nature of the dielectric medium in which the forces are observed.

But Faraday has observed that the quantity of electricity induced by a given electromotive force on the surface of a conductor which bounds a dielectric is not the same for all dielectrics. The induced electricity is greater for most solid and liquid dielectrics than for air and gases. Hence these bodies are said to have a greater specific inductive capacity than air, which he adopted as the standard medium.

We may express the theory of Faraday in mathematical language by saying that in a dielectric medium the induction across any surface is the product of the normal electric force into the coefficient of specific inductive capacity of that medium. If we denote this coefficient by K, then in every part of the investigation of surface-integrals we must multiply X, Y, and Z by K, so that the equation of Poisson will become

$$\frac{d}{dx}.K\frac{dV}{dx}+\frac{d}{dy}.K\frac{dV}{dy}+\frac{d}{dz}.K\frac{dV}{dz}+4\pi\rho=0. \qquad (1)$$

At the surface of separation of two media whose inductive capacities are K_1 and K_2, and in which the potentials are V_1 and V_2, the characteristic equation may be written

$$K_1\frac{dV_1}{d\nu_1}+K_2\frac{dV_2}{d\nu_2}+4\pi\sigma=0; \qquad (2)$$

where ν_1, ν_2 are the normals drawn in the two media, and σ is the true surface-density on the surface of separation; that is to say, the quantity of electricity which is actually on the surface in the form of a charge, and which can be altered only by conveying electricity to or from the spot.

Apparent distribution of Electricity.

83*b*.] If we begin with the actual distribution of the potential and deduce from it the volume density ρ' and the surface density σ' on the hypothesis that K is everywhere equal to unity, we may call ρ' the apparent volume density and σ' the apparent surface density, because a distribution of electricity thus defined would account for the actual distribution of potential, on the hypothesis that the law of electric force as given in Art. 66 requires no modification on account of the different properties of dielectrics.

The apparent charge of electricity within a given region may increase or diminish without any passage of electricity through the bounding surface of the region. We must therefore distinguish it from the true charge, which satisfies the equation of continuity.

In a heterogeneous dielectric in which K varies continuously, if ρ' be the apparent volume-density,

$$\frac{d^2V}{dx^2}+\frac{d^2V}{dy^2}+\frac{d^2V}{dz^2}+4\pi\rho'=0. \tag{3}$$

Comparing this with the equation above, we find

$$4\pi(\rho-K\rho')+\frac{dK}{dx}\frac{dV}{dx}+\frac{dK}{dy}\frac{dV}{dy}+\frac{dK}{dz}\frac{dV}{dz}=0. \tag{4}$$

The true electrification, indicated by ρ, in the dielectric whose variable inductive capacity is denoted by K, will produce the same potential at every point as the apparent electrification, denoted by ρ', would produce in a dielectric whose inductive capacity is everywhere equal to unity.

The apparent surface charge, σ', is that deduced from the electrical forces in the neighbourhood of the surface, using the ordinary characteristic equation

$$\frac{dV_1}{d\nu_1}+\frac{dV_2}{d\nu_2}+4\pi\sigma'=0. \tag{5}$$

If a solid dielectric of any form is a perfect insulator, and if its surface receives no charge, then the true electrification remains zero, whatever be the electrical forces acting on it.

Hence
$$K_1\frac{dV_1}{d\nu_1}+K_2\frac{dV_2}{d\nu_2}=0,$$

$$\frac{dV_1}{d\nu_1}=\frac{4\pi\sigma' K_2}{K_1-K_2}, \qquad \frac{dV_2}{d\nu_2}=\frac{4\pi\sigma' K_1}{K_2-K_2}.$$

The surface-density σ' is that of the apparent electrification produced at the surface of the solid dielectric by induction. It disappears entirely when the inducing force is removed, but if during the action of the inducing force the apparent electrification of the surface is discharged by passing a flame over the surface, then, when the inducing force is taken away, there will appear a true electrification opposite to σ' *.

* See Faraday's 'Remarks on Static Induction,' *Proceedings of the Royal Institution*, Feb. 12, 1858.

CHAPTER III.

ON ELECTRICAL WORK AND ENERGY IN A SYSTEM OF CONDUCTORS.

84.] *On the Work which must be done by an external agent in order to charge an electrified system in a given manner.*

The work spent in bringing a quantity of electricity δe from an infinite distance (or from any place where the potential is zero) to a given part of the system where the potential is V, is, by the definition of potential (Art. 70), $V\delta e$.

The effect of this operation is to increase the charge of the given part of the system by δe, so that if it was e before, it will become $e+\delta e$ after the operation.

We may therefore express the work done in producing a given alteration in the charges of the system by the integral

$$W = \Sigma\left(\int V\delta e\right); \tag{1}$$

where the summation, (Σ), is to be extended to all parts of the electrified system.

It appears from the expression for the potential in Art. 73, that the potential at a given point may be considered as the sum of a number of parts, each of these parts being the potential due to a corresponding part of the charge of the system.

Hence if V is the potential at a given point due to a system of charges which we may call $\Sigma(e)$, and V' the potential at the same point due to another system of charges which we may call $\Sigma(e')$, the potential at the same point due to both systems of charges existing together would be $V+V'$.

If, therefore, every one of the charges of the system is altered in the ratio of n to 1, the potential at any given point in the system will also be altered in the ratio of n to 1.

Let us, therefore, suppose that the operation of charging the system is conducted in the following manner. Let the system be originally free from charge and at potential zero, and let the different portions of the system be charged simultaneously, each at a rate proportional to its final charge.

Thus if e is the final charge, and V the final potential of any part of the system, then, if at any stage of the operation the charge is ne, the potential will be nV, and we may represent the process of charging by supposing n to increase continuously from 0 to 1.

While n increases from n to $n+\delta n$, any portion of the system whose final charge is e, and whose final potential is V, receives an increment of charge $e\delta n$, its potential being nV, so that the work done on it during this operation is $eVn\delta n$.

Hence the whole work done in charging the system is

$$\Sigma eV\int_0^1 n\,dn = \tfrac{1}{2}\Sigma(eV), \tag{2}$$

or half the sum of the products of the charges of the different portions of the system into their respective potentials.

This is the work which must be done by an external agent in order to charge the system in the manner described, but since the system is a conservative system, the work required to bring the system into the same state by any other process must be the same.

We may therefore call

$$W = \tfrac{1}{2}\Sigma(eV) \tag{3}$$

the electric energy of the system, expressed in terms of the charges of the different parts of the system and their potentials.

85 *a*.] Let us next suppose that the system passes from the state (e, V) to the state (e', V') by a process in which the different charges increase simultaneously at rates proportional for each to its total increment $e'-e$.

If at any instant the charge of a given portion of the system is $e+n(e'-e)$, its potential will be $V+n(V'-V)$, and the work done in altering the charge of this portion will be

$$\int_0^1 (e'-e)[V+n(V'-V)]\,dn = \tfrac{1}{2}(e'-e)(V+V');$$

so that if we denote by W' the energy of the system in the state (e', V')

$$W'-W = \tfrac{1}{2}\Sigma(e'-e)(V'+V). \tag{4}$$

But $$W = \tfrac{1}{2}\Sigma(eV),$$

and $$W' = \tfrac{1}{2}\Sigma(e'V').$$

Substituting these values in equation (4) we find

$$\Sigma(eV') = \Sigma(e'V). \tag{5}$$

Hence if, in the same fixed system of electrified conductors, we consider two different states of electrification, the sum of the products of the charges in the first state into the potentials of the corresponding portions of the conductors in the second state, is equal to the sum of the products of the charges in the second state into the potentials of the corresponding conductors in the first state.

This result corresponds, in the elementary theory of electricity, to Green's Theorem in the analytical theory. By properly choosing the initial and final state of the system, we may deduce a number of useful results.

85 *b*.] From (4) and (5) we find another expression for the increment of the energy, in which it is expressed in terms of the increments of potential,

$$W' - W = \tfrac{1}{2}\Sigma(e'+e)(V'-V). \tag{6}$$

If the increments are infinitesimal, we may write (4) and (6)

$$dW = \Sigma(V\delta e) = \Sigma(e\delta V), \tag{7}$$

and if we denote by W_e and W_V the expressions for W in terms of the charges and the potentials of the system respectively, and by A_r, e_r, and V_r a particular conductor of the system, its charge, and its potential, then

$$V_r = \frac{dW_e}{de_r}, \tag{8}$$

$$e_r = \frac{dW_V}{dV_r}. \tag{9}$$

86.] If in any fixed system of conductors, any one of them, which we may denote by A_t, is without charge, both in the initial and final state, then for that conductor $e_t = 0$, and $e_t' = 0$, so that the terms depending on A_t vanish from both members of equation (5).

If another conductor, say A_u, is at potential zero in both states of the system, then $V_u = 0$ and $V_u' = 0$, so that the terms depending on A_u vanish from both members of equation (5).

If, therefore, all the conductors except two, A_r and A_s, are either insulated and without charge, or else connected to the earth, equation (5) is reduced to the form

$$e_r V_r' + e_s V_s' = e_r' V_r + e_s' V_s. \tag{10}$$

If in the initial state

$$e_r = 1 \quad \text{and} \quad e_s = 0,$$

and in the final state

$$e_r' = 0 \quad \text{and} \quad e_s' = 1,$$

equation (10) becomes

$$V_r' = V_s; \tag{11}$$

or if a unit charge communicated to A_r raises A_s to a potential V, then a unit charge communicated to A_s will raise A_r to the same potential V, provided that every one of the other conductors of the system is either insulated and without charge, or else connected to earth so that its potential is zero.

This is the first instance we have met with in electricity of a reciprocal relation. Such reciprocal relations occur in every branch of science, and often enable us to deduce the solution of new problems from those of simpler problems already solved.

Thus from the fact that at a point outside a conducting sphere whose charge is 1 the potential is r^{-1}, where r is the distance from the centre, we conclude that if a small body whose charge is 1 is placed at a distance r from the centre of a conducting sphere without charge, it will raise the potential of the sphere to r^{-1}.

Let us next suppose that in the initial state

$$V_r = 1 \quad \text{and} \quad V_s = 0,$$

and in the final state

$$V_r' = 0 \quad \text{and} \quad V_s' = 1,$$

equation (10) becomes

$$e_s = e_r'; \tag{12}$$

or if, when A_r is raised to unit potential, a charge e is induced on A_s, then if A_s is raised to unit potential, an equal charge e will be induced on A_r.

Let us suppose in the third place, that in the initial state

$$V_r = 1 \quad \text{and} \quad e_s = 0,$$

and that in the final state

$$V_r' = 0 \quad \text{and} \quad e_s' = 1,$$

equation (10) becomes in this case

$$e_r' + V_s = 0. \tag{13}$$

Hence if when A_s is without charge, the operation of charging A_r to potential unity raises A_s to potential V, then if A_r is kept at potential zero, a unit charge communicated to A_s will induce on A_r a negative charge, the numerical value of which is V.

In all these cases we may suppose some of the other conductors to be insulated and without charge, and the rest to be connected to earth.

The third case is an elementary form of one of Green's theorems. As an example of its use let us suppose that we have ascertained the distribution of electric charge on the different elements of a conducting system at potential zero, induced by a charge unity communicated to a given body A_s of the system.

Let η_r be the charge of A_r under these circumstances. Then if we suppose A_s without charge, and the other bodies raised each to a different potential, the potential of A_s will be

$$V_s = -\Sigma(\eta_r V_r). \tag{14}$$

Thus if we have ascertained the surface density at any given point of a hollow conducting vessel due to a unit charge placed at a given point within it, then, if we know the value of the potential at every point of a surface of the same size and form as the interior surface of the vessel, we can deduce the potential at a point within it the position of which corresponds to that of the unit charge.

Hence if the potential is known for all points of a closed surface it may be determined for any point within the surface, if there be no electrified body within it, and for any point outside, if there be no electrified body outside.

Theory of a system of conductors.

87.] Let $A_1, A_2, \ldots A_n$ be n conductors of any form; let $e_1, e_2, \ldots e_n$ be their charges; and $V_1, V_2, \ldots V_n$ their potentials.

Let us suppose that the dielectric medium which separates the conductors remains the same, and does not become charged with electricity during the operations to be considered.

We have shown in Art. 84 that the potential of each conductor is a homogeneous linear function of the n charges.

Hence since the electric energy of the system is half the sum of the products of the potential of each conductor into its charge, the electric energy must be a homogeneous quadratic function of the n charges, of the form

$$W_e = \tfrac{1}{2}p_{11}e_1^2 + p_{12}e_1e_2 + \tfrac{1}{2}p_{22}e_2^2 + p_{13}e_1e_3 + p_{23}e_2e_3 + \tfrac{1}{2}p_{33}e_3^2 + \&c. \tag{15}$$

The suffix e indicates that W is to be expressed as a function of the charges. When W is written without a suffix it denotes the expression (3), in which both charges and potentials occur.

From this expression we can deduce the potential of any one of the conductors. For since the potential is defined as the work which must be done to bring a unit of electricity from potential zero to the given potential, and since this work is spent in increasing W, we have only to differentiate W_e with respect to the charge of the given conductor to obtain its potential. We thus obtain

$$\left.\begin{array}{l} V_1 = p_{11} e_1 \dots + p_{r1} e_r \dots + p_{n1} e_n, \\ \text{-} \quad \text{-} \quad \text{-} \quad \text{-} \quad \text{-} \quad \text{-} \quad \text{-} \quad \text{-} \\ V_s = p_{1s} e_1 \dots + p_{rs} e_r \dots + p_{ns} e_n, \\ \text{-} \quad \text{-} \quad \text{-} \quad \text{-} \quad \text{-} \quad \text{-} \quad \text{-} \quad \text{-} \\ V_n = p_{1n} e_1 \dots + p_{rn} e_r \dots + p_{nn} e_n, \end{array}\right\} \qquad (16)$$

a system of n linear equations which express the n potentials in terms of the n charges.

The coefficients p_{rs} &c., are called coefficients of potential. Each has two suffixes, the first corresponding with that of the charge, and the second with that of the potential.

The coefficient p_{rr}, in which the two suffixes are the same, denotes the potential of A_r when its charge is unity, that of all the other conductors being zero. There are n coefficients of this kind, one for each conductor.

The coefficient p_{rs}, in which the two suffixes are different, denotes the potential of A_s when A_r receives a charge unity, the charge of each of the other conductors, except A_r, being zero.

We have already proved in Art. 86 that $p_{rs} = p_{sr}$, but we may prove it more briefly by considering that

$$p_{rs} = \frac{dV_s}{de_r} = \frac{d}{de_r}\frac{dW_e}{de_s} = \frac{d}{de_s}\frac{dW_e}{de_r} = \frac{dV_r}{de_s} = p_{sr}. \qquad (17)$$

The number of *different* coefficients with double suffix is therefore $\frac{1}{2} n (n-1)$, being one for each pair of conductors.

By solving the equations (16) for e_1, e_2 &c., we obtain n equations giving the charges in terms of the potentials

$$\left.\begin{array}{l} e_1 = q_{11} V_1 \dots + q_{1s} V_s \dots + q_{1n} V_n, \\ \text{-} \quad \text{-} \quad \text{-} \quad \text{-} \quad \text{-} \quad \text{-} \quad \text{-} \quad \text{-} \\ e_r = q_{r1} V_1 \dots + q_{rs} V_s \dots + q_{rn} V_n, \\ \text{-} \quad \text{-} \quad \text{-} \quad \text{-} \quad \text{-} \quad \text{-} \quad \text{-} \quad \text{-} \\ e_n = q_{n1} V_1 \dots + q_{ns} V_s \dots + q_{n} V_{nn}, \end{array}\right\} \qquad (18)$$

We have in this case also $q_{rs} = q_{sr}$, for

$$q_{rs} = \frac{de_r}{dV_s} = \frac{d}{dV_s}\frac{dW_V}{dV_r} = \frac{d}{dV_r}\frac{dW_V}{dV_s} = \frac{de_s}{dV_r} = q_{sr}. \qquad (19)$$

By substituting the values of the charges in the equation for the electric energy

$$W = \tfrac{1}{2}\,[e_1 V_1 + \ldots + e_r V_r \ldots + e_n V_n], \qquad (20)$$

we obtain an expression for the energy in terms of the potentials

$$W_V = \tfrac{1}{2}\,q_{11} V_1^2 + q_{12} V_1 V_2 + \tfrac{1}{2}\,q_{22} V_2^2 + q_{13} V_1 V_3 + q_{23} V_2 V_3 + \tfrac{1}{2}\,q_{33} V_3^2 + \&c. \qquad (21)$$

A coefficient in which the two suffixes are the same is called the Electric Capacity of the conductor to which it belongs.

Definition. The Capacity of a conductor is its charge when its own potential is unity, and that of all the other conductors is zero.

This is the proper definition of the capacity of a conductor when no further specification is made. But it is sometimes convenient to specify the condition of some or all of the other conductors in a different manner, as for instance to suppose that the charge of certain of them is zero, and we may then define the capacity of the conductor under these conditions as its charge when its potential is unity.

The other coefficients are called coefficients of induction. Any one of them, as q_{rs} denotes the charge of A_r when A_s is raised to potential unity, the potential of all the conductors except A_s being zero.

The mathematical calculation of the coefficients of potential and of capacity is in general difficult. We shall afterwards prove that they have always determinate values, and in certain special cases we shall calculate these values. We shall also shew how they may be determined by experiment.

When the capacity of a conductor is spoken of without specifying the form and position of any other conductor in the same system, it is to be interpreted as the capacity of the conductor when no other conductor or electrified body is within a finite distance of the conductor referred to.

It is sometimes convenient, when we are dealing with capacities and coefficients of induction only, to write them in the form $[A\,.\,P]$, this symbol being understood to denote the charge on A when P is raised to unit potential.

In like manner $[(A+B)\,.\,(P+Q)]$ would denote the charge on

$A+B$ when P and Q are both raised to potential 1, and it is manifest that since

$$[(A+B)(P+Q)] = [AP]+[AQ]+[BP]+[BQ]$$
$$= [(P+Q)(A+B)],$$

the compound symbols may be combined by addition and multiplication as if they were symbols of quantity.

The symbol $[A.A]$ denotes the charge on A when the potential of A is 1, that is to say, the capacity of A.

In like manner $[(A+B)(A+Q)]$ denotes the sum of the charges on A and B when A and Q are raised to potential 1, the potential of all the conductors except A and Q being zero.

It may be decomposed into

$$[A.A]+[A.B]+[A.Q]+[B.Q].$$

The coefficients of potential cannot be dealt with in this way. The coefficients of induction represent charges, and these charges can be combined by addition, but the coefficients of potential represent potentials, and if the potential of A is V_1 and that of B is V_2, the sum V_1+V_2 has no physical meaning bearing on the phenomena, though V_1-V_2 represents the electromotive force from A to B.

The coefficients of induction between two conductors may be expressed in terms of the capacities of the conductors and that of the two conductors together, thus:

$$[A.B] = \tfrac{1}{2}[(A+B)(A+B)]-\tfrac{1}{2}[A.A]-\tfrac{1}{2}[B.B].$$

Dimensions of the coefficients.

88.] Since the potential of a charge e at a distance r is $\frac{e}{r}$, the dimensions of a charge of electricity are equal to those of the product of a potential into a line.

The coefficients of capacity and induction have therefore the same dimensions as a line, and each of them may be represented by a straight line, the length of which is independent of the system of units which we employ.

For the same reason, any coefficient of potential may be represented as the reciprocal of a line.

On certain conditions which the coefficients must satisfy.

89 a.] In the first place, since the electric energy of a system is an essentially positive quantity, its expression as a quadratic

function of the charges or of the potentials must be positive, whatever values, positive or negative, are given to the charges or the potentials.

Now the conditions that a homogeneous quadratic function of n variables shall be always positive are n in number, and may be written

$$\left.\begin{array}{l} p_{11} > 0, \\ \begin{vmatrix} p_{11}, & p_{12} \\ p_{21}, & p_{22} \end{vmatrix} > 0, \\ - - - - - - - \\ \begin{vmatrix} p_{11} & \cdots & p_{1n} \\ - & - - & - \\ p_{n1} & \cdots & p_{nn} \end{vmatrix} > 0. \end{array}\right\} \qquad (22)$$

These n conditions are necessary and sufficient to ensure that W shall be essentially positive *.

But since in equation (16) we may arrange the conductors in any order, every determinant must be positive which is formed symmetrically from the coefficients belonging to any combination of the n conductors, and the number of these combinations is $2^n - 1$.

Only n, however, of the conditions so found can be independent.

The coefficients of capacity and induction are subject to conditions of the same form.

89 *b.*] *The coefficients of potential are all positive, but none of the coefficients* p_{rs} *is greater than* p_{rr} *or* p_{ss}.

For let a charge unity be communicated to A_r, the other conductors being uncharged. A system of equipotential surfaces will be formed. Of these one will be the surface of A_r, and its potential will be p_{rr}. If A_s is placed in a hollow excavated in A_r so as to be completely enclosed by it, then the potential of A_s will also be p_{rr}.

If, however, A_s is outside of A_r its potential p_{rs} will lie between p_{rr} and zero.

For consider the lines of force issuing from the charged conductor A_r. The charge is measured by the excess of the number of lines which issue from it over those which terminate in it. Hence, if the conductor has no charge, the number of lines which enter the conductor must be equal to the number which issue from it. The lines which enter the conductor come from places of greater potential, and those which issue from it go to places of less poten-

* See Williamson's *Differential Calculus*, 3rd edition, p. 407.

tial. Hence the potential of an uncharged conductor must be intermediate between the highest and lowest potentials in the field, and therefore the highest and lowest potentials cannot belong to any of the uncharged bodies.

The highest potential must therefore be p_{rr}, that of the charged body A_r, the lowest must be that of space at an infinite distance, which is zero, and all the other potentials such as p_{rs} must lie between p_{rr} and zero.

If A_s completely surrounds A_t, then $p_{rs} = p_{rt}$.

89 *c.*] *None of the coefficients of induction are positive, and the sum of all those belonging to a single conductor is not numerically greater than the coefficient of capacity of that conductor, which is always positive.*

For let A_r be maintained at potential unity while all the other conductors are kept at potential zero, then the charge on A_r is q_{rr}, and that on any other conductor A_s is q_{rs}.

The number of lines of force which issue from A_r is q_{rr}. Of these some terminate in the other conductors, and some may proceed to infinity, but no lines of force can pass between any of the other conductors or from them to infinity, because they are all at potential zero.

No line of force can issue from any of the other conductors such as A_s, because no part of the field has a lower potential than A_s. If A_s is completely cut off from A_r by the closed surface of one of the conductors, then q_{rs} is zero. If A_s is not thus cut off, q_{rs} is a negative quantity.

If one of the conductors A_t completely surrounds A_r, then all the lines of force from A_r fall on A_t and the conductors within it, and the sum of the coefficients of induction of these conductors with respect to A_r will be equal to q_{rr} with its sign changed. But if A_r is not completely surrounded by a conductor the arithmetical sum of the coefficients of induction q_{rs}, &c. will be less than q_{rr}.

We have deduced these two theorems independently by means of electrical considerations. We may leave it to the mathematical student to determine whether one is a mathematical consequence of the other.

89 *d.*] When there is only one conductor in the field its coefficient of potential on itself is the reciprocal of its capacity.

The centre of mass of the electricity when there are no external forces is called the electric centre of the conductor. If the conductor is symmetrical about a centre of figure, this point is the electric centre. If the dimensions of the conductor are small compared with the distances considered, the position of the electric centre may be estimated sufficiently nearly by conjecture.

The potential at a distance c from the electric centre must be between

$$\frac{e}{c}\left(1+\frac{a^2}{c^2}\right) \text{ and } \frac{e}{c}\left(1-\tfrac{1}{2}\frac{a^2}{c^2}\right);$$

where e is the charge, and a is the greatest distance of any part of the surface of the body from the electric centre.

For if the charge be concentrated in two points at distances a on opposite sides of the electric centre, the first of these expressions is the potential at a point in the line joining the charges, and the second at a point in a line perpendicular to the line joining the charges. For all other distributions within the sphere whose radius is a the potential is intermediate between those values.

If there are two conductors in the field, their mutual coefficient of potential is $\frac{1}{c'}$, where c' cannot differ from c, the distance between the electric centres, by more than $\frac{a^2+b^2}{c}$; a and b being the greatest distances of any part of the surfaces of the bodies from their respective electric centres.

89 *e*.] If a new conductor is brought into the field the coefficient of potential of any one of the others on itself is diminished.

For let the new body, B, be supposed at first to be a non-conductor free from charge in any part, then when one of the conductors, A_1, receives a charge e_1, the distribution of the electricity on the conductors of the system will not be disturbed by B, as B is still without charge in any part, and the electric energy of the system will be simply

$$\tfrac{1}{2}e_1 V_1 = \tfrac{1}{2}e_1^2 p_{11}.$$

Now let B become a conductor. Electricity will flow from places of higher to places of lower potential, and in so doing will diminish the electric energy of the system, so that the quantity $\frac{1}{2}e_1^2 p_{11}$ must diminish.

But e_1 remains constant, therefore p_{11} must diminish.

Also if B increases by another body b being placed in contact with it, p_{11} will be further diminished.

For let us first suppose that there is no electric communication between B and b; the introduction of the new body b will diminish p_{11}. Now let a communication be opened between B

and *b*. If any electricity flows through it, it flows from a place of higher to a place of lower potential, and therefore, as we have shown, still further diminishes p_{11}.

Hence the diminution of p_{11} by the body *B* is greater than that which would be produced by any body the surface of which can be inscribed in *B*, and less than that produced by any body the surface of which can be described about *B*.

We shall shew in Chapter XI, that a sphere of diameter *b* at a distance *r* diminishes the value of p_{11} by a quantity which is approximately $\frac{1}{8}\frac{b^3}{r^4}$.

Hence if the body *B* is of any other figure, and if *b* is its greatest diameter, the diminution of the value of p_{11} must be less than $\frac{1}{8}\frac{b^3}{r^4}$.

Hence if the greatest diameter of *B* is so small compared with its distance from A_1 that we may neglect quantities of the order $\frac{1}{8}\frac{b^3}{r^4}$, we may consider the reciprocal of the capacity of A_1 when alone in the field as a sufficient approximation to p_{11}.

90 *a*.] Let us therefore suppose that the capacity of A_1 when alone in the field is K_1, and that of A_2, K_2, and let the mean distance between A_1 and A_2 be *r*, where *r* is very great compared with the dimensions of A_1 and A_2, then we may write

$$p_{11} = \frac{1}{K_1}, \quad p_{12} = \frac{1}{r}, \quad p_{22} = \frac{1}{K_2};$$

$$V_1 = e_1 K_1^{-1} + e_2 r^{-1},$$

$$V_2 = e_1 r^{-1} + e_2 K_2^{-1}.$$

Hence

$$q_{11} = K_1 (1 - K_1 K_2 r^{-2})^{-1},$$

$$q_{12} = -K_1 K_2 r^{-1} (1 - K_1 K_2 r^{-2})^{-1},$$

$$q_{22} = K_2 (1 - K_1 K_2 r^{-2})^{-1}.$$

Of these coefficients q_{11} and q_{22} are the capacities of A_1 and A_2 when, instead of being each alone at an infinite distance from any other body, they are brought so as to be at a distance *r* from each other.

90 *b*.] When two conductors are placed so near together that their coefficient of mutual induction is large, the combination is called a Condenser.

Let *A* and *B* be the two conductors or electrodes of a condenser.

Let L be the capacity of A, N that of B, and M the coefficient of mutual induction. (We must remember that M is essentially negative, so that the numerical value of $L+M$ and $M+N$ is less than L or N.)

Let us suppose that a and b are the electrodes of another condenser at a distance R from the first, R being very great compared with the dimensions of either condenser, and let the coefficients of capacity and induction of the condenser ab when alone be l, m, n. Let us calculate the effect of one of the condensers on the coefficients of the other.

Let $$D = LN - M^2 \quad \text{and} \quad d = ln - m^2;$$

then the coefficients of potential for each condenser by itself are

$$\begin{aligned} p_{AA} &= D^{-1}N, & p_{aa} &= d^{-1}n, \\ p_{AB} &= -D^{-1}M, & p_{ab} &= -d^{-1}m, \\ p_{BB} &= D^{-1}L, & p_{bb} &= d^{-1}l. \end{aligned}$$

The values of these coefficients will not be sensibly altered when the two condensers are at a distance R.

The coefficient of potential of any two conductors at distance R is R^{-1}, so that

$$p_{Aa} = p_{Ab} = p_{Ba} = p_{Bb} = R^{-1}.$$

The equations of potential are therefore

$$\begin{aligned} V_A &= D^{-1}Ne_A - D^{-1}Me_B + R^{-1}e_a + R^{-1}e_b, \\ V_B &= -D^{-1}Me_A + D^{-1}Le_B + R^{-1}e_a + R^{-1}e_b, \\ V_a &= R^{-1}e_A + R^{-1}e_B + d^{-1}ne_a - d^{-1}me_b, \\ V_b &= R^{-1}e_A + R^{-1}e_b - d^{-1}me_a + d^{-1}le_b. \end{aligned}$$

Solving these equations for the charges, we find

$$q_{AA} = L' = L + \frac{(L+M)^2(l+2m+n)}{R^2-(L+2M+N)(l+2m+n)},$$

$$q_{AB} = M' = M + \frac{(L+M)(M+N)[l+2m+n]}{R^2-(L+2M+N)(l+2m+n)},$$

$$q_{Aa} = -\frac{R(L+M)(l+m)}{R^2-(L+2M+N)(l+2m+n)},$$

$$q_{Ab} = -\frac{R(L+M)(m+n)}{R^2-(L+2M+N)(l+2m+n)};$$

where L', M', N' are what L, M, N become when the second condenser is brought into the field.

If only one conductor, a, is brought into the field, $m=n=0$, and

$$q_{AA}=L'=L+\frac{(L+M)^2 l}{R^2-l(L+2M+N)},$$

$$q_{AB}=M'=M+\frac{(L+M)(M+N)l}{R^2-l(L+2M+N)},$$

$$q_{Aa}=-\frac{Rl(L+M)}{R^2-l(L+2M+N)}.$$

If there are only the two simple conductors, A and a,

$$M=N=m=n=0,$$

and $$q_{AA}=L+\frac{L^2 l}{R^2-Ll},\qquad q_{Aa}=-\frac{RLl}{R^2-Ll};$$

expressions which are the same as those found in Art. 90 a.

The quantity $L+2M+N$ is the total charge of the condenser when its electrodes are at potential 1. It cannot exceed half the greatest diameter of the condenser.

$L+M$ is the charge of the first electrode, and $M+N$ that of the second when both are at potential 1. These quantities must be each of them positive and less than the capacity of the electrode by itself. Hence the corrections to be applied to the coefficients of capacity of a condenser are much smaller than those for a simple conductor of equal capacity.

Approximations of this kind are often useful in estimating the capacities of conductors of irregular form placed at a finite distance from other conductors.

91.] When a round conductor, A_3, of small size compared with the distances between the conductors, is brought into the field, the coefficient of potential of A_1 on A_2 will be increased when A_3 is inside and diminished when A_3 is outside of a sphere whose diameter is the straight line $A_1 A_2$.

For if A_1 receives a unit charge there will be a distribution of electricity on A_3, $+e$ being on the side furthest from A_1, and $-e$ on the side nearest A_1. The potential at A_2 due to this distribution on A_3 will be positive or negative as $+e$ or $-e$ is nearest to A_2, and if the form of A_3 is not very elongated this will depend on whether the angle $A_1 A_3 A_2$ is obtuse or acute, and therefore on whether A_3 is inside or outside the sphere described on $A_1 A_2$ as diameter.

If A_3 is of an elongated form it is easy to see that if it is placed with its longest axis in the direction of the tangent to the circle drawn through the points A_1, A_3, A_2 it may increase the potential of A_2, even when it is entirely outside the sphere, and how by placing it with its longest axis in the direction of the radius of the sphere, it may diminish the potential of A_2, even when entirely within the sphere. But this proposition is only intended for forming a rough estimate of the phenomena to be expected in a given arrangement of apparatus.

92.] If a new conductor, A_3, is introduced into the field, the capacities of all the conductors already there are increased, and the numerical values of the coefficients of induction between every pair of them are diminished.

Let us suppose that A_1 is at potential unity and all the rest at potential zero. Since the charge of the new conductor is negative it will induce a positive charge on every other conductor, and will therefore increase the positive charge of A_1 and diminish the negative charge of each of the other conductors.

93 *a.*] *Work done by the electric forces during the displacement of a system of insulated charged conductors.*

Since the conductors are insulated, their charges remain constant during the displacement. Let their potentials be V_1, V_2, ... V_n before and V_1', V_2', ... V_n' after the displacement. The electrical energy is

$$W = \tfrac{1}{2}\Sigma(eV)$$

before the displacement, and

$$W' = \tfrac{1}{2}\Sigma(eV')$$

after the displacement.

The work done by the electric forces during the displacement is the excess of the initial energy W over the final energy W', or

$$W - W' = \tfrac{1}{2}\Sigma\left[e(V - V')\right].$$

This expression gives the work done during any displacement, small or large, of an insulated system.

To find the force tending to produce a particular kind of displacement, let ϕ be the variable whose variation corresponds to the kind of displacement, and let Φ be the corresponding force, reckoned positive when the electric force tends to increase ϕ, then

$$\Phi\, d\phi = -dW_e,$$

or

$$\Phi = -\frac{dW_e}{d\phi};$$

where W_e denotes the expression for the electric energy as a quadratic function of the charges.

93 b.] To prove that $\frac{dW_e}{d\phi}+\frac{dW_V}{d\phi}=0.$

We have three different expressions for the energy of the system,

(1) $$W=\tfrac{1}{2}\Sigma(eV),$$

a definite function of the n charges and n potentials

(2) $$W_e=\tfrac{1}{2}\Sigma\Sigma(e_r\,e_s\,p_{rs}),$$

where r and s may be the same or different, and both rs and sr are to be included in the summation.

This is a function of the n charges and of the variables which define the configuration. Let ϕ be one of these.

(3) $$W_V=\tfrac{1}{2}\Sigma\Sigma(V_r\,V_s\,q_{rs}),$$

where the summation is to be taken as before. This is a function of the n potentials and of the variables which define the configuration of which ϕ is one.

Since $$W=W_e=W_V,$$
$$W_e+W_V-2W=0.$$

Now let the n charges, the n potentials, and ϕ vary in any consistent manner, and we must have

$$\Sigma\left[\left(\frac{dW_e}{de_r}-V_r\right)\delta e_r\right]+\Sigma\left[\left(\frac{dW_V}{dV_s}-e_s\right)\delta V_s\right]+\left(\frac{dW_e}{d\phi}+\frac{dW_V}{d\phi}\right)\delta\phi=0.$$

Now the n charges, the n potentials, and ϕ are not all independent of each other, for in fact only $n+1$ of them can be independent. But we have already proved that

$$\frac{dW_e}{de_r}=V_r,$$

so that the first sum of terms vanishes identically, and it follows from this, even if we had not already proved it that

$$\frac{dW_V}{dV_s}=e_s,$$

and that lastly,

$$\frac{dW_e}{d\phi}+\frac{dW_V}{d\phi}=0.$$

Work done by the electric forces during the displacement of a system whose potentials are maintained constant.

93 c.] It follows from the last equation that the force $\Phi=\frac{dW_V}{d\phi}$,

and if the system is displaced under the condition that all the potentials remain constant, the work done by the electric forces is

$$\int \Phi\, d\phi = \int dW'_V = W_V - W_V;$$

or the work done by the electric forces in this case is equal to the *increment* of the electric energy.

Here, then, we have an increase of energy together with a quantity of work done by the system. The system must therefore be supplied with energy from some external source, such as a voltaic battery, in order to maintain the potentials constant during the displacement.

The work done by the battery is therefore equal to the sum of the work done by the system and the increment of energy, or, since these are equal, the work done by the battery is twice the work done by the system of conductors during the displacement.

On the comparison of similar electrified systems.

94.] If two electrified systems are similar in a geometrical sense, so that the lengths of corresponding lines in the two systems are as L to L', then if the dielectric which separates the conducting bodies is the same in both systems, the coefficients of induction and of capacity will be in the proportion of L to L'. For if we consider corresponding portions, A and A', of the two systems, and suppose the quantity of electricity on A to be e, and that on A' to be e', then the potentials V and V' at corresponding points B and B', due to this electrification, will be

$$V = \frac{e}{AB}, \text{ and } V' = \frac{e'}{A'B'}.$$

But AB is to $A'B'$ as L to L', so that we must have

$$e : e' :: LV : L'V'.$$

But if the inductive capacity of the dielectric is different in the two systems, being K in the first and K' in the second, then if the potential at any point of the first system is to that at the corresponding point of the second as V to V', and if the quantities of electricity on corresponding parts are as E to E', we shall have

$$e : e' :: LVK : L'V'K'.$$

By this proportion we may find the relation between the total charges of corresponding parts of two systems, which are in the first place geometrically similar, in the second place composed of dielectric media of which the specific inductive capacity at corresponding points is in the proportion of K to K', and in the third place so electrified that the potentials of corresponding points are as V to V'.

From this it appears that if q be any coefficient of capacity or induction in the first system, and q' the corresponding one in the second,

$$q : q' :: LK : L'K';$$

and if p and p' denote corresponding coefficients of potential in the two systems,

$$p : p' :: \frac{1}{LK} : \frac{1}{L'K'}.$$

If one of the bodies be displaced in the first system, and the corresponding body in the second system receive a similar displacement, then these displacements are in the proportion of L to L', and if the forces acting on the two bodies are as F to F', then the work done in the two systems will be as FL to $F'L'$.

But the total electrical energy is half the sum of the charges of electricity multiplied each by the potential of the charged body, so that in the similar systems, if W and W' be the total electrical energy in the two systems respectively,

$$W : W' :: eV : e'V',$$

and the difference of energy after similar displacements in the two systems will be in the same proportion. Hence, since FL is proportional to the electrical work done during the displacement,

$$FL : F'L' :: eV : e'V'.$$

Combining these proportions, we find that the ratio of the resultant force on any body of the first system to that on the corresponding body of the second system is

$$F : F' :: V^2K : V'^2K',$$

or

$$F : F' :: \frac{e^2}{L^2K} : \frac{e'^2}{L'^2K'}.$$

The first of these proportions shews that in similar systems the force is proportional to the square of the electromotive force and to the inductive capacity of the dielectric, but is independent of the actual dimensions of the system.

Hence two conductors placed in a liquid whose inductive capacity is greater than that of air, and electrified to given potentials, will attract each other more than if they had been electrified to the same potentials in air.

The second proportion shews that if the quantity of electricity on each body is given, the forces are proportional to the squares of the charges and inversely to the squares of the distances, and also inversely to the inductive capacities of the media.

Hence, if two conductors with given charges are placed in a liquid whose inductive capacity is greater than that of air, they will attract each other less than if they had been surrounded with air and charged with the same quantities of electricity.

CHAPTER IV.

GENERAL THEOREMS.

95 *a*.] IN the second chapter we have calculated the potential function and investigated some of its properties on the hypothesis that there is a direct action at a distance between electrified bodies, which is the resultant of the direct actions between the various electrified parts of the bodies.

If we call this the direct method of investigation, the inverse method will consist in assuming that the potential is a function characterised by properties the same as those which we have already established, and investigating the form of the function.

In the direct method the potential is calculated from the distribution of electricity by a process of integration, and is found to satisfy certain partial differential equations. In the inverse method the partial differential equations are supposed given, and we have to find the potential and the distribution of electricity.

It is only in problems in which the distribution of electricity is given that the direct method can be used. When we have to find the distribution on a conductor we must make use of the inverse method.

We have now to shew that the inverse method leads in every case to a determinate result, and to establish certain general theorems deduced from Poisson's partial differential equation

$$\frac{d^2V}{dx^2}+\frac{d^2V}{dy^2}+\frac{d^2V}{dz^2}+4\pi\rho=0.$$

The mathematical ideas expressed by this equation are of a different kind from those expressed by the definite integral

$$V=\int_{-\infty}^{+\infty}\int_{-\infty}^{+\infty}\int_{-\infty}^{+\infty}\frac{\rho}{r}\,dx'\,dy'\,dz'.$$

In the differential equation we express that the sum of the second derivatives of V in the neighbourhood of any point is related to the density at that point in a certain manner, and no relation is expressed between the value of V at that point and the value of ρ at any point at a finite distance from it.

In the definite integral, on the other hand, the distance of the point (x', y', z'), at which ρ exists, from the point (x, y, z), at which V exists, is denoted by r, and is distinctly recognised in the expression to be integrated.

The integral, therefore, is the appropriate mathematical expression for a theory of action between particles at a distance, whereas the differential equation is the appropriate expression for a theory of action exerted between contiguous parts of a medium.

We have seen that the result of the integration satisfies the differential equation. We have now to shew that it is the only solution of that equation satisfying certain conditions.

We shall in this way not only establish the mathematical equivalence of the two expressions, but prepare our minds to pass from the theory of direct action at a distance to that of action between contiguous parts of a medium.

95 *b*.] The theorems considered in this chapter relate to the properties of certain volume-integrals taken throughout a finite region of space which we may refer to as the electric field.

The element of these integrals, that is to say, the quantity under the integral sign, is either the square of a certain vector quantity whose direction and magnitude varies from point to point in the field, or the product of one vector into the resolved part of another in its own direction.

Of the different modes in which a vector quantity may be distributed in space, two are of special importance.

The first is that in which the vector may be represented as the space-variation [Art. 17] of a scalar function called the Potential.

Such a distribution may be called an Irrotational distribution. The resultant force arising from the attraction or repulsion of any combination of centres of force, the law of each being any given function of the distance, is distributed irrotationally.

The second mode of distribution is that in which the convergence [Art. 25] is zero at every point. Such a distribution may be called a Solenoidal distribution. The velocity of an incompressible fluid is distributed in a solenoidal manner.

When the central forces which, as we have said, give rise to an irrotational distribution of the resultant force, vary according to the inverse square of the distance, then, if these centres are outside the field, the distribution within the field will be solenoidal as well as irrotational.

When the motion of an incompressible fluid which, as we have said, is solenoidal, arises from the action of central forces depending on the distance, or of surface pressures, on a frictionless fluid originally at rest, the distribution of velocity is irrotational as well as solenoidal.

When we have to specify a distribution which is at once irrotational and solenoidal, we shall call it a Laplacian distribution; Laplace having pointed out some of the most important properties of such a distribution.

The volume integrals discussed in this chapter are, as we shall see, expressions for the energy of the electric field. In the first group of theorems, beginning with Green's Theorem, the energy is expressed in terms of the electromotive intensity, a vector which is distributed irrotationally in all cases of electric equilibrium. It is shewn that if the surface-potential be given, then of all irrotational distributions, that which is also solenoidal has the least energy; whence it also follows that there can be only one Laplacian distribution consistent with the surface potentials.

In the second group of theorems, including Thomson's Theorem, the energy is expressed in terms of the electric displacement, a vector of which the distribution is solenoidal. It is shewn that if the surface-charges are given, then of all solenoidal distributions that has least energy which is also irrotational, whence it also follows that there can be only one Laplacian distribution consistent with the given surface-charges.

The demonstration of all these theorems is conducted in the same way. In order to avoid the repetition in every case of the steps of a surface integration conducted with reference to rectangular axes, we make use in each case of the result of Theorem III, Art. 21,* where the relation between a volume-integral and the corresponding surface-integral is fully worked out. All that we have to do, therefore, is to substitute for X, Y, and Z in that Theorem the components of the vector on which the particular theorem depends.

In the first edition of this book the statement of each theorem was cumbered with a multitude of alternative conditions which

* This theorem seems to have been first given by Ostrogradsky in a paper read in 1828. but published in 1831 in the *Mém. de l'Acad. de St. Pétersbourg*, T. I. p. 39. It may be regarded, however, as a form of the equation of continuity.

were intended to shew the generality of the theorem and the variety of cases to which it might be applied, but which tended rather to confuse in the mind of the reader what was assumed with what was to be proved.

In the present edition each theorem is at first stated in a more definite, if more restricted, form, and it is afterwards shewn what further degree of generality the theorem admits of.

We have hitherto used the symbol V for the potential, and we shall continue to do so whenever we are dealing with electrostatics only. In this chapter, however, and in those parts of the second volume in which the electric potential occurs in electro-magnetic investigations, we shall use Ψ as a special symbol for the electric potential.

Green's Theorem.

96 *a*.] The following important theorem was given by George Green, in his 'Essay on the Application of Mathematics to Electricity and Magnetism.'

The theorem relates to the space bounded by the closed surface s. We may refer to this finite space as the Field. Let ν be a normal drawn from the surface s into the field, and let l, m, n be the direction cosines of this normal, then

$$l\frac{d\Psi}{dx}+m\frac{d\Psi}{dy}+n\frac{d\Psi}{dz}=\frac{d\Psi}{d\nu} \tag{1}$$

will be the rate of variation of the function Ψ in passing along the normal ν. Let it be understood that the value of $\frac{d\Psi}{d\nu}$ is to be taken at the surface itself, where $\nu = 0$.

Let us also write, as in Arts. 26 and 77,

$$\frac{d^2\Psi}{dx^2}+\frac{d^2\Psi}{dy^2}+\frac{d^2\Psi}{dz^2}=-\nabla^2\Psi, \tag{2}$$

and when there are two functions, Ψ and Φ, let us write

$$\frac{d\Psi}{dx}\frac{d\Phi}{dx}+\frac{d\Psi}{dy}\frac{d\Phi}{dy}+\frac{d\Psi}{dz}\frac{d\Phi}{dz}=-S.\nabla\Psi\nabla\Phi. \tag{3}$$

The reader who is not acquainted with the method of Quaternions may, if it pleases him, regard the expressions $\nabla^2\Psi$ and $S.\nabla\Psi\nabla\Phi$ as mere conventional abbreviations for the quantities to which they are equated above, and as in what follows we shall employ ordinary Cartesian methods, it will not be necessary to remember the Quaternion interpretation of these expressions. The reason, however, why we use as our abbreviations these expressions and not single letters arbitrarily chosen, is, that in the language of Quaternions they represent fully the quantities to which they are equated. The operator ∇ applied to the scalar function Ψ gives the space-variation of that function, and the expression $-S.\nabla\Psi\nabla\Phi$ is the scalar part of the product of two space-variations, or the product of either space-variation into the resolved part of the other in its own direction. The expression $\frac{d\Psi}{d\nu}$ is usually written in Quaternions $S.U\nu\nabla\Psi$, $U\nu$ being a unit-vector in the direction of the normal. There does not seem much advantage in using this notation here, but we shall find the advantage of doing so when we come to deal with anisotropic media.

Statement of Green's Theorem.

Let Ψ and Φ be two functions of x, y, z, which, with their first derivatives, are finite and continuous within the acyclic region s, bounded by the closed surface s, then

$$\iint \Psi \frac{d\Phi}{d\nu}\, ds - \iiint \Psi\nabla^2\Phi ds = \iiint S.\nabla\Psi\nabla\Phi ds$$

$$= \iint \Phi \frac{d\Psi}{d\nu}\, ds - \iiint \Phi\nabla^2\Psi ds; \qquad (4)$$

where the double integrals are to be extended over the whole closed surface s, and the triple integrals throughout the field, s, enclosed by that surface.

To prove this, let us write, in Art. 21, Theorem III,

$$X = \Psi\frac{d\Phi}{dx}, \quad Y = \Psi\frac{d\Phi}{dy}, \quad Z = \Psi\frac{d\Phi}{dz}, \qquad (5)$$

$$\text{then } R\cos\epsilon = \Psi\left(l\frac{d\Phi}{dx} + m\frac{d\Phi}{dy} + n\frac{d\Phi}{dz}\right)$$

$$= \Psi\frac{d\Phi}{d\nu}, \text{ by (1)}; \qquad (6)$$

$$\text{and } \frac{dX}{dx} + \frac{dY}{dy} + \frac{dZ}{dz} = \Psi\left(\frac{d^2\Phi}{dx^2} + \frac{d^2\Phi}{dy^2} + \frac{d^2\Phi}{dz^2}\right)$$

$$+ \frac{d\Psi}{dx}\frac{d\Phi}{dx} + \frac{d\Psi}{dy}\frac{d\Phi}{dy} + \frac{d\Psi}{dz}\frac{d\Phi}{dz}$$

$$= -\Psi\nabla^2\Phi - S.\nabla\Psi\nabla\Phi, \text{ by (2) and (3).} \qquad (7)$$

But by Theorem III

$$\iint R\cos\epsilon\, ds = -\iiint \left(\frac{dX}{dx} + \frac{dY}{dy} + \frac{dZ}{dz}\right) ds;$$

or by (6) and (7)

$$\iint \Psi \frac{d\Phi}{d\nu} ds - \iiint \Psi \nabla^2 \Phi d\varsigma = \iiint S.\nabla\Psi\nabla\Phi d\varsigma. \qquad (8)$$

Since in the second member of this equation Ψ and Φ may be interchanged, we may do so in the first, and we thus obtain the complete statement of Green's Theorem, as given in equation (4).

96 *b*.] We have next to shew that Green's Theorem is true when one of the functions, say Ψ, is a many-valued one, provided that its first derivatives are single-valued, and do not become infinite within the acyclic region ς.

Since $\nabla\Psi$ and $\nabla\Phi$ are single-valued, the second member of equation (4) is single-valued; but since Ψ is many-valued, any one element of the first member, as $\Psi \nabla^2 \Phi$, is many-valued. If, however, we select one of the many values of Ψ, as Ψ_0, at the point A within the region ς, then the value of Ψ at any other point, P, will be definite. For, since the selected value of Ψ is continuous within the region, the value of Ψ at P must be that which is arrived at by continuous variation along any path from A to P, beginning with the value Ψ_0 at A. If the value at P were different for two paths between A and P, then these two paths must embrace between them a closed curve at which the first derivatives of Ψ become infinite. Now this is contrary to the specification, for since the first derivatives do not become infinite within the region ς, the closed curve must be entirely without the region; and since the region is acyclic, two paths within the region cannot embrace anything outside the region.

Hence, if Ψ_0 is given as the value of Ψ at the point A, the value at P is definite.

If any other value of Ψ, say $\Psi_0 + n\kappa$, had been chosen as the value at A, then the value at P would have been $\Psi + n\kappa$. But the value of the first member of equation (4) would be the same as before, for the change amounts to increasing the first member by

$$n\kappa \left[\iint \frac{d\Phi}{d\nu} ds - \iiint \nabla^2 \Phi d\varsigma \right];$$

and this, by Theorem III, is zero.

96 *c*.] If the region ς is doubly or multiply connected, we may reduce it to an acyclic region by closing each of its circuits with a diaphragm.

Let s_1 be one of these diaphragms, and κ_1 the corresponding cyclic constant, that is to say, the increment of Ψ in going once round the circuit in the positive direction. Since the region s lies on both sides of the diaphragm s_1, every element of s_1 will occur twice in the surface integral.

If we suppose the normal ν_1 drawn towards the positive side of ds_1, and ν_1' drawn towards the negative side,

$$\frac{d\Phi}{d\nu_1'} = -\frac{d\Phi}{d\nu_1},$$

and

$$\Psi_1' = \Psi_1 + \kappa,$$

so that the element of the surfaee-integral arising from ds_1 will be

$$\Psi_1 \frac{d\Phi}{d\nu_1} ds_1 + \Psi_1' \frac{d\Phi}{d\nu_1'} ds_1 = -\kappa_1 \frac{d\Phi}{d\nu_1} ds_1.$$

Hence if the region s is multiply connected, the first term of equation (4) must be written

$$\iint \Psi \frac{d\Phi}{d\nu} ds - \kappa_1 \iint \frac{d\Phi}{d\nu_1} ds_1 - \&c. - \kappa_n \iint \frac{d\Phi}{d\nu_n} ds_n - \iiint \Psi \nabla^2 \Phi ds; \quad (4_a)$$

where the first surface-integral is to be taken over the bounding surface, and the others over the different diaphragms, each element of surface of a diaphragm being taken once only, and the normal being drawn in the positive direction of the circuit.

This modification of the theorem in the case of multiply-connected regions was first shewn to be necessary by Helmholtz *, and was first applied to the theorem by Thomson †.

96 *d.*] Let us now suppose, with Green, that one of the functions, say Φ, does not satisfy the condition that it and its first derivatives do not become infinite within the given region, but that it becomes infinite at the point P, and at that point only, in that region, and that very near to P the value of Φ is $\Phi_0 + e/r$ ‡, where Φ_0 is a finite and continuous quantity, and r is the distance from P. This will be the case if Φ is the potential of a quantity of electricity e concentrated at the point P, together with any distribution of electricity the volume density of which is nowhere infinite within the region considered.

Let us now suppose a very small sphere whose radius is a to be described about P as centre; then since in the region outside this sphere, but within the surface s, Φ presents no singularity, we

* 'Ueber Integrale der hydrodynamischen Gleichungen welche den Wirbelbewegungen entsprechen,' *Crelle*, 1858. Translated by Prof. Tait, *Phil. Mag.*, 1867 (I).

† 'On Vortex Motion,' *Trans. R. S Edin.* xxv part i. p. 241 (1867).

‡ The mark / separates the numerator from the denominator of a fraction.

may apply Green's Theorem to this region, remembering that the surface of the small sphere is to be taken account of in forming the surface-integral.

In forming the volume-integrals we have to subtract from the volume-integral arising from the whole region that arising from the small sphere.

Now $\iiint \Phi \nabla^2 \Psi \, dx\, dy\, dz$ for the sphere cannot be numerically greater than

$$(\nabla^2\Psi)_g \iiint \Phi \, dx\, dy\, dz,$$

or

$$(\nabla^2 \Psi)_g \, \{2\pi e a^2 + \tfrac{4}{3}\pi a^3 \Phi_0\},$$

where the suffix, $_g$, attached to any quantity, indicates that the greatest numerical value of that quantity within the sphere is to be taken.

This volume-integral, therefore, is of the order a^2, and may be neglected when a diminishes and ultimately vanishes.

The other volume-integral

$$\iiint \Psi \, \nabla^2 \, \Phi \, dx\, dy\, dz$$

cannot be numerically greater than

$$\Psi_g \, (\nabla^2 \Phi_0)_g \, \tfrac{4}{3}\pi a^3,$$

and is of the order a^3, and may be neglected when a vanishes.

The surface-integral $\iint \Phi \frac{d\Psi}{d\nu} \, ds$ cannot be numerically greater than $\Phi_g \iint \frac{d\Psi}{d\nu} \, ds$.

Now by Theorem III

$$\iint \frac{d\Psi}{d\nu} \, ds = -\iiint \nabla^2 \Psi \, dx\, dy\, dz,$$

and this cannot be numerically greater than $(\nabla^2 \Psi)_g \, \frac{4}{3}\pi a^3$, and Φ_g at the surface is approximately $\frac{e}{a}$, so that $\iint \Phi \frac{d\Psi}{d\nu} \, ds$ cannot be numerically greater than

$$\tfrac{4}{3}\pi a^2 e (\nabla^2 \Psi)_g,$$

and is therefore of the order a^2, and may be neglected when a vanishes.

But the surface-integral on the other side of the equation, namely

$$\iint \Psi \frac{d\Phi}{d\nu} \, ds,$$

does not vanish, for $\iint \frac{d\Phi}{d\nu} ds = -4\pi e$;

and if Ψ_0 be the value of Ψ at the point P,

$$\iint \Psi \frac{d\Phi}{d\nu} ds = -4\pi e \Psi_0.$$

Equation (4) therefore becomes in this case

$$\iint \Psi \frac{d\Phi}{d\nu} ds - \iiint \Psi \nabla^2 \Phi ds - 4\pi e \Psi_0 = \iint \Phi \frac{d\Psi}{d\nu} ds - \iiint \Phi \nabla^2 \Psi ds. \quad (4_b)$$

97 *a*.] We may illustrate this case of Green's Theorem by employing it as Green does to determine the surface-density of a distribution which will produce a potential whose values inside and outside a given closed surface are given. These values must coincide at the surface, also within the surface $\nabla^2 \Psi = 0$, and outside $\nabla^2 \Psi' = 0$.

Green begins with the direct process, that is to say, the distribution of the surface density, σ, being given, the potentials at an internal point P and an external point P' are found by integrating the expressions

$$\Psi_P = \iint \frac{\sigma}{r} ds, \qquad \Psi'_{P'} = \iint \frac{\sigma}{r'} ds; \quad (9)$$

where r and r' are measured from the points P and P' respectively.

Now let $\Phi = 1/r$, then applying Green's Theorem to the space within the surface, and remembering that $\nabla^2 \Phi = 0$ and $\nabla^2 \Psi = 0$, we find

$$\iint \Psi \frac{d\frac{1}{r}}{d\nu} ds - 4\pi \Psi_P = \iint \frac{1}{r} \frac{d\Psi}{d\nu} ds, \quad (10)$$

where Ψ_P is the value of Ψ at P.

Again, if we apply the theorem to the space between the surface s and a surface surrounding it at an infinite distance a, the part of the surface-integral belonging to the latter surface will be of the order $1/a$ and may be neglected, and we have

$$\iint \Psi' \frac{d\frac{1}{r}}{d\nu'} ds = \iint \frac{1}{r} \frac{d\Psi'}{d\nu'} ds. \quad (11)$$

Now at the surface, $\Psi = \Psi'$, and since the normals ν and ν' are drawn in opposite directions,

$$\frac{d\frac{1}{r}}{d\nu} + \frac{d\frac{1}{r}}{d\nu'} = 0.$$

Hence on adding equations (10) and (11), the left-hand members destroy each other, and we have

$$-4\pi\Psi_P = \iint \frac{1}{r}\left(\frac{d\Psi}{d\nu} + \frac{d\Psi'}{d\nu'}\right) ds. \tag{12}$$

97 *b*.] Green also proves that if the value of the potential at every point of a closed surface s be given arbitrarily, the potential at any point inside or outside the surface may be determined.

For this purpose he supposes the function Φ to be such that near the point P its value is sensibly $1/r$, while at the surface s its value is zero, and at every point within the surface $\nabla^2\Phi = 0$.

That such a function must exist, Green proves from the physical consideration that if s is a conducting surface connected to the earth, and if a unit of electricity is placed at the point P, the potential within s must satisfy the above conditions. For since s is connected to the earth the potential must be zero at every point of s, and since the potential arises from the electricity at P and the electricity induced on s, $\nabla^2\Phi = 0$ at every point within the surface.

Applying Green's Theorem to this case, we find

$$4\pi\Psi_P = \iint \Psi \frac{d\Phi}{d\nu} ds, \tag{13}$$

where, in the surface-integral, Ψ is the given value of the potential at the element of surface ds; and since, if σ_P is the density of the electricity induced on s by unit of electricity at P,

$$4\pi\sigma_P + \frac{d\Phi}{d\nu} = 0, \tag{14}$$

we may write equation (13)

$$\Psi_P = \iint \Psi\sigma\, ds, \tag{15}$$

where σ is the surface-density of the electricity induced on ds by a charge equal to unity at the point P.

Hence if the value of σ is known at every point of the surface for a particular position of P, then we can calculate by ordinary integration the potential at the point P, supposing the potential at every point of the surface to be given, and the potential within the surface to be subject to the condition

$$\nabla^2\Psi = 0.$$

We shall afterwards prove that if we have obtained a value of Ψ which satisfies these conditions, it is the only value of Ψ which satisfies them.

Green's Function.

98.] Let a closed surface s be maintained at potential zero. Let P and Q be two points on the positive side of the surface s (we may suppose either the inside or the outside positive), and let a small body charged with unit of electricity be placed at P; the potential at the point Q will consist of two parts, of which one is due to the direct action of the electricity at P, while the other is due to the action of the electricity induced on s by P. The latter part of the potential is called Green's Function, and is denoted by G_{pq}.

This quantity is a function of the positions of the two points P and Q, the form of the function depending on the surface s. It has been calculated for the case in which s is a sphere, and for a very few other cases. It denotes the potential at Q due to the electricity induced on s by unit of electricity at P.

The actual potential at any point Q due to the electricity at P and to the electricity induced on s is $1/r_{pq}+G_{pq}$, where r_{pq} denotes the distance between P and Q.

At the surface s, and at all points on the negative side of s, the potential is zero, therefore

$$G_{pa} = -\frac{1}{r_{pa}}, \tag{1}$$

where the suffix $_a$ indicates that a point A on the surface s is taken instead of Q.

Let $\sigma_{pa'}$ denote the surface-density induced by P at a point A' of the surface s, then, since G_{pq} is the potential at Q due to the superficial distribution,

$$G_{pq} = \iint \frac{\sigma_{pa'}}{r_{qa'}}\, ds', \tag{2}$$

where ds' is an element of the surface s at A', and the integration is to be extended over the whole surface s.

But if unit of electricity had been placed at Q, we should have had by equation (1),

$$\frac{1}{r_{qa'}} = -G_{qa'} \tag{3}$$

$$= -\iint \frac{\sigma_{qa}}{r_{aa'}}\, ds; \tag{4}$$

where σ_{qa} is the density at A of the electricity induced by Q, ds is an element of surface, and $r_{aa'}$ is the distance between A and A'.

Substituting this value of $1/r_{qd'}$ in the expression for G_{pq}, we find

$$G_{pq} = -\iiiint \frac{\sigma_{qd}\sigma_{pd'}}{r_{dd'}}\, ds\, ds', \tag{5}$$

Since this expression is not altered by changing $_p$ into $_q$ and $_q$ into $_p$, we find that

$$G_{pq} = G_{qp}; \tag{6}$$

a result which we have already shewn to be necessary in Art. 87, but which we now see to be deducible from the mathematical process by which Green's function may be calculated.

If we assume any distribution of electricity whatever, and place in the field a point charged with unit of electricity, and if the surface of potential zero completely separates the point from the assumed distribution, then if we take this surface for the surface s, and the point for P, Green's function, for any point on the same side of the surface as P, will be the potential of the assumed distribution on the other side of the surface. In this way we may construct any number of cases in which Green's function can be found for a particular position of P. To find the form of the function when the form of the surface is given and the position of P is arbitrary, is a problem of far greater difficulty, though, as we have proved, it is mathematically possible.

Let us suppose the problem solved, and that the point P is taken within the surface. Then for all external points the potential of the superficial distribution is equal and opposite to that of P. The superficial distribution is therefore *centrobaric**, and its action on all external points is the same as that of a unit of negative electricity placed at P.

99 *a*.] If in Green's Theorem we make $\Psi = \Phi$, we find

$$\iint \Psi \frac{d\Psi}{d\nu}\, ds - \iiint \Psi \nabla^2 \Psi\, d\varsigma = \iiint (\nabla\Psi)^2\, d\varsigma. \tag{16}$$

If Ψ is the potential of a distribution of electricity in space with a volume-density ρ and on conductors whose surfaces are s_1, s_2, &c., and whose potentials are Ψ_1, Ψ_2, &c., with surface-densities σ_1, σ_2, &c., then

$$\nabla^2 \Psi = 4\pi\rho, \tag{17}$$

$$\frac{d\Psi}{d\nu} = -4\pi\sigma, \tag{18}$$

$$\text{and} \iint \frac{d\Psi}{d\nu_1}\, ds_1 = -4\pi e_1, \tag{19}$$

where e_1 is the charge of the surface s_1.

* Thomson and Tait's *Natural Philosophy*, § 526.

Dividing (16) by -8π, we find

$$\frac{1}{2}(\Psi_1 e_1 + \Psi_2 e_2 + \&c.) + \frac{1}{2}\iiint \Psi\rho\, dx\, dy\, dz$$

$$= \frac{1}{8\pi}\iiint\left[\left(\frac{d\Psi}{dx}\right)^2 + \left(\frac{d\Psi}{dy}\right)^2 + \left(\frac{d\Psi}{dz}\right)^2\right] dx\, dy\, dz. \qquad (20)$$

The first term is the electric energy of the system arising from the surface-distributions, and the second is that arising from the distribution of electricity through the field, if such a distribution exists.

Hence the second member of the equation expresses the whole electric energy of the system, the potential Ψ being a given function of x, y, z.

As we shall often have occasion to employ this volume-integral, we shall denote it by the abbreviation W_{Ψ}, so that

$$W_{\Psi} = \frac{1}{8\pi}\iiint\left[\left(\frac{d\Psi}{dx}\right)^2 + \left(\frac{d\Psi}{dy}\right)^2 + \left(\frac{d\Psi}{dz}\right)^2\right] dx\, dy\, dz. \qquad (21)$$

If the only charges are those on the surfaces of the conductors, $\rho = 0$, and the second term of the first member of equation (20) disappears.

The first term is the expression for the energy of the charged system expressed, as in Art. 84, in terms of the charges and the potentials of the conductors, and this expression for the energy we denote by W.

99 *b*.] Let Ψ be a function of x, y, z, subject to the condition that its value at the closed surface s is $\bar{\Psi}$, a known quantity for every point of the surface. The value of Ψ at points not on the surface s is perfectly arbitrary.

Let us also write

$$W = \frac{1}{8\pi}\iiint\left[\left(\frac{d\Psi}{dx}\right)^2 + \left(\frac{d\Psi}{dy}\right)^2 + \left(\frac{d\Psi}{dz}\right)^2\right] dx\, dy\, dz, \qquad (22)$$

the integration being extended throughout the space within the surface; then we shall prove that if Ψ_1 is a particular form of Ψ which satisfies the surface condition and also satisfies Laplace's Equation

$$\nabla^2 \Psi_1 = 0 \qquad (23)$$

at every point within the surface, then W_1, the value of W corresponding to Ψ_1, is less than that corresponding to any function which differs from Ψ_1 at any point within the surface.

For let Ψ be any function coinciding with Ψ_1 at the surface but not at every point within it, and let us write

$$\Psi = \Psi_1 + \Psi_2; \qquad (24)$$

then Ψ_2 is a function which is zero at every point of the surface.

The value of W for Ψ will be evidently

$$W = W_1 + W_2 + \frac{1}{4\pi}\iiint\left(\frac{d\Psi_1}{dx}\frac{d\Psi_2}{dx} + \frac{d\Psi_1}{dy}\frac{d\Psi_2}{dy} + \frac{d\Psi_1}{dz}\frac{d\Psi_2}{dz}\right)dx\,dy\,dz. \quad (25)$$

By Green's Theorem the last term may be written

$$\frac{1}{4\pi}\iiint \Psi_2 \nabla^2 \Psi_1 ds - \frac{1}{4\pi}\iint \Psi_2 \frac{d\Psi_1}{d\nu} ds. \quad (26)$$

The volume-integral vanishes because $\nabla^2\Psi_1 = 0$ within the surface, and the surface-integral vanishes because at the surface $\Psi_2 = 0$. Hence equation (25) is reduced to the form

$$W = W_1 + W_2. \quad (27)$$

Now the elements of the integral W_2 being sums of three squares, are incapable of negative values, so that the integral itself can only be positive or zero. Hence if W_2 is not zero it must be positive, and therefore W greater than W_1. But if W_2 is zero, every one of its elements must be zero, and therefore

$$\frac{d\Psi_2}{dx} = 0, \quad \frac{d\Psi_2}{dy} = 0, \quad \frac{d\Psi_2}{dz} = 0$$

at every point within the surface, and Ψ_2 must be a constant within the surface. But at the surface $\Psi_2 = 0$, therefore $\Psi_2 = 0$ at every point within the surface, and $\Psi = \Psi_1$, so that if W is not greater than W_1, Ψ must be identical with Ψ_1 at every point within the surface.

It follows from this that Ψ_1 is the only function of x, y, z which becomes equal to $\bar{\Psi}$ at the surface, and which satisfies Laplace's Equation at every point within the surface.

For if these conditions are satisfied by any other function Ψ_3, then W_3 must be less than any other value of W. But we have already proved that W_1 is less than any other value, and therefore than W_3. Hence no function different from Ψ_1 can satisfy the conditions.

The case which we shall find most useful is that in which the field is bounded by one exterior surface, s, and any number of interior surfaces, s_1, s_2, &c., and when the conditions are that the value of Ψ shall be zero at s, Ψ_1 at s_1, Ψ_2 at s_2, and so on, where Ψ_1, Ψ_2, &c. are constant for each surface, as in a system of conductors, the potentials of which are given.

Of all values of Ψ satisfying these conditions, that gives the minimum value of W_Ψ for which $\nabla^2\Psi = 0$ at every point in the field.

Thomson's Theorem.

Lemma.

100 *a*.] Let Ψ be any function of x, y, z which is finite and continuous within the closed surface s, and which at certain closed surfaces, s_1, s_2, s_p, &c., has the values Ψ_1, Ψ_2, Ψ_p, &c. constant for each surface.

Let u, v, w be functions of x, y, z, which we may consider as the components of a vector $\mathfrak{C}$ subject to the solenoidal condition

$$-S.\nabla\mathfrak{C} = \frac{du}{dx} + \frac{dv}{dy} + \frac{dw}{dz} = 0, \tag{28}$$

and let us put in Theorem III

$$X = \Psi u, \quad Y = \Psi v, \quad Z = \Psi w; \tag{29}$$

we find as the result of these substitutions

$$\Sigma_p \iint \Psi_p (l_p u + m_p v + n_p w)\, ds_p + \iiint \Psi \left(\frac{du}{dx} + \frac{dv}{dy} + \frac{dw}{dz}\right) dx\, dy\, dz$$
$$+ \iiint \left(u\frac{d\Psi}{dx} + v\frac{d\Psi}{dy} + w\frac{d\Psi}{dz}\right) dx\, dy\, dz = 0, \tag{30}$$

the surface-integrals being extended over the different surfaces and the volume-integrals being taken throughout the whole field. Now the first volume-integral vanishes in virtue of the solenoidal condition for u, v, w, and the surface-integrals vanish in the following cases:—

(1) When at every point of the surface $\Psi = 0$.

(2) When at every point of the surface $lu + mv + nw = 0$.

(3) When the surface is entirely made up of parts which satisfy either (1) or (2).

(4) When Ψ is constant over the whole closed surface, and

$$\iint (lu + mv + nw)\, ds = 0.$$

Hence in these four cases the volume-integral

$$M = \iiint \left(u\frac{d\Psi}{dx} + v\frac{d\Psi}{dy} + w\frac{d\Psi}{dz}\right) dx\, dy\, dz = 0. \tag{31}$$

100 *b*.] Now consider a field bounded by the external closed surface s, and the internal closed surfaces s_1, s_2, &c.

Let Ψ be a function of x, y, z, which within the field is finite and continuous and satisfies Laplace's Equation

$$\nabla^2\Psi = 0, \tag{32}$$

and has the constant, but not given, values Ψ_1, Ψ_2, &c. at the surfaces s_1, s_2, &c. respectively, and is zero at the external surface s.

The charge of any of the conducting surfaces, as s_1, is given by the surface-integral

$$e_1 = -\frac{1}{4\pi}\iint \frac{d\Psi}{d\nu_1} ds_1, \qquad (33)$$

the normal ν_1 being drawn from the surface s_1 into the electric field.

100 *c*.] Now let f, g, h be functions of x, y, z, which we may consider as the components of a vector $\mathfrak{D}$, subject only to the conditions that at every point of the field they must satisfy the solenoidal equation

$$\frac{df}{dx} + \frac{dg}{dy} + \frac{dh}{dz} = 0, \qquad (34)$$

and that at any one of the internal closed surfaces, as s_1, the surface-integral

$$\iint (l_1 f + m_1 g + n_1 h)\, ds = e_1, \qquad (35)$$

where l, m, n are the direction cosines of the normal ν_1 drawn outwards from the surface s_1 into the electric field, and e_1 is the same quantity as in equation (33), being, in fact, the electric charge of the conductor whose surface is s_1.

We have to consider the value of the volume-integral

$$W_{\mathfrak{D}} = 2\pi \iiint (f^2 + g^2 + h^2)\, dx\, dy\, dz, \qquad (36)$$

extended throughout the whole of the field within s and without s_1, s_2, &c., and to compare it with

$$W_\Psi = \frac{1}{8\pi}\iiint \left[\left(\frac{d\Psi}{dx}\right)^2 + \left(\frac{d\Psi}{dy}\right)^2 + \left(\frac{d\Psi}{dz}\right)^2\right] dx\, dy\, dz, \qquad (37)$$

the limits of integration being the same.

Let us write

$$u = f + \frac{1}{4\pi}\frac{d\Psi}{dx}, \quad v = g + \frac{1}{4\pi}\frac{d\Psi}{dy}, \quad w = h + \frac{1}{4\pi}\frac{d\Psi}{dz}, \qquad (38)$$

$$\text{and} \quad W_{\mathfrak{S}} = 2\pi \iiint (u^2 + v^2 + w^2)\, dx\, dy dz; \qquad (39)$$

then since

$$f^2 + g^2 + h^2 = \frac{1}{16\pi^2}\left[\left(\frac{d\Psi}{dx}\right)^2 + \left(\frac{d\Psi}{dy}\right)^2 + \left(\frac{d\Psi}{dz}\right)^2\right]$$

$$+ u^2 + v^2 + w^2 - \frac{1}{2\pi}\left[u\frac{d\Psi}{dx} + v\frac{d\Psi}{dy} + w\frac{d\Psi}{dz}\right],$$

$$W_{\mathfrak{D}} = W_\Psi + W_{\mathfrak{S}} - \iiint \left(u\frac{d\Psi}{dx} + v\frac{d\Psi}{dy} + w\frac{d\Psi}{dz}\right) dx\, dy\, dz. \qquad (40)$$

Now in the first place, u, v, w satisfy the solenoidal condition at every point of the field, for by equations (38)

$$\frac{du}{dx}+\frac{dv}{dy}+\frac{dw}{dz}=\frac{df}{dx}+\frac{dg}{dy}+\frac{dh}{dz}-\frac{1}{4\pi}\nabla^2\Psi, \qquad (41)$$

and by the conditions expressed in equations (34) and (32), both parts of the second member of (41) are zero.

In the second place, the surface-integral

$$\iint (l_1 u+m_1 v+n_1 w)\,ds_1$$
$$=\iint (l_1 f+m_1 g+n_1 h)\,ds_1+\frac{1}{4\pi}\iint \frac{d\Psi}{d\nu_1}ds_1, \qquad (42)$$

but by (35) the first term of the second member is e, and by (33) the second term is $-e$, so that

$$\iint (l_1 u+m_1 v+n_1 w)\,ds_1=0. \qquad (43)$$

Hence, since Ψ_1 is constant, the fourth condition of Art. 100*a* is satisfied, and the last term of equation (40) is zero, so that the equation is reduced to the form

$$W_{\mathfrak{D}}=W_{\Psi}+W_{\mathfrak{S}}. \qquad (44)$$

Now since the element of the integral $W_{\mathfrak{S}}$ is the sum of three squares, $u^2+v^2+w^2$, it must be either positive or zero. If at any point within the field u, v, and w are not each of them equal to zero, the integral $W_{\mathfrak{S}}$ must have a positive value, and $W_{\mathfrak{D}}$ must therefore be greater than W_{Ψ}. But the values $u=v=w=0$ at every point satisfy the conditions.

Hence, if at every point

$$f=-\frac{1}{4\pi}\frac{d\Psi}{dn}, \quad g=-\frac{1}{4\pi}\frac{d\Psi}{dy}, \quad h=-\frac{1}{4\pi}\frac{d\Psi}{dz}; \qquad (45)$$

then

$$W_{\mathfrak{D}}=W_{\Psi}, \qquad (46)$$

and the value of $W_{\mathfrak{D}}$ corresponding to these values of f, g, h, is less than the value corresponding to any values of f, g, h, differing from these.

Hence the problem of determining the displacement and potential, at every point of the field, when the charge on each conductor is given, has one and only one solution.

This theorem in one of its more general forms was first stated by Sir W. Thomson*. We shall afterwards show of what generalization it is capable.

* *Cambridge and Dublin Mathematical Journal, February*, 1848.

100 *d*.] This theorem may be modified by supposing that the vector $\mathfrak{D}$, instead of satisfying the solenoidal condition at every point of the field, satisfies the condition

$$\frac{df}{dx} + \frac{dg}{dy} + \frac{dh}{dz} = \rho, \tag{47}$$

where ρ is a finite quantity, whose value is given at every point in the field, and may be positive or negative, continuous or discontinuous, its volume-integral within a finite region being, however, finite.

We may also suppose that at certain surfaces in the field

$$lf + mg + nh + l'f' + m'g' + n'h' = \sigma, \tag{48}$$

when l, m, n and l', m', n' are the direction cosines of the normals drawn from a point of the surface towards those regions in which the components of the displacement are f, g, h and f', g', h' respectively, and σ is a quantity given at all points of the surface, the surface-integral of which, over a finite surface, is finite.

100 *e*.] We may also alter the condition at the bounding surfaces by supposing that at every point of these surfaces

$$lf + mg + nh = \sigma, \tag{49}$$

where σ is given for every point.

(In the original statement we supposed only the value of the *integral* of σ over each of the surfaces to be given. Here we suppose its value given for every element of surface, which comes to the same thing as if, in the original statement, we had considered every element as a separate surface.)

None of these modifications will affect the truth of the theorem provided we remember that Ψ must satisfy the corresponding conditions, namely, the general condition,

$$\frac{d^2\Psi}{dx^2} + \frac{d^2\Psi}{dy^2} + \frac{d^2\Psi}{dz^2} + 4\pi\rho = 0, \tag{50}$$

and the surface condition

$$\frac{d\Psi}{d\nu} + \frac{d\Psi'}{d\nu'} + 4\pi\sigma = 0. \tag{51}$$

For if, as before,

$$f + \frac{1}{4\pi}\frac{d\Psi}{dx} = u, \quad g + \frac{1}{4\pi}\frac{d\Psi}{dy} = v, \quad h + \frac{1}{4\pi}\frac{d\Psi}{dz} = w,$$

then u, v, w will satisfy the general solenoidal condition

$$\frac{du}{dx} + \frac{dv}{dy} + \frac{dw}{dz} = 0,$$

and the surface condition

$$lu + mv + nw + l'u' + m'v' + n'w' = 0,$$

and at the bounding surface

$$lu + mv + nw = 0,$$

whence we find as before that

$$M = \iiint \left(u\frac{d\Psi}{dx} + v\frac{d\Psi}{dy} + w\frac{d\Psi}{dz}\right) = 0,$$

and that

$$W_{\mathfrak{D}} = W_{\Psi} + W_{\mathfrak{E}}.$$

Hence as before it is shewn that $W_{\mathfrak{D}}$ is a unique minimum when $W_{\mathfrak{E}} = 0$, which implies that $\mathfrak{E}$ is everywhere zero, and therefore

$$f = -\frac{1}{4\pi}\frac{d\Psi}{dx}, \quad g = -\frac{1}{4\pi}\frac{d\Psi}{dy}, \quad h = -\frac{1}{4\pi}\frac{d\Psi}{dz}.$$

101 *a.*] In our statement of these theorems we have hitherto confined ourselves to that theory of electricity which assumes that the properties of an electric system depend on the form and relative position of the conductors, and on their charges, but takes no account of the nature of the dielectric medium between the conductors.

According to that theory, for example, there is an invariable relation between the surface density of a conductor and the electromotive intensity just outside it, as expressed in the law of Coulomb

$$R = 4\pi\sigma.$$

But this is true only in the standard medium, which we may take to be air. In other media the relation is different, as was proved experimentally, though not published, by Cavendish, and afterwards rediscovered independently by Faraday.

In order to express the phenomenon completely, we find it necessary to consider two vector quantities, the relation between which is different in different media. One of these is the electromotive intensity, the other is the electric displacement. The electromotive intensity is connected by equations of invariable form with the potential, and the electric displacement is connected by equations of invariable form with the distribution of electricity, but the relation between the electromotive intensity and the electric displacement depends on the nature of the dielectric medium, and must be expressed by equations, the most general form of which is as yet not fully determined, and can be determined only by experiments on dielectrics.

101 *b.*] The electromotive intensity is a vector defined in Art. 68, as the mechanical force on a small quantity e of electricity divided by e. We shall denote its components by the letters P, Q, R, and the vector itself by $\mathfrak{E}$.

In electrostatics, the line integral of $\mathfrak{E}$ is always independent of the path of integration, or in other words $\mathfrak{E}$ is the space-variation of a potential. Hence

$$P = -\frac{d\Psi}{dx}, \quad Q = -\frac{d\Psi}{dy}, \quad R = -\frac{d\Psi}{dz},$$

or more briefly, in the language of Quaternions

$$\mathfrak{E} = -\nabla\Psi.$$

101 *c.*] The electric displacement in any direction is defined in Art. 68, as the quantity of electricity carried through a small area A, the plane of which is normal to that direction, divided by A. We shall denote the rectangular components of the electric displacement by the letters f, g, h, and the vector itself by $\mathfrak{D}$.

The volume-density at any point is determined by the equation

$$\rho = \frac{df}{dx} + \frac{dg}{dy} + \frac{dh}{dz},$$

or in the language of Quaternions

$$\rho = -S.\nabla\mathfrak{D}.$$

The surface-density at any point of a charged surface is determined by the equation

$$\sigma = lf + mg + nh + l'f' + m'g' + n'h',$$

where f, g, h are the components of the displacement on one side of the surface, the direction cosines of the normal drawn from the surface on that side being l, m, n, and f', g', h' and l', m', n' are the components of the displacements, and the direction cosines of the normal on the other side.

This is expressed in Quaternions by the equation

$$\sigma = -[S\,.\,U\nu\,\mathfrak{D} + S\,.\,U\nu'\,\mathfrak{D}'],$$

where $U\nu$, $U\nu'$ are unit normals on the two sides of the surface, and S indicates that the scalar part of the product is to be taken.

When the surface is that of a conductor, ν being the normal drawn outwards, then since f', g', h' and $\mathfrak{D}'$ are zero, the equation is reduced to the form

$$\sigma = (lf + mg + nh);$$
$$= -S\,.\,U\nu\,\mathfrak{D}.$$

The whole charge of the conductor is therefore

$$e = \iint (lf + mg + nh)\,ds;$$
$$= -\iint S\,.\,U\nu\,\mathfrak{D}\,ds.$$

101 *d.*] The electric energy of the system is, as was shown in Art. 84, half the sum of the products of the charges into their respective potentials. Calling this energy W,

$$W = \tfrac{1}{2}\Sigma e\Psi$$
$$= \frac{1}{2}\iiint \rho\Psi\, dx\, dy\, dz + \frac{1}{2}\iint \sigma\Psi\, ds,$$
$$= \frac{1}{2}\iiint \Psi\left(\frac{df}{dx} + \frac{dg}{dy} + \frac{dh}{dz}\right) dx\, dy\, dz$$
$$+ \frac{1}{2}\iint \Psi\,(lf + mg + nh)\, ds;$$

where the volume-integral is to be taken throughout the electric field, and the surface-integral over the surfaces of the conductors.

Writing in Theorem III, Art. 21,

$$X = \Psi f, \quad Y = \Psi g, \quad Z = \Psi h,$$

we find

$$\iint \Psi\,(lf + mg + nh)\, ds = -\iiint \Psi\left(\frac{df}{dx} + \frac{dg}{dy} + \frac{dh}{dz}\right) dx\, dy\, dz;$$
$$-\iiint \left(f\frac{d\Psi}{dx} + g\frac{d\Psi}{dy} + h\frac{d\Psi}{dz}\right) dx\, dy\, dz.$$

Substituting this value for the surface-integral in W we find

$$W = -\frac{1}{2}\iiint \left(f\frac{d\Psi}{dx} + g\frac{d\Psi}{dy} + h\frac{d\Psi}{dz}\right) dx\, dy\, dz;$$

or $$W = \frac{1}{2}\iiint (fP + gQ + hR)\, dx\, dy\, dz.$$

101 *e*.] We now come to the relation between 𝔇 and 𝔈.

The unit of electricity is usually defined with reference to experiments conducted in air. We now know from the experiments of Boltzmann that the dielectric constant of air is somewhat greater than that of a vacuum, and that it varies with the density. Hence, strictly speaking, all measurements of electric quantity require to be corrected to reduce them either to air of standard pressure and temperature, or, what would be more scientific, to a vacuum, just as indices of refraction measured in air require a similar correction, the correction in both cases being so small that it is sensible only in measurements of extreme accuracy.

In the standard medium

$$4\pi\mathfrak{D} = \mathfrak{E},$$

or $$4\pi f = P, \quad 4\pi g = Q, \quad 4\pi h = R.$$

In an isotropic medium whose dielectric constant is K

$$4\pi D = K\mathfrak{E},$$
$$4\pi f = KP, \quad 4\pi g = KQ, \quad 4\pi h = KR.$$

There are some media, however, of which glass has been the most carefully investigated, in which the relation between 𝔇 and 𝔈 is more complicated, and involves the time variation of one or both of these quantities, so that the relation must be of the form

$$F(\mathfrak{D}, \mathfrak{E}, \dot{\mathfrak{D}}, \dot{\mathfrak{E}}, \ddot{\mathfrak{D}}, \ddot{\mathfrak{E}}, \&c.) = 0.$$

We shall not attempt to discuss relations of this more general kind at present, but shall confine ourselves to the case in which $\mathfrak{D}$ is a linear and vector function of $\mathfrak{E}$.

The most general form of such a relation may be written

$$4\pi\mathfrak{D} = \phi(\mathfrak{E}),$$

where ϕ during the present investigation always denotes a linear and vector function. The components of $\mathfrak{D}$ are therefore homogeneous linear functions of those of $\mathfrak{E}$, and may be written in the form

$$4\pi f = K_{xx}P + K_{xy}Q + K_{xz}R;$$
$$4\pi g = K_{yx}P + K_{yy}Q + K_{yz}R;$$
$$4\pi h = K_{zx}P + K_{zy}Q + K_{zz}R;$$

where the first suffix of each coefficient K indicates the direction of the displacement, and the second, that of the electromotive intensity.

The most general form of a linear and vector function involves nine independent coefficients. When the coefficients which have the same pair of suffixes are equal, the function is said to be self-conjugate.

If we express $\mathfrak{E}$ in terms of $\mathfrak{D}$ we shall have

$$\mathfrak{E} = 4\pi\phi^{-1}\mathfrak{D},$$

or

$$P = 4\pi(k_{xx}f + k_{yx}g + k_{zx}h),$$
$$Q = 4\pi(k_{xy}f + k_{yy}g + k_{zy}h),$$
$$R = 4\pi(k_{xz}f + k_{yz}g + k_{zz}h).$$

101 *f*.] The work done by the electromotive intensity whose components are P, Q, R, in producing a displacement whose components are df, dg, and dh, in unit of volume of the medium, is

$$dW = P\,df + Q\,dg + R\,dh.$$

Since a dielectric under electric displacement is a conservative system, W must be a function of f, g, h, and since f, g, h may vary independently, we have

$$P = \frac{dW}{df}, \qquad Q = \frac{dW}{dg}, \qquad R = \frac{dW}{dh}.$$

Hence

$$\frac{dP}{dg} = \frac{d^2W}{df\,dg} = \frac{d^2W}{dg\,df} = \frac{dQ}{df}.$$

But $\dfrac{dP}{dg} = 4\pi k_{yx}$, the coefficient of g in the expression for P, and $\dfrac{dQ}{df} = 4\pi k_{xy}$, the coefficient of f in the expression for Q.

Hence if a dielectric is a conservative system, (and we know that it is so, because it can retain its energy for an indefinite time),

$$k_{xy} = k_{yx},$$

and ϕ^{-1} is a self-conjugate function.

Hence it follows that ϕ also is self-conjugate, and

$$K_{xy} = K_{yx}.$$

101 *g.*] The expression for the energy may therefore be written in either of the forms

$$W_{\mathfrak{E}} = \frac{1}{8\pi}\iiint [K_{xx}P^2 + K_{yy}Q^2 + K_{zz}R^2 + 2K_{yz}QR + 2K_{zx}RP + 2K_{xy}PQ]\,dx\,dy\,dz,$$

or

$$W_{\mathfrak{D}} = 2\pi\iiint [k_{xx}f^2 + k_{yy}g^2 + k_{zz}h^2 + 2k_{yz}gh + 2k_{zx}hf + 2k_{xy}fg]\,dx\,dy\,dz,$$

where the suffix denotes the vector in terms of which W is to be expressed. When there is no suffix, the energy is understood to be expressed in terms of both vectors.

We have thus, in all, six different expressions for the energy of the electric field. Three of these involve the charges and potentials of the surfaces of conductors, and are given in Art. 87.

The other three are volume-integrals taken throughout the electric field, and involve the components of electromotive intensity or of electric displacement, or of both.

The first three therefore belong to the theory of action at a distance, and the last three to the theory of action by means of the intervening medium.

These three expressions for W may be written,

$$W = -\frac{1}{2}\iiint S.\mathfrak{D}\mathfrak{E}\,d\varsigma,$$

$$W_{\mathfrak{E}} = -\frac{1}{8\pi}\iiint S.\mathfrak{E}\phi\mathfrak{E}\,d\varsigma,$$

$$W_{\mathfrak{D}} = -2\pi\iiint S.\mathfrak{D}\phi\mathfrak{D}\,d\varsigma.$$

101 *h.*] To extend Green's Theorem to the case of a heterogeneous anisotropic medium, we have only to write in Theorem III,

$$X = \Psi\left[K_{xx}\frac{d\Phi}{dx} + K_{xy}\frac{d\Phi}{dy} + K_{xz}\frac{d\Phi}{dz}\right],$$

$$Y = \Psi\left[K_{yx}\frac{d\Phi}{dx} + K_{yy}\frac{d\Phi}{dy} + K_{yz}\frac{d\Phi}{dz}\right],$$

$$Z = \Psi\left[K_{zx}\frac{d\Phi}{dx} + K_{zy}\frac{d\Phi}{dy} + K_{zz}\frac{d\Phi}{dz}\right],$$

and we obtain (remembering that the order of the suffixes of the coefficients is indifferent),

$$\iint \Psi \Big[(K_{xx} l + K_{yx} m + K_{zx} n) \frac{d\Phi}{dx} + (K_{xy} l + K_{yy} m + K_{zy} n) \frac{d\Phi}{dy} + (K_{xz} l + K_{yz} m + K_{zz} n) \frac{d\Phi}{dz} \Big] ds$$

$$+ \iiint \Psi \Big[\frac{d}{dx}\Big(K_{xx}\frac{d\Phi}{dx} + K_{xy}\frac{d\Phi}{dy} + K_{xz}\frac{d\Phi}{dz}\Big) + \frac{d}{dy}\Big(K_{yx}\frac{d\Phi}{dx} + K_{yy}\frac{d\Phi}{dy} + K_{yz}\frac{d\Phi}{dz}\Big) + \frac{d}{dz}\Big(K_{zx}\frac{d\Phi}{dx} + K_{zy}\frac{d\Phi}{dy} + K_{zz}\frac{d\Phi}{dz}\Big) \Big] dx\, dy\, dz$$

$$= -\iiint \Big[K_{xx}\frac{d\Psi}{dx}\frac{d\Phi}{dx} + K_{yy}\frac{d\Psi}{dy}\frac{d\Phi}{dy} + K_{zz}\frac{d\Psi}{dz}\frac{d\Phi}{dz} + K_{yz}\Big(\frac{d\Psi}{dy}\frac{d\Phi}{dz} + \frac{d\Psi}{dz}\frac{d\Phi}{dy}\Big) + K_{zx}\Big(\frac{d\Psi}{dz}\frac{d\Phi}{dx} + \frac{d\Psi}{dx}\frac{d\Phi}{dz}\Big) + K_{xy}\Big(\frac{d\Phi}{dx}\frac{d\Phi}{dy} + \frac{d\Psi}{dy}\frac{d\Phi}{dx}\Big) \Big] dx\, dy\, dz$$

$$= \iint \Phi \Big[(K_{xx} l + K_{yx} m + K_{zx} n) \frac{d\Psi}{dx} + (K_{xy} l + K_{yy} m + K_{zy} n) \frac{d\Psi}{dy} + (K_{xz} l + K_{yz} m + K_{zz} n) \frac{d\Psi}{dz} \Big] ds$$

$$+ \iiint \Phi \Big[\frac{d}{dx}\Big(K_{xx}\frac{d\Psi}{dx} + K_{xy}\frac{d\Psi}{dy} + K_{xz}\frac{d\Psi}{dz}\Big) + \frac{d}{dy}\Big(K_{yx}\frac{d\Psi}{dx} + K_{yy}\frac{d\Psi}{dy} + K_{yz}\frac{d\Psi}{dz}\Big) + \frac{d}{dz}\Big(K_{zx}\frac{d\Psi}{dx} + K_{zy}\frac{d\Psi}{dy} + K_{zz}\frac{d\Psi}{dz}\Big) \Big] dx\, dy\, dz.$$

Using quaternion notation the result may be written more briefly,

$$\iint \Psi\, S.\, U\nu\, \phi(\nabla\Phi)\, ds - \iiint \Psi\, S.(\nabla\phi\nabla)\,\Phi\, d\varsigma$$

$$= -\iiint S.\nabla\Psi\,\phi\nabla\Phi\, d\varsigma = \iiint S.\nabla\Phi\,\phi\,\nabla\Psi\, d\varsigma$$

$$= \iint \Phi\, S.\, U\nu\, \phi(\nabla\Psi)\, ds - \iiint \Phi\, S.(\nabla\phi\nabla)\,\Psi\, d\varsigma.$$

Limits between which the electric capacity of a conductor must lie.

102 *a*.] The capacity of a conductor or system of conductors has been already defined as the charge of that conductor or system

of conductors when raised to potential unity, all the other conductors in the field being at potential zero.

The following method of determining limiting values between which the capacity must lie, was suggested by a paper 'On the Theory of Resonance,' by the Hon. J. W. Strutt, Phil. Trans. 1871. See Art. 308.

Let s_1 denote the surface of the conductor, or system of conductors, whose capacity is to be determined, and s_0 the surface of all other conductors. Let the potential of s_1 be Ψ_1, and that of s_0, Ψ_0. Let the charge of s_1 be e_1. That of s_0 will be $-e_1$.

Then if q is the capacity of s_1,

$$q = \frac{e_1}{\Psi_1 - \Psi_0}, \tag{1}$$

and if W is the energy of the system with its actual distribution of electricity

$$W = \tfrac{1}{2} e_1 (\Psi_1 - \Psi_0), \tag{2}$$

and

$$q = \frac{2W}{(\Psi_1 - \Psi_0)^2} = \frac{e_1^2}{2W}. \tag{3}$$

To find an upper limit of the value of the capacity. Assume any value of Ψ which is equal to 1 at s_1 and equal to zero at s_0, and calculate the value of the volume-integral

$$W_\Psi = \frac{1}{8\pi} \iiint \left[\left(\frac{d\Psi}{dx}\right)^2 + \left(\frac{d\Psi}{dy}\right)^2 + \left(\frac{d\Psi}{dz}\right)^2 \right] dx\, dy\, dz \tag{4}$$

extended over the whole field.

Then as we have proved (Art. 99 *b*) that W cannot be greater than W_Ψ, the capacity, q, cannot be greater than $2W_\Psi$.

To find a lower limit of the value of the capacity. Assume any system of values of f, g, h, which satisfies the equation

$$\frac{df}{dx} + \frac{dg}{dy} + \frac{dh}{dz} = 0, \tag{5}$$

and let it make

$$\iint (l_1 f + m_1 g + n_1 h)\, ds_1 = e_1. \tag{6}$$

Calculate the value of the volume-integral

$$W_{\mathfrak{D}} = 2\pi \iiint (f^2 + g^2 + h^2)\, dx\, dy\, dz, \tag{7}$$

extended over the whole field; then as we have proved (Art. 100 *c*) that W cannot be greater than $W_{\mathfrak{D}}$, the capacity, q, cannot be less than

$$\frac{e_1^2}{2W_{\mathfrak{D}}}. \tag{8}$$

The simplest method of obtaining a system of values of f, g, h, which will satisfy the solenoidal condition, is to assume a distribution of electricity on the surface of s_1, and another on s_0, the sum of the charges being zero, then to calculate the potential, Ψ, due to this distribution, and the electric energy of the system thus arranged, which we may call W_σ.

If we then make

$$f = -\frac{1}{4\pi}\frac{d\Psi}{dx}, \quad g = -\frac{1}{4\pi}\frac{d\Psi}{dy}, \quad h = -\frac{1}{4\pi}\frac{d\Psi}{dz},$$

these values of f, g, h will satisfy the solenoidal condition.

But in this case we can determine $W_{\mathfrak{D}}$ without going through the process of finding the volume-integral. For since this solution makes $\nabla^2\Psi = 0$ at all points in the field, we can obtain $W_{\mathfrak{D}}$ in the form of the surface-integrals,

$$W_{\mathfrak{D}} = \frac{1}{2}\iint \Psi\sigma_1\, ds_1 + \frac{1}{2}\iint \Psi\sigma_0\, ds_0, \tag{9}$$

where the first integral is extended over the surface s_1 and the second over the surface s_0.

If the surface s_0 is at an infinite distance from s_1, the potential at s_0 is zero and the second term vanishes.

102 *b*.] An approximation to the solution of any problem of the distribution of electricity on conductors whose potentials are given may be made in the following manner:—

Let s_1 be the surface of a conductor or system of conductors maintained at potential 1, and let s_0 be the surface of all the other conductors, including the hollow conductor which surrounds the rest, which last, however, may in certain cases be at an infinite distance from the others.

Begin by drawing a set of lines, straight or curved, from s_1 to s_0.

Along each of these lines, assume Ψ so that it is equal to 1 at s_1, and equal to 0 at s_0. Then if P is a point on one of these lines we may take $\Psi_1 = \dfrac{Ps_0}{s_1 s_0}$ as a first approximation.

We shall thus obtain a first approximation to Ψ which satisfies the condition of being equal to unity at s_1 and equal to zero at s_0.

The value of W_Ψ calculated from Ψ_1 would be greater than W.

Let us next assume as a second approximation to the lines of force

$$f = -p\frac{d\Psi_1}{dx}, \quad g = -p\frac{d\Psi_1}{dy}, \quad h = -p\frac{d\Psi_1}{dz}. \tag{10}$$

The vector whose components are a, b, c is normal to the surfaces for which Ψ_1 is constant. Let us determine p so as to make a, b, c satisfy the solenoidal condition. We thus get

$$p\left(\frac{d^2\Psi_1}{dx^2}+\frac{d^2\Psi_1}{dy^2}+\frac{d^2\Psi_1}{dz^2}\right)+\frac{dp}{dx}\frac{d\Psi}{dx}+\frac{dp}{dy}\frac{d\Psi}{dy}+\frac{dp}{dz}\frac{d\Psi}{dz}=0. \qquad (11)$$

If we draw a line from s_1 to s_0 whose direction is always normal to the surfaces for which Ψ is constant, and if we denote the length of this line measured from s_0 by s, then

$$R\frac{dx}{ds}=-\frac{d\Psi_1}{dx}, \quad R\frac{dy}{ds}=-\frac{d\Psi_1}{dy}, \quad R\frac{dz}{ds}=-\frac{d\Psi_1}{dz}, \qquad (12)$$

where R is the resultant intensity $=-\frac{d\Psi}{ds}$, so that

$$\frac{dp}{dx}\frac{d\Psi}{dx}+\frac{dp}{dy}\frac{d\Psi}{dy}+\frac{dp}{dz}\frac{d\Psi}{dz}=-R\frac{dp}{ds},$$

$$=R^2\frac{dp}{d\Psi}, \qquad (13)$$

and equation (11) becomes

$$p\nabla^2\Psi=R^2\frac{dp}{d\Psi}, \qquad (14)$$

whence

$$p=C\,exp.\int_0^{\Psi_1}\frac{\nabla^2\Psi_1}{R^2}d\Psi_1, \qquad (15)$$

the integral being a line integral taken along the line s.

Let us next assume that along the line s,

$$-\frac{d\Psi_2}{ds}=a\frac{dx}{ds}+b\frac{dy}{ds}+c\frac{dz}{ds},$$

$$=-p\frac{d\Psi_1}{ds}, \qquad (16)$$

then

$$\Psi_2=C\int_0^{\Psi_1}\left(exp.\int\frac{\nabla^2\Psi_1}{R^2}d\Psi_1\right)d\Psi_1, \qquad (17)$$

the integration being always understood to be performed along the line s.

The constant C is now to be determined from the condition that $\Psi_2=1$ at s_1 when also $\Psi_1=1$, so that

$$C\int_0^1 exp.\int_0^{\Psi}\frac{\nabla^2\Psi}{R^2}d\Psi\,d\Psi=1. \qquad (18)$$

This gives a second approximation to Ψ, and the process may be repeated.

The results obtained from calculating W_{Ψ_1}, $W_{\mathfrak{D}_2}$, W_{Ψ_2}, &c., give capacities alternately above and below the true capacity and continually approximating thereto.

The process as indicated above involves the calculation of the form of the line s and integration along this line, operations which are in general too difficult for practical purposes.

In certain cases however we may obtain an approximation by a simpler process.

102 *c*.] As an illustration of this method, let us apply it to obtain successive approximations to the equipotential surfaces and lines of induction in the electric field between two surfaces which are nearly but not exactly plane and parallel, one of which is maintained at potential zero, and the other at potential unity.

Let the equations of the two surfaces be

$$z_1 = f_1(x,y) = a \tag{19}$$

for the surface whose potential is zero, and

$$z_2 = f_2(x,y) = b \tag{20}$$

for the surface whose potential is unity, a and b being given functions of x and y, of which b is always greater than a. The first derivatives of a and b with respect to x and y are small quantities of which we may neglect powers and products of more than two dimensions.

We shall begin by supposing that the lines of induction are parallel to the axis of z, in which case

$$f = 0, \quad g = 0, \quad \frac{dh}{dz} = 0. \tag{21}$$

Hence h is constant along each individual line of induction, and

$$\Psi = -4\pi \int_a^z h\, dz = -4\pi h(z-a). \tag{22}$$

When $z = b$, $\Psi = 1$, hence

$$h = -\frac{1}{4\pi(b-a)}, \tag{23}$$

and

$$\Psi = \frac{z-a}{b-a}, \tag{24}$$

which gives a first approximation to the potential, and indicates a series of equipotential surfaces the intervals between which, measured parallel to z, are equal.

To obtain a second approximation to the lines of induction, let us assume that they are everywhere normal to the equipotential surfaces as given by equation (24).

This is equivalent to the conditions

$$4\pi f = \lambda \frac{d\Psi}{dx}, \qquad 4\pi g = \lambda \frac{d\Psi}{dy}, \qquad 4\pi h = \lambda \frac{d\Psi}{dz},$$

where λ is to be determined so that at every point of the field

$$\frac{df}{dx} + \frac{dg}{dy} + \frac{dh}{dz} = 0, \tag{26}$$

and also so that the line-integral

$$4\pi\int\left(f\frac{dx}{ds}+g\frac{dy}{ds}+h\frac{dz}{ds}\right)ds, \tag{27}$$

taken along any line of induction from the surface a to the surface b, shall be equal to -1.

Let us assume

$$\lambda = 1+A+B(z-a)+C(z-a)^2, \tag{28}$$

and let us neglect powers and products of A, B, C, and at this stage of our work powers and products of the first derivatives of a and b.

The solenoidal condition then gives

$$B=-\nabla^2 a, \qquad C=-\tfrac{1}{2}\frac{\nabla^2(b-a)}{b-a}, \tag{29}$$

where

$$\nabla^2 = -\left(\frac{d^2}{dx^2}+\frac{d^2}{dy^2}\right). \tag{30}$$

If instead of taking the line-integral along the new line of induction, we take it along the old line of induction, parallel to z, the second condition gives

$$1 = 1+A+\tfrac{1}{2}B(b-a)+\tfrac{1}{3}C(b-a)^2.$$

Hence

$$A=\tfrac{1}{6}(b-a)\nabla^2(2a+b), \tag{31}$$

and

$$\lambda = 1+\tfrac{1}{6}(b-a)\nabla^2(2a+b)-(z-a)\nabla^2 a-\tfrac{1}{2}\frac{(z-a)^2}{b-a}\nabla^2(b-a). \tag{32}$$

We thus find for the second approximation to the components of displacement,

$$\begin{aligned}
-4\pi f &= \frac{\lambda}{b-a}\left[\frac{da}{dx}+\frac{d(b-a)}{dx}\,\frac{z-a}{b-a}\right],\\
-4\pi g &= \frac{\lambda}{b-a}\left[\frac{da}{dy}+\frac{d(b-a)}{dy}\,\frac{z-a}{b-a}\right],\\
-4\pi h &= \frac{\lambda}{b-a},
\end{aligned} \tag{33}$$

and for the second approximation to the potential,

$$\Psi = \frac{z-a}{b-a}+\tfrac{1}{6}\nabla^2(2a+b)(z-a)-\tfrac{1}{2}\nabla^2 a\frac{(z-a)^2}{b-a} -\tfrac{1}{6}\nabla^2(b-a)\frac{(z-a)^3}{(b-a)^2}. \tag{34}$$

If σ_a and σ_b are the surface-densities and Ψ_a and Ψ_b the potentials of the surfaces a and b respectively,

$$\sigma_a = \frac{1}{4\pi}(\Psi_a-\Psi_b)\left[\frac{1}{b-a}+\tfrac{1}{3}\nabla^2 a+\tfrac{1}{6}\nabla^2 b\right]$$

$$\sigma_b = \frac{1}{4\pi}(\Psi_b-\Psi_a)\left[\frac{1}{b-a}-\tfrac{1}{6}\nabla^2 a-\tfrac{1}{3}\nabla^2 b\right].$$

CHAPTER V.

MECHANICAL ACTION BETWEEN TWO ELECTRICAL SYSTEMS.

103.] Let E_1 and E_2 be two electrical systems, the mutual action between which we propose to investigate. Let the distribution of electricity in E_1 be defined by the volume-density, ρ_1, of the element whose coordinates are x_1, y_1, z_1. Let ρ_2 be the volume-density of the element of E_2, whose coordinates are x_2, y_2, z_2.

Then the x-component of the force acting on the element of E_1 on account of the repulsion of the element of E_2 will be

$$\rho_1 \rho_2 \frac{x_1 - x_2}{r^3} dx_1\, dy_1\, dz_1\, dx_2\, dy_2\, dz_2,$$

where $$r^2 = (x_1-x_2)^2+(y_1-y_2)^2+(z_1-z_2)^2,$$
and if A denotes the x component of the whole force acting on E_1 on account of the presence of E_2

$$A = \iiiiiint \frac{x_1-x_2}{r^3} \rho_1 \rho_2\, dx_1\, dy_1\, dz_1\, dx_2\, dy_2\, dz_2, \qquad (1)$$

where the integration with respect to x_1, y_1, z_1 is extended throughout the region occupied by E_1, and the integration with respect to x_2, y_2, z_2 is extended throughout the region occupied by E_2.

Since, however, ρ_1 is zero except in the system E_1, and ρ_2 is zero except in the system E_2, the value of the integral will not be altered by extending the limits of the integrations, so that we may suppose the limits of every integration to be $\pm\, \infty$.

This expression for the force is a literal translation into mathematical symbols of the theory which supposes the electric force to act directly between bodies at a distance, no attention being bestowed on the intervening medium.

If we now define Ψ_2, the potential at the point x_1, y_1, z_1, arising from the presence of the system E_2, by the equation

$$\Psi_2 = \iiint \frac{\rho_2}{r}\, dx_2\, dy_2\, dz_2, \qquad (2)$$

Ψ_2 will vanish at an infinite distance, and will everywhere satisfy the equation

$$\nabla^2 \Psi_2 = 4\pi\rho_2. \qquad (3)$$

We may now express A in the form of a triple integral

$$A = -\iiint \frac{d\Psi_2}{dx} \rho_1 \, dx_1 \, dy_1 \, dz_1 . \tag{4}$$

Here the potential Ψ_2 is supposed to have a definite value at every point of the field, and in terms of this, together with the distribution, ρ_1, of electricity in the first system E_1, the force A is expressed, no explicit mention being made of the distribution of electricity in the second system E_2.

Now let Ψ_1 be the potential arising from the first system, expressed as a function of x, y, z, and defined by the equation

$$\Psi_1 = \iiint \frac{\rho_1}{r} \, dx_1 \, dy_1 \, dz_1 , \tag{5}$$

Ψ_1 will vanish at an infinite distance, and will everywhere satisfy the equation

$$\nabla^2 \Psi_1 = 4\pi\rho_1 . \tag{6}$$

We may now eliminate ρ_1 from A and obtain

$$A = -\frac{1}{4\pi} \iiint \frac{d\Psi_2}{dx} \nabla^2 \Psi_1 \, dx_1 \, dy_1 \, dz_1 , \tag{7}$$

in which the force is expressed in terms of the two potentials only.

104.] In all the integrations hitherto considered, it is indifferent what limits are prescribed, provided they include the whole of the system E_1. In what follows we shall suppose the systems E_1 and E_2 to be such that a certain closed surface s contains within it the whole of E_1 but no part of E_2.

Let us also write

$$\rho = \rho_1 + \rho_2 , \quad \Psi = \Psi_1 + \Psi_2 , \tag{8}$$

then within s, $\qquad \rho_2 = 0, \; \rho = \rho_1 ,$

and without s $\qquad \rho_1 = 0, \; \rho = \rho_2 . \qquad (9)$

Now

$$A_{11} = -\iiint \frac{d\Psi_1}{dx} \rho_1 \, dx_1 \, dy_1 \, dz_1 \tag{10}$$

represents the resultant force, in the direction x, on the system E_1 arising from the electricity in the system itself. But on the theory of direct action this must be zero, for the action of any particle P on another Q is equal and opposite to that of Q on P, and since the components of both actions enter into the integral, they will destroy each other.

We may therefore write

$$A = -\frac{1}{4\pi} \iiint \frac{d\Psi}{dx} \nabla^2 \Psi \, dx_1 \, dy_1 \, dz_1 , \tag{11}$$

where Ψ is the potential arising from both systems, the integration being now limited to the space within the closed surface s, which includes the whole of the system E_1 but none of E_2.

105.] If the action of E_2 on E_1 is effected, not by direct action at a distance, but by means of a distribution of stress in a medium extending continuously from E_2 to E_1, it is manifest that if we know the stress at every point of any closed surface s which completely separates E_1 from E_2, we shall be able to determine completely the mechanical action of E_2 on E_1. For if the force on E_1 is not completely accounted for by the stress through s, there must be direct action between something outside of s and something inside of s.

Hence if it is possible to account for the action of E_2 on E_1 by means of a distribution of stress in the intervening medium, it must be possible to express this action in the form of a surface-integral extended over any surface s which completely separates E_2 from E_1.

Let us therefore endeavour to express

$$A = \frac{1}{4\pi}\iiint \frac{d\Psi}{dx}\left[\frac{d^2\Psi}{dx^2} + \frac{d^2\Psi}{dy^2} + \frac{d^2\Psi}{dz^2}\right] dx\, dy\, dz \tag{12}$$

in the form of a surface integral.

By Theorem III we may do so if we can determine X, Y and Z, so that

$$\frac{d\Psi}{dx}\left(\frac{d^2\Psi}{dx^2} + \frac{d^2\Psi}{dy^2} + \frac{d^2\Psi}{dz^2}\right) = \frac{dX}{dx} + \frac{dY}{dy} + \frac{dZ}{dz}. \tag{13}$$

Taking the terms separately,

$$\frac{d\Psi}{dx}\frac{d^2\Psi}{dx^2} = \frac{1}{2}\frac{d}{dx}\left(\frac{d\Psi}{dx}\right)^2,$$

$$\frac{d\Psi}{dx}\frac{d^2\Psi}{dy^2} = \frac{d}{dy}\left(\frac{d\Psi}{dx}\frac{d\Psi}{dy}\right) - \frac{d\Psi}{dy}\frac{d^2\Psi}{dx\,dy},$$

$$= \frac{d}{dy}\left(\frac{d\Psi}{dx}\frac{d\Psi}{dy}\right) - \frac{1}{2}\frac{d}{dx}\left(\frac{d\Psi}{dy}\right)^2.$$

Similarly $$\frac{d\Psi}{dx}\frac{d^2\Psi}{dz^2} = \frac{d}{dz}\left(\frac{d\Psi}{dx}\frac{d\Psi}{dz}\right) - \frac{1}{2}\frac{d}{dx}\left(\frac{d\Psi}{dz}\right)^2.$$

If, therefore, we write

$$\left.\begin{aligned}
\left(\frac{d\Psi}{dx}\right)^2-\left(\frac{d\Psi}{dy}\right)^2-\left(\frac{d\Psi}{dz}\right)^2 &= 8\pi p_{xx},\\
\left(\frac{d\Psi}{dy}\right)^2-\left(\frac{d\Psi}{dz}\right)^2-\left(\frac{d\Psi}{dx}\right)^2 &= 8\pi p_{yy},\\
\left(\frac{d\Psi}{dz}\right)^2-\left(\frac{d\Psi}{dx}\right)^2-\left(\frac{d\Psi}{dy}\right)^2 &= 8\pi p_{zz},\\
\frac{d\Psi}{dy}\frac{d\Psi}{dz} &= 4\pi p_{yz} = 4\pi p_{zy},\\
\frac{d\Psi}{dz}\frac{d\Psi}{dx} &= 4\pi p_{zx} = 4\pi p_{xz},\\
\frac{d\Psi}{dx}\frac{d\Psi}{dy} &= 4\pi p_{xy} = 4\pi p_{yx},
\end{aligned}\right\} \tag{14}$$

then $A=\iiint\left(\frac{dp_{xx}}{dx}+\frac{dp_{yx}}{dy}+\frac{dp_{zx}}{dz}\right)dx\,dy\,dz,$ (15)

the integration being extended throughout the space within s. Transforming the volume-integral by Theorem III, Art. 21,

$$A=\iint(lp_{xx}+mp_{yx}+np_{zx})\,ds, \tag{16}$$

where ds is an element of any closed surface including the whole of E_1 but none of E_2, and $l\,m\,n$ are the direction cosines of the normal drawn from ds outwards.

For the components of the force on E_1 in the directions of y and z, we obtain in the same way

$$B=\iint(lp_{xy}+mp_{yy}+np_{zy})\,ds, \tag{17}$$

$$C=\iint(lp_{xz}+mp_{yz}+np_{zz})\,ds. \tag{18}$$

If the action of the system E_2 on E_1 does in reality take place by direct action at a distance, without the intervention of any medium, we must consider the quantities p_{xx} &c. as mere abbreviated forms for certain symbolical expressions, and as having no physical significance.

But if we suppose that the mutual action between E_2 and E_1 is kept up by means of stress in the medium between them, then since the equations (16), (17), (18) give the components of the resultant force arising from the action, on the outside of the surface s, of the stress whose six components are p_{xx} &c., we must consider p_{xx} &c. as the components of a stress actually existing in the medium.

106.] To obtain a clearer view of the nature of this stress let us alter the form of part of the surface s so that the element ds may become part of an equipotential surface. (This alteration of the surface is legitimate provided we do not thereby exclude any part of E_1 or include any part of E_2).

Let ν be a normal to ds drawn outwards.

Let $R = -\dfrac{d\Psi}{d\nu}$ be the intensity of the electromotive force in the direction of ν, then

$$\frac{d\Psi}{dx} = -Rl, \quad \frac{d\Psi}{dy} = -Rm, \quad \frac{d\Psi}{dz} = -Rn.$$

Hence the six components of stress are

$$p_{xx} = \frac{1}{8\pi} R^2 (l^2 - m^2 - n^2), \quad p_{yz} = \frac{1}{4\pi} R^2 mn,$$

$$p_{yy} = \frac{1}{8\pi} R^2 (m^2 - n^2 - l^2), \quad p_{zx} = \frac{1}{4\pi} R^2 nl,$$

$$p_{zz} = \frac{1}{8\pi} R^2 (n^2 - l^2 - m^2), \quad p_{xy} = \frac{1}{4\pi} R^2 lm.$$

If a, b, c are the components of the force on ds per unit of area

$$a = lp_{xx} + mp_{xy} + np_{zx} = \frac{1}{8\pi} R^2 l,$$

$$b = \frac{1}{8\pi} R^2 m,$$

$$c = \frac{1}{8\pi} R^2 n.$$

Hence the force exerted by the part of the medium outside ds on the part of the medium inside ds is normal to the element and directed outwards, that is to say, it is a tension like that of a rope, and its value per unit of area is $\dfrac{1}{8\pi} R^2$.

Let us next suppose that the element ds is at right angles to the equipotential surfaces which cut it, in which case

$$l\frac{d\Psi}{dx} + m\frac{d\Psi}{dy} + n\frac{d\Psi}{dz} = 0. \tag{19}$$

Now
$$8\pi(lp_{xx} + mp_{xy} + np_{zx}) = l\left[\left(\frac{d\Psi}{dx}\right)^2 - \left(\frac{d\Psi}{dy}\right)^2 - \left(\frac{d\Psi}{dz}\right)^2\right] + 2m\frac{d\Psi}{dx}\frac{d\Psi}{dy} + 2n\frac{d\Psi}{dx}\frac{d\Psi}{dz}. \tag{20}$$

Multiplying (19) by $2\dfrac{d\Psi}{dx}$ and subtracting from (20), we find

$$8\pi(lp_{xx}+mp_{xy}+np_{xz}) = -l\left[\left(\frac{d\Psi}{dx}\right)^2+\left(\frac{d\Psi}{dy}\right)^2+\left(\frac{d\Psi}{dz}\right)^2\right] = -lR^2. \qquad (21)$$

Hence the components of the tension per unit of area of ds are

$$a = -\frac{1}{8\pi}R^2 l,$$

$$b = -\frac{1}{8\pi}R^2 m,$$

$$c = -\frac{1}{8\pi}R^2 n.$$

Hence if the element ds is at right angles to an equipotential surface, the force which acts on it is normal to the surface, and its numerical value per unit of area is the same as in the former case, but the direction of the force is different, for it is a pressure instead of a tension.

We have thus completely determined the type of the stress at any given point of the medium.

The direction of the electromotive intensity at the point is a principal axis of stress, and the stress in this direction is a tension whose numerical value is

$$p = \frac{1}{8\pi}R^2, \qquad (22)$$

where R is the electromotive intensity.

Any direction at right angles to this is also a principal axis of stress, and the stress along this axis is a pressure whose numerical magnitude is also p.

The stress as thus defined is not of the most general type, for it has two of its principal stresses equal to each other, and the third has the same value with the sign reversed.

These conditions reduce the number of independent variables which determine the stress from six to three, and accordingly it is completely determined by the three components of the electromotive intensity

$$-\frac{d\Psi}{dx}, \quad -\frac{d\Psi}{dy}, \quad -\frac{d\Psi}{dz}.$$

The three relations between the six components of stress are

$$\left.\begin{aligned} p^2_{yz} &= (p_{xx}+p_{yy})(p_{zz}+p_{xx}),\\ p^2_{zx} &= (p_{yy}+p_{zz})(p_{xx}+p_{yy}),\\ p^2_{xy} &= (p_{zz}+p_{xx})(p_{yy}+p_{zz}). \end{aligned}\right\} \qquad (23)$$

107.] Let us now examine whether the results we have obtained will require modification when a finite quantity of electricity is collected on a finite surface so that the volume-density becomes infinite at the surface.

In this case, as we have shown in Art. 78, the components of the electromotive intensity are discontinuous at the surface. Hence the components of stress will also be discontinuous at the surface.

Let $l\ m\ n$ be the direction cosines of the normal to ds. Let P, Q, R be the components of the electromotive intensity on the side on which the normal is drawn, and $P'\ Q'\ R'$ their values on the other side.

Then by Art. (78 a) if σ is the surface-density

$$\left.\begin{aligned} P-P' &= 4\pi\sigma l, \\ Q-Q' &= 4\pi\sigma m, \\ R-R' &= 4\pi\sigma n. \end{aligned}\right\} \tag{24}$$

Let a be the x-component of the resultant force acting on the surface per unit of area, arising from the stress on both sides, then

$$\begin{aligned} a &= l\,(p_{xx}-p'_{xx})+m\,(p_{xy}-p'_{xy})+n(p_{xz}-p'_{xz}), \\ &= \frac{1}{8\pi}\,l\,\{(P^2-P'^2)-(Q^2-Q'^2)-(R^2-R'^2)\}, \\ &\qquad +\frac{1}{4\pi}\,m\,(PQ-P'Q')+\frac{1}{4\pi}\,n\,(PR-P'R'), \\ &= \frac{1}{8\pi}\,l\,\{(P-P')(P+P')-(Q-Q')(Q+Q')-(R-R')(R+R')\}, \\ &\qquad +\frac{1}{8\pi}\,m\,\{(P-P')(Q+Q')+(P+P')(Q-Q')\}, \\ &\qquad +\frac{1}{8\pi}\,n\,\{(P-P')(R+R')+(P+P')(R-R')\}, \\ &= \frac{1}{2}\,l\sigma\,\{\,l(P+P')-m\,(Q+Q')-n\,(R+R')\}, \\ &\quad +\frac{1}{2}\,m\sigma\,\{\,l(Q+Q')+m\,(P+P')\}+\frac{1}{2}\,n\sigma\,\{\,l(R+R')+n\,(P+P')\}, \\ &= \frac{1}{2}\,\sigma\,(P+P'). \end{aligned} \tag{25}$$

Hence, assuming that the stress at any point is given by equations (14), we find that the resultant force in the direction of x on a charged surface per unit of volume is equal to the surface-density multiplied into the arithmetical mean of the x-components of the electromotive intensity on the two sides of the surface.

This is the same result as we obtained in Art. 79 by a process essentially similar.

Hence the hypothesis of stress in the surrounding medium is applicable to the case in which a finite quantity of electricity is collected on a finite surface.

The resultant force on an element of surface is usually deduced from the theory of action at a distance by considering a portion of the surface, the dimensions of which are very small compared with the radii of curvature of the surface.*

On the normal to the middle point of this portion of the surface take a point P whose distance from the surface is very small compared with the dimensions of the portion of the surface. The electromotive intensity at this point, due to the small portion of the surface, will be approximately the same as if the surface had been an infinite plane, that is to say $2\pi\sigma$ in the direction of the normal drawn from the surface. For a point P' just on the other side of the surface the intensity will be the same, but in the opposite direction.

Now consider the part of the electromotive intensity arising from the rest of the surface and from other electrified bodies at a finite distance from the element of surface. Since the points P and P' are infinitely near one another, the components of the electromotive intensity arising from electricity at a finite distance will be the same for both points.

Let P_0 be the x-component of the electromotive intensity on A or A' arising from electricity at a finite distance, then the total value of the x-component for A will be

$$P = P_0 + 2\pi\sigma l,$$

and for A'

$$P' = P_0 - 2\pi\sigma l.$$

Hence

$$P_0 = \tfrac{1}{2}(P+P').$$

Now the resultant mechanical force on the element of surface must arise entirely from the action of electricity at a finite distance, since the action of the element on itself must have a resultant zero. Hence the x-component of this force per unit of area must be

$$a = \sigma P_0, \quad = \tfrac{1}{2}\sigma(P+P').$$

108.] If we define the potential (as in equation (2)) in terms of a distribution of electricity supposed to be given, then it follows

* This method is due to Laplace. See Poisson, 'Sur la Distribution de l'électricité &c.' *Mém. de l'Institut*, 1811, p. 30.

from the fact that the action and reaction between any pair of electric particles are equal and opposite, that the x-component of the force arising from the action of a system on itself must be zero, and we may write this in the form

$$\frac{1}{4\pi}\iiint\frac{d\Psi}{dx}\,\nabla^2\Psi\,dx\,dy\,dz = 0. \tag{26}$$

But if we define Ψ as a function of x, y, z which satisfies the equation
$$\nabla^2\Psi = 0$$
at every point outside the closed surface s, and is zero at an infinite distance, the fact, that the volume-integral extended throughout any space including s is zero, would seem to require proof.

One method of proof is founded on the theorem (Art. 100 *a*), that if $\nabla^2\Psi$ is given at every point, and $\Psi = 0$ at an infinite distance, then the value of Ψ at every point is determinate and equal to

$$\Psi' = \frac{1}{4\pi}\iiint\frac{1}{r}\,\nabla^2\Psi\,dx\,dy\,dz, \tag{27}$$

where r is the distance between the element $dx\,dy\,dz$ at which the concentration of Ψ is given $= \nabla^2\Psi$ and the point $x'\,y'\,z'$ at which Ψ' is to be found.

This reduces the theorem to what we deduced from the first definition of Ψ.

But when we consider Ψ as the primary function of x, y, z, from which the others are derived, it is more appropriate to reduce (26) to the form of a surface-integral,

$$A = \iint(l p_{xx} + m p_{xy} + n p_{xz})\,dS, \tag{28}$$

and if we suppose the surface S to be everywhere at a great distance a from the surface s, which includes every point where $\nabla^2\Psi$ differs from zero, then we know that Ψ cannot be numerically greater than e/a, where $4\pi e$ is the volume-integral of $\nabla^2\Psi$, and that R cannot be greater than $d\Psi/da$ or $-e/a^2$, and that the quantities p_{xx}, p_{xy}, p_{xz} can none of them be greater than p or $R^2/8\pi$ or $e^2/8\pi a^4$. Hence the surface-integral taken over a sphere whose radius is very great and equal to a cannot exceed $e^2/2a^2$, and when a is increased without limit, the surface-integral must become ultimately zero.

But this surface-integral is equal to the volume-integral (26), and the value of this volume-integral is the same whatever be the size of the space enclosed within S, provided S encloses every point at which $\nabla^2\Psi$ differs from zero. Hence, since the integral is zero when a is infinite, it must also be zero when the limits of integration are defined by any surface which includes every point at which $\nabla^2\Psi$ differs from zero.

109.] The distribution of stress considered in this chapter is precisely that to which Faraday was led in his investigation of induction through dielectrics. He sums up in the following words:—

'(1297) The direct inductive force, which may be conceived to be exerted in lines between the two limiting and charged conducting surfaces, is accompanied by a lateral or transverse force equivalent to a dilatation or repulsion of these representative lines (1224.); or the attracting force which exists amongst the particles of the dielectric in the direction of the induction is accompanied by a repulsive or a diverging force in the transverse direction.

'(1298) Induction appears to consist in a certain polarized state of the particles, into which they are thrown by the electrified body sustaining the action, the particles assuming positive and negative points or parts, which are symmetrically arranged with respect to each other and the inducting surfaces or particles. The state must be a forced one, for it is originated and sustained only by force, and sinks to the normal or quiescent state when that force is removed. It can be *continued* only in insulators by the same portion of electricity, because they only can retain this state of the particles.'

This is an exact account of the conclusions to which we have been conducted by our mathematical investigation. At every point of the medium there is a state of stress such that there is tension along the lines of force and pressure in all directions at right angles to these lines, the numerical magnitude of the pressure being equal to that of the tension, and both varying as the square of the resultant force at the point.

The expression 'electric tension' has been used in various senses by different writers. I shall always use it to denote the tension along the lines of force, which, as we have seen, varies from point to point, and is always proportional to the square of the resultant force at the point.

110.] The hypothesis that a state of stress of this kind exists in a fluid dielectric, such as air or turpentine, may at first sight appear at variance with the established principle that at any point in a fluid the pressures in all directions are equal. But in the deduction of this principle from a consideration of the mobility and equilibrium of the parts of the fluid it is taken for granted that no action such as that which we here suppose to take place along the lines of force exists in the fluid. The state of stress which we have been studying is perfectly consistent with the mobility and equilibrium of the fluid, for we have seen that, if any portion of the fluid is devoid of electric charge, it experiences no resultant force from the stresses on its surface, however intense these may be. It is only when a portion of the fluid becomes charged that its equilibrium is disturbed by the stresses on its surface, and we know that in this case it actually tends to move. Hence the supposed state of stress is not inconsistent with the equilibrium of a fluid dielectric.

The quantity W, which was investigated in Chapter IV, Art. 99, may be interpreted as the energy in the medium due to the distribution of stress. It appears from the theorems of that chapter that the distribution of stress which satisfies the conditions there given also makes W an absolute minimum. Now when the energy is a minimum for any configuration, that configuration is one of equilibrium, and the equilibrium is stable. Hence the dielectric, when subjected to the inductive action of electrified bodies, will of itself take up a state of stress distributed in the way we have described.

It must be carefully borne in mind that we have made only one step in the theory of the action of the medium. We have supposed it to be in a state of stress, but we have not in any way accounted for this stress, or explained how it is maintained. This step, however, seems to me to be an important one, as it explains, by the action of the consecutive parts of the medium, phenomena which were formerly supposed to be explicable only by direct action at a distance.

111.] I have not been able to make the next step, namely, to account by mechanical considerations for these stresses in the dielectric. I therefore leave the theory at this point, merely stating what are the other parts of the phenomenon of induction in dielectrics.

I. *Electric Displacement.* When induction is transmitted through a dielectric, there is in the first place a displacement of electricity in the direction of the induction. For instance, in a Leyden jar, of which the inner coating is charged positively and the outer coating negatively, the displacement of positive electricity in the substance of the glass is from within outwards.

Any increase of this displacement is equivalent, during the time of increase, to a current of positive electricity from within outwards, and any diminution of the displacement is equivalent to a current in the opposite direction.

The whole quantity of electricity displaced through any area of a surface fixed in the dielectric is measured by the quantity which we have already investigated (Art. 75) as the surface-integral of induction through that area, multiplied by $K/4\pi$, where K is the specific inductive capacity of the dielectric.

II. *Surface charge of the particles of the dielectric.* Conceive any portion of the dielectric, large or small, to be separated (in imagination) from the rest by a closed surface, then we must suppose that on every elementary portion of this surface there is a charge measured by the total displacement of electricity through that element of surface *reckoned inwards.*

In the case of the Leyden jar of which the inner coating is charged positively, any portion of the glass will have its inner side charged positively and its outer side negatively. If this portion be entirely in the interior of the glass, its surface charge will be neutralized by the opposite charge of the parts in contact with it, but if it be in contact with a conducting body, which is incapable of maintaining in itself the inductive state, the surface charge will not be neutralized, but will constitute that apparent charge which is commonly called the Charge of the Conductor.

The charge therefore at the bounding surface of a conductor and the surrounding dielectric, which on the old theory was called the charge of the conductor, must be called in the theory of induction the surface charge of the surrounding dielectric.

According to this theory, all charge is the residual effect of the polarization of the dielectric. This polarization exists throughout the interior of the substance, but it is there neutralized by the juxtaposition of oppositely charged parts, so that it is only at the surface of the dielectric that the effects of the charge become apparent.

The theory completely accounts for the theorem of Art. 77, that the total induction through a closed surface is equal to the total quantity of electricity within the surface multiplied by 4π. For what we have called the induction through the surface is simply the electric displacement multiplied by 4π, and the total displacement outwards is necessarily equal to the total charge within the surface.

The theory also accounts for the impossibility of communicating an 'absolute charge' to matter. For every particle of the dielectric has equal and opposite charges on its opposite sides, if it would not be more correct to say that these charges are only the manifestations of a single phenomenon, which we may call Electric Polarization.

A dielectric medium, when thus polarized, is the seat of electrical energy, and the energy in unit of volume of the medium is numerically equal to the electric tension on unit of area, both quantities being equal to half the product of the displacement and the resultant electromotive intensity, or

$$p = \tfrac{1}{2}\mathfrak{D}\mathfrak{E} = \frac{1}{8\pi}K\mathfrak{E}^2 = \frac{2\pi}{K}\mathfrak{D}^2,$$

where p is the electric tension, $\mathfrak{D}$ the displacement, $\mathfrak{E}$ the electromotive intensity, and K the specific inductive capacity.

If the medium is not a perfect insulator, the state of constraint, which we call electric polarization, is continually giving way. The medium yields to the electromotive force, the electric stress is relaxed, and the potential energy of the state of constraint is converted into heat. The rate at which this decay of the state of polarization takes place depends on the nature of the medium. In some kinds of glass, days or years may elapse before the polarization sinks to half its original value. In copper, a similar change is effected in less than the billionth of a second.

We have supposed the medium after being polarized to be simply left to itself. In the phenomenon called the electric current the constant passage of electricity through the medium tends to restore the state of polarization as fast as the conductivity of the medium allows it to decay. Thus the external agency which maintains the current is always doing work in restoring the polarization of the medium, which is continually becoming relaxed, and the potential energy of this polarization is continually becoming transformed into heat, so that the final result of the energy expended in maintaining the current is to gradually raise the temperature of the conductor, till as much heat is lost by conduction and radiation from its surface as is generated in the same time by the electric current.

CHAPTER VI.

ON POINTS AND LINES OF EQUILIBRIUM.

112.] If at any point of the electric field the resultant force is zero, the point is called a Point of equilibrium.

If every point on a certain line is a point of equilibrium, the line is called a Line of equilibrium.

The conditions that a point shall be a point of equilibrium are that at that point

$$\frac{dV}{dx} = 0, \quad \frac{dV}{dy} = 0, \quad \frac{dV}{dz} = 0.$$

At such a point, therefore, the value of V is a maximum, or a minimum, or is stationary, with respect to variations of the coordinates. The potential, however, can have a maximum or a minimum value only at a point charged with positive or with negative electricity, or throughout a finite space bounded by a surface charged positively or negatively. If, therefore, a point of equilibrium occurs in an uncharged part of the field it must be a stationary point, and not a maximum or a minimum.

In fact, the first condition of a maximum or minimum is that

$$\frac{d^2V}{dx^2}, \quad \frac{d^2V}{dy^2}, \quad \text{and} \quad \frac{d^2V}{dz^2}$$

must be all negative or all positive, if they have finite values.

Now, by Laplace's equation, at a point where there is no charge, the sum of these three quantities is zero, and therefore this condition cannot be satisfied.

Instead of investigating the analytical conditions for the cases in which the components of the force simultaneously vanish, we shall give a general proof by means of the equipotential surfaces.

If at any point, P, there is a true maximum value of V, then, at all other points in the immediate neighbourhood of P, the value of V is less than at P. Hence P will be surrounded by a series of closed equipotential surfaces, each outside the one before it, and at all points of any one of these surfaces the electrical force will be directed outwards. But we have proved, in Art. 76, that the surface-integral of the electromotive intensity taken over any closed surface gives the total charge within that surface multiplied by 4π. Now, in this case the force is everywhere outwards, so that the surface-integral is necessarily positive, and therefore there is positive charge within the surface, and, since we may take the surface as near to P as we please, there is positive charge at the point P.

In the same way we may prove that if V is a minimum at P, then P is negatively charged.

Next, let P be a point of equilibrium in a region devoid of charge, and let us describe a sphere of very small radius round P, then, as we have seen, the potential at this surface cannot be everywhere greater or everywhere less than at P. It must therefore be greater at some parts of the surface and less at others. These portions of the surface are bounded by lines in which the potential is equal to that at P. Along lines drawn from P to points at which the potential is less than that at P the electrical force is from P, and along lines drawn to points of greater potential the force is towards P. Hence the point P is a point of stable equilibrium for some displacements, and of unstable equilibrium for other displacements.

113.] To determine the number of the points and lines of equilibrium, let us consider the surface or surfaces for which the potential is equal to C, a given quantity. Let us call the regions in which the potential is less than C the negative regions, and those in which it is greater than C the positive regions. Let V_0 be the lowest, and V_1 the highest potential existing in the electric field. If we make $C = V_0$, the negative region will include only the point or conductor of lowest potential, and this is necessarily charged negatively. The positive region consists of the rest of space, and since it surrounds the negative region it is periphractic. See Art. 18.

If we now increase the value of C, the negative region will expand, and new negative regions will be formed round negatively charged bodies. For every negative region thus formed the surrounding positive region acquires one degree of periphraxy.

As the different negative regions expand, two or more of them may meet in a point or a line. If $n+1$ negative regions meet, the positive region loses n degrees of periphraxy, and the point or the line in which they meet is a point or line of equilibrium of the nth degree.

When C becomes equal to V_1 the positive region is reduced to the point or the conductor of highest potential, and has therefore lost all its periphraxy. Hence, if each point or line of equilibrium counts for one, two, or n, according to its degree, the number so made up by the points or lines now considered will be less by one than the number of negatively charged bodies.

There are other points or lines of equilibrium which occur where the positive regions become separated from each other, and the negative region acquires periphraxy. The number of these, reckoned according to their degrees, is less by one than the number of positively charged bodies.

If we call a point or line of equilibrium positive when it is the meeting-place of two or more positive regions, and negative when the regions which unite there are negative, then, if there are p bodies positively and n bodies negatively charged, the sum of the degrees of the positive points and lines of equilibrium will be $p-1$, and that of the negative ones $n-1$. The surface which surrounds the electrical system at an infinite distance from it is to be reckoned as a body whose charge is equal and opposite to the sum of the charges of the system.

But, besides this definite number of points and lines of equilibrium arising from the junction of different regions, there may be others, of which we can only affirm that their number must be even. For if, as any one of the negative regions expands, it meets itself, it becomes a cyclic region, and it may acquire, by repeatedly meeting itself, any number of degrees of cyclosis, each of which corresponds to the point or line of equilibrium at which the cyclosis was established. As the negative region continues to expand till it fills all space, it loses every degree of cyclosis it has acquired, and becomes at last acyclic. Hence there is a set of points or lines of equilibrium at which cyclosis is lost, and these are equal in number of degrees to those at which it is acquired.

If the form of the charged bodies or conductors is arbitrary, we can only assert that the number of these additional points or lines is even, but if they are charged points or spherical conductors, the number arising in this way cannot exceed $(n-1)(n-2)$, where n is the number of bodies.

114.] The potential close to any point P may be expanded in the series

$$V = V_0 + H_1 + H_2 + \&c.;$$

where H_1, H_2, &c. are homogeneous functions of x, y, z, whose dimensions are 1, 2, &c. respectively.

Since the first derivatives of V vanish at a point of equilibrium, $H_1 = 0$, if P be a point of equilibrium.

Let H_n be the first function which does not vanish, then close to the point P we may neglect all functions of higher degrees as compared with H_n.

Now
$$H_n = 0$$
is the equation of a cone of the degree n, and this cone is the cone of closest contact with the equipotential surface at P.

It appears, therefore, that the equipotential surface passing through P has, at that point, a conical point touched by a cone of the second or of a higher degree. The intersection of this cone with a sphere whose centre is the vertex is called the Nodal line.

If the point P is not on a line of equilibrium the nodal line does not intersect itself, but consists of n or some smaller number of closed curves.

If the nodal line intersects itself, then the point P is on a line of equilibrium, and the equipotential surface through P cuts itself in that line.

If there are intersections of the nodal line not on opposite points of the sphere, then P is at the intersection of three or more lines of equilibrium. For the equipotential surface through P must cut itself in each line of equilibrium.

115.] If two sheets of the same equipotential surface intersect, they must intersect at right angles.

For let the tangent to the line of intersection be taken as the axis of z, then $d^2V/dz^2 = 0$. Also let the axis of x be a tangent to one of the sheets, then $d^2V/dx^2 = 0$. It follows from this, by Laplace's equation, that $d^2V/dy^2 = 0$, or the axis of y is a tangent to the other sheet.

This investigation assumes that H_2 is finite. If H_2 vanishes, let the tangent to the line of intersection be taken as the axis of z, and let $x = r\cos\theta$, and $y = r\sin\theta$, then, since
$$\frac{d^2V}{dz^2} = 0, \quad \frac{d^2V}{dx^2} + \frac{d^2V}{dy^2} = 0;$$
or
$$\frac{d^2V}{dr^2} + \frac{1}{r}\frac{dV}{dr} + \frac{1}{r^2}\frac{d^2V}{d\theta^2} = 0;$$
the solution of which equation in ascending powers of r is
$$V = V_0 + A_1 r\cos(\theta + a) + A_2 r^2 \cos(2\theta + a_2) + \&c. + A_n r^n \cos(n\theta + a_n).$$
At a point of equilibrium A_1 is zero. If the first term that does not vanish is that in r^n, then
$$V - V_0 = A_n r^n \cos(n\theta + a_n) + \text{terms in higher powers of } r.$$

This equation shews that n sheets of the equipotential surface $V=V_0$ intersect at angles each equal to π/n. This theorem was given by Rankine*.

It is only under certain conditions that a line of equilibrium can exist in free space, but there must be a line of equilibrium on the surface of a conductor whenever the surface density of the conductor is positive in one portion and negative in another.

In order that a conductor may be charged oppositely on different portions of its surface, there must be in the field some places where the potential is higher than that of the body and others where it is lower.

Let us begin with two conductors electrified positively to the same potential. There will be a point of equilibrium between the two bodies. Let the potential of the first body be gradually diminished. The point of equilibrium will approach it, and, at a certain stage of the process, will coincide with a point on its surface. During the next stage of the process, the equipotential surface round the second body which has the same potential as the first body will cut the surface of the second body at right angles in a closed curve, which is a line of equilibrium. This closed curve, after sweeping over the entire surface of the conductor, will again contract to a point; and then the point of equilibrium will move off on the other side of the first body, and will be at an infinite distance when the charges of the two bodies are equal and opposite.

Earnshaw's Theorem.

116.] A charged body placed in a field of electric force cannot be in stable equilibrium.

First, let us suppose the electricity of the moveable body (A), and also that of the system of surrounding bodies (B), to be fixed in those bodies.

Let V be the potential at any point of the moveable body due to the action of the surrounding bodies (B), and let e be the electricity on a small portion of the moveable body A surrounding this point. Then the potential energy of A with respect to B will be

$$M = \Sigma\,(Ve),$$

where the summation is to be extended to every charged portion of A.

* 'Summary of the Properties of certain Stream Lines,' *Phil. Mag.*, Oct. 1864. See also, Thomson and Tait's *Natural Philosophy*, § 780; and Rankine and Stokes, in the *Proc. R. S.*, 1867, p. 468; also W. R. Smith, *Proc. R. S. Edin.*, 1869-70, p. 79.

Let a, b, c be the coordinates of any charged part of A with respect to axes fixed in A, and parallel to those of x, y, z. Let the absolute coordinates of the origin of these axes be ξ, η, ζ.

Let us suppose for the present that the body A is constrained to move parallel to itself, then the absolute coordinates of the point a, b, c will be

$$x = \xi + a, \qquad y = \eta + b, \qquad z = \zeta + c.$$

The potential of the body A with respect to B may now be expressed as the sum of a number of terms, in each of which V is expressed in terms of a, b, c and ξ, η, ζ and the sum of these terms is a function of the quantities a, b, c, which are constant for each point of the body, and of ξ, η, ζ which vary when the body is moved.

Since Laplace's equation is satisfied by each of these terms it is satisfied by their sum, or

$$\frac{d^2M}{d\xi^2} + \frac{d^2M}{d\eta^2} + \frac{d^2M}{d\zeta^2} = 0.$$

Now let a small displacement be given to A, so that

$$d\xi = l\,dr, \qquad d\eta = m\,dr, \qquad d\zeta = n\,dr;$$

and let dM be the increment of the potential of A with respect to the surrounding system B.

If this be positive, work will have to be done to increase r, and there will be a force $R = dM/dr$ tending to diminish r and to restore A to its former position, and for this displacement therefore the equilibrium will be stable. If, on the other hand, this quantity is negative, the force will tend to increase r, and the equilibrium will be unstable.

Now consider a sphere whose centre is the origin and whose radius is r, and so small that when the point fixed in the body lies within this sphere no part of the moveable body A can coincide with any part of the external system B. Then, since within the sphere $\nabla^2 M = 0$, the surface-integral

$$\iint \frac{dM}{dr}\,dS,$$

taken over the surface of the sphere, is zero.

Hence, if at any part of the surface of the sphere dM/dr is positive, there must be some other part of the surface where it is negative, and if the body A be displaced in a direction in which dM/dr is negative, it will tend to move from its original position, and its equilibrium is therefore necessarily unstable.

The body therefore is unstable even when constrained to move parallel to itself, and *à fortiori* it is unstable when altogether free.

Now let us suppose that the body A is a conductor. We might treat this as a case of equilibrium of a system of bodies, the moveable electricity being considered as part of that system, and we might argue that as the system is unstable when deprived of so many degrees of freedom by the fixture of its electricity, it must *à fortiori* be unstable when this freedom is restored to it.

But we may consider this case in a more particular way, thus—

First, let the electricity be fixed in A, and let A move through the small distance dr. The increment of the potential of A due to this cause has been already considered.

Next, let the electricity be allowed to move within A into its position of equilibrium, which is always stable. During this motion the potential will necessarily be *diminished* by a quantity which we may call Cdr.

Hence the total increment of the potential when the electricity is free to move will be

$$\left(\frac{dM}{dr} - C\right)dr;$$

and the force tending to bring A back towards its original position will be

$$\frac{dM}{dr} - C,$$

where C is always positive.

Now we have shewn that dM/dr is negative for certain directions of r, hence when the electricity is free to move the instability in these directions will be increased.

CHAPTER VII.

FORMS OF THE EQUIPOTENTIAL SURFACES AND LINES OF INDUCTION IN SIMPLE CASES.

117.] We have seen that the determination of the distribution of electricity on the surface of conductors may be made to depend on the solution of Laplace's equation

$$\frac{d^2V}{dx^2} + \frac{d^2V}{dy^2} + \frac{d^2V}{dz^2} = 0,$$

V being a function of x, y, and z, which is always finite and continuous, which vanishes at an infinite distance, and which has a given constant value at the surface of each conductor.

It is not in general possible by known mathematical methods to solve this equation so as to fulfil arbitrarily given conditions, but it is easy to write down any number of expressions for the function V which shall satisfy the equation, and to determine in each case the forms of the conducting surfaces, so that the function V shall be the true solution.

It appears, therefore, that what we should naturally call the inverse problem of determining the forms of the conductors when the expression for the potential is given is more manageable than the direct problem of determining the potential when the form of the conductors is given.

In fact, every electrical problem of which we know the solution has been constructed by this inverse process. It is therefore of great importance to the electrician that he should know what results have been obtained in this way, since the only method by which he can expect to solve a new problem is by reducing it to one of the cases in which a similar problem has been constructed by the inverse process.

This historical knowledge of results can be turned to account in two ways. If we are required to devise an instrument for making electrical measurements with the greatest accuracy, we may select those forms for the electrified surfaces which correspond to cases of which we know the accurate solution. If, on the other hand, we are required to estimate what will be the electrification of bodies whose forms are given, we may begin with some case in which one of the equipotential surfaces takes a form somewhat resembling the given form, and then by a tentative method we may modify the problem till it more nearly corresponds to the given case. This method is evidently very imperfect considered from a mathematical point of view, but it is the only one we have, and if we are not allowed to choose our conditions, we can make only an approximate calculation of the electrification. It appears, therefore, that what we want is a knowledge of the forms of equipotential surfaces and lines of induction in as many different cases as we can collect together and remember. In certain classes of cases, such as those relating to spheres, there are known mathematical methods by which we may proceed. In other cases we cannot afford to despise the humbler method of actually drawing tentative figures on paper, and selecting that which appears least unlike the figure we require.

This latter method I think may be of some use, even in cases in which the exact solution has been obtained, for I find that an eye-knowledge of the forms of the equipotential surfaces often leads to a right selection of a mathematical method of solution.

I have therefore drawn several diagrams of systems of equipotential surfaces and lines of induction, so that the student may make himself familiar with the forms of the lines. The methods by which such diagrams may be drawn will be explained in Art. 123.

118.] In the first figure at the end of this volume we have the sections of the equipotential surfaces surrounding two points charged with quantities of electricity of the same kind and in the ratio of 20 to 5.

Here each point is surrounded by a system of equipotential surfaces which become more nearly spheres as they become smaller, though none of them are accurately spheres. If two of these surfaces, one surrounding each point, be taken to represent the surfaces of two conducting bodies, nearly but not quite spherical, and if these bodies be charged with the same kind of electricity, the charges being as 4 to 1, then the diagram will represent the equipotential surfaces, provided we expunge all those which are drawn inside the two bodies. It appears from the diagram that the action between the bodies will be the same as that between two points having the same charges, these points being not exactly in the middle of the axis of each body, but each somewhat more remote than the middle point from the other body.

The same diagram enables us to see what will be the distribution of electricity on one of the oval figures, larger at one end than the other, which surround both centres. Such a body, if charged with 25 units of electricity and free from external influence, will have the surface-density greatest at the small end, less at the large end, and least in a circle somewhat nearer the smaller than the larger end.

There is one equipotential surface, indicated by a dotted line, which consists of two lobes meeting at the conical point *P*. That point is a point of equilibrium, and the surface-density on a body of the form of this surface would be zero at this point.

The lines of force in this case form two distinct systems, divided from one another by a surface of the sixth degree, indicated by a dotted line, passing through the point of equilibrium, and somewhat resembling one sheet of the hyperboloid of two sheets.

This diagram may also be taken to represent the lines of force and equipotential surfaces belonging to two spheres of gravitating matter whose masses are as 4 to 1.

119.] In the second figure we have again two points whose charges are as 20 to 5, but the one positive and the other negative. In this case one of the equipotential surfaces, that, namely, corresponding to potential zero, is a sphere. It is marked in the diagram by the dotted circle *Q*. The importance of this spherical surface will be seen when we come to the theory of Electrical Images.

We may see from this diagram that if two round bodies are charged with opposite kinds of electricity they will attract each other as much as two points having the same charges but placed somewhat nearer together than the middle points of the round bodies.

Here, again, one of the equipotential surfaces, indicated by a dotted line, has two lobes, an inner one surrounding the point whose charge is 5 and an outer one surrounding both bodies, the two lobes meeting in a conical point *P* which is a point of equilibrium.

If the surface of a conductor is of the form of the outer lobe, a roundish body having, like an apple, a conical dimple at one end of its axis, then, if this conductor be electrified, we shall be able to determine the surface-density at any point. That at the bottom of the dimple will be zero.

Surrounding this surface we have others having a rounded dimple which flattens and finally disappears in the equipotential surface passing through the point marked *M*.

The lines of force in this diagram form two systems divided by a surface which passes through the point of equilibrium.

If we consider points on the axis on the further side of the point B, we find that the resultant force diminishes to the double point P, where it vanishes. It then changes sign, and reaches a maximum at M, after which it continually diminishes.

This maximum, however, is only a maximum relatively to other points on the axis, for if we consider a surface through M perpendicular to the axis, M is a point of minimum force relatively to neighbouring points on that surface.

120.] Figure III represents the equipotential surfaces and lines of induction due to a point whose charge is 10 placed at A, and surrounded by a field of force, which, before the introduction of the charged point, was uniform in direction and magnitude at every part.

The equipotential surfaces have each of them an asymptotic plane. One of them, indicated by a dotted line, has a conical point and a lobe surrounding the point A. Those below this surface have one sheet with a depression near the axis. Those above have a closed portion surrounding A and a separate sheet with a slight depression near the axis.

If we take one of the surfaces below A as the surface of a conductor, and another a long way below A as the surface of another conductor at a different potential, the system of lines and surfaces between the two conductors will indicate the distribution of electric force. If the lower conductor is very far from A its surface will be very nearly plane, so that we have here the solution of the distribution of electricity on two surfaces, both of them nearly plane and parallel to each other, except that the upper one has a protuberance near its middle point, which is more or less prominent according to the particular equipotential surface we choose.

121.] Figure IV represents the equipotential surfaces and lines of induction due to three points A, B and C, the charge of A being 15 units of positive electricity, that of B 12 units of negative electricity, and that of C 20 units of positive electricity. These points are placed in one straight line, so that

$$AB = 9, \quad BC = 16, \quad AC = 25.$$

In this case, the surface for which the potential is zero consists of two spheres whose centres are A and C and their radii 15 and 20. These spheres intersect in the circle which cuts the plane of the paper at right angles in D and D', so that B is the centre of this circle and its radius is 12. This circle is an example of a line of equilibrium, for the resultant force vanishes at every point of this line.

If we suppose the sphere whose centre is A to be a conductor with a charge of 3 units of positive electricity, and placed under the influence of 20 units of positive electricity at C, the state of the case will be represented by the diagram if we leave out all the lines within the sphere A. The part of this spherical surface within the small circle DD' will be negatively charged by the influence of C. All the rest of the sphere will be positively charged, and the small circle DD' itself will be a line of no charge.

We may also consider the diagram to represent the sphere whose centre is C, charged with 8 units of positive electricity, and influenced by 15 units of positive electricity placed at A.

The diagram may also be taken to represent a conductor whose surface consists of the larger segments of the two spheres meeting in DD', charged with 23 units of positive electricity.

We shall return to the consideration of this diagram as an illustration of Thomson's *Theory of Electrical Images*. See Art. 168.

122.] These diagrams should be studied as illustrations of the language of Faraday in speaking of 'lines of force,' the 'forces of an electrified body,' &c.

The word Force denotes a restricted aspect of that action between two material bodies by which their motions are rendered different from what they would have been in the absence of that action. The whole phenomenon, when both bodies are contemplated at once, is called Stress, and may be described as a transference of momentum from one body to the other. When we restrict our attention to the first of the two bodies, we call the stress acting on it the Moving Force, or simply the Force on that body, and it is measured by the momentum which that body is receiving per unit of time.

The mechanical action between two charged bodies is a stress, and that on one of them is a force. The force on a small charged body is proportional to its own charge, and the force per unit of charge is called the Intensity of the force.

The word Induction was employed by Faraday to denote the mode in which the charges of electrified bodies are related to each other, every unit of positive charge being connected with a unit of negative charge by a line, the direction of which, in fluid dielectrics, coincides at every part of its course with that of the electric intensity. Such a line is often called a line of Force, but it is more correct to call it a line of Induction.

Now the quantity of electricity in a body is measured, according to Faraday's ideas, by the *number* of lines of force, or rather of induction, which proceed from it. These lines of force must all terminate somewhere, either on bodies in the neighbourhood, or on the walls and roof of the room, or on the earth, or on the heavenly bodies, and wherever they terminate there is a quantity of electricity exactly equal and opposite to that on the part of the body from which they proceeded. By examining the diagrams this will be seen to be the case. There is therefore no contradiction between Faraday's views and the mathematical results of the old theory, but, on the contrary, the idea of lines of force throws great light on these results, and seems to afford the means of rising by a continuous process from the somewhat rigid conceptions of the old theory to notions which may be capable of greater expansion, so as to provide room for the increase of our knowledge by further researches.

123.] These diagrams are constructed in the following manner:—

First, take the case of a single centre of force, a small electrified body with a charge e. The potential at a distance r is $V = e/r$; hence, if we make $r = e/V$, we shall find r, the radius of the sphere for which the potential is V. If we now give to V the values 1, 2, 3, &c., and draw the corresponding spheres, we shall obtain a series of equipotential surfaces, the potentials corresponding to which are measured by the natural numbers. The sections of these spheres by a plane passing through their common centre will be circles, each of which we may mark with the number denoting its potential. These are indicated by the dotted semi-circles on the right hand of Fig. 6.

If there be another centre of force, we may in the same way draw the equipotential surfaces belonging to it, and if we now wish to find the form of the equipotential surfaces due to both centres together, we must remember that if V_1 be the potential due to one centre, and V_2 that due to the other, the potential due to both will be $V_1 + V_2 = V$. Hence, since at every intersection of the equipotential surfaces belonging to the two series we know both V_1 and V_2, we also know the value of V. If therefore we draw a surface which passes through all those intersections for which the value of V is the same, this surface will coincide with a true equipotential surface at all these intersections, and if the original systems of surfaces are drawn sufficiently close, the new surface may be drawn with any required degree of accuracy. The equipotential surfaces due to two points whose charges are equal and opposite are represented by the continuous lines on the right hand side of Fig. 6.

This method may be applied to the drawing of any system of equipotential surfaces when the potential is the sum of two potentials, for which we have already drawn the equipotential surfaces.

The lines of force due to a single centre of force are straight lines radiating from that centre. If we wish to indicate by these lines the intensity as well as the direction of the force at any point, we must draw them so that they mark out on the equipotential surfaces portions over which the surface-integral of induction has definite values. The best way of doing this is to suppose our plane figure to be the section of a figure in space formed by the revolution of the plane figure about an axis passing through the centre of force. Any straight line radiating from the centre and making an angle θ with the axis will then trace out a cone, and the surface-integral of the induction through that part of any surface which is cut off by this cone on the side next the positive direction of the axis is $2\pi e\,(1-\cos\theta)$.

If we further suppose this surface to be bounded by its intersection with two planes passing through the axis, and inclined at the angle whose arc is equal to half the radius, then the induction through the surface so bounded is

$$e\,(1-\cos\theta) = 2\,\Phi, \text{ say;}$$

$$\text{and} \quad \theta = \cos^{-1}\left(1-2\,\frac{\Phi}{e}\right).$$

If we now give to Φ a series of values 1, 2, 3 ... e, we shall find a corresponding series of values of θ, and if e be an integer, the number of corresponding lines of force, including the axis, will be equal to e.

We have thus a method of drawing lines of force so that the charge of any centre is indicated by the number of lines which diverge from it, and the induction through any surface cut off in the way described is measured by the number of lines of force which pass through it. The dotted straight lines on the left hand side of Fig. 6 represent the lines of force due to each of two electrified points whose charges are 10 and -10 respectively.

If there are two centres of force on the axis of the figure we may draw the lines of force for each axis corresponding to values

Fig 6.

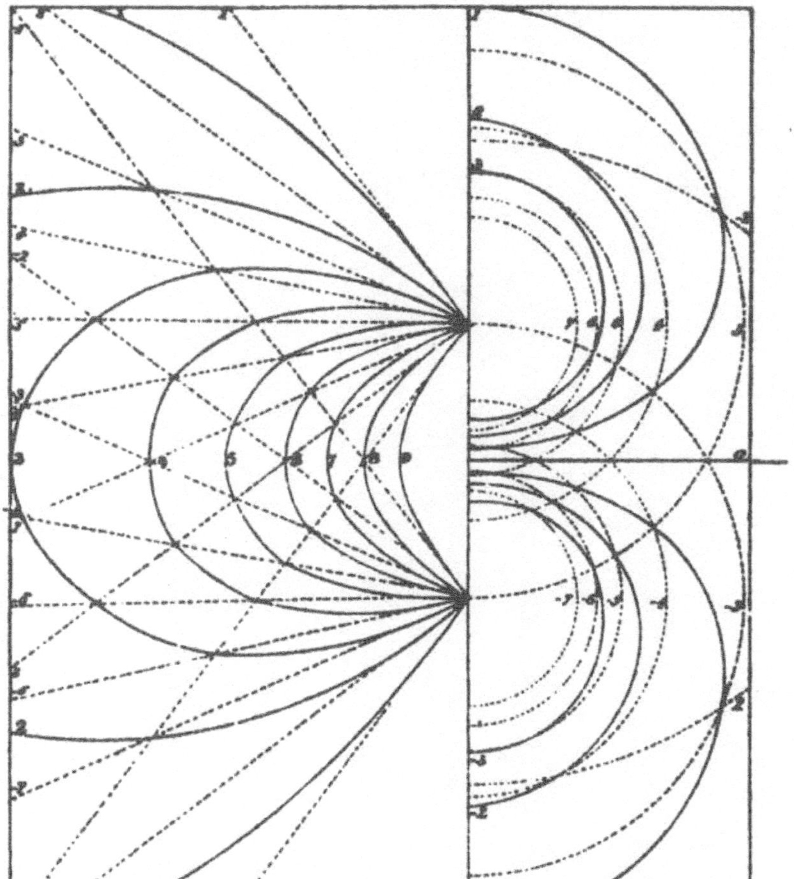

Method of drawing

Lines of Force and Equipotential Surfaces.

of Φ_1 and Φ_2, and then, by drawing lines through the consecutive intersections of these lines for which the value of $\Phi_1 + \Phi_2$ is the same, we may find the lines of force due to both centres, and in the same way we may combine any two systems of lines of force which are symmetrically situated about the same axis. The continuous curves on the left hand side of Fig. 6 represent the lines of force due to the two charged points acting at once.

After the equipotential surfaces and lines of force have been constructed by this method the accuracy of the drawing may be tested by observing whether the two systems of lines are everywhere orthogonal, and whether the distance between consecutive equipotential surfaces is to the distance between consecutive lines of force as half the mean distance from the axis is to the assumed unit of length.

In the case of any such system of finite dimensions the line of force whose index number is Φ has an asymptote which passes through the electric centre (Art. 89 *d*) of the system, and is inclined to the axis at an angle whose cosine is $1 - 2\,\Phi/e$, where e is the total electrification of the system, provided Φ is less than e. Lines of force whose index is greater than e are finite lines. If e is zero, they are all finite.

The lines of force corresponding to a field of uniform force parallel to the axis are lines parallel to the axis, the distances from the axis being the square roots of an arithmetical series.

The theory of equipotential surfaces and lines of force in two dimensions will be given when we come to the theory of conjugate functions*.

* See a paper 'On the Flow of Electricity in Conducting Surfaces,' by Prof. W. R. Smith, *Proc. R. S. Edin.*, 1869–70, p. 79.

CHAPTER VIII.

SIMPLE CASES OF ELECTRIFICATION.

Two Parallel Planes.

124.] WE shall consider, in the first place, two parallel plane conducting surfaces of infinite extent, at a distance c from each other, maintained respectively at potentials A and B.

It is manifest that in this case the potential V will be a function of the distance z from the plane A, and will be the same for all points of any parallel plane between A and B, except near the boundaries of the electrified surfaces, which by the supposition are at an infinitely great distance from the point considered.

Hence, Laplace's equation becomes reduced to

$$\frac{d^2 V}{dz^2} = 0,$$

the integral of which is

$$V = C_1 + C_2 z;$$

and since when $z = 0$; $V = A$, and when $z = c$, $V = B$,

$$V = A + (B - A)\frac{z}{c}.$$

For all points between the planes, the resultant intensity is normal to the planes, and its magnitude is

$$R = \frac{A - B}{c}.$$

In the substance of the conductors themselves, $R = 0$. Hence the distribution of electricity on the first plane has a surface-density σ, where

$$4\pi\sigma = R = \frac{A - B}{c}.$$

On the other surface, where the potential is B, the surface-density σ' will be equal and opposite to σ, and

$$4\pi\sigma' = -R = \frac{B - A}{c}.$$

Let us next consider a portion of the first surface whose area is S, taken so that no part of S is near the boundary of the surface.

The quantity of electricity on this surface is $e_1 = S\sigma$, and, by Art. 79, the force acting on every unit of electricity is $\frac{1}{2}R$, so that the whole force acting on the area S, and attracting it towards the other plane, is

$$F = \tfrac{1}{2} R S \sigma = \frac{1}{8\pi} R^2 S = \frac{S}{8\pi} \frac{(B-A)^2}{c^2}.$$

Here the attraction is expressed in terms of the area S, the difference of potentials of the two surfaces $(A-B)$, and the distance between them c. The attraction, expressed in terms of the charge e_1, on the area S, is

$$F = \frac{2\pi}{S} e_1^2.$$

The electrical energy due to the distribution of electricity on the area S, and that on the corresponding area S' on the surface B defined by projecting S on the surface B by a system of lines of force, which in this case are normals to the planes, is

$$\begin{aligned} W &= \tfrac{1}{2}(e_1 A + e_2 B), \\ &= \tfrac{1}{2} \frac{S}{4\pi} \frac{(A-B)^2}{c}, \\ &= \frac{R^2}{8\pi} Sc, \\ &= \frac{2\pi}{S} e_1^2 c, \\ &= Fc. \end{aligned}$$

The first of these expressions is the general expression of electrical energy (Art. 84).

The second gives the energy in terms of the area, the distance, and the difference of potentials.

The third gives it in terms of the resultant force R, and the volume Sc included between the areas S and S', and shews that the energy in unit of volume is p where $8\pi p = R^2$.

The attraction between the planes is pS, or in other words, there is an electrical tension (or negative pressure) equal to p on every unit of area.

The fourth expression gives the energy in terms of the charge.

The fifth shews that the electrical energy is equal to the work which would be done by the electric force if the two surfaces were to be brought together, moving parallel to themselves, with their electric charges constant.

To express the charge in terms of the difference of potentials, we have

$$e_1 = \frac{1}{4\pi}\frac{S}{c}(A-B) = q(A-B).$$

The coefficient q represents the charge due to a difference of potentials equal to unity. This coefficient is called the Capacity of the surface S, due to its position relatively to the opposite surface.

Let us now suppose that the medium between the two surfaces is no longer air but some other dielectric substance whose specific inductive capacity is K, then the charge due to a given difference of potentials will be K times as great as when the dielectric is air, or

$$e_1 = \frac{KS}{4\pi c}(A-B).$$

The total energy will be

$$W = \frac{KS}{8\pi c}(A-B)^2,$$

$$= \frac{2\pi}{KS}e_1^2 c.$$

The force between the surfaces will be

$$F = pS = \frac{KS}{8\pi}\frac{(A-B)^2}{c^2},$$

$$= \frac{2\pi}{KS}e_1^2.$$

Hence the force between two surfaces kept at given potentials varies directly as K, the specific capacity of the dielectric, but the force between two surfaces charged with given quantities of electricity varies inversely as K.

Two Concentric Spherical Surfaces.

125.] Let two concentric spherical surfaces of radii a and b, of which b is the greater, be maintained at potentials A and B respectively, then it is manifest that the potential V is a function of r the distance from the centre. In this case, Laplace's equation becomes

$$\frac{d^2V}{dr^2} + \frac{2}{r}\frac{dV}{dr} = 0.$$

The solution of this is

$$V = C_1 + C_2 r^{-1};$$

and the condition that $V=A$ when $r=a$, and $V=B$ when $r=b$, gives for the space between the spherical surfaces,

$$V = \frac{Aa - Bb}{a - b} + \frac{A - B}{a^{-1} - b^{-1}} r^{-1};$$

$$R = -\frac{dV}{dr} = \frac{A - B}{a^{-1} - b^{-1}} r^{-2}.$$

If σ_1, σ_2 are the surface-densities on the opposed surfaces of a solid sphere of radius a, and a spherical hollow of radius b, then

$$\sigma_1 = \frac{1}{4\pi a^2} \frac{A - B}{a^{-1} - b^{-1}}, \qquad \sigma_2 = \frac{1}{4\pi b^2} \frac{B - A}{a^{-1} - b^{-1}}.$$

If e_1 and e_2 are the whole charges of electricity on these surfaces,

$$e_1 = 4\pi a^2 \sigma_1 = \frac{A - B}{a^{-1} - b^{-1}} = -e_2.$$

The capacity of the enclosed sphere is therefore $\dfrac{ab}{b - a}$.

If the outer surface of the shell be also spherical and of radius c, then, if there are no other conductors in the neighbourhood, the charge on the outer surface is

$$e_3 = Bc.$$

Hence the whole charge on the inner sphere is

$$e_1 = \frac{ab}{b - a}(A - B),$$

and that of the outer shell

$$e_2 + e_3 = \frac{ab}{b - a}(B - A) + Bc.$$

If we put $b = \infty$, we have the case of a sphere in an infinite space. The electric capacity of such a sphere is a, or it is numerically equal to its radius.

The electric tension on the inner sphere per unit of area is

$$p = \frac{1}{8\pi} \frac{b^2}{a^2} \frac{(A - B)^2}{(b - a)^2},$$

The resultant of this tension over a hemisphere is $\pi a^2 p = F$ normal to the base of the hemisphere, and if this is balanced by a surface tension exerted across the circular boundary of the hemisphere, the tension on unit of length being T, we have

$$F = 2\pi a T.$$

Hence

$$F = \frac{b^2}{8} \frac{(A - B)^2}{(b - a)^2} = \frac{e_1^2}{8a^2},$$

$$T = \frac{b^2}{16\pi a} \frac{(A - B)^2}{(b - a)^2}.$$

If a spherical soap bubble is electrified to a potential A, then, if its radius is a, the charge will be Aa, and the surface-density will be

$$\sigma = \frac{1}{4\pi}\frac{A}{a}.$$

The resultant intensity just outside the surface will be $4\pi\sigma$, and inside the bubble it is zero, so that by Art. 79 the electrical force on unit of area of the surface will be $2\pi\sigma^2$, acting outwards. Hence the electrification will diminish the pressure of the air within the bubble by $2\pi\sigma^2$, or

$$\frac{1}{8\pi}\frac{A^2}{a^2}.$$

But it may be shewn that if T_0 is the tension which the liquid film exerts across a line of unit length, then the pressure from within required to keep the bubble from collapsing is $2T_0/a$. If the electrical force is just sufficient to keep the bubble in equilibrium when the air within and without is at the same pressure,

$$A^2 = 16\pi a T_0.$$

Two Infinite Coaxal Cylindric Surfaces.

126.] Let the radius of the outer surface of a conducting cylinder be a, and let the radius of the inner surface of a hollow cylinder, having the same axis with the first, be b. Let their potentials be A and B respectively. Then, since the potential V is in this case a function of r, the distance from the axis, Laplace's equation becomes

$$\frac{d^2V}{dr^2} + \frac{1}{r}\frac{dV}{dr} = 0,$$

whence

$$V = C_1 + C_2 \log r.$$

Since $V = A$ when $r = a$, and $V = B$ when $r = b$,

$$V = \frac{A\log\frac{b}{r} + B\log\frac{r}{a}}{\log\frac{b}{a}}.$$

If σ_1, σ_2 are the surface-densities on the inner and outer surfaces,

$$4\pi\sigma_1 = \frac{A-B}{a\log\frac{b}{a}}, \qquad 4\pi\sigma_2 = \frac{B-A}{b\log\frac{b}{a}}.$$

If e_1 and e_2 are the charges on the portions of the two cylinders between two sections transverse to the axis at a distance l from each other,

$$e_1 = 2\pi a l \sigma_1 = \frac{1}{2}\frac{A-B}{\log\frac{b}{a}} l = -e_2.$$

The capacity of a length l of the interior cylinder is therefore

$$\frac{1}{2}\frac{l}{\log\frac{b}{a}}.$$

If the space between the cylinders is occupied by a dielectric of specific capacity K instead of air, then the capacity of the inner cylinder is

$$\frac{1}{2}\frac{lK}{\log\frac{b}{a}}.$$

The energy of the electrical distribution on the part of the infinite cylinder which we have considered is

$$\frac{1}{4}\frac{lK(A-B)^2}{\log\frac{b}{a}}.$$

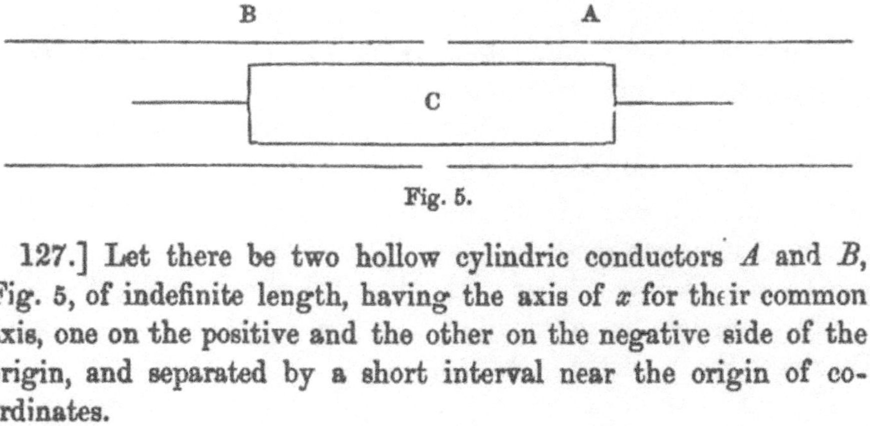

Fig. 5.

127.] Let there be two hollow cylindric conductors A and B, Fig. 5, of indefinite length, having the axis of x for their common axis, one on the positive and the other on the negative side of the origin, and separated by a short interval near the origin of co-ordinates.

Let a hollow cylinder C of length $2l$ be placed with its middle point at a distance x on the positive side of the origin, so as to extend into both the hollow cylinders.

Let the potential of the positive hollow cylinder be A, that of the negative one B, and that of the internal one C, and let us put α for the capacity per unit of length of C with respect to A, and β for the same quantity with respect to B.

The surface densities of the parts of the cylinders at fixed points near the origin and at points at given small distances from the ends of the inner cylinder will not be affected by the value of x provided a considerable length of the inner cylinder enters each of the hollow cylinders. Near the ends of the hollow cylinders, and near the ends of the inner cylinder, there will be distributions of electricity which we are not yet able to calculate, but the distribution near the origin will not be altered by the motion of the inner cylinder provided neither of its ends comes near the origin, and the distributions at the ends of the inner cylinder will move with it, so that the only effect of the motion will be to increase or diminish the length of those parts of the inner cylinder where the distribution is similar to that on an infinite cylinder.

Hence the whole energy of the system will be, so far as it depends on x,

$$Q = \tfrac{1}{2}\alpha(l+x)(C-A)^2 + \tfrac{1}{2}\beta(l-x)(C-B)^2 + \text{quantities independent of } x;$$

and the resultant force parallel to the axis of the cylinders will be

$$X = \frac{dQ}{dx} = \tfrac{1}{2}\alpha(C-A)^2 - \tfrac{1}{2}\beta(C-B)^2.$$

If the cylinders A and B are of equal section, $\alpha = \beta$, and

$$X = \alpha(B-A)(C - \tfrac{1}{2}(A+B)).$$

It appears, therefore, that there is a constant force acting on the inner cylinder tending to draw it into that one of the outer cylinders from which its potential differs most.

If C be numerically large and $A+B$ comparatively small, then the force is approximately $X = \alpha(B-A)C$;

so that the difference of the potentials of the two cylinders can be measured if we can measure X, and the delicacy of the measurement will be increased by raising C, the potential of the inner cylinder.

This principle in a modified form is adopted in Thomson's Quadrant Electrometer, Art. 219.

The same arrangement of three cylinders may be used as a measure of capacity by connecting B and C. If the potential of A is zero, and that of B and C is V, then the quantity of electricity on A will be

$$E_3 = (q_{13} + \alpha(l+x))V;$$

so that by moving C to the right till x becomes $x+\xi$ the capacity of the cylinder C becomes increased by the definite quantity $\alpha\xi$, where

$$\alpha = \frac{1}{2\log\frac{b}{a}},$$

a and b being the radii of the opposed cylindric surfaces.

CHAPTER IX.

SPHERICAL HARMONICS.

128.] The mathematical theory of spherical harmonics has been made the subject of several special treatises. The *Handbuch der Kugelfunctionen* of Dr. E. Heine, which is the most elaborate work on the subject, has now (1878) reached a second edition in two volumes, and Dr. F. Neumann has published his *Beiträge zur Theorie der Kugelfunctionen* (Leipzig, Teubner, 1878). The treatment of the subject in Thomson and Tait's *Natural Philosophy* is considerably improved in the second edition (1879), and Mr. Todhunter's *Elementary Treatise on Laplace's Functions, Lamé's Functions, and Bessel's Functions*, together with Mr. Ferrers' *Elementary Treatise on Spherical Harmonics and subjects connected with them*, have rendered it unnecessary to devote much space in a book on electricity to the purely mathematical development of the subject.

I have retained however the specification of a spherical harmonic in terms of its poles.

On Singular Points at which the Potential becomes Infinite.

129 *a*] If a charge, A_0, of electricity is uniformly spread over the surface of a sphere the coordinates of whose centre are (a, b, c) the potential at any point (x, y, z) outside the sphere is, by Art. 125,

$$V = \frac{A_0}{r}, \tag{1}$$

where $$r^2 = (x-a)^2 + (y-b)^2 + (z-c)^2. \tag{2}$$

As the expression for V is independent of the radius of the sphere, the form of the expression will be the same if we suppose the radius infinitely small. The physical interpretation of the expression would be that the charge A_0 is placed on the surface of an infinitely small sphere, which is sensibly the same as a mathematical point. We have already (Arts. 55, 81) shewn that there is a limit to the surface-density of electricity, so that it is physically impossible to place a finite charge of electricity on a sphere of less than a certain radius.

Nevertheless as the equation (1) represents a possible distribution of potential in the space surrounding a sphere, we may for mathematical purposes treat it as if it arose from a charge A_0 condensed at the mathematical point (a, b, c) and we may call the point an infinite point of order zero.

There are other kinds of singular points, the properties of which we shall now investigate, but before doing so we must define certain expressions which we shall find useful in dealing with directions in space, and with the points on a sphere which correspond to them.

129 *b*.] An *axis* is any definite direction in space. We may suppose it defined by a mark made on the surface of a sphere at the point where the radius drawn *from* the centre in the direction of the axis meets the surface. This point is called the Pole of the axis. An axis has therefore one pole only, not two.

If μ is the cosine of the angle between the axis h and any vector r, and if

$$p = \mu r, \tag{3}$$

p is the resolved part of r in the direction of the axis h.

Different axes are distinguished by different suffixes, and the cosine of the angle between two axes is denoted by λ_{mn}, where m and n are the suffixes specifying the axes.

Differentiation with respect to an axis, h, whose direction cosines are L, M, N, is denoted by

$$\frac{d}{dh} = L\frac{d}{dx} + M\frac{d}{dy} + N\frac{d}{dz}. \tag{4}$$

From these definitions it is evident that

$$\frac{dr}{dh_m} = \frac{p_m}{r} = \mu_m, \tag{5}$$

$$\frac{dp_n}{dh_m} = \lambda_{mn} = \frac{dp_m}{dh_n}, \tag{6}$$

$$\frac{d\mu_m}{dh_n} = \frac{\lambda_{mn} - \mu_m \mu_n}{r}. \tag{7}$$

If we now suppose that the potential at the point (x, y, z) due to a singular point of any order placed at the origin is

$$Af(x, y, z),$$

then if such a point be placed at the extremity of the axis h, the potential at (x, y, z) will be

$$Af[(x-Lh), \quad (y-Mh), \quad (z-Nh)],$$

and if a point in all respects the same, except that the sign of A is reversed, be placed at the origin, the potential due to the pair of points will be

$$V = Af[(x-Lh), \quad (y-Mh), \quad (z-Nh)] - Af(x, y, z),$$

$$= -Ah\frac{d}{dh}f(x, y, z) + \text{terms containing } h^2.$$

If we now diminish h and increase A without limit, their product continuing finite and equal to A', the ultimate value of the potential of the pair of points will be

$$V' = -A'\frac{d}{dh}f(x, y, z). \qquad (8)$$

If $f(x, y, z)$ satisfies Laplace's equation, then, since this equation is linear, V', which is the difference of two functions, each of which separately satisfies the equation, must itself satisfy it.

129 c.] Now the potential due to an infinite point of order zero

$$V_0 = A_0\frac{1}{r}, \qquad (9)$$

satisfies Laplace's equation, therefore every function formed from this by differentiation with respect to any number of axes in succession must also satisfy that equation.

A point of the first order may be formed by taking two points of order zero, having equal and opposite charges $-A_0$ and A_0, and placing the first at the origin and the second at the extremity of the axis h_1. The value of h_1 is then diminished and that of A_0 increased indefinitely, but so that the product $A_0 h_1$ is always equal to A_1. The ultimate result of this process, when the two points coincide, is a point of the first order whose moment is A_1 and whose axis is h_1. A point of the first order is therefore a double point. Its potential is

$$V_1 = -h_1\frac{d}{dh_1}V_0$$

$$= A_1\frac{\mu_1}{r^2}. \qquad (10)$$

By placing a point of the first order at the origin, whose moment is $-A_1$, and another at the extremity of the axis h_2 whose moment is A_1, and then diminishing h_2 and increasing A_1, so that

$$A_1 h_2 = \tfrac{1}{2}A_2, \qquad (11)$$

we obtain a point of the second order, whose potential is

$$V_2 = -\tfrac{1}{2} h_2 \frac{d}{dh_2} V_1$$

$$= A_2 \tfrac{1}{2} \frac{3\mu_1\mu_2 - \lambda_{12}}{r^3}. \qquad (12)$$

We may call a point of the second order a quadruple point because it is constructed by making four points of order zero approach each other. It has two axes h_1 and h_2 and a moment A_2. The directions of these axes and the magnitude of the moment completely define the nature of the point.

By differentiating with respect to n axes in succession we obtain the potential due to a point of the nth order. It will be the product of three factors, a constant, a certain combination of cosines, and $r^{-(n+1)}$. It is convenient, for reasons which will appear as we go on, to make the numerical value of the constant such that when all the axes coincide with the vector, the coefficient of the moment is $r^{-(n+1)}$. We therefore divide by n when we differentiate with respect to h_n.

In this way we obtain a definite numerical value for a particular potential, to which we restrict the name of The Solid Harmonic of degree $-(n+1)$, namely

$$V_n = (-)^n \frac{1}{1.2.3\dots n} \frac{d}{dh_1} \cdot \frac{d}{dh_2} \cdots \frac{d}{dh_n} \cdot \frac{1}{r}. \qquad (13)$$

If this quantity is multiplied by a constant it is still the potential due to a certain point of the nth order.

129 *d*.] The result of the operation (13) is of the form

$$V = Y_n r^{-(n+1)}, \qquad (14)$$

where Y_n is a function of the n cosines $\mu_1 \dots \mu_n$ of the angles between r and the n axes, and of the $\tfrac{1}{2} n(n-1)$ cosines λ_{12}, &c. of the angles between pairs of the axes.

If we consider the directions of r and the n axes as determined by points on a spherical surface, we may regard Y_n as a quantity varying from point to point on that surface, being a function of the $\tfrac{1}{2} n(n+1)$ distances between the n poles of the axes and the pole of the vector. We therefore call Y_n the Surface Harmonic of order n.

130 *a*.] We have next to shew that to every surface-harmonic of order n there corresponds not only a solid harmonic of degree $-(n+1)$ but another of degree n, or that

$$H_n = Y_n r^n = V_n r^{2n+1} \qquad (15)$$

satisfies Laplace's equation.

For
$$\frac{dH_n}{dx} = (2n+1)\, r^{2n-1} x V_n + r^{2n+1} \frac{dV_n}{dx},$$
$$\frac{d^2H_n}{dx^2} = (2n+1)\left[(2n-1)x^2 + r^2\right] r^{2n-3} V_n + 2(2n+1)\, r^{2n-1} x \frac{dV_n}{dx}$$
$$+ r^{2n+1} \frac{d^2V_n}{dx^2}.$$

Hence
$$\frac{d^2H_n}{dx^2} + \frac{d^2H_n}{dy^2} + \frac{d^2H_n}{dz^2} = (2n+1)(2n+2)\, r^{2n-1} V_n$$
$$+ 2(2n+1)\, r^{2n-1}\left(x\frac{dV_n}{dx} + y\frac{dV_n}{dy} + z\frac{dV_n}{dz}\right)$$
$$+ r^{2n+1}\left(\frac{d^2V_n}{dx^2} + \frac{d^2V_n}{dy^2} + \frac{d^2V_n}{dz^2}\right). \quad (16)$$

Now, since V_n is a homogeneous function of x, y, and z, of negative degree $n+1$,
$$x\frac{dV_n}{dx} + y\frac{dV_n}{dy} + z\frac{dV_n}{dz} = -(n+1)V_n. \quad (17)$$

The first two terms therefore of the right-hand member of equation (16) destroy each other, and, since V_n satisfies Laplace's equation, the third term is zero, so that H_n also satisfies Laplace's equation, and is therefore a solid harmonic of degree n.

This is a particular case of the more general theorem of electrical inversion, which asserts that if $F(x, y, z)$ is a function of x, y, and z which satisfies Laplace's equation, then there exists another function,
$$\frac{a}{r} F\left(\frac{a^2 x}{r^2}, \frac{a^2 y}{r^2}, \frac{a^2 z}{r^2}\right),$$
which also satisfies Laplace's equation. See Art. 162.

130 *b*.] The surface harmonic Y_n contains $2n$ arbitrary variables, for it is defined by the positions of its n poles on the sphere, and each of these is defined by two coordinates.

Hence the solid harmonics V_n and H_n also contain $2n$ arbitrary variables. Each of these quantities, however, when multiplied by a constant, will still satisfy Laplace's equation.

To prove that AH_n is the most general rational homogeneous function of degree n which can satisfy Laplace's equation, we observe that K, the general rational homogeneous function of degree n, contains $\frac{1}{2}(n+1)(n+2)$ terms. But $\nabla^2 K$ is a homogeneous function of degree $n-2$, and therefore contains $\frac{1}{2}n(n-1)$ terms, and the condition $\nabla^2 K = 0$ requires that each of these must vanish. There are therefore $\frac{1}{2}n(n-1)$ equations between the coefficients of the $\frac{1}{2}(n+1)(n+2)$ terms of the function K, leaving $2n+1$ independent constants in the most general form of the homogeneous function of degree n which satisfies Laplace's equation. But H_n, when multiplied by an arbitrary constant, satisfies the required conditions, and has $2n+1$ arbitrary constants. It is therefore of the most general form.

131 *a*.] We are now able to form a distribution of potential such that neither the potential itself nor its first derivatives become infinite at any point.

The function $V_n = Y_n r^{-(n+1)}$ satisfies the condition of vanishing at infinity, but becomes infinite at the origin.

The function $H_n = Y_n r^n$ is finite and continuous at finite distances from the origin, but does not vanish at an infinite distance.

But if we make $a^n Y_n r^{-(n+1)}$ the potential at all points outside a sphere whose centre is the origin, and whose radius is a, and $a^{-(n+1)} Y_n r^n$ the potential at all points within the sphere, and if on the sphere itself we suppose electricity spread with a surface density σ such that

$$4\pi\sigma a^2 = (2n+1) Y_n, \tag{18}$$

then all the conditions will be satisfied for the potential due to a shell charged in this manner.

For the potential is everywhere finite and continuous, and vanishes at an infinite distance; its first derivatives are everywhere finite and are continuous except at the charged surface, where they satisfy the equation

$$\frac{dV}{d\nu} + \frac{dV'}{d\nu'} + 4\pi\sigma = 0, \tag{19}$$

and Laplace's equation is satisfied at all points both inside and outside of the sphere.

This, therefore, is a distribution of potential which satisfies the conditions, and by Art. 100 *a* it is the only distribution which can satisfy them.

131 *b*.] The potential due to a sphere of radius a whose surface density is given by the equation

$$4\pi a^2 \sigma = (2n+1) Y_n, \tag{20}$$

is, at all points external to the sphere, identical with that due to the corresponding singular point of order n.

Let us now suppose that there is an electrical system which we may call E, external to the sphere, and that Ψ is the potential due to this system, and let us find the value of $\Sigma(\Psi e)$ for the singular point. This is the part of the electric energy depending on the action of the external system on the singular point.

If A_0 is the charge of a single point of order zero, then the potential energy in question is

$$W_0 = A_0 \Psi. \tag{21}$$

If there are two such points, a negative one at the origin and a positive one of equal numerical value at the extremity of the axis h_1, then the potential energy will be

$$-A_0 \Psi + A_0 \left(\Psi + h_1 \frac{d\Psi}{dh_1} + \tfrac{1}{2} h_1^2 \frac{d^2 \Psi}{dh_1^2} + \&c.\right),$$

and when A_0 increases and h_1 diminishes indefinitely, but so that $A_0 h_1 = A_1$, the value of the potential energy for a point of the first order will be

$$W_1 = A_1 \frac{d\Psi}{dh_1}. \tag{22}$$

Similarly for a point of order n the potential energy will be

$$W_n = \frac{1}{1.2\ldots n} A_n \frac{d^n \Psi}{dh_1 \ldots dh_n} \text{*}. \tag{23}$$

131 c.] If we suppose the external system to be made up of parts, any one of which is denoted by dE, and the singular point to be made up of parts any one of which is de, then

$$\Psi = \Sigma \left(\frac{1}{r} dE\right). \tag{24}$$

But if V_n is the potential due to the singular point,

$$V_n = \Sigma \left(\frac{1}{r} de\right), \tag{25}$$

and the potential energy due to the action of E on e is

$$W = \Sigma (\Psi\, de) = \Sigma\Sigma \left(\frac{1}{r} dE\, de\right) = \Sigma V_n\, dE, \tag{26}$$

the last expression being the potential energy due to the action of e on E.

Similarly, if σds is an element of electricity on the shell, since the potential due to the shell at the external system E is V_n, we have

$$W = \Sigma (V_n dE) = \Sigma\Sigma \left(\frac{1}{r} dE\, \sigma\, ds\right) = \Sigma (\Psi\, \sigma\, ds). \tag{27}$$

The last term contains a summation to be extended over the

* We shall find it convenient, in what follows, to denote the product of the positive integral numbers $1.2.3\ldots n$ by $n!$

surface of the sphere. Equating it to the first expression for W, we have

$$\iint \Psi\sigma\, ds = \Sigma\,(\Psi \dot{d}e),$$

$$= \frac{1}{n\,!} A_n \frac{d^n\,\Psi}{dh_1 \ldots dh_n}. \qquad (28)$$

If we remember that $4\,\pi\sigma a^2 = (2\,n+1)\,Y_n$, and that $A_n = a^n$, this becomes

$$\iint \Psi\, Y_n\, ds = \frac{4\,\pi}{n\,!\,(2\,n+1)}\; a^{n+2}\; \frac{d^n\,\Psi}{dh_1 \ldots dh_n}. \qquad (29)$$

This equation reduces the operation of taking the surface integral of $\Psi\, Y_n\, ds$ over every element of the surface of a sphere of radius a, to that of differentiating Ψ with respect to the n axes of the harmonic and taking the value of the differential coefficient at the centre of the sphere, provided that Ψ satisfies Laplace's equation at all points within the sphere, and Y_n is a surface harmonic of order n.

132.] Let us now suppose that Ψ is a solid harmonic of positive degree m of the form

$$\Psi = a^{-m}\, Y_m r^m. \qquad (30)$$

At the spherical surface, $r = a$, and $\Psi = Y_m$, so that equation (29) becomes in this case

$$\iint Y_m Y_n ds = \frac{4\,\pi}{n\,!\,(2\,n+1)}\, a^{n-m+2}\, \frac{d^n\,(Y_m r^m)}{dh_1 \ldots dh_n}, \qquad (31)$$

where the value of the differential coefficient is to be taken at the centre of the sphere.

When n is less than m, the result of the differentiation is a homogeneous function of x, y and z of degree $m-n$, the value of which at the centre of the sphere is zero. If n is equal to m the result of the differentiation is a constant, the value of which we shall determine in Art. 134 b. If the differentiation is carried further, the result is zero. Hence the surface-integral $\iint Y_m\, Y_n\, ds$ vanishes whenever m and n are different.

The steps by which we have arrived at this result are all of them purely mathematical, for though we have made use of terms having a physical meaning, such as electrical energy, each of these terms is regarded not as a physical phenomenon to be investigated, but as a definite mathematical expression. A mathematician has as much right to make use of these as of any other mathematical functions which he may find useful, and a physicist, when he has to follow a mathematical calculation, will understand it all the better if each of the steps of the calculation admits of a physical interpretation.

133.] We shall now determine the form of the surface harmonic Y_n as a function of the position of a point P on the sphere with respect to the n poles of the harmonic.

We have

$$\left.\begin{array}{l} Y_0 = 1, \quad Y_1 = \mu_1, \quad Y_2 = \frac{3}{2}\mu_1\mu_2 - \frac{1}{2}\lambda_{12}, \\ Y_3 = \frac{5}{2}\mu_1\mu_2\mu_3 - \frac{1}{2}(\mu_1\lambda_{23} + \mu_2\lambda_{31} + \mu_3\lambda_{12}), \end{array}\right\} \tag{32}$$

and so on.

Every term of Y_n therefore consists of products of cosines, those of the form μ, with a single suffix, being cosines of the angles between P and the different poles, and those of the form λ, with double suffixes, being cosines of the angles between the poles.

Since each axis is introduced by one of the n differentiations, the symbol of that axis must occur once and only once among the suffixes of the cosines of each term.

Hence if in any term there are s cosines with double suffixes, there must be $n-2s$ cosines with single suffixes.

Let the sum of all products of cosines in which s of them have double suffixes be written in the abbreviated form

$$\Sigma(\mu^{n-2s}\lambda^s).$$

In every one of the products all the suffixes occur once, and none is repeated.

If we wish to express that a particular suffix, m, occurs among the μ's only or among the λ's only, we write it as a suffix to the μ or the λ. Thus the equation

$$\Sigma(\mu^{n-2s}\lambda^s) = \Sigma(\mu_m{}^{n-2s}\lambda^s) + \Sigma(\mu^{n-2s}\lambda_m{}^s) \tag{33}$$

expresses that the whole set of products may be divided into two parts, in one of which the suffix m occurs among the direction cosines of the variable point P, and in the other among the cosines of the angles between the poles.

Let us now assume that for a particular value of n

$$\begin{aligned} Y_n = A_{n.0}\Sigma(\mu^n) + A_{n.1}\Sigma(\mu^{n-2}\lambda^1) + \&c. \\ + A_{n.s}\Sigma(\mu^{n-2s}\lambda^s) + \&c., \end{aligned} \tag{34}$$

when the A's are numerical coefficients. We may write the series in the abbreviated form

$$Y_n = S[A_{n.s}\Sigma(\mu^{n-2s}\lambda^s)], \tag{35}$$

when S indicates a summation in which all values of s, including zero, not greater than $\frac{1}{2}n$, are to be taken.

To obtain the corresponding solid harmonic of negative degree $(n+1)$ and order n, we multiply by $r^{-(n+1)}$, and obtain

$$V_n = S[A_{n.s} r^{2s-2n-1} \Sigma (p^{n-2s} \lambda^s)]; \qquad (36)$$

putting $r\mu = p$, as in equation (3).

If we differentiate V_n with respect to a new axis h_m we obtain $-(n+1) V_{n+1}$, and therefore

$$(n+1) V_{n+1} = S[A_{n.s}(2n+1-2s) r^{2s-2n-3} \Sigma (p_m^{\ n-2s+1} \lambda^s)$$
$$- A_{n.s} r^{2s-2n-1} \Sigma (p^{n-2s-1} . \lambda_m^{\ s+1})]. \qquad (37)$$

If we wish to obtain the terms containing s cosines with double suffixes, we must diminish s by unity in the last term, and we find

$$(n+1) V_{n+1} = S[r^{2s-2n-3} \{A_{n.s}(2n-2s+1) \Sigma (p_m^{\ n-2s+1} \lambda^s)$$
$$- A_{n.s-1} \Sigma (p^{n-2s+1} \lambda_m^{\ s})\}]. \qquad (38)$$

Now the two classes of products are not distinguished from each other in any way except that the suffix m occurs among the p's in one and among the λ's in the other. Hence their coefficients must be the same, and since we ought to be able to obtain the same result by putting $n+1$ for n in the expression for V_n and multiplying by $n+1$, we obtain the following equations,

$$(n+1) A_{n+1.s} = (2n-2s+1) A_{n.s} = -A_{n.s-1}. \qquad (39)$$

If we put $s = 0$, we obtain

$$(n+1) A_{n+1} = (2n+1) A_n; \qquad (40)$$

and therefore, since $A_{1.0} = 1$,

$$A_{n.0} = \frac{2n!}{2^n (n!)^2}; \qquad (41)$$

and from this we obtain the general value of the coefficient

$$A_{n.s} = (-)^s \frac{(2n-2s)!}{2^{n-s} n! (n-s)!}; \qquad (42)$$

and finally the trigonometrical expression for the surface harmonic, as

$$Y_n = S[(-)^s \frac{(2n-2s)!}{2^{n-s} n! (n-s)!} \Sigma (\mu^{n-2s} \lambda^s)]. \qquad (43)$$

This expression gives the value of the surface harmonic at any point P of the spherical surface in terms of the cosines of the distances of P from the different poles and of the distances of the poles from each other.

It is easy to see that if any one of the poles be removed to the opposite point of the spherical surface, the value of the harmonic will have its sign reversed. For any cosine involving the index of this pole will have its sign reversed, and in each term of the harmonic the index of the pole occurs once and only once.

Hence if two or any even number of poles are removed to the points respectively opposite to them, the value of the harmonic will be unaltered.

Professor Sylvester, however, has shewn (*Phil. Mag.*, Oct. 1876) that when the harmonic is given, the problem of finding the n lines which coincide with the axes has one and only one solution, though, as we have just seen, the directions to be reckoned positive along these axes may be reversed in pairs.

134.] We are now able to determine the value of the surface integral $\iint Y_m Y_n ds$ when the order of the two surface harmonics is the same, though the directions of their axes may be in general different.

For this purpose we have to form the solid harmonic $Y_m r^m$ and to differentiate it with respect to each of the n axes of Y_n.

Any term of $Y_m r_m$ of the form $r^m \mu^{m-2s} \lambda^s$ may be written $r^{2s} p_m^{m-2s} \lambda_{mm}^s$. Differentiating this n times in succession with respect to the n axes of Y_n, we find that in differentiating r^{2s} with respect to s of these axes we introduce s of the p_n's, and the numerical factor

$$2s(2s-2)\dots 2, \text{ or } 2^s s!$$

In continuing the differentiation with respect to the next s axes, the p_n's become converted into λ_{nn}'s, but no numerical factor is introduced, and in differentiating with respect to the remaining $n-2s$ axes, the p_m's become converted into λ_{mn}'s, so that the result is $2^s s!\, \lambda_{nn}^s \lambda_{mm}^s \lambda_{mn}^{m-2s}$.

We have therefore, by equation (31),

$$\iint Y_m Y_n ds = \frac{4\pi}{n!(2n+1)} a^{n-m+2} \frac{d^n (Y_m r^m)}{dh_1 \dots dh_n}, \tag{44}$$

and by equation (43),

$$Y_m r^m = S\left[(-)^s \frac{(2m-2s)!}{2^{m-s} m!(m-s)!} \Sigma (r^{2s} p_m^{m-2s} \lambda_{nn}^s)\right]. \tag{45}$$

Hence performing the differentiations and remembering that $m = n$, we find

$$\iint Y_m Y_n ds =$$

$$\frac{4\pi a^2}{(2n+1)(n!)^2} S\left[(-)^s \frac{(2n-2s)!\, s!}{2^{n-2s}(n-s)!} \Sigma (\lambda^s_{mm} \lambda^s_{nn} \lambda^{n-2s}_{mn})\right]. \tag{46}$$

135 *a*.] The expression (46) for the surface-integral of the product of two surface-harmonics assumes a remarkable form if we suppose all the axes of one of the harmonics, Y_m, to coincide with each other, so that Y_m becomes what we shall afterwards define as the zonal harmonic of order m, denoted by the symbol P_m.

In this case all the cosines of the form λ_{nm} may be written μ_n, where μ_n denotes the cosine of the angle between the common axis of P_m and one of the axes of Y_n. The cosines of the form λ_{mm} will all become equal to unity, so that for $\Sigma\lambda^s_{mm}$ we must put the number of combinations of s symbols, each of which is distinguished by two suffixes out of n, no suffix being repeated. Hence

$$\Sigma\lambda^s_{mm} = \frac{n!}{2^s\, s!\,(n-2s)!}. \tag{47}$$

The number of permutations of the remaining $n-2s$ indices of the axes of P_m is $(n-2s)!$ Hence

$$\Sigma\,(\lambda_{mn}^{n-2s}) = (n-2s)!\,\mu^{n-2s}. \tag{48}$$

Equation (46) therefore becomes, when all the axes of Y_m coincide with each other,

$$\iint Y_n P_m\, ds = \frac{4\pi a^2}{(2n+1)\,n!}\, S\left[(-)^s \frac{(2n-2s)!}{2^{n-s}\,(n-s)!}\,\Sigma\,(\mu^{n-2s}\lambda^s)\right] \tag{49}$$

$$= \frac{4\pi a^2}{2n+1}\, Y_{n(m)}, \text{ by equation (43)}, \tag{50}$$

where $Y_{n(m)}$ denotes the value of Y_n at the pole of P_m.

We may obtain the same result by the following shorter process:—

Let a system of rectangular coordinates be taken so that the axis of z coincides with the axis of P_m, and let $Y_n r^n$ be expanded as a homogeneous function of x, y, z of degree n.

At the pole of P_m, $x = y = 0$ and $z = r$, so that if Cz^n is the term not involving x or y, C is the value of Y_n at the pole of P_m.

Equation (31) becomes in this case

$$\iint Y_n P_m\, ds = \frac{4\pi a^2}{2n+1}\,\frac{1}{n!}\,\frac{d^m}{dz^m}\,(Y_n r^n).$$

If m is equal to n, the result of differentiating Cz^n is $n!\,C$, and is zero for the other terms. Hence

$$\iint Y_n P_m\, ds = \frac{4\pi a^2}{2n+1}\, C,$$

C being the value of Y_n at the pole of P_m.

135 *b*.] This result is a very important one in the theory of

spherical harmonics, as it shews how to determine a series of spherical harmonics which expresses the value of a quantity having any arbitrarily assigned finite and continuous value at each point of a spherical surface.

For let F be the value of the quantity and ds the element of surface at a point Q of the spherical surface, then if we multiply Fds by P_n, the zonal harmonic whose pole is the point P of the same surface, and integrate over the surface, the result, since it depends on the position of the point P, may be considered as a function of the position of P.

But since the value at P of the zonal harmonic whose pole is Q is equal to the value at Q of the zonal harmonic of the same order whose pole is P, we may suppose that for every element ds of the surface a zonal harmonic is constructed having its pole at Q and having a coefficient Fds.

We shall thus have a system of zonal harmonics superposed on each other with their poles at every point of the sphere where F has a value. Since each of these is a multiple of a surface harmonic of order n, their sum is a multiple of a surface harmonic (not necessarily zonal) of order n.

The surface integral $\iint FP_n ds$ considered as a function of the point P is therefore a multiple of a surface harmonic Y_n; so that

$$\frac{2n+1}{4\pi a^2}\iint FP_n ds$$

is also that particular surface harmonic of the nth order which belongs to the series of harmonics which expresses F, provided F can be so expressed.

For if F can be expressed in the form

$$F = A_0 Y_0 + A_1 Y_1 + \&c. + A_n Y_n + \&c.,$$

then if we multiply by $P_n ds$ and take the surface integral over the whole sphere, all terms involving products of harmonics of different orders will vanish, leaving

$$\iint FP_n ds = \frac{4\pi a^2}{2n+1} A_n Y_n.$$

Hence the only possible expansion of F in spherical harmonics is

$$F = \frac{1}{4\pi a^2}\left[\iint FP_0 ds + \&c. + (2n+1)\iint FP_n ds + \&c.\right]. \qquad (51)$$

Conjugate Harmonics.

136.] We have seen that the surface integral of the product of two harmonics of different orders is always zero. But even when the two harmonics are of the same order, the surface integral of their product may be zero. The two harmonics are then said to be conjugate to each other. The condition of two harmonics of the same order being conjugate to each other is expressed in terms of equation (46) by making its members equal to zero.

If one of the harmonics is zonal, the condition of conjugacy is that the value of the other harmonic at the pole of the zonal harmonic must be zero.

If we begin with a given harmonic of the nth order, then, in order that a second harmonic may be conjugate to it, its $2n$ variables must satisfy one condition.

If a third harmonic is to be conjugate to both, its $2n$ variables must satisfy two conditions. If we go on constructing harmonics, each of which is conjugate to all those before it, the number of conditions for each will be equal to the number of harmonics already in existence, so that the $(2n+1)^{\text{th}}$ harmonic will have $2n$ conditions to satisfy by means of its $2n$ variables, and will therefore be completely determined.

Any multiple AY_n of a surface harmonic of the nth order can be expressed as the sum of multiples of any set of $2n+1$ conjugate harmonics of the same order, for the coefficients of the $2n+1$ conjugate harmonics are a set of disposable quantities equal in number to the $2n$ variables of Y_n and the coefficient A.

In order to find the coefficient of any one of the conjugate harmonics, say Y_n^σ, suppose that

$$AY_n = A_0 Y_n^\sigma + \&c. + A_\sigma Y_n^\sigma + \&c.$$

Multiply by $Y_n^\sigma ds$ and find the surface integral over the sphere. All the terms involving products of harmonics conjugate to each other will vanish, leaving

$$A\iint Y_n Y_n^\sigma ds = A_\sigma \iint \left(Y_n^\sigma\right)^2 ds, \tag{52}$$

an equation which determines A_σ.

Hence if we suppose a set of $2n+1$ conjugate harmonics given, any other harmonic of the nth order can be expressed in terms of them, and this only in one way. Hence no other harmonic can be conjugate to all of them.

137.] We have seen that if a complete system of $2n+1$ harmonics of the nth order, all conjugate to each other, be given, any other harmonic of that order can be expressed in terms of these. In such a system of $2n+1$ harmonics there are $2n(2n+1)$ variables connected by $n(2n+1)$ equations, $n(2n+1)$ of the variables may therefore be regarded as arbitrary.

We might, as Thomson and Tait have suggested, select as a system of conjugate harmonics one in which each harmonic has its n poles distributed so that j of them coincide at the pole of the axis of x, k at the pole of y, and $l(=n-j-k)$ at the pole of z. The $n+1$ distributions for which $l=0$ and the n distributions for which $l=1$ being given, all the others may be expressed in terms of these.

The system which has been actually adopted by all mathematicians (including Thomson and Tait) is that in which $n-\sigma$ of the poles are made to coincide at a point which we may call the Positive Pole of the sphere, and the remaining σ poles are placed at equal distances round the equator when their number is odd, or at equal distances round one half of the equator when their number is even.

In this case $\mu_1, \mu_2, \ldots \mu_{n-\sigma}$ are each of them equal to $\cos\theta$, which we shall denote by μ. If we also write ν for $\sin\theta$, $\mu_{n-\sigma+1} \ldots \mu_n$ are of the form $\nu\cos(\phi-\beta)$, where β is the azimuth of one of the poles on the equator.

Also the value of λ_{pq} is unity, if p and q are both less than $n-\sigma$, zero when one is greater and the other less than this number, and $\cos r\pi/\sigma$ when both are greater, r being an integral number less than σ.

138.] When all the poles coincide at the pole of the sphere, $\sigma=0$, and the harmonic is called a Zonal harmonic. As the zonal harmonic is of great importance we shall reserve for it the symbol P_n.

We may obtain its value either from the trigonometrical expression (43) or more directly by differentiation, thus

$$P_n = (-)^n \frac{r^{n+1}}{n!} \frac{d^n}{dz^n}\left(\frac{1}{r}\right), \tag{52}$$

$$P_n = \frac{1.3.5\ldots(2n-1)}{1.2.3\ldots n}\left[\mu^n - \frac{n(n-1)}{2.(2n-1)}\mu^{n-2} + \frac{n(n-1)(n-2)(n-3)}{2.4.(2n-1)(2n-3)}\mu^{n-4} - \&c.\right]$$

$$= \Sigma\left[(-)^p \frac{(2n-2p)!}{2^n p!(n-p)!(n-2p)!}\mu^{n-2p}\right], \tag{53}$$

where we must give to p every integral value from zero to the greatest integer which does not exceed $\frac{1}{2}n$.

It is sometimes convenient to express P_n as a homogeneous function of $\cos\theta$ and $\sin\theta$, or, as we write them, μ and ν,

$$P_n = \mu^n - \frac{n(n-1)}{2.2}\mu^{n-2}\nu^2 + \frac{n(n-1)(n-2)(n-3)}{2.2.4.4}\mu^{n-4}\nu^4 - \&c.,$$

$$= \Sigma\left[(-)^p \frac{n!}{2^{2p}(p!)^2(n-2p)!}\mu^{n-2p}\nu^{2p}\right]. \tag{54}$$

It is shewn in the mathematical treatises on this subject that $P_n(\mu)$ is the coefficient of h^n in the expansion of $(1-2\mu h+h^2)^{-\frac{1}{2}}$.

The surface integral of the square of the zonal harmonic, or

$$\iint (P_n)^2\,ds = 2\pi a^2 \int_{-1}^{+1} (P_n(\mu))^2 d\mu = \frac{4\pi a^2}{2n+1}. \tag{55}$$

Hence
$$\int_{-1}^{+1} (P_n(\mu))^2\,d\mu = \frac{2}{2n+1}. \tag{56}$$

139.] If we consider a zonal harmonic simply as a function of μ, and without any explicit reference to a spherical surface, it may be called a Legendre's Coefficient.

If we consider the zonal harmonic as existing on a spherical surface the points of which are defined by the coordinates θ and ϕ, and if we suppose the pole of the zonal harmonic to be at the point (θ', ϕ'), then the value of the zonal harmonic at the point (θ, ϕ) is a function of the four angles $\theta', \phi'; \theta, \phi$, and because it is a function of μ, the cosine of the arc joining the points (θ, ϕ) and (θ', ϕ'), it will be unchanged in value if θ and θ', and also ϕ and ϕ', are made to change places. The zonal harmonic so expressed has been called Laplace's Coefficient. Thomson and Tait call it the Biaxal Harmonic.

Any homogeneous function of x, y, z which satisfies Laplace's equation may be called a Solid harmonic, and the value of a solid harmonic at the surface of a sphere whose centre is the origin may be called a Surface harmonic. In this book we have defined a surface harmonic by means of its n poles, so that it has only $2n$ variables. The more general surface harmonic, which has $2n+1$ variables, is the more restricted surface harmonic multiplied by an arbitrary constant. The more general surface harmonic, when expressed in terms of θ and ϕ, is called a Laplace's Function.

140 *a*.] To obtain the other harmonics of the symmetrical system, we have to differentiate with respect to σ axes in the plane of xy inclined to each other at angles equal to π/σ. This may be most

conveniently done by means of the system of imaginary coordinates given in Thomson and Tait's *Natural Philosophy*, vol. I, p. 148 (or p. 185 of 2nd edition).

If we write

$$\xi = x + iy, \qquad \eta = x - iy, \tag{57}$$

where i denotes $\sqrt{-1}$, the operation of differentiating with respect to the σ axes may be written

$$i\left(\frac{d^\sigma}{d\xi^\sigma} - \frac{d^\sigma}{d\eta^\sigma}\right) = \overset{(\sigma)}{Ds} \tag{58}$$

if one of the axes coincides with the axis of y, and

$$\left(\frac{d^\sigma}{d\xi^\sigma} + \frac{d^\sigma}{d\eta^\sigma}\right) = \overset{(\sigma)}{Dc} \tag{59}$$

if the axis of y bisects the angle between two of the axes.

We shall find it convenient to express these operations by the abbreviated symbols of operation $\overset{(\sigma)}{Ds}$ and $\overset{(\sigma)}{Dc}$, respectively. They are, of course, real operations, and may be expressed without the use of imaginary symbols thus—

$$2^{\sigma-1}\overset{(\sigma)}{Ds} = \sigma\frac{d^{\sigma-1}}{dx^{\sigma-1}}\frac{d}{dy} - \frac{\sigma(\sigma-1)(\sigma-2)}{1.2.3}\frac{d^{\sigma-3}}{dx^{\sigma-3}}\frac{d^3}{dy^3} + \&c., \tag{60}$$

$$2^{\sigma-1}\overset{(\sigma)}{Dc} = \frac{d^\sigma}{dx^\sigma} - \frac{\sigma(\sigma-1)}{1.2}\frac{d^{\sigma-2}}{dx^{\sigma-2}}\frac{d^2}{dy^2} + \&c. \tag{61}$$

We shall also write

$$\frac{d^{n-\sigma}}{dz^{n-\sigma}}\overset{(\sigma)}{Ds} = \underset{n}{\overset{(\sigma)}{Ds}} \quad \text{and} \quad \frac{d^{n-\sigma}}{dz^{n-\sigma}}\overset{(\sigma)}{Dc} = \underset{n}{\overset{(\sigma)}{Dc}}; \tag{62}$$

so that $\underset{n}{\overset{(\sigma)}{Ds}}$ and $\underset{n}{\overset{(\sigma)}{Dc}}$ denote the operations of differentiating with respect to n axes, $n-\sigma$ of which coincide with the axis of z, while the remaining σ make angles π/σ with each other in the plane of xy, $\underset{n}{\overset{(\sigma)}{Ds}}$ being used when the axis of y coincides with one of the axes, and $\underset{n}{\overset{(\sigma)}{Dc}}$ when the axis of y bisects the angle between two of the axes.

The two tesseral surface harmonics of order n and type σ may now be written

$$\underset{n}{\overset{(\sigma)}{Ys}} = (-)^n\frac{1}{n!}r^{n+1}\underset{n}{\overset{(\sigma)}{Ds}}\frac{1}{r}, \tag{63}$$

$$\underset{n}{\overset{(\sigma)}{Yc}} = (-)^n\frac{1}{n!}r^{n+1}\underset{n}{\overset{(\sigma)}{Dc}}\frac{1}{r}. \tag{64}$$

Writing $\mu = \cos\theta, \quad \nu = \sin\theta, \quad \rho^2 = x^2 + y^2,$

so that $z = \mu r, \quad \rho = \nu r, \quad x = \rho\cos\phi, \quad y = \rho\sin\phi,$

we have
$$\overset{(\sigma)}{D_s}\frac{1}{r} = (-1)^\sigma \frac{(2\sigma)!}{2^{2\sigma}\sigma!} i(\eta^\sigma - \xi^\sigma)\frac{1}{r^{2\sigma+1}}, \tag{65}$$

$$\overset{(\sigma)}{D_c}\frac{1}{r} = (-)^\sigma \frac{(2\sigma)!}{2^{2\sigma}\sigma!}(\xi^\sigma + \eta^\sigma)\frac{1}{r^{2\sigma+1}}, \tag{66}$$

in which we may write
$$\frac{i}{2}(\eta^\sigma - \xi^\sigma) = \rho^\sigma \sin\sigma\phi, \qquad \frac{1}{2}(\xi^\sigma + \eta^\sigma) = \rho^\sigma\cos\sigma\phi. \tag{67}$$

We have now only to differentiate with respect to z, which we may do either so as to obtain the result in terms of r and z, or as a homogeneous function of z and ρ divided by a power of r,

$$\frac{d^{n-\sigma}}{dz^{n-\sigma}}\frac{1}{r^{2\sigma+1}} = (-1)^{n-\sigma}\frac{(2n)!}{2^n n!}\frac{2^\sigma\sigma!}{(2\sigma)!}\frac{1}{r^{2n+1}}\times$$
$$\left[z^{n-\sigma} - \frac{(n-\sigma)(n-\sigma-1)}{2(2n-1)}z^{n-\sigma-2}r^2 + \&c.\right], \tag{68}$$

or
$$\frac{d^{n-\sigma}}{dz^{n-\sigma}}\frac{1}{r^{2\sigma+1}} = (-1)^{n-\sigma}\frac{(n+\sigma)!}{(2\sigma)!}\frac{1}{r^{2n+1}}\times$$
$$\left[z^{n-\sigma} - \frac{(n-\sigma)(n-\sigma-1)}{4(\sigma+1)}z^{n-\sigma-2}\rho^2 + \&c.\right]. \tag{69}$$

If we write

$$\Theta_n^{(\sigma)} = \nu^\sigma\left[\mu^{n-\sigma} - \frac{(n-\sigma)(n-\sigma-1)}{2\,.\,(2n-1)}\mu^{n-\sigma-2}\right.$$
$$\left. + \frac{(n-\sigma)(n-\sigma-1)(n-\sigma-2)(n-\sigma-3)}{2\,.\,4\,.\,(2n-1)(2n-3)}\mu^{n-\sigma-4} - \&c.\right], \tag{70}$$

and

$$\underset{\sim}{\vartheta}_n^{(\sigma)} = \nu^\sigma\left[\mu^{n-\sigma} - \frac{(n-\sigma)(n-\sigma-1)}{4(\sigma+1)}\mu^{n-\sigma-2}\nu^2\right.$$
$$\left. + \frac{(n-\sigma)(n-\sigma-1)(n-\sigma-2)(n-\sigma-3)}{4\,.\,8\,.\,(\sigma+1)(\sigma+2)}\mu^{n-\sigma-4}\nu^4 - \&c.\right], \tag{71}$$

then
$$\Theta_n^{(\sigma)} = \frac{2^{n-\sigma}n!(n+\sigma)!}{(2n)!\,\sigma!}\underset{\sim}{\vartheta}_n^{(\sigma)}, \tag{72}$$

so that these two functions differ only by a constant factor.

We may now write the expressions for the two tesseral harmonics of order n and type σ in terms either of Θ or ϑ,

$$\overset{(\sigma)}{\underset{n}{Y_s}} = \frac{(2n)!}{2^{n+\sigma}n!\,n!}\Theta_n^{(\sigma)}\,2\sin\sigma\phi = \frac{(n+\sigma)!}{2^{2\sigma}n!\,\sigma!}\underset{\sim}{\vartheta}_n^{(\sigma)}\,2\sin\sigma\phi, \tag{73}$$

$$\overset{(\sigma)}{\underset{n}{Y_c}} = \frac{(2n)!}{2^{n+\sigma}n!\,n!}\Theta_n^{(\sigma)}\,2\cos\sigma\phi = \frac{(n+\sigma)!}{2^{2\sigma}n!\,\sigma!}\underset{\sim}{\vartheta}_n^{(\sigma)}\,2\cos\sigma\phi. \tag{74}$$

We must remember that when $\sigma = 0$, $\sin\sigma\phi = 0$ and $\cos\sigma\phi = 1$.

For every value of σ from 1 to n inclusive there is a pair of harmonics, but when $\sigma = 0$, $\overset{(0)}{\underset{n}{Ys}} = 0$ and $\overset{(0)}{\underset{n}{Yc}} = 2P_n$, the zonal harmonic. The whole number of harmonics of order n is therefore $2n+1$, as it ought to be.

140 *b*.] The numerical value of Y adopted in this treatise is that which we find by differentiating r^{-1} with respect to the n axes and dividing by $n!$ It is the product of four factors, the sine or cosine of $\sigma\phi$, ν^σ, a function of μ (or of μ and ν), and a numerical coefficient.

The product of the second and third factors, that is to say, the part depending on θ, has been expressed in terms of three different symbols which differ from each other only by their numerical factors. When it is expressed as the product of ν^σ into a series of descending powers of μ, the first term being $\mu^{n-\sigma}$, it is the function which we, following Thomson and Tait, denote by Θ.

The function which Heine (*Handbuch der Kugelfunctionen*, § 47) denotes by $P_\sigma^{(n)}$, and calls *eine zugeordnete Function erster Art*, or, as Todhunter translates it, an 'Associated Function of the First Kind,' is related to $\Theta_n^{(\sigma)}$ by the equation

$$\Theta_n^{(\sigma)} = (-)^{\frac{\sigma}{2}} P_\sigma^{(n)}. \qquad (75)$$

The series of descending powers of μ, beginning with $\mu^{n-\sigma}$, is expressed by Heine by the symbol $\mathfrak{P}_\sigma^{(n)}$, and by Todhunter by the symbol $\varpi(\sigma, n)$.

This series may also be expressed in two other forms,

$$\mathfrak{P}_\sigma^{(n)} = \varpi(\sigma, n) = \frac{(n-\sigma)!}{(2n)!}\frac{d^{n+\sigma}}{d\mu^{n+\sigma}}(\mu^2-1)^n$$

$$= \frac{2^n (n-\sigma)!\, n!}{(2n)!}\frac{d^\sigma}{d\mu^\sigma} P_n. \qquad (76)$$

The last of these, in which the series is obtained by differentiating the zonal harmonic with respect to μ, seems to have suggested the symbol adopted by Ferrers, who defines it thus

$$T_n^{(\sigma)} = \nu^\sigma \frac{d^\sigma}{d\mu^\sigma} P_n = \frac{(2n)!}{2^n (n-\sigma)!\, n!}\Theta_n^{(\sigma)}. \qquad (77)$$

When the same quantity is expressed as a homogeneous function of μ and ν, and divided by the coefficient of $\mu^{n-\sigma}\nu^\sigma$, it is what we have already denoted by $\mathfrak{Z}_n^{(\sigma)}$.

140 *c*.] The harmonics of the symmetrical system have been classified by Thomson and Tait with reference to the form of the spherical curves at which they become zero.

The value of the zonal harmonic at any point of the sphere is a function of the cosine of the polar distance, which if equated to zero gives an equation of the nth degree, all whose roots lie between -1 and $+1$, and therefore correspond to n parallels of latitude on the sphere.

The zones included between these parallels are alternately positive and negative, the circle surrounding the pole being always positive.

The zonal harmonic is therefore suitable for expressing a function which becomes zero at certain parallels of latitude on the sphere, or at certain conical surfaces in space.

The other harmonics of the symmetrical system occur in pairs, one involving the cosine and the other the sine of $\sigma\phi$. They therefore become zero at σ meridian circles on the sphere and also at $n-\sigma$ parallels of latitude, so that the spherical surface is divided into $2\sigma(n-\sigma-1)$ quadrilaterals or tesserae, together with 4σ triangles at the poles. They are therefore useful in investigations relating to quadrilaterals or tesserae on the sphere bounded by meridian circles and parallels of latitude.

They are all called Tesseral harmonics except the last pair, which becomes zero at n meridian circles only, which divide the spherical surface into $2n$ sectors. This pair are therefore called Sectorial harmonics.

141.] We have next to find the surface integral of the square of any tesseral harmonic taken over the sphere. This we may do by the method of Art. 134. We convert the surface harmonic $Y_n^{(\sigma)}$ into a solid harmonic of positive degree by multiplying it by r^n, we differentiate this solid harmonic with respect to the n axes of the harmonic itself, and then make $x=y=z=0$, and we multiply the result by $\frac{4\pi a^2}{n!\,(2n+1)}$.

These operations are indicated in our notation by

$$\iint \left(Y_n^{(\sigma)}\right)^2 ds = \frac{4\pi a^2}{n!\,(2n+1)} D_n^{(\sigma)}\left(r^n Y_n^{(\sigma)}\right). \tag{78}$$

Writing the solid harmonic in the form of a homogeneous function of z and ξ, η, viz.,

$$r^n Y_{s\,n}^{(\sigma)} = \frac{(n+\sigma)!}{2^{2\sigma}\,n!\,\sigma!}\, i\,(\eta^\sigma - \xi^\sigma)\left(z^{n-\sigma} - \frac{(n-\sigma)(n-\sigma-1)}{4\,.\,(\sigma+1)} z^{n-\sigma-2}\xi\eta + \&c.\right); \tag{79}$$

we find that on performing the differentiations with respect to z, all the terms of the series except the first disappear, and the factor $(n-\sigma)!$ is introduced.

Continuing the differentiation with respect to ξ and η we get rid also of these variables and introduce the factor $\sigma!$, so that the final result is

$$\iint \left(Y_{s\,n}^{(\sigma)}\right)^2 ds = \frac{8\pi a^2}{2n+1} \frac{(n+\sigma)!\,(n-\sigma)!}{2^{2\sigma}\, n!\, n!}. \qquad (80)$$

We shall denote the second member of this equation by the abbreviated symbol $[n, \sigma]$.

This expression is correct for all values of σ from 1 to n inclusive, but there is no harmonic in $\sin\sigma\phi$ corresponding to $\sigma = 0$.

In the same way we can shew that

$$\iint \left(Y_{c\,n}^{(\sigma)}\right)^2 ds = \frac{8\pi a^2}{2n+1} \frac{(n+\sigma)!\,(n-\sigma)!}{2^{2\sigma}\, n!\, n!} \qquad (81)$$

for all values of σ from 1 to n inclusive.

When $\sigma = 0$, the harmonic becomes the zonal harmonic, and

$$\iint \left(Y_{c\,n}^{(0)}\right)^2 ds = \iint (P_n)^2\, ds = \frac{4\pi a^2}{2n+1}. \qquad (82)$$

a result which may be obtained directly from equation (50) by putting $Y_n = P_m$ and remembering that the value of the zonal harmonic at its pole is unity.

142 *a*.] We can now apply the method of Art. 136 to determine the coefficient of any given tesseral surface harmonic in the expansion of any arbitrary function of the position of a point on a sphere. For let F be the arbitrary function, and let A_n^σ be the coefficient of $Y_n^{(\sigma)}$ in the expansion of this function in surface harmonics of the symmetrical system

$$\iint F Y_n^{(\sigma)}\, ds = A_n^{(\sigma)} \iint \left(Y_n^{(\sigma)}\right)^2 ds = A_n^{(\sigma)}\, [n, \sigma], \qquad (83)$$

where $[n, \sigma]$ is the abbreviation for the value of the surface integral given in equation (80).

142 *b*.] Let Ψ be any function which satisfies Laplace's equation, and which has no singular values within a distance a of a point O, which we may take as the origin of coordinates. It is always possible to expand such a function in a series of solid harmonics of positive degree, having their origin at O.

One way of doing this is to describe a sphere about O as centre with a radius less than a, and to expand the value of the potential at the surface of the sphere in a series of surface harmonics. Multiplying each of these harmonics by r/a raised to a power equal to the order of the surface harmonic, we obtain the solid harmonics of which the given function is the sum.

But a more convenient method, and one which does not involve integration, is by differentation with respect to the axes of the harmonics of the symmetrical system.

For instance, let us suppose that in the expression of Ψ, there is a term of the form $\overset{(\sigma)}{\underset{n}{Ac}}\,\overset{(\sigma)}{\underset{n}{Yc}}\,r^n$.

If we perform on Ψ and on its expansion the operation

$$\frac{d^{n-\sigma}}{dz^{n-\sigma}}\left(\frac{d^\sigma}{d\xi^\sigma}+\frac{d^\sigma}{d\eta^\sigma}\right),$$

and put x, y, z equal to zero after differentiating, all the terms of the expansion vanish except that containing $\overset{(\sigma)}{\underset{n}{Ac}}$.

Expressing the operator on Ψ in terms of differentiators with respect to the real axes, we obtain the equation

$$\frac{d^{n-\sigma}}{dz^{n-\sigma}}\left(\frac{d^\sigma}{dx^\sigma}-\frac{\sigma(\sigma-1)}{1.2}\frac{d^{\sigma-2}}{dx^{\sigma-2}}\frac{d^2}{dy^2}+\&c.\right)\Psi$$
$$=\overset{(\sigma)}{\underset{n}{Ac}}\,\frac{(n+\sigma)!\,(n-\sigma)!}{2^\sigma n!}, \quad (84)$$

from which we can determine the coefficient of any harmonic of the series in terms of the differential coefficients of Ψ with respect to x, y, z at the origin.

143.] It appears from equation (50) that it is always possible to express a harmonic as the sum of a system of zonal harmonics of the same order, having their poles distributed over the surface of the sphere. The simplification of this system, however, does not appear easy. I have, however, for the sake of exhibiting to the eye some of the features of spherical harmonics, calculated the zonal harmonics of the third and fourth orders, and drawn, by the method already described for the addition of functions, the equipotential lines on the sphere for harmonics which are the sums of two zonal harmonics. See Figures VI to IX at the end of this volume.

Fig. VI represents the difference of two zonal harmonics of the third order whose axes are inclined 120° in the plane of the paper, and this difference is the harmonic of the second type in which $\sigma = 1$, the axis being perpendicular to the paper.

In Fig. VII the harmonic is also of the third order, but the axes of the zonal harmonics of which it is the sum are inclined 90°, and the result is not of any type of the symmetrical system. One of the nodal lines is a great circle, but the other two which are intersected by it are not circles.

Fig. VIII represents the difference of two zonal harmonics of the fourth order whose axes are at right angles. The result is a tesseral harmonic for which $n = 4$, $\sigma = 2$.

Fig. IX represents the sum of the same zonal harmonics. The result gives some notion of one type of the more general harmonic of the fourth order. In this type the nodal line on the sphere consists of six ovals not intersecting each other. Within these ovals the harmonic is positive, and in the sextuply connected part of the spherical surface which lies outside the ovals, the harmonic is negative.

All these figures are orthogonal projections of the spherical surface.

I have also drawn in Fig. V a plane section through the axis of a sphere, to shew the equipotential surfaces and lines of force due to a spherical surface electrified according to the values of a spherical harmonic of the first order.

Within the sphere the equipotential surfaces are equidistant planes, and the lines of force are straight lines parallel to the axis, their distances from the axis being as the square roots of the natural numbers. The lines outside the sphere may be taken as a representation of those which would be due to the earth's magnetism if it were distributed according to the most simple type.

144 *a.*] We are now able to determine the distribution of electricity on a spherical conductor under the action of electric forces whose potential is given.

By the methods already given we expand Ψ, the potential due to the given forces, in a series of solid harmonics of positive degree having their origin at the centre of the sphere.

Let $A_n r^n Y_n$ be one of these, then since within the conducting sphere the potential is uniform, there must be a term $-A_n r^n Y_n$ arising from the distribution of electricity on the surface of the sphere, and therefore in the expansion of $4\pi\sigma$ there must be a term

$$4\pi\sigma_n = (2n+1)a^{n-1}A_n Y_n.$$

In this way we can determine the coefficients of the harmonics of all orders except zero in the expression for the surface density. The coefficient corresponding to order zero depends on the charge, e, of the sphere, and is given by $4\pi\sigma_0 = a^{-2}e$.

The potential of the sphere is

$$V = \Psi_0 + \frac{e}{a}.$$

144 *b.*] Let us next suppose that the sphere is placed in the neighbourhood of conductors connected with the earth, and that

Green's Function, G, has been determined in terms of x, y, z and x', y', z', the coordinates of any two points in the region in which the sphere is placed.

If the surface density on the sphere is expressed in a series of spherical harmonics, then the electrical phenomena outside the sphere, arising from this charge on the sphere, are identical with those arising from an imaginary series of singular points all at the centre of the sphere, the first of which is a single point having a charge equal to that of the sphere and the others are multiple points of different orders corresponding to the harmonics which express the surface density.

Let Green's function be denoted by $G_{pp'}$, where p indicates the point whose coordinates are x, y, z, and p' the point whose coordinates are x', y', z'.

If a charge A_0 is placed at the point p', then, considering x', y', z' as constants, $G_{pp'}$ becomes a function of x, y, z and the potential arising from the electricity induced on surrounding bodies by A_0 is

$$\Psi = A_0 G_{pp'}. \tag{1}$$

If, instead of placing the charge A_0 at the point p', it were distributed uniformly over a sphere of radius a having its centre at p', the value of Ψ at points outside the sphere would be the same.

If the charge on the sphere is not uniformly distributed, let its surface density be expressed, as it always can, in a series of spherical harmonics, thus

$$4\pi a^2 \sigma = A_0 + 3A_1 Y_1 + \&c. + (2n+1) A_n Y_n. \tag{2}$$

The potential arising from any term of this distribution, say

$$4\pi a^2 \sigma_n = (2n+1) A_n Y_n, \tag{3}$$

will be $\dfrac{r^n}{a^{n+1}} A_n Y_n$ for points inside the sphere, and $\dfrac{a^n}{r^{n+1}} A_n Y_n$ for points outside the sphere.

Now the latter expression, by equations (13), (14), Art. 129, is equal to

$$(-1)^n A_n \frac{a^n}{n!} \frac{d^n}{dh_1 \ldots dh_n} \frac{1}{r};$$

or the potential outside the sphere, due to the charge on the surface of the sphere, is equivalent to that due to a certain multiple point whose axes are $h_1 \ldots h_n$ and whose moment is $A_n a^n$.

Hence the distribution of electricity on the surrounding conductors and the potential due to this distribution is the same as that which would be due to such a multiple point.

The potential, therefore, at the point p, or (x, y, z), due to the induced electrification of surrounding bodies, is

$$\Psi_n = A_n \frac{a^n}{n!} \frac{d'^n}{d'h_1 \dots d'h_n} G, \tag{4}$$

where the accent over the d's indicates that the differentiations are to be performed with respect to x', y', z'. These coordinates are afterwards to be made equal to those of the centre of the sphere.

It is convenient to suppose Y_n broken up into its $2n+1$ constituents of the symmetrical system. Let $A_n^{(\sigma)} Y_n^{(\sigma)}$ be one of these, then

$$\frac{d'^n}{d'h_1 \dots d'h_n} = D_n'^{(\sigma)}. \tag{5}$$

It is unnecessary here to supply the affix s or c, which indicates whether $\sin \sigma\phi$ or $\cos \sigma\phi$ occurs in the harmonic

We may now write the complete expression for Ψ,

$$\Psi = A_0 G + \Sigma\Sigma \left[A_n^{(\sigma)} \frac{a^n}{n!} D_n'^{(\sigma)} G \right]. \tag{6}$$

But within the sphere the potential is constant, or

$$\Psi + \frac{1}{a} A_0 + \Sigma\Sigma \left[\frac{r^{n_1}}{a^{n_1+1}} A_{n_1}^{(\sigma_1)} Y_{n_1}^{(\sigma_1)} \right] = \text{constant}. \tag{7}$$

Now perform on this expression the operation $D_{n_1}^{(\sigma_1)}$, where the differentiations are to be with respect to x y, z, and the values of n_1 and σ_1 are independent of those of n and σ. All the terms of (7) will disappear except that in $Y_{n_1}^{(\sigma_1)}$, and we find

$$-2 \frac{(n_1+\sigma_1)!\,(n_1-\sigma_1)!}{2^{2\sigma_1} n_1!} \frac{1}{a^{n_1+1}} A_{n_1}^{(\sigma_1)}$$

$$= A_0 D_{n_1}^{(\sigma_1)} G + \Sigma\Sigma \left[A_n^{\sigma} \frac{a^n}{n!} D_{n_1}^{(\sigma_1)} D_n'^{(\sigma)} G \right]. \tag{8}$$

We thus obtain a set of equations, the first member of each of which contains one of the coefficients which we wish to determine. The first term of the second member contains A_0, the charge of the sphere, and we may regard this as the principal term.

Neglecting, for the present, the other terms, we obtain as a first approximation

$$A_{n_1}^{(\sigma_1)} = -\frac{1}{2} \frac{2^{2\sigma} n_1!}{(n_1+\sigma_1)!\,(n_1-\sigma_1)!} A_0 a^{n_1+1} D_{n_1}^{(\sigma_1)} G. \tag{9}$$

If the shortest distance from the centre of the sphere to the nearest of the surrounding conductors is denoted by b,

$$a^{n_1+1} D_{n_1}^{(\sigma_1)} G < \left(\frac{a}{b}\right)^{n_1+1}.$$

If, therefore, b is large compared with a, the radius of the sphere, the coefficients of the other spherical harmonics are very small compared with A_0. The terms after the first on the right-hand side of equation (8) will therefore be of an order of magnitude similar to $\left(\frac{a}{b}\right)^{2n+n_1+1}$.

We may therefore neglect them in a first approximation, and in a second approximation we may insert in these terms the values of the coefficients obtained by the first approximation, and so on till we arrive at the degree of approximation required.

Distribution of electricity on a nearly spherical conductor.

145 *a*.] Let the equation of the surface of the conductor be

$$r = a(1+F), \tag{1}$$

where F is a function of the direction of r, that is to say of θ and ϕ, and is a quantity the square of which may be neglected in this investigation.

Let F be expanded in the form of a series of surface harmonics

$$F = f_0 + f_1 Y_1 + f_2 Y_2 + \&c. + f_n Y_n. \tag{2}$$

Of these terms, the first depends on the excess of the mean radius above a. If therefore we assume that a is the mean radius, that is to say, approximately the radius of a sphere whose volume is equal to that of the given conductor, the coefficient f_0 will disappear.

The second term, that in f_1, depends on the distance of the centre of mass of the conductor, supposed of uniform density, from the origin. If therefore we take that centre for origin, the coefficient f_1 will also disappear.

We shall begin by supposing that the conductor has a charge A_0, and that no external electrical force acts on it. The potential outside the conductor must therefore be of the form

$$V = A_0 \frac{1}{r} + A_1 Y_1' \frac{1}{r^2} + \&c. + A_n Y_n' \frac{1}{r^{n+1}}, \tag{3}$$

where the surface harmonics are not assumed to be of the same types as in the expansion of F.

At the surface of the conductor the potential is that of the conductor, namely, the constant quantity a.

Hence, expanding the powers of r in terms of a and F, and neglecting the square and higher powers of F, we have

$$\alpha = A_0 \frac{1}{a}(1-F) + A_1 \frac{1}{a^2} Y_1'(1-2F) + \&c.$$

$$+ A_n \frac{1}{a^{n+1}} Y_n'(1-(n+1)F). \quad (4)$$

Since the coefficients A_1, &c. are evidently small compared with A_0, we may begin by neglecting products of these coefficients into F.

If we then write for F in its first term its expansion in spherical harmonics, and equate to zero the terms involving harmonics of the same order, we find

$$\alpha = A_0 \frac{1}{a}, \quad (5)$$

$$A_1 Y_1' = A_0 a f_1 Y_1 = 0, \quad (6)$$

. . . .

$$A_n Y_n' = A_0 a^n f_n Y_n. \quad (7)$$

It follows from these equations that the Y''s must be of the same type as the Y's, and therefore identical with them, and that $A_1 = 0$ and $A_n = A_0 a^n f_n$.

To determine the density at any point of the surface, we have the equation

$$4\pi\sigma = -\frac{dV}{d\nu} = -\frac{dV}{dr}\cos\epsilon; \quad (8)$$

where ν is the normal and ϵ is the angle which the normal makes with the radius. Since in this investigation we suppose F and its first differential coefficients with respect to θ and ϕ to be small, we may put $\cos\epsilon = 1$, so that

$$4\pi\sigma = -\frac{dV}{dr} = A_0 \frac{1}{r^2} + \&c. + (n+1) A_n Y_n \frac{1}{r^{n+2}}. \quad (9)$$

Expanding the powers of r in terms of a and F, and neglecting products of F into A_n, we find

$$4\pi\sigma = A_0 \frac{1}{a^2}(1-2F) + \&c. + (n+1) A_n \frac{1}{a^{n+2}} Y_n. \quad (10)$$

Expanding F in spherical harmonics and giving A_n its value as already found, we obtain

$$4\pi\sigma = A_0 \frac{1}{a^2}[1 + f_2 Y_2 + 2 f_3 Y_3 + \&c. + (n-1) f_n Y_n]. \quad (11)$$

Hence, if the surface differs from that of a sphere by a thin stratum whose depth varies according to the values of a spherical harmonic of order n, the ratio of the difference of the surface densities at any two points to their sum will be $n-1$ times the ratio of the difference of the radii at the same two points to their sum.

145 b.] If a nearly spherical conductor is acted on by external electric forces, let the potential, U, arising from these forces be expanded in a series of spherical harmonics of positive degree, having their origin at the centre of volume of the conductor

$$U = B_0 + B_1 r Y_1' + B_2 r^2 Y_2' + \&c. + B_n r^n Y_n', \tag{12}$$

where the accent over Y indicates that this harmonic is not necessarily of the same type as the harmonic of the same order in the expansion of F.

If the conductor had been accurately spherical, the potential arising from its surface charge at a point outside the conductor would have been

$$V = A_0 \frac{1}{r} - B_1 \frac{a^3}{r^2} Y_1' - \&c. - B_n \frac{a^{2n+1}}{r^{n+1}} Y_n'. \tag{13}$$

Let the actual potential arising from the surface charge be $V + W$, where

$$W = C_0 \frac{1}{r} + C_1 \frac{1}{r^2} Y_1'' + \&c. + C_m \frac{1}{r^{m+1}} Y_m'' + \dots; \tag{14}$$

the harmonics with a double accent being different from those occurring either in F or in U, and the coefficients C being small because F is small.

The condition to be fulfilled is that, when $r = a(1+F)$,

$$U + V + W = \text{constant} = A_0 \frac{1}{a} + B_0,$$

the potential of the conductor.

Expanding the powers of r in terms of a and F, and retaining the first power of F when it is multiplied by A or B, but neglecting it when it is multiplied by the small quantity C, we find

$$F\left[- A_0 \frac{1}{a} + 3 B_1 a^3 Y_1' + 5 B_2 a^5 Y_2' + \&c. + (2n+1) B_n a^{2n+1} Y_n'\right]$$
$$+ C_0 \frac{1}{r} + C_1 \frac{1}{r^2} Y_1'' + \&c. + C_m \frac{1}{r^{m+1}} Y_m'' = 0. \tag{15}$$

To determine the coefficients C, we must perform the multiplication indicated in the first term, and express the result in a series of spherical harmonics. This series, with the signs reversed, will be the series for W at the surface of the conductor.

The product of two spherical harmonics of orders n and m, is a rational function of degree $n+m$ in x/r, y/r, and z/r, and can therefore be expanded in a series of spherical harmonics of orders not exceeding $m+n$. If, therefore, F can be expanded in spherical harmonics of orders not exceeding m, and if the potential due to external forces can be expanded in spherical harmonics of orders not exceeding n, the potential arising from the surface charge will involve spherical harmonics of orders not exceeding $m+n$.

This surface density can then be found from the potential by the equation

$$4\pi\sigma + \frac{d}{dr}(U+V+W) = 0. \tag{16}$$

145 c.] *A nearly spherical conductor enclosed in a nearly spherical and nearly concentric vessel.*

Let the equation of the surface of the conductor be

$$r = a(1+F), \tag{17}$$

where
$$F = f_1 Y_1 + \&c. + f_n^{(\sigma)} Y_n^{(\sigma)}. \tag{18}$$

Let the equation of the inner surface of the vessel be

$$r = b(1+G), \tag{19}$$

where
$$G = g_1 Y_1 + \&c. + g_n^{(\sigma)} Y_n^{(\sigma)}, \tag{20}$$

the f's and g's being small compared with unity, and $Y_n^{(\sigma)}$ being the surface harmonic of order n and type σ.

Let the potential of the conductor be α, and that of the vessel β. Let the potential at any point between the conductor and the vessel be expanded in spherical harmonics, thus

$$\Psi = h_0 + h_1 Y_1 r + \&c. + h_n^{(\sigma)} Y_n^{(\sigma)} r^n$$
$$+ k_0\frac{1}{r} + k_1 Y_1 \frac{1}{r^2} + \&c. + k_n^{(\sigma)} Y_n^{(\sigma)} \frac{1}{r^{n+1}}, \tag{21}$$

then we have to determine the constants of the forms h and k so that when $r = a(1+F)$, $\Psi = \alpha$, and when $r = b(1+G)$, $\Psi = \beta$.

It is manifest, from our former investigation, that all the h's and k's except h_0 and k_0 will be small quantities, the products of which into F may be neglected. We may, therefore, write

$$\alpha = h_0 + k_0\frac{1}{a}(1-F) + \&c. + \left(h_n^{(\sigma)} a^n + k_n^{(\sigma)} \frac{1}{a^{n+1}}\right) Y_n^{(\sigma)}, \tag{22}$$

$$\beta = h_0 + k_0\frac{1}{b}(1-G) + \&c. + \left(h_n^{(\sigma)} b^n + k_n^{(\sigma)} \frac{1}{b^{n+1}}\right) Y_n^{(\sigma)}. \tag{23}$$

We have therefore
$$\alpha = h_0 + k_0\frac{1}{a}, \tag{24}$$

$$\beta = h_0 + k_0\frac{1}{b}, \tag{25}$$

$$k_0 \frac{1}{a} f_n^{(\sigma)} = h_n^{(\sigma)} a^n + k_n^{(\sigma)} \frac{1}{a^{n+1}}, \tag{26}$$

$$k_0 \frac{1}{a} g_n^{(\sigma)} = h_n^{(\sigma)} b^n + k_n^{(\sigma)} \frac{1}{b^{n+1}}, \tag{27}$$

whence we find for the charge of the inner conductor

$$k_0 = (\alpha - \beta) \frac{ab}{b-a}, \tag{28}$$

and for the coefficients of the harmonics of order n

$$h_n = k_0 \frac{b^n g_n - a^n f_n}{b^{2n+1} - a^{2n+1}}, \tag{29}$$

$$k_n = k_0 a^n b^n \frac{b^{n+1} f_n - a^{n+1} g_n}{b^{2n+1} - a^{2n+1}}, \tag{30}$$

where we must remember that the coefficients f_n, g_n, h_n, k_n are those belonging to the same type as well as order.

The surface density on the inner conductor is given by the equation

$$4\pi\sigma a^2 = k_0 (1 + \dots + A_n Y_n^{(\sigma)} + \dots)$$

where $$A_n = \frac{f_n\{na^{2n+1} + (n+1) b^{2n+1}\} - g_n (2n+1) a^{n+1} b^n}{b^{2n+1} - a^{2n+1}}. \tag{31}$$

146.] As an example of the application of zonal harmonics, let us investigate the equilibrium of electricity on two spherical conductors.

Let a and b be the radii of the spheres, and c the distance between their centres. We shall also, for the sake of brevity, write $a = cx$, and $b = cy$, so that x and y are numerical quantities less than unity.

Let the line joining the centres of the spheres be taken as the axis of the zonal harmonics, and let the pole of the zonal harmonics belonging to either sphere be the point of that sphere nearest to the other.

Let r be the distance of any point from the centre of the first sphere, and s the distance of the same point from that of the second sphere.

Let the surface density, σ_1, of the first sphere be given by the equation

$$4\pi\sigma_1 a^2 = A + A_1 P_1 + 3 A_2 P_2 + \text{\&c.} + (2m+1) A_m P_m, \tag{1}$$

so that A is the total charge of the sphere, and A_1, &c. are the coefficients of the zonal harmonics P_1, &c.

The potential due to this distribution of charge may be represented by

$$U' = \frac{1}{a}\left[A + A_1 P_1 \frac{r}{a} + A_2 P_2 \frac{r^2}{a^2} + \&c. + A_m P_m \frac{r^m}{a^m}\right] \tag{2}$$

for points inside the sphere, and by

$$U = \frac{1}{r}\left[A + A_1 P_1 \frac{a}{r} + A_2 P_2 \frac{a^2}{r^2} + \&c. + A_m P_m \frac{a^m}{r^m}\right] \tag{3}$$

for points outside.

Similarly, if the surface density on the second sphere is given by the equation

$$4\pi\sigma_2 b^2 = B + B_1 P_1 + \&c. + (2n+1) B_n P_n, \tag{4}$$

the potential inside and outside this sphere may be represented by equations of the form

$$V' = \frac{1}{b}\left[B + B_1 P_1 \frac{s}{b} + \&c. + B_n P_n \frac{s^n}{b^n}\right], \tag{5}$$

$$V = \frac{1}{s}\left[B + B_1 P_1 \frac{b}{s} + \&c. + B_n P_n \frac{b^n}{s^n}\right], \tag{6}$$

where the general harmonics are related to the second sphere.

The charges of the sphere are A and B respectively.

The potential at every point within the first sphere is constant and equal to α, the potential of that sphere, so that within the first sphere $\qquad U' + V = \alpha. \qquad (7)$

Similarly, if the potential of the second sphere is β, for points within that sphere, $\qquad U + V' = \beta. \qquad (8)$

For points outside both spheres the potential is Ψ, where

$$U + V = \Psi. \tag{9}$$

On the axis, between the centres of the spheres,

$$r + s = c. \tag{10}$$

Hence, differentiating with respect to r, and after differentiation making $r = 0$, and remembering that at the pole each of the zonal harmonics is unity, we find

$$\begin{aligned} A_1 \frac{1}{a^2} - \frac{dV}{ds} &= 0, \\ A_2 \frac{2!}{a^3} + \frac{d^2 V}{ds^2} &= 0, \\ \cdot\quad\cdot\quad\cdot\quad&\cdot\quad\cdot\quad\cdot\quad\cdot \\ A_m \frac{m!}{a^{m+1}} + (-1)^m \frac{d^m V}{ds^m} &= 0, \end{aligned} \tag{11}$$

where, after differentiation, s is to be made equal to c.

If we perform the differentiations, and write $a/c = x$ and $b/c = y$, these equations become

$$\left.\begin{aligned}
0 &= A_1 + Bx^2 + 2B_1 x^2 y + 3B_2 x^2 y^2 + \&c. + (n+1) B_n x^2 y^n, \\
0 &= A_2 + Bx^3 + 3B_1 x^3 y + 6B_2 x^3 y^2 + \&c. + \tfrac{1}{2}(n+1)(n+2) B_n x^3 y^n, \\
&\quad \cdots \\
0 &= A_m + Bx^{m+1} + (m+1) B_1 x^{m+1} y + \tfrac{1}{2}(m+1)(m+2) B_2 x^{m+1} y^2 \\
&\qquad + \&c. + \frac{(m+n)!}{m!\,n!} B_n x^{m+1} y^n.
\end{aligned}\right\} \quad (12)$$

By the corresponding operations for the second sphere we find,

$$\left.\begin{aligned}
0 &= B_1 + Ay^2 + 2A_1 xy^2 + 3A_2 x^2 y^2 + \&c. + (m+1) A_m x^m y^2, \\
0 &= B_2 + Ay^3 + 3A_1 xy^3 + 6A_2 x^2 y^3 + \&c. + \tfrac{1}{2}(m+1)(m+2) A_m x^m y^3, \\
&\quad \cdots \\
0 &= B_n + Ay^{n+1} + (n+1) A_1 xy^{n+1} + \tfrac{1}{2}(n+1)(n+2) x^2 y^{n+1} + \&c. \\
&\qquad + \frac{(m+n)!}{m!\,n!} A_m x^m y^{n+1}.
\end{aligned}\right\} \quad (13)$$

To determine the potentials, α and β, of the two spheres we have the equations (7) and (8), which we may now write

$$c\alpha = A\frac{1}{x} + B + B_1 y + B_2 y^2 + \&c. + B_n y^n, \qquad (14)$$

$$c\beta = B\frac{1}{y} + A + A_1 x + A_2 x^2 + \&c. + A_m x^m. \qquad (15)$$

If, therefore, we confine our attention to the coefficients A_1 to A_m and B_1 to B_n, we have $m+n$ equations from which to determine these quantities in terms of A and B, the charges of the two spheres, and by inserting the values of these coefficients in (14) and (15) we may express the potentials of the spheres in terms of their charges.

These operations may be expressed in the form of determinants, but for purposes of calculation it is more convenient to proceed as follows.

Inserting in equations (12) the values of $B_1 \ldots B_n$ from equations (13), we find

$$\begin{aligned}
A_1 = -Bx^2 &+ A\,x^2y^3[2.1 + 3.1y^2 + 4.1y^4 + 5.1y^6 + 6.1y^8 \\
&+ A_1 x^3 y^3 [2.2 + 3.3y^2 + 4.4y^4 + 5.5y^6 \\
&+ A_2 x^4 y^3 [2.3 + 3.6y^2 + 4.10y^4 \\
&+ A_3 x^5 y^3 [2.4 + 3.10y^2 \\
&+ A_4 x^6 y^3 [2.5.
\end{aligned} \qquad (16)$$

$$\begin{aligned} A_2 = -Bx^3 &+ A\ x^3y^3[3\cdot 1+6\cdot 1y^2+10\cdot 1y^4+15\cdot 1y^6 \\ &+A_1x^4y^3[3\cdot 2+6\cdot 3y^2+10\cdot 4y^4 \\ &+A_2x^5y^3[3\cdot 3+6\cdot 6y^2 \\ &+A_3x^6y^3[3\cdot 4. \end{aligned} \tag{17}$$

$$\begin{aligned} A_3 = -Bx^4 &+ A\ x^4y^3[4\cdot 1+10\cdot 1y^2+20\cdot 1y^4 \\ &+A_1x^5y^3[4\cdot 2+10\cdot 3y^2 \\ &+A_2x^6y^3[4\cdot 3. \end{aligned} \tag{18}$$

$$\begin{aligned} A_4 = -Bx^5 &+ A\ x^5y^3[5+15y^2 \\ &+A_1x^6y^3[5\cdot 2. \end{aligned} \tag{19}$$

By substituting in the second members of these equations the approximate values of A_1 &c., and repeating the process for further approximations, we may carry the approximation to the coefficient to any extent in ascending powers and products of x and y. If we write

$$A_n = \ \ p_nA - q_nB,$$
$$B_n = -r_nA + s_nB;$$

we find

$$\begin{aligned} p_1 = x^2y^3&[2+3y^2+4y^4+5y^6+6y^8+7y^{10}+8y^{12}+9y^{14}+\&c. \\ &+x^5y^6[8+30y^2+75y^4+154y^6+280y^8+\&c. \\ &+x^7y^6[18+90y^2+288y^4+735y^6+\&c. \\ &+x^9y^6[32+200y^2+780y^4+\&c. \\ &+x^{11}y^6[50+375y^2+\&c. \\ &+x^{13}y^6[72+\&c. \\ &\ \cdot\ \cdot\ \cdot\ \cdot\ \cdot\ \cdot\ \cdot \\ &+x^8y^9[32+192y^2 \\ &+x^{10}y^9[144+\&c. \\ &\ \cdot\ \cdot\ \cdot\ \cdot\ \cdot\ \cdot\ \cdot \end{aligned} \tag{20}$$

$$\begin{aligned} q_1 = x^3& \\ &+x^5y^3[4+9y^2+16y^4+25y^6+36y^8+49y^{10}+64y^{12}+\&c. \\ &+x^7y^3[6+18y^2+40y^4+75y^6+126y^8+196y^{10}+\&c. \\ &+x^9y^3[8+30y^2+80y^4+175y^6+336y^8+\&c. \\ &+x^{11}y^3[10+45y^2+140y^4+350y^6+\&c. \\ &+x^{13}y^3[12+63y^2+224y^4+\&c. \\ &+x^{15}y^3[14+84y^2+\&c. \\ &+x^{17}y^3[16+\&c. \\ &\ \cdot\ \cdot\ \cdot\ \cdot\ \cdot\ \cdot\ \cdot \end{aligned}$$

$$+x^8y^6[\ 16+\ \ 72y^2+\ 209y^4+488y^6+\&c.$$
$$+x^{10}y^6[\ 60+\ 342y^2+1222y^4+\&c.$$
$$+x^{12}y^6[150+1050y^2+\&c.$$
$$+x^{14}y^6[308+\&c.$$
. . . .
$$+{}^{11}xy^9[\ 64+\&c. \qquad (21)$$

It will be more convenient in subsequent operations to write these coefficients in terms of a, b, and c, and to arrange the terms according to their dimensions in c. This will make it easier to differentiate with respect to c. We thus find

$$\begin{aligned} p_1 = {} & 2a^2b^3c^{-5}+3a^2b^5c^{-7}+4a^2b^7c^{-9}+(5a^2b^9+8a^5b^6)c^{-11} \\ & +(6a^2b^{11}+30a^5b^8+18a^7b^6)c^{-13} \\ & +(7a^2b^{13}+75a^5b^{10}+90a^7b^8+32a^9b^6)c^{-15} \\ & +(8a^2b^{15}+154a^5b^{12}+288a^7b^{10}+32a^8b^9+200a^9b^8+50a^{11}b^6)c^{-17} \\ & +(9a^2b^{17}+280a^5b^{14}+735a^7b^{12}+192a^8b^{11}+780a^9b^{10} \\ & \qquad +144a^{10}b^9+375a^{11}b^8+72a^{13}b^6)c^{-19}. \end{aligned} \qquad (22)$$

$$\begin{aligned} q_1 = {} & a^2c^{-2}+4a^5b^3c^{-8}+(6a^7b^3+9a^5b^5)c^{-10} \\ & +(8a^9b^3+18a^7b^5+16a^5b^7)c^{-12} \\ & +(10a^{11}b^3+30a^9b^5+16a^8b^6+40a^7b^7+25a^5b^9)c^{-14} \\ & +(12a^{13}b^3+45a^{11}b^5+60a^{10}b^6+80a^9b^7 \\ & \qquad +72a^8b^8+75a^7b^9+36a^5b^{11})c^{-16} \\ & +(14a^{15}b^3+63a^{13}b^5+150a^{12}b^6+140a^{11}b^7+342a^{10}b^8 \\ & \qquad +175a^9b^9+209a^8b^{10}+126a^7b^{11}+49a^5b^{13})c^{-18} \\ & +(16a^{17}b^3+84a^{15}b^5+308a^{14}b^6+224a^{13}b^7+1050a^{12}b^8 \\ & \qquad +414a^{11}b^9+1222a^{10}b^{10}+336a^9b^{11}+488a^8b^{12}+196a^7b^{13} \\ & \qquad +64a^5b^{15})c^{-20}. \end{aligned} \qquad (23)$$

$$\begin{aligned} p_2 = {} & 3a^3b^3c^{-6}+6a^3b^5c^{-8}+10a^3b^7c^{-10}+(12a^6b^6+15a^3b^9)c^{-12} \\ & +(27a^8b^6+54a^6b^8+21a^3b^{11})c^{-14} \\ & +(48a^{10}b^6+162a^8b^8+158a^6b^{10}+28a^3b^{13})c^{-16} \\ & +(75a^{12}b^6+360a^{10}b^8+48a^9b^9+606a^8b^{10} \\ & \qquad +372a^6b^{12}+36a^3b^{15})c^{-18}. \end{aligned} \qquad (24)$$

$$\begin{aligned} q_2 = {} & a^3c^{-3}+6a^6b^3c^{-9}+(9a^8b^3+18a^6b^5)c^{-11} \\ & +(12a^{10}b^3+36a^8b^5+40a^6b^7)c^{-13} \\ & +(15a^{12}b^3+60a^{10}b^5+24a^9b^6+100a^8b^7+75a^6b^9)c^{-15} \\ & +(18a^{14}b^3+90a^{12}b^5+90a^{11}b^6+200a^{10}b^7 \\ & \qquad +126a^9b^8+225a^8b^9+126a^6b^{11})c^{-17} \\ & +(21a^{16}b^3+126a^{14}b^5+225a^{13}b^6+350a^{12}b^7+594a^{11}b^8 \\ & \qquad +525a^{10}b^9+418a^9b^{10}+441a^8b^{11}+196a^6b^{13})c^{-19}. \end{aligned} \qquad (25)$$

$$p_3 = 4a^4b^3c^{-7} + 10a^4b^5c^{-9} + 20a^4b^7c^{-11} + (16a^7b^6 + 35a^4b^9)c^{-13}$$
$$+ (36a^9b^6 + 84a^7b^8 + 56a^4b^{11})c^{-15}$$
$$+ (64a^{11}b^6 + 252a^9b^8 + 282a^7b^{10} + 84a^4b^{13})c^{-17}. \tag{26}$$

$$q_3 = a^4c^{-4} + 8a^7b^3c^{-10} + (12a^9b^3 + 30a^7b^5)c^{-12}$$
$$+ (16a^{11}b^3 + 60a^9b^5 + 80a^7b^7)c^{-14}$$
$$+ (20a^{13}b^3 + 100a^{11}b^5 + 32a^{10}b^6 + 200a^9b^7 + 175a^7b^9)c^{-16}$$
$$+ (24a^{15}b^3 + 150a^{13}b^5 + 120a^{12}b^6 + 400a^{11}b^7 + 192a^{10}b^8$$
$$+ 525a^9b^9 + 336a^7b^{11})c^{-18}. \tag{27}$$

$$p_4 = 5a^5b^3c^{-8} + 15a^5b^5c^{-10} + 35a^5b^7c^{-12} + (20a^8b^6 + 70a^5b^9)c^{-14}$$
$$+ (45a^{10}b^6 + 120a^8b^8 + 126a^5b^{11})c^{-16}. \tag{28}$$

$$q_4 = a^5c^{-5} + 10a^8b^3c^{-11} + (15a^{10}b^3 + 45a^8b^5)c^{-13}$$
$$+ (20a^{12}b^3 + 90a^{10}b^5 + 140a^8b^7)c^{-15}$$
$$+ (25a^{14}b^3 + 150a^{12}b^5 + 40a^{11}b^6 + 350a^{10}b^7 + 350a^8b^9)c^{-17}. \tag{29}$$

$$p_5 = 6a^6b^3c^{-9} + 21a^6b^5c^{-11} + 56a^6b^7c^{-13}$$
$$+ (24a^9b^6 + 126a^6b^9)c^{-15}. \tag{30}$$

$$q_5 = a^6c^{-6} + 12a^9b^3c^{-12} + (18a^{11}b^3 + 63a^9b^5)c^{-14}$$
$$+ (24a^{13}b^3 + 126a^{11}b^5 + 224a^9b^7)c^{-16}. \tag{31}$$

$$p_6 = 7a^7b^3c^{-10} + 28a^7b^5c^{-12} + 84a^7b^7c^{-14}. \tag{32}$$

$$q_6 = a^7c^{-7} + 14a^{10}b^3c^{-13} + (21a^{12}b^3 + 84a^{10}b^5)c^{-15}. \tag{33}$$

$$p_7 = 8a^8b^3c^{-11} + 36a^8b^5c^{-13}. \tag{34}$$

$$q_7 = a^8c^{-8} + 16a^{11}b^3c^{-14}. \tag{35}$$

$$p_8 = 9a^9b^3c^{-12}. \tag{36}$$

$$q_8 = a^9c^{-9}. \tag{37}$$

The values of the r's and s's may be written down by exchanging a and b in the q's and p's respectively.

If we now calculate the potentials of the two spheres in terms of these coefficients in the form

$$\alpha = lA + mB, \tag{38}$$

$$\beta = mA + nB, \tag{39}$$

then l, m, n are the coefficients of potential (Art. 87), and of these

$$m = c^{-1} + p_1ac^{-2} + p_2a^2c^{-3} + \&c., \tag{40}$$

$$n = b^{-1} - q_1ac^{-2} - q_2a^2c^{-3} - \&c., \tag{41}$$

or, expanding in terms of a, b, c,

$$\begin{aligned}
m = {} & c^{-1} + 2a^3b^3c^{-7} + 3a^3b^3(a^2+b^2)c^{-9} + a^3b^3(4a^4+6a^2b^2+4b^4)c^{-11} \\
& + a^3b^3[5a^6+10a^4b^2+8a^3b^3+10a^2b^4+5b^6]c^{-13} \\
& + a^3b^3[6a^8+15a^6b^2+30a^5b^3+20a^4b^4 \\
& \qquad +30a^3b^5+15a^2b^6+6b^8]c^{-15} \\
& + a^3b^3[7a^{10}+21a^8b^2+75a^7b^3+35a^6b^4+144a^5b^5 \\
& \qquad +35a^4b^6+75a^3b^7+21a^2b^8+7b^{10}]c^{-17} \\
& + a^3b^3[8a^{12}+28a^{10}b^2+154a^9b^3+56a^8b^4+446a^7b^5+102a^6b^6 \\
& \qquad +446a^5b^7+56a^4b^8+154a^3b^9+28a^2b^{10}+8b^{12}]c^{-19} \\
& + a^3b^3[9a^{14}+36a^{12}b^2+280a^{11}b^3+84a^{10}b^4+1107a^9b^5+318a^8b^6 \\
& \qquad +1668a^7b^7+318a^6b^8+1107a^5b^9+84a^4b^{10}+280a^3b^{11} \\
& \qquad +36a^2b^{12}+9b^{14}]c^{-21}.
\end{aligned} \tag{42}$$

$$\begin{aligned}
n = {} & b^{-1} - a^3c^{-4} - a^5c^{-6} - a^7c^{-8} - (a^3+4b^3)a^6c^{-10} \\
& -(a^5+12a^2b^3+9b^5)a^6c^{-12} - (a^7+25a^4b^3+36a^2b^5+16b^7)a^6c^{-14} \\
& -(a^9+44a^6b^3+96a^4b^5+16a^3b^6+80a^2b^7+25b^9)a^6c^{-16} \\
& -(a^{11}+70a^8b^3+210a^6b^5+84a^5b^6+260a^4b^7 \\
& \qquad +72a^3b^8+150a^2b^9+36b^{11})a^6c^{-18} \\
& -(a^{13}+104a^{10}b^3+406a^8b^5+272a^7b^6+680a^6b^7+468a^5b^8 \\
& \qquad +575a^4b^9+209a^3b^{10}+252a^2b^{11}+49b^{13})a^6c^{-20} \\
& -(a^{15}+147a^{12}b^3+720a^{10}b^5+693a^9b^6+1548a^8b^7+1836a^7b^8 \\
& \qquad +1814a^6b^9+1640a^5b^{10}+1113a^4b^{11}+488a^3b^{12} \\
& \qquad +392a^2b^{13}+64b^{15})a^6c^{-22}.
\end{aligned} \tag{43}$$

The value of l can be obtained from that of n by exchanging a and b.

The potential energy of the system is, by Art. 87,

$$W = \tfrac{1}{2}lA^2 + mAB + \tfrac{1}{2}nB^2, \tag{44}$$

and the repulsion between the two spheres is, by Art. 93 a,

$$-\frac{dW}{dc} = \tfrac{1}{2}A^2\frac{dl}{dc} + AB\frac{dm}{dc} + \tfrac{1}{2}B^2\frac{dn}{dc}. \tag{45}$$

The surface density at any point of either sphere is given by equations (1) and (4) in terms of the coefficients A_n and B_n.

CHAPTER X.

CONFOCAL QUADRIC SURFACES *.

147.] Let the general equation of a confocal system be

$$\frac{x^2}{\lambda^2-a^2}+\frac{y^2}{\lambda^2-b^2}+\frac{z^2}{\lambda^2-c^2}=1, \tag{1}$$

where λ is a variable parameter, which we shall distinguish by a suffix for the species of quadric, viz. we shall take λ_1 for the hyperboloids of two sheets, λ_2 for the hyperboloids of one sheet, and λ_3 for the ellipsoids. The quantities

$$a,\ \lambda_1,\ b,\ \lambda_2,\ c,\ \lambda_3$$

are in ascending order of magnitude. The quantity a is introduced for the sake of symmetry, but in our results we shall always suppose $a=0$.

If we consider the three surfaces whose parameters are $\lambda_1, \lambda_2, \lambda_3$, we find, by elimination between their equations, that the value of x^2 at their point of intersection satisfies the equation

$$x^2(b^2-a^2)(c^2-a^2)=(\lambda_1^2-a^2)(\lambda_2^2-a^2)(\lambda_3^2-a^2). \tag{2}$$

The values of y^2 and z^2 may be found by transposing a, b, c symmetrically.

Differentiating this equation with respect to λ_1, we find

$$\frac{dx}{d\lambda_1}=\frac{\lambda_1}{\lambda_1^2-a^2}x. \tag{3}$$

If ds_1 is the length of the intercept of the curve of intersection of λ_2 and λ_3 cut off between the surfaces λ_1 and $\lambda_1+d\lambda_1$, then

$$\overline{\frac{ds_1}{d\lambda_1}}\Big|^2=\overline{\frac{dx}{d\lambda_1}}\Big|^2+\overline{\frac{dy}{d\lambda_1}}\Big|^2+\overline{\frac{dz}{d\lambda_1}}\Big|^2=\frac{\lambda_1^2(\lambda_2^2-\lambda_1^2)(\lambda_3^2-\lambda_1^2)}{(\lambda_1^2-a^2)(\lambda_1^2-b^2)(\lambda_1^2-c^2)}. \tag{4}$$

* This investigation is chiefly borrowed from a very interesting work,—*Leçons sur les Fonctions Inverses des Transcendantes et les Surfaces Isothermes.* Par G. Lamé. Paris, 1857.

The denominator of this fraction is the product of the squares of the semi-axes of the surface λ_1.

If we put

$$D_1^2 = \lambda_3^2 - \lambda_2^2, \quad D_2^2 = \lambda_3^2 - \lambda_1^2, \quad \text{and} \quad D_3^2 = \lambda_2^2 - \lambda_1^2, \tag{5}$$

and if we make $a = 0$, then

$$\frac{ds_1}{d\lambda_1} = \frac{D_2 D_3}{\sqrt{b^2 - \lambda_1^2}\sqrt{c^2 - \lambda_1^2}}. \tag{6}$$

It is easy to see that D_2 and D_3 are the semi-axes of the central section of λ_1 which is conjugate to the diameter passing through the given point, and that D_2 is parallel to ds_2, and D_3 to ds_3.

If we also substitute for the three parameters $\lambda_1, \lambda_2, \lambda_3$ their values in terms of three functions α, β, γ, defined by the equations

$$\begin{aligned} \alpha &= \int_0^{\lambda_1} \frac{c\,d\lambda_1}{\sqrt{(b^2 - \lambda_1^2)(c^2 - \lambda_1^2)}}, \\ \beta &= \int_b^{\lambda_2} \frac{c\,d\lambda_2}{\sqrt{(\lambda_2^2 - b^2)(c^2 - \lambda_2^2)}}, \\ \gamma &= \int_c^{\lambda_3} \frac{c\,d\lambda_3}{\sqrt{(\lambda_3^2 - b^2)(\lambda_3^2 - c^2)}}; \end{aligned} \tag{7}$$

then $$ds_1 = \frac{1}{c} D_2 D_3 \, d\alpha, \quad ds_2 = \frac{1}{c} D_3 D_1 \, d\beta, \quad ds_3 = \frac{1}{c} D_1 D_2 \, d\gamma. \tag{8}$$

148.] Now let V be the potential at any point α, β, γ, then the resultant force in the direction of ds_1 is

$$R_1 = -\frac{dV}{ds_1} = -\frac{dV}{d\alpha}\frac{d\alpha}{ds_1} = -\frac{dV}{d\alpha}\frac{c}{D_2 D_3}. \tag{9}$$

Since ds_1, ds_2, and ds_3 are at right angles to each other, the surface-integral over the element of area $ds_2\, ds_3$ is

$$\begin{aligned} R_1\, ds_2\, ds_3 &= -\frac{dV}{d\alpha}\frac{c}{D_2 D_3} \cdot \frac{D_3 D_1}{c} \cdot \frac{D_1 D_2}{c} \cdot d\beta\, d\gamma \\ &= -\frac{dV}{d\alpha}\frac{D_1^2}{c}\, d\beta\, d\gamma. \end{aligned} \tag{10}$$

Now consider the element of volume intercepted between the surfaces α, β, γ, and $\alpha + d\alpha$, $\beta + d\beta$, $\gamma + d\gamma$. There will be eight such elements, one in each octant of space.

We have found the surface-integral of the normal component of the force (measured inwards) for the element of surface intercepted from the surface α by the surfaces β and $\beta + d\beta$, γ and $\gamma + d\gamma$.

The surface-integral for the corresponding element of the surface $a+da$ will be

$$+\frac{dV}{da}\frac{D_1^2}{c}\,d\beta\,d\gamma+\frac{d^2V}{da^2}\frac{D_1^2}{c}\,da\,d\beta\,d\gamma$$

since D_1 is independent of a. The surface-integral for the two opposite faces of the element of volume will be the sum of these quantities, or

$$\frac{d^2V}{da^2}\frac{D_1^2}{c}\,da\,d\beta\,d\gamma.$$

Similarly the surface-integrals for the other two pairs of faces will be

$$\frac{d^2V}{d\beta^2}\frac{D_2^2}{c}\,da\,d\beta\,d\gamma \quad\text{and}\quad \frac{d^2V}{d\gamma^2}\frac{D_3^2}{c}\,da\,d\beta\,d\gamma.$$

These six faces enclose an element whose volume is

$$ds_1\,ds_2\,ds_3=\frac{D_1^2\,D_2^2\,D_3^2}{c^3}\,da\,d\beta\,d\gamma,$$

and if ρ is the volume-density within that element, we find by Art. 77 that the total surface-integral of the element, together with the quantity of electricity within it, multiplied by 4π is zero, or, dividing by $da\,d\beta\,d\gamma$,

$$\frac{d^2V}{da^2}D_1^2+\frac{d^2V}{d\beta^2}D_2^2+\frac{d^2V}{d\gamma^2}D_3^2+4\pi\rho\frac{D_1^2\,D_2^2\,D_3^2}{c^2}=0, \tag{11}$$

which is the form of Poisson's extension of Laplace's equation referred to ellipsoidal coordinates.

If $\rho=0$ the fourth term vanishes, and the equation is equivalent to that of Laplace.

For the general discussion of this equation the reader is referred to the work of Lamé already mentioned.

149.] To determine the quantities a, β, γ, we may put them in the form of ordinary elliptic integrals by introducing the auxiliary angles θ, ϕ, and ψ, where

$$\lambda_1=b\sin\theta, \tag{12}$$

$$\lambda_2=\sqrt{c^2\sin^2\phi+b^2\cos^2\phi}, \tag{13}$$

$$\lambda_3=c\sec\psi. \tag{14}$$

If we put $b=kc$, and $k^2+k'^2=1$, we may call k and k' the two complementary moduli of the confocal system, and we find

$$a=\int_0^\theta\frac{d\theta}{\sqrt{1-k^2\sin^2\theta}}, \tag{15}$$

an elliptic integral of the first kind, which we may write according to the usual notation $F(k,\theta)$.

In the same way we find

$$\beta = \int_0^\phi \frac{d\phi}{\sqrt{1-k'^2\cos^2\phi}} = F(k') - F(k',\phi), \tag{16}$$

where $F(k')$ is the complete function for modulus k',

$$\gamma = \int_0^\psi \frac{d\psi}{\sqrt{1-k^2\cos^2\psi}} = F(k) - F(k,\psi). \tag{17}$$

Here α is represented as a function of the angle θ, which is accordingly a function of the parameter λ_1, β as a function of ϕ and thence of λ_2, and γ as a function of ψ and thence of λ_3.

But these angles and parameters may be considered as functions of α, β, γ. The properties of such inverse functions, and of those connected with them, are explained in the treatise of M. Lamé on that subject.

It is easy to see that since the parameters are periodic functions of the auxiliary angles, they will be periodic functions of the quantities α, β, γ: the periods of λ_1 and λ_3 are $4F(k)$, and that of λ_2 is $2F(k')$.

Particular Solutions.

150.] If V is a linear function of α, β, or γ, the equation is satisfied. Hence we may deduce from the equation the distribution of electricity on any two confocal surfaces of the same family maintained at given potentials, and the potential at any point between them.

The Hyperboloids of Two Sheets.

When α is constant the corresponding surface is a hyperboloid of two sheets. Let us make the sign of α the same as that of x in the sheet under consideration. We shall thus be able to study one of these sheets at a time.

Let α_1, α_2 be the values of α corresponding to two single sheets, whether of different hyperboloids or of the same one, and let V_1, V_2 be the potentials at which they are maintained. Then, if we make

$$V = \frac{\alpha_1 V_2 - \alpha_2 V_1 + \alpha(V_1 - V_2)}{\alpha_1 - \alpha_2}, \tag{18}$$

the conditions will be satisfied at the two surfaces and throughout the space between them. If we make V constant and equal to V_1 in the space beyond the surface α_1, and constant and equal to V_2

in the space beyond the surface a_2, we shall have obtained the complete solution of this particular case.

The resultant force at any point of either sheet is

$$R_1 = -\frac{dV}{ds_1} = -\frac{dV}{da}\frac{da}{ds_1}, \tag{19}$$

$$\text{or} \quad R_1 = \frac{V_1 - V_2}{a_1 - a_2}\frac{c}{D_2 D_3}. \tag{20}$$

If p_1 be the perpendicular from the centre on the tangent plane at any point, and P_1 the product of the semi-axes of the surface, then $p_1 D_2 D_3 = P_1$.

Hence we find

$$R_1 = \frac{V_1 - V_2}{a_1 - a_2}\frac{c p_1}{P_1}, \tag{21}$$

or the force at any point of the surface is proportional to the perpendicular from the centre on the tangent plane.

The surface-density σ may be found from the equation

$$4\pi\sigma = R_1. \tag{22}$$

The total quantity of electricity on a segment cut off by a plane whose equation is $x = a$ from one sheet of the hyperboloid is

$$Q = \frac{c}{2}\frac{V_1 - V_2}{a_1 - a_2}\left(\frac{a}{\lambda_1} - 1\right). \tag{23}$$

The quantity on the whole infinite sheet is therefore infinite.

The limiting forms of the surface are:—

(1) When $a = F(k)$ the surface is the part of the plane of xz on the positive side of the positive branch of the hyperbola whose equation is

$$\frac{x^2}{b^2} - \frac{z^2}{c^2 - b^2} = 1. \tag{24}$$

(2) When $a = 0$ the surface is the plane of yz.

(3) When $a = -F(k)$ the surface is the part of the plane of xz on the negative side of the negative branch of the same hyperbola.

The Hyperboloid of One Sheet.

By making β constant we obtain the equation of the hyperboloid of one sheet. The two surfaces which form the boundaries of the electric field must therefore belong to two different hyperboloids. The investigation will in other respects be the same as for the hyperboloids of two sheets, and when the difference of potentials is given the density at any point of the surface will be proportional to the perpendicular from the centre on the tangent plane, and the whole quantity on the infinite sheet will be infinite.

Limiting Forms.

(1) When $\beta = 0$ the surface is the part of the plane of xz between the two branches of the hyperbola whose equation is written above, (24).

(2) When $\beta = F(k')$ the surface is the part of the plane of xy which is on the outside of the focal ellipse whose equation is

$$\frac{x^2}{c^2} + \frac{y^2}{c^2 - b^2} = 1. \tag{25}$$

The Ellipsoids.

For any given ellipsoid γ is constant. If two ellipsoids, γ_1 and γ_2, be maintained at potentials V_1 and V_2, then, for any point γ in the space between them, we have

$$V = \frac{\gamma_1 V_2 - \gamma_2 V_1 + \gamma(V_1 - V_2)}{\gamma_1 - \gamma_2}. \tag{26}$$

The surface-density at any point is

$$\sigma = -\frac{1}{4\pi}\frac{V_1 - V_2}{\gamma_1 - \gamma_2}\frac{cp_3}{P_3}, \tag{27}$$

where p_3 is the perpendicular from the centre on the tangent plane, and P_3 is the product of the semi-axes.

The whole charge of electricity on either surface is given by

$$Q_2 = c\frac{V_1 - V_2}{\gamma_1 - \gamma_2} = -Q_1, \tag{28}$$

and is finite.

When $\gamma = F(k)$ the surface of the ellipsoid is at an infinite distance in all directions.

If we make $V_2 = 0$ and $\gamma_2 = F(k)$, we find for the quantity of electricity on an ellipsoid maintained at potential V in an infinitely extended field,

$$Q = c\frac{V}{F(k) - \gamma}. \tag{29}$$

The limiting form of the ellipsoids occurs when $\gamma = 0$, in which case the surface is the part of the plane of xy within the focal ellipse, whose equation is written above, (25).

The surface-density on either side of the elliptic plate whose equation is (25), and whose eccentricity is k, is

$$\sigma = \frac{V}{4\pi\sqrt{c^2 - b^2}}\frac{1}{F(k)}\frac{1}{\sqrt{1 - \frac{x^2}{c^2} - \frac{y^2}{c^2 - b^2}}}, \tag{30}$$

and its charge is

$$Q = c\frac{V}{F(k)}. \tag{31}$$

Particular Cases.

151.] If c remains finite, while b and therefore k is diminished till it becomes ultimately zero, the system of surfaces becomes transformed in the following manner :—

The real axis and one of the imaginary axes of each of the hyperboloids of two sheets are indefinitely diminished, and the surface ultimately coincides with two planes intersecting in the axis of z.

The quantity a becomes identical with θ, and the equation of the system of meridional planes to which the first system is reduced is

$$\frac{x^2}{(\sin a)^2} - \frac{y^2}{(\cos a)^2} = 0. \tag{32}$$

As regards the quantity β, if we take the definition given in page 216 (7) we shall be led to an infinite value of the integral at the lower limit. In order to avoid this we define β in this particular case as the value of the integral

$$\int_{\lambda_2}^{c} \frac{c\,d\lambda_2}{\lambda_2\sqrt{c^2-\lambda_2{}^2}}.$$

If we now put $\lambda_2 = c \sin\phi$, β becomes

$$\int_{\phi}^{\frac{\pi}{2}} \frac{d\phi}{\sin\phi}, \qquad \text{i.e. } \log\cot\tfrac{1}{2}\phi.$$

Whence $$\cos\phi = \frac{e^{\beta}-e^{-\beta}}{e^{\beta}+e^{-\beta}},$$

and therefore $$\sin\phi = \frac{2}{e^{\beta}+e^{-\beta}}.$$

If we call the exponential quantity $\frac{1}{2}(e^{\beta}+e^{-\beta})$ the hyperbolic cosine of β, or more concisely the hypocosine of β, or $\cosh\beta$, and if we call $\frac{1}{2}(e^{\beta}-e^{-\beta})$ the hyposine of β, or $\sinh\beta$, and if in the same way we employ functions of a similar character analogous to the other simple trigonometrical ratios, then $\lambda_2 = c\,\text{sech}\,\beta$, and the equation of the system of hyperboloids of one sheet is

$$\frac{x^2+y^2}{(\text{sech}\,\beta)^2} - \frac{z^2}{(\tanh\beta)^2} = c^2. \tag{35}$$

The quantity γ is reduced to ψ, so that $\lambda_3 = c\,\text{cosec}\,\gamma$, and the equation of the system of ellipsoids is

$$\frac{x^2+y^2}{(\sec\gamma)^2} + \frac{z^2}{(\tan\gamma)^2} = c^2. \tag{36}$$

Ellipsoids of this kind, which are figures of revolution about their conjugate axes, are called planetary ellipsoids.

The quantity of electricity on a planetary ellipsoid maintained at potential V in an infinite field, is

$$Q = c\frac{V}{\frac{1}{2}\pi - \gamma}, \tag{37}$$

where $c \sec\gamma$ is the equatorial radius, and $c \tan\gamma$ is the polar radius.

If $\gamma = 0$, the figure is a circular disk of radius c, and

$$\sigma = \frac{V}{2\pi^2\sqrt{c^2 - r^2}}, \tag{38}$$

$$Q = c\frac{V}{\frac{1}{2}\pi}. \tag{39}$$

152.] *Second Case.* Let $b = c$, then $k = 1$ and $k' = 0$,

$$a = \log\tan\frac{\pi + 2\theta}{4}, \text{ whence } \lambda_1 = c\tanh a, \tag{40}$$

and the equation of the hyperboloids of revolution of two sheets becomes

$$\frac{x^2}{(\tanh a)^2} - \frac{y^2 + z^2}{(\operatorname{sech} a)^2} = c^2. \tag{41}$$

The quantity β becomes reduced to ϕ, and each of the hyperboloids of one sheet is reduced to a pair of planes intersecting in the axis of x whose equation is

$$\frac{y^2}{(\sin\beta)^2} - \frac{z^2}{(\cos\beta)^2} = 0. \tag{42}$$

This is a system of meridional planes in which β is the longitude.

The quantity γ as defined in page 216, (7) becomes in this case infinite at the lower limit. To avoid this let us define it as the value of the integral $\displaystyle\int_{\lambda_3}^{\infty}\frac{c\,d\lambda_3}{\lambda_3^2 - c^2}$.

If we then put $\lambda_3 = c\sec\psi$, we find $\gamma = \displaystyle\int_{\psi}^{\frac{\pi}{2}}\frac{d\psi}{\sin\psi}$, whence $\lambda_3 = c\coth\gamma$, and the equation of the family of ellipsoids is

$$\frac{x^2}{(\coth\gamma)^2} + \frac{y^2 + z^2}{(\operatorname{cosech}\gamma)^2} = c^2. \tag{43}$$

These ellipsoids, in which the transverse axis is the axis of revolution, are called ovary ellipsoids.

The quantity of electricity on an ovary ellipsoid maintained at potential V in an infinite field becomes in this case, by (29),

$$cV \div \int_{\psi_0}^{\frac{\pi}{2}}\frac{d\psi}{\sin\psi}, \tag{44}$$

where $c\sec\psi_0$ is the polar radius.

If we denote the polar radius by A and the equatorial by B, the result just found becomes

$$V\frac{\sqrt{A^2-B^2}}{\log\frac{A+\sqrt{A^2-B^2}}{B}}. \qquad (45)$$

If the equatorial radius is very small compared to the polar radius, as in a wire with rounded ends,

$$Q=\frac{AV}{\log 2A-\log B}. \qquad (46)$$

When both b and c become zero, their ratio remaining finite, the system of surfaces becomes two systems of confocal cones, and a system of spherical surfaces of which the radius is inversely proportional to γ.

If the ratio of b to c is zero or unity, the system of surfaces becomes one system of meridian planes, one system of right cones having a common axis, and a system of concentric spherical surfaces of which the radius is inversely proportional to γ. This is the ordinary system of spherical polar coordinates.

Cylindric Surfaces.

153.] When c is infinite the surfaces are cylindric, the generating lines being parallel to the axes of z. One system of cylinders is hyperbolic, viz. that into which the hyperboloids of two sheets degenerate. Since, when c is infinite, k is zero, and therefore $\theta=a$, it follows that the equation of this system is

$$\frac{x^2}{\sin^2 a}-\frac{y^2}{\cos^2 a}=b^2. \qquad (47)$$

The other system is elliptic, and since when $k=0$, β becomes

$$\int_b^{\lambda_2}\frac{d\lambda_2}{\sqrt{\lambda_2^{\,2}-b^2}}, \text{ or } \lambda_2=b\cosh\beta,$$

the equation of this system is

$$\frac{x^2}{(\cosh\beta)^2}+\frac{y^2}{(\sinh\beta)^2}=b^2. \qquad (48)$$

These two systems are represented in Fig. X at the end of this volume.

Confocal Paraboloids.

154.] If in the general equations we transfer the origin of coordinates to a point on the axis of x distant t from the centre of the system, and if for x, λ, b, and c we substitute $t+x$, $t+\lambda$, $t+b$, and $t+c$ respectively, and then make t increase indefinitely, we obtain, in the limit, the equation of a system of paraboloids whose foci are at the points $x=b$ and $x=c$, viz. the equation is

$$4(x-\lambda)+\frac{y^2}{\lambda-b}+\frac{z^2}{\lambda-c}=0. \qquad (49)$$

If the variable parameter is λ for the first system of elliptic paraboloids, μ for the hyperbolic paraboloids, and ν for the second system of elliptic paraboloids, we have λ, b, μ, c, ν in ascending order of magnitude, and

$$\left.\begin{aligned} x &= \lambda+\mu+\nu-c-b, \\ y^2 &= 4\frac{(b-\lambda)(\mu-b)(\nu-b)}{c-b}, \\ z^2 &= 4\frac{(c-\lambda)(c-\mu)(\nu-c)}{c-b}. \end{aligned}\right\} \tag{50}$$

In order to avoid infinite values in the integrals (7) the corresponding integrals in the paraboloidal system are taken between different limits.

We write in this case

$$\alpha = \int_\lambda^b \frac{d\lambda}{\sqrt{(b-\lambda)(c-\lambda)}},$$

$$\beta = \int_b^\mu \frac{d\mu}{\sqrt{(\mu-b)(c-\mu)}},$$

$$\gamma = \int_c^\nu \frac{d\nu}{\sqrt{(\nu-b)(\nu-c)}}.$$

From these we find

$$\left.\begin{aligned} \lambda &= \tfrac{1}{2}(c+b)-\tfrac{1}{2}(c-b)\cosh\alpha, \\ \mu &= \tfrac{1}{2}(c+b)-\tfrac{1}{2}(c-b)\cos\beta, \\ \nu &= \tfrac{1}{2}(c+b)+\tfrac{1}{2}(c-b)\cosh\gamma; \end{aligned}\right\} \tag{51}$$

$$\left.\begin{aligned} x &= \tfrac{1}{2}(c+b)+\tfrac{1}{2}(c-b)(\cosh\gamma-\cos\beta-\cosh\alpha), \\ y &= 2(c-b)\sinh\frac{\alpha}{2}\sin\frac{\beta}{2}\cosh\frac{\gamma}{2}, \\ z &= 2(c-b)\cosh\frac{\alpha}{2}\cos\frac{\beta}{2}\sinh\frac{\gamma}{2}. \end{aligned}\right\} \tag{52}$$

When $b=c$ we have the case of paraboloids of revolution about the axis of x, and

$$\begin{aligned} x &= a(e^{2\alpha}-e^{2\gamma}), \\ y &= 2ae^{\alpha+\gamma}\cos\beta, \\ z &= 2ae^{\alpha+\gamma}\sin\beta. \end{aligned} \tag{53}$$

The surfaces for which β is constant are planes through the axis, β being the angle which such a plane makes with a fixed plane through the axis.

The surfaces for which α is constant are confocal paraboloids. When $\alpha=-\infty$ the paraboloid is reduced to a straight line terminating at the origin.

We may also find the values of α, β, γ in terms of r, θ, and ϕ, the spherical polar coordinates referred to the focus as orgin, and the axis of the parabolas as axis of the sphere,

$$\begin{aligned} \alpha &= \log(r^{\frac{1}{2}} \cos \tfrac{1}{2}\theta), \\ \beta &= \phi, \\ \gamma &= \log(r^{\frac{1}{2}} \sin \tfrac{1}{2}\theta). \end{aligned} \tag{54}$$

We may compare the case in which the potential is equal to α, with the zonal solid harmonic $r_i Q_i$. Both satisfy Laplace's equation, and are homogeneous functions of x, y, z, but in the case derived from the paraboloid there is a discontinuity at the axis, and i has a value not differing by any finite quantity from zero.

The surface-density on an electrified paraboloid in an infinite field (including the case of a straight line infinite in one direction) is inversely as the square root of the distance from the focus, or, in the case of the line, from the extremity of the line.

CHAPTER XI.

THEORY OF ELECTRIC IMAGES AND ELECTRIC INVERSION.

155.] We have already shewn that when a conducting sphere is under the influence of a known distribution of electricity, the distribution of electricity on the surface of the sphere can be determined by the method of spherical harmonics.

For this purpose we require to expand the potential of the influencing system in a series of solid harmonics of positive degree, having the centre of the sphere as origin, and we then find a corresponding series of solid harmonics of negative degree, which express the potential due to the electrification of the sphere.

By the use of this very powerful method of analysis, Poisson determined the electrification of a sphere under the influence of a given electrical system, and he also solved the more difficult problem to determine the distribution of electricity on two conducting spheres in presence of each other. These investigations have been pursued at great length by Plana and others, who have confirmed the accuracy of Poisson.

In applying this method to the most elementary case of a sphere under the influence of a single electrified point, we require to expand the potential due to the electrified point in a series of solid harmonics, and to determine a second series of solid harmonics which express the potential, due to the electrification of the sphere, in the space outside.

It does not appear that any of these mathematicians observed that this second series expresses the potential due to an imaginary electrified point, which has no physical existence as an electrified point, but which may be called an electrical image, because the action of the surface on external points is the same as that which would be produced by the imaginary electrified point if the spherical surface were removed.

This discovery seems to have been reserved for Sir W. Thomson, who has developed it into a method of great power for the solution of electrical problems, and at the same time capable of being presented in an elementary geometrical form.

His original investigations, which are contained in the *Cambridge and Dublin Mathematical Journal*, 1848, are expressed in terms of the ordinary theory of attraction at a distance, and make no use of the method of potentials and of the general theorems of Chapter IV, though they were probably discovered by these methods. Instead, however, of following the method of the author, I shall make free use of the idea of the potential and of equipotential surfaces, whenever the investigation can be rendered more intelligible by such means.

Theory of Electric Images.

156.] Let A and B, Figure 7, represent two points in a uniform dielectric medium of infinite extent. Let the charges of A and B be e_1 and e_2 respectively. Let P be any point in space whose distances from A and B are r_1 and r_2 respectively. Then the value of the potential at P will be

$$V = \frac{e_1}{r_1} + \frac{e_2}{r_2}. \qquad (1)$$

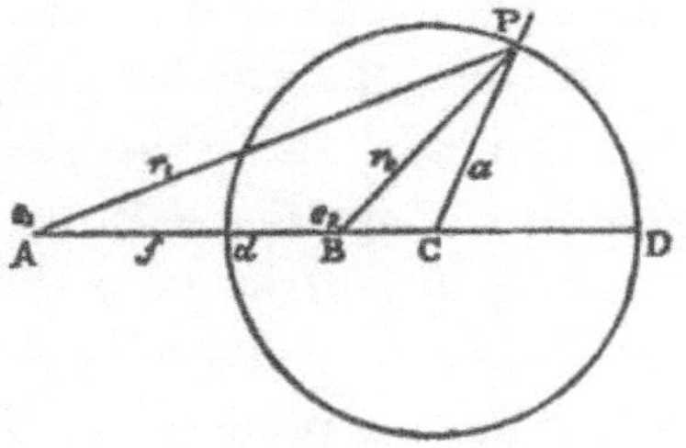

Fig. 7.

The equipotential surfaces due to this distribution of electricity are represented in Fig. I (at the end of this volume) when e_1 and e_2 are of the same sign, and in Fig. II when they are of opposite signs. We have now to consider that surface for which $V = 0$, which is the only spherical surface in the system. When e_1 and e_2 are of the same sign, this surface is entirely at an infinite distance, but when they are of opposite signs there is a plane or spherical surface at a finite distance for which the potential is zero.

The equation of this surface is

$$\frac{e_1}{r_1} + \frac{e_2}{r_2} = 0. \qquad (2)$$

Its centre is at a point C in AB produced, such that

$$AC : BC :: e_1^2 : e_2^2,$$

and the radius of the sphere is

$$AB \frac{e_1 e_2}{e_1^2 - e_2^2}.$$

The two points A and B are inverse points with respect to this sphere, that is to say, they lie in the same radius, and the radius is a mean proportional between their distances from the centre.

Since this spherical surface is at potential zero, if we suppose it constructed of thin metal and connected with the earth, there will be no alteration of the potential at any point either outside or inside, but the electrical action everywhere will remain that due to the two electrified points A and B.

If we now keep the metallic shell in connection with the earth and remove the point B, the potential within the sphere will become everywhere zero, but outside it will remain the same as before. For the surface of the sphere still remains at the same potential, and no change has been made in the exterior electrification.

Hence, if an electrified point A be placed outside a spherical conductor which is at potential zero, the electrical action at all points outside the sphere will be that due to the point A together with another point B within the sphere, which we may call the electrical image of A.

In the same way we may shew that if B is a point placed inside the spherical shell, the electrical action within the sphere is that due to B, together with its image A.

157.] *Definition of an Electrical Image.* An electrical image is an electrified point or system of points on one side of a surface which would produce on the other side of that surface the same electrical action which the actual electrification of that surface really does produce.

In Optics a point or system of points on one side of a mirror or lens which if it existed would emit the system of rays which actually exists on the other side of the mirror or lens, is called a *virtual* image.

Electrical images correspond to virtual images in Optics in being related to the space on the other side of the surface. They do not correspond to them in actual position, or in the merely approximate character of optical foci.

There are no *real* electrical images, that is, imaginary electrified points which would produce, in the region on the same side of the electrified surface, an effect equivalent to that of the electrified surface.

For if the potential in any region of space is equal to that due to a certain electrification in the same region it must be actually produced by that electrification. In fact, the electrification at any point may be found from the potential near that point by the application of Poisson's equation.

Let a be the radius of the sphere.

Let f be the distance of the electrified point A from the centre C.

Let e be the charge of this point.

Then the image of the point is at B, on the same radius of the sphere at a distance $\frac{a^2}{f}$, and the charge of the image is $-e\frac{a}{f}$.

We have shewn that this image will produce the same effect on the opposite side of the surface as the actual electrification of the surface does. We shall next determine the surface-density of this electrification at any point P of the spherical surface, and for this purpose we shall make use of the theorem of Coulomb, Art. 80, that if R is the resultant force at the surface of a conductor, and σ the superficial density,

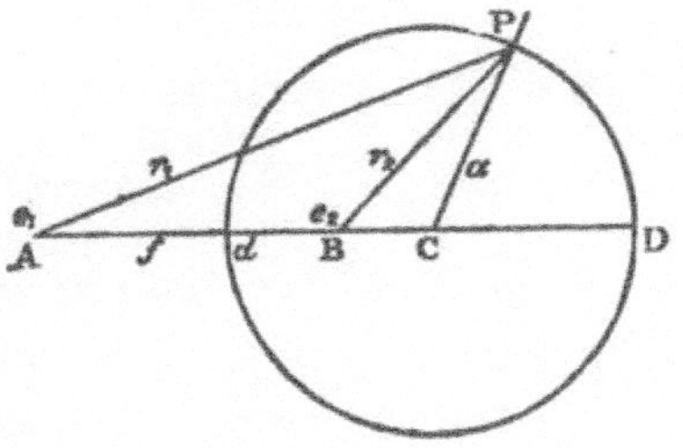

Fig. 7.

$$R = 4\pi\sigma,$$

R being measured away from the surface.

We may consider R as the resultant of two forces, a repulsion $\frac{e}{AP^2}$ acting along AP, and an attraction $e\frac{a}{f}\frac{1}{PB^2}$ acting along PB.

Resolving these forces in the directions of AC and CP, we find that the components of the repulsion are

$$\frac{ef}{AP^3} \text{ along } AC, \text{ and } \frac{ea}{AP^3} \text{ along } CP.$$

Those of the attraction are

$$-e\frac{a}{f}\frac{1}{BP^3}BC \text{ along } AC, \text{ and } -e\frac{a^2}{f}\frac{1}{BP^3} \text{ along } CP.$$

Now $BP = \frac{a}{f}AP$, and $BC = \frac{a^2}{f}$, so that the components of the attraction may be written

$$-ef\frac{1}{AP^3} \text{ along } AC, \text{ and } -e\frac{f^2}{a}\frac{1}{AP^3} \text{ along } CP.$$

The components of the attraction and the repulsion in the direction of AC are equal and opposite, and therefore the resultant force is entirely in the direction of the radius CP. This only confirms what we have already proved, that the sphere is an equipotential surface, and therefore a surface to which the resultant force is everywhere perpendicular.

The resultant force measured along CP, the normal to the surface in the direction towards the side on which A is placed, is

$$R = -e\frac{f^2-a^2}{a}\frac{1}{AP^3}. \tag{3}$$

If A is taken inside the sphere f is less than a, and we must measure R inwards. For this case therefore

$$R = -e\frac{a^2-f^2}{a}\frac{1}{AP^3}. \tag{4}$$

In all cases we may write

$$R = -e\frac{AD.Ad}{CP}\frac{1}{AP^3}, \tag{5}$$

where AD, Ad are the segments of any line through A cutting the sphere, and their product is to be taken positive in all cases.

158.] From this it follows, by Coulomb's theorem, Art. 80, that the surface-density at P is

$$\sigma = -e\frac{AD.Ad}{4\pi.CP}\frac{1}{AP^3}. \tag{6}$$

The density of the electricity at any point of the sphere varies inversely as the cube of its distance from the point A.

The effect of this superficial distribution, together with that of the point A, is to produce on the same side of the surface as the point A a potential equivalent to that due to e at A, and its image $-e\frac{a}{f}$ at B, and on the other side of the surface the potential is everywhere zero. Hence the effect of the superficial distribution by itself is to produce a potential on the side of A equivalent to that due to the image $-e\frac{a}{f}$ at B, and on the opposite side a potential equal and opposite to that of e at A.

The whole charge on the surface of the sphere is evidently $-e\frac{a}{f}$ since it is equivalent to the image at B.

We have therefore arrived at the following theorems on the action of a distribution of electricity on a spherical surface, the surface-density being inversely as the cube of the distance from a point A either without or within the sphere.

Let the density be given by the equation

$$\sigma = \frac{C}{AP^3}, \tag{7}$$

where C is some constant quantity, then by equation (6)

$$C = -e\frac{AD.Ad}{4\pi a}. \tag{8}$$

The action of this superficial distribution on any point separated from A by the surface is equal to that of a quantity of electricity $-e$, or

$$\frac{4\pi a C}{AD.Ad}$$

concentrated at A.

Its action on any point on the same side of the surface with A is equal to that of a quantity of electricity

$$\frac{4\pi C a^2}{f\,AD.Ad}$$

concentrated at B the image of A.

The whole quantity of electricity on the sphere is equal to the first of these quantities if A is within the sphere, and to the second if A is without the sphere.

These propositions were established by Sir W. Thomson in his original geometrical investigations with reference to the distribution of electricity on spherical conductors, to which the student ought to refer.

159.] If a system in which the distribution of electricity is known is placed in the neighbourhood of a conducting sphere of radius a, which is maintained at potential zero by connection with the earth, then the electrifications due to the several parts of the system will be superposed.

Let A_1, A_2, &c. be the electrified points of the system, f_1, f_2, &c. their distances from the centre of the sphere, e_1, e_2, &c. their charges, then the images B_1, B_2, &c. of these points will be in the same radii as the points themselves, and at distances $\frac{a^2}{f_1}$, $\frac{a^2}{f_2}$, &c. from the centre of the sphere, and their charges will be

$$-e_1\frac{a}{f_1}, \quad -e_2\frac{a}{f_2}, \text{ \&c.}$$

The potential on the outside of the sphere due to the superficial electrification will be the same as that which would be produced by the system of images B_1, B_2, &c. This system is therefore called the electrical image of the system A_1, A_2, &c.

If the sphere instead of being at potential zero is at potential V, we must superpose a distribution of electricity on its outer surface having the uniform surface-density

$$\sigma = \frac{V}{4\pi a}.$$

The effect of this at all points outside the sphere will be equal to that of a quantity Va of electricity placed at its centre, and at all points inside the sphere the potential will be simply increased by V.

The whole charge on the sphere due to an external system of influencing points, A_1, A_2, &c. is

$$E = Va - e_1\frac{a}{f_1} - e_2\frac{a}{f_2} - \&c., \qquad (9)$$

from which either the charge E or the potential V may be calculated when the other is given.

When the electrified system is within the spherical surface the induced charge on the surface is equal and of opposite sign to the inducing charge, as we have before proved it to be for every closed surface, with respect to points within it.

*160.] The energy due to the mutual action between an electrified point e, at a distance f from the centre of the sphere greater than a the radius, and the electrification of the spherical surface due to the influence of the electrified point and the charge of the sphere, is

$$M = \frac{Ee}{f} - \frac{1}{2}\frac{e^2 a^3}{f^2(f^2-a^2)}, \qquad (10)$$

where V is the potential, and E the charge of the sphere.

The repulsion between the electrified point and the sphere is therefore, by Art. 92,

* [The discussion in the text will perhaps be more easily understood if the problem be regarded as an example of Art. 86. Let us then suppose that what is described as an electrified point is really a small spherical conductor, the radius of which is b and the potential v. We have thus a particular case of the problem of two spheres of which one solution has already been given in Art. 146, and another will be given in Art. 173. In the case before us however the radius b is so small that we may consider the electricity of the small conductor to be uniformly distributed over its surface and all the electric images except the first image of the small conductor to be disregarded.

We thus have
$$V = \frac{E}{a} + \frac{e}{f},$$

$$v = \frac{E + e\dfrac{a}{f}}{f} - \frac{ea}{f^2-a^2} + \frac{e}{b}.$$

The energy of the system is therefore, Art. 85,

$$\frac{E^2}{2a} + \frac{Ee}{f} + \frac{e^2}{2}\left(\frac{1}{b} - \frac{a^3}{f^2(f^2-a^2)}\right).$$

By means of the above equations we may also express the energy in terms of the potentials: to the same order of approximation it is

$$\frac{aV^2}{2} - \frac{ab}{f}Vv + \frac{1}{2}\left(b + \frac{ab^2}{f^2-a^2}\right)v^2.]$$

$$F = ea\left(\frac{V}{f^2} - \frac{ef}{(f^2-a^2)^2}\right)$$

$$= \frac{e}{f^2}\left(E - e\frac{a^3(2f^2-a^2)}{f(f^2-a^2)^2}\right). \qquad (11)$$

Hence the force between the point and the sphere is always an attraction in the following cases—

(1) When the sphere is uninsulated.

(2) When the sphere has no charge.

(3) When the electrified point is very near the surface.

In order that the force may be repulsive, the potential of the sphere must be positive and greater than $e\frac{f^3}{(f^2-a^2)^2}$, and the charge of the sphere must be of the same sign as e and greater than $e\frac{a^3(2f^2-a^2)}{f(f^2-a^2)^2}$.

At the point of equilibrium the equilibrium is unstable, the force being an attraction when the bodies are nearer and a repulsion when they are farther off.

When the electrified point is within the spherical surface the force on the electrified point is always away from the centre of the sphere, and is equal to

$$\frac{e^2af}{(a^2-f^2)^2}.$$

The surface-density at the point of the sphere nearest to the electrified point where it lies outside the sphere is

$$\sigma_1 = \frac{1}{4\pi a^2}\left\{Va - e\frac{a(f+a)}{(f-a)^2}\right\}$$

$$= \frac{1}{4\pi a^2}\left\{E - e\frac{a^2(3f-a)}{f(f-a)^2}\right\}. \qquad (12)$$

The surface-density at the point of the sphere farthest from the electrified point is

$$\sigma_2 = \frac{1}{4\pi a^2}\left\{Va - e\frac{a(f-a)}{(f+a)^2}\right\}$$

$$= \frac{1}{4\pi a^2}\left\{E + e\frac{a^2(3f+a)}{f(f+a)^2}\right\}. \qquad (13)$$

When E, the charge of the sphere, lies between

$$e\frac{a^2(3f-a)}{f(f-a)^2} \quad \text{and} \quad -e\frac{a^2(3f+a)}{f(f+a)^2}$$

the electrification will be negative next the electrified point and positive on the opposite side. There will be a circular line of division between the positively and the negatively electrified parts of the surface, and this line will be a line of equilibrium.

If
$$E = ea\left(\frac{1}{\sqrt{f^2-a^2}} - \frac{1}{f}\right), \tag{14}$$

the equipotential surface which cuts the sphere in the line of equilibrium is a sphere whose centre is the electrified point and whose radius is $\sqrt{f^2-a^2}$.

The lines of force and equipotential surfaces belonging to a case of this kind are given in Figure IV at the end of this volume.

Images in an Infinite Plane Conducting Surface.

161.] If the two electrified points A and B in Art. 156 are electrified with equal charges of electricity of opposite signs, the surfaces of zero potential will be the plane, every point of which is equidistant from A and B.

Hence, if A be an electrified point whose charge is e, and AD a perpendicular on the plane, produce AD to B so that $DB = AB$, and place at B a charge equal to $-e$, then this charge at B will be the image of A, and will produce at all points on the same side of the plane as A, an effect equal to that of the actual electrification of the plane. For the potential on the side of A due to A and B fulfils the conditions that $\nabla^2 V = 0$ everywhere except at A, and that $V = 0$ at the plane, and there is only one form of V which can fulfil these conditions.

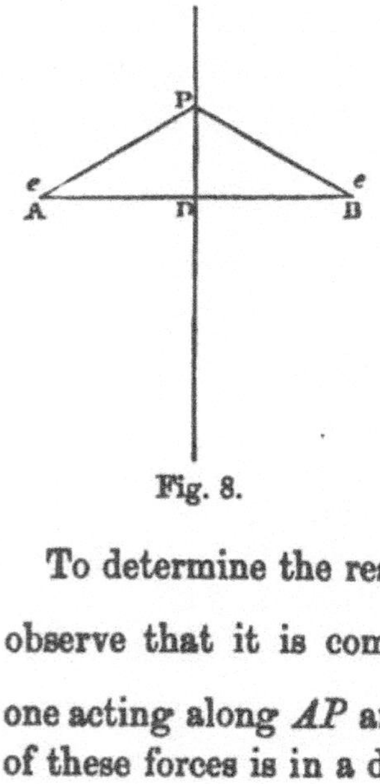

Fig. 8.

To determine the resultant force at the point P of the plane, we observe that it is compounded of two forces each equal to $\frac{e}{AP^2}$, one acting along AP and the other along PB. Hence the resultant of these forces is in a direction parallel to AB and equal to

$$\frac{e}{AP^2} \cdot \frac{AB}{AP}.$$

Hence R, the resultant force measured from the surface towards the space in which A lies, is

$$R = -\frac{2eAD}{AP^3}, \tag{15}$$

and the density at the point P is

$$\sigma = -\frac{eAD}{2\pi AP^3}. \tag{16}$$

On Electrical Inversion.

162.] The method of electrical images leads directly to a method of transformation by which we may derive from any electrical problem of which we know the solution any number of other problems with their solutions.

We have seen that the image of a point at a distance r from the centre of a sphere of radius R is in the same radius and at a distance r' such that $rr' = R^2$. Hence the image of a system of points, lines, or surfaces is obtained from the original system by the method known in pure geometry as the method of inversion, and described by Chasles, Salmon, and other mathematicians.

If A and B are two points, A' and B' their images, O being the centre of inversion, and R the radius of the sphere of inversion,

$$OA.OA' = R^2 = OB.OB'.$$

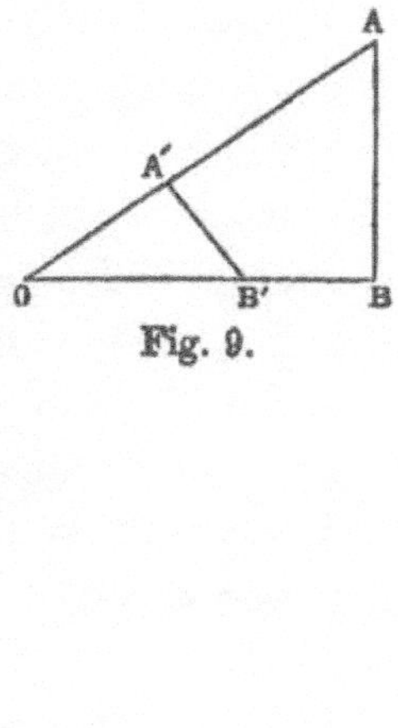

Fig. 9.

Hence the triangles OAB, $OB'A'$ are similar, and $AB : A'B' :: OA : OB' :: OA.OB : R^2$.

If a quantity of electricity e be placed at A, its potential at B will be $V = \frac{e}{AB}$.

If e' be placed at A' its potential at B' will be

$$V' = \frac{e'}{A'B'}.$$

In the theory of electrical images

$$e : e' :: OA : R :: R : OA'.$$

Hence $$V : V' :: R : OB, \tag{17}$$

or the potential at B due to the electricity at A is to the potential at the image of B due to the electrical image of A as R is to OB.

Since this ratio depends only on OB and not on OA, the potential at B due to any system of electrified bodies is to that at B' due to the image of the system as R is to OB.

If r be the distance of any point A from the centre, and r' that of its image A', and if e be the electrification of A, and e' that of A', also if L, S, K be linear, superficial, and solid elements at A, and L', S', K' their images at A', and λ, σ, ρ, λ', σ', ρ' the corresponding line surface and volume densities of electricity at the two points,

V the potential at A due to the original system, and V' the potential at A' due to the inverse system, then

$$\left.\begin{aligned}
\frac{r'}{r} = \frac{L'}{L} = \frac{R^2}{r^2} = \frac{r'^2}{R^2}, \quad \frac{S'}{S} = \frac{R^4}{r^4} = \frac{r'^4}{R^4}, \quad \frac{K'}{K} = \frac{R^6}{r^6} = \frac{r'^6}{R^6};\\
\frac{e'}{e} = \frac{R}{r} = \frac{r'}{R}, \quad \frac{\lambda'}{\lambda} = \frac{r}{R} = \frac{R}{r'},\\
\frac{\sigma'}{\sigma} = \frac{r^3}{R^3} = \frac{R^3}{r'^3}, \quad \frac{\rho'}{\rho} = \frac{r^5}{R^5} = \frac{R^5}{r'^5},\\
\frac{V'}{V} = \frac{r}{R} = \frac{R}{r'}.
\end{aligned}\right\} \text{*}(18)$$

If in the original system a certain surface is that of a conductor, and has therefore a constant potential P, then in the transformed system the image of the surface will have a potential $P\frac{R}{r'}$. But by placing at O, the centre of inversion, a quantity of electricity equal to $-PR$, the potential of the transformed surface is reduced to zero.

Hence, if we know the distribution of electricity on a conductor when insulated in open space and charged to the potential P, we can find by inversion the distribution on a conductor whose form is the image of the first under the influence of an electrified point with a charge $-PR$ placed at the centre of inversion, the conductor being in connexion with the earth.

163.] The following geometrical theorems are useful in studying cases of inversion.

Every sphere becomes, when inverted, another sphere, unless it passes through the centre of inversion, in which case it becomes a plane.

If the distances of the centres of the spheres from the centre of inversion are a and a', and if their radii are α and α', and if we define the *power* of a sphere with respect to the centre of inversion to be the product of the segments cut off by the sphere from a line through the centre of inversion, then the power of the first sphere is $a^2-\alpha^2$, and that of the second is $a'^2-\alpha'^2$. We have in this case

$$\frac{a'}{a} = \frac{\alpha'}{\alpha} = \frac{R^2}{a^2-\alpha^2} = \frac{a'^2-\alpha'^2}{R^2}, \tag{19}$$

or the ratio of the distances of the centres of the first and second spheres is equal to the ratio of their radii, and to the ratio of the

* See Thomson and Tait's *Natural Philosophy*, § 515.

power of the sphere of inversion to the power of the first sphere, or of the power of the second sphere to the power of the sphere of inversion.

The image of the centre of inversion with regard to one sphere is the inverse point of the centre of the other sphere.

In the case in which the inverse surfaces are a plane and a sphere, the perpendicular from the centre of inversion on the plane is to the radius of inversion as this radius is to the diameter of the sphere, and the sphere has its centre on this perpendicular and passes through the centre of inversion.

Every circle is inverted into another circle unless it passes through the centre of inversion, in which case it becomes a straight line.

The angle between two surfaces, or two lines at their intersection, is not changed by inversion.

Every circle which passes through a point, and the image of that point with respect to a sphere, cuts the sphere at right angles.

Hence, any circle which passes through a point and cuts the sphere at right angles passes through the image of the point.

164.] We may apply the method of inversion to deduce the distribution of electricity on an uninsulated sphere under the influence of an electrified point from the uniform distribution on an insulated sphere not influenced by any other body.

If the electrified point be at A, take it for the centre of inversion, and if A is at a distance f from the centre of the sphere whose radius is a, the inverted figure will be a sphere whose radius is a' and whose centre is distant f', where

$$\frac{a'}{a} = \frac{f'}{f} = \frac{R^2}{f^2-a^2}. \tag{20}$$

The centre of either of these spheres corresponds to the inverse point of the other with respect to A, or if C is the centre and B the inverse point of the first sphere, C' will be the inverse point, and B' the centre of the second.

Now let a quantity e' of electricity be communicated to the second sphere, and let it be uninfluenced by external forces. It will become uniformly distributed over the sphere with a surface-density

$$\sigma' = \frac{e'}{4\pi a'^2}. \tag{21}$$

Its action at any point outside the sphere will be the same as that of a charge e' placed at B' the centre of the sphere.

At the spherical surface and within it the potential is

$$P' = \frac{e'}{a'}, \tag{22}$$

a constant quantity.

Now let us invert this system. The centre B' becomes in the inverted system the inverse point B, and the charge e' at B' becomes $e'\frac{R}{f'}$ at B, and at any point separated from B by the surface the potential is that due to this charge at B.

The potential at any point P on the spherical surface, or on the same side as B, is in the inverted system

$$\frac{e'}{a'}\frac{R}{AP}.$$

If we now superpose on this system a charge e at A, where

$$e = -\frac{e'}{a'}R, \tag{23}$$

the potential on the spherical surface, and at all points on the same side as B, will be reduced to zero. At all points on the same side as A the potential will be that due to a charge e at A, and a charge $e'\frac{R}{f'}$ at B.

But
$$e'\frac{R}{f'} = -e\frac{a'}{f'} = -e\frac{a}{f}, \tag{24}$$

as we found before for the charge of the image at B.

To find the density at any point of the first sphere we have

$$\sigma = \sigma'\frac{R^3}{AP^3}. \tag{25}$$

Substituting for the value of σ' in terms of the quantities belonging to the first sphere, we find the same value as in Art. 158,

$$\sigma = -\frac{e(f^2 - a^2)}{4\pi a AP^3}. \tag{26}$$

On Finite Systems of Successive Images.

165.] If two conducting planes intersect at an angle which is a submultiple of two right angles, there will be a finite system of images which will completely determine the electrification.

For let AOB be a section of the two conducting planes perpendicular to their line of intersection, and let the angle of intersection $AOB = \frac{\pi}{n}$, let P be an electrified point, and let $PO = r$, and $POB = \theta$. Then, if we draw a circle with centre O and radius

OP, and find points which are the successive images of P in the two planes beginning with OB, we shall find Q_1 for the image of P in OB, P_2 for the image of Q_1 in OA, Q_3 for that of P_2 in OB, P_3 for that of Q_3 in OA, and Q_2 for that of P_3 in OB.

If we had begun with the image of P in AO we should have found the same points in the reverse order Q_2, P_3, Q_3, P_2, Q_1, provided AOB is a submultiple of two right angles.

For the alternate images P_1, P_2, P_3 are ranged round the circle at angular intervals equal to $2AOB$, and the intermediate images Q_1, Q_2, Q_3 are at intervals of the same magnitude. Hence, if $2AOB$ is a submultiple of 2π, there will be a finite number of images, and none of these will fall within the angle AOB. If, however, AOB is not a submultiple of π, it will be impossible to represent the actual electrification as the result of a finite series of electrified points.

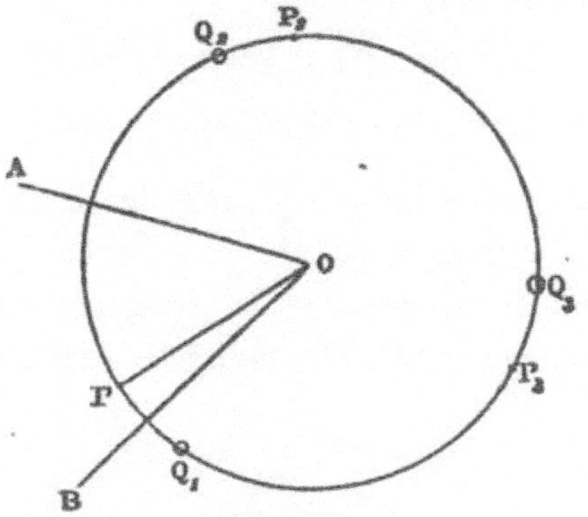

Fig. 10.

If $AOB = \frac{\pi}{n}$, there will be n negative images Q_1, Q_2, &c., each equal and of opposite sign to P, and $n-1$ positive images P_2, P_3, &c., each equal to P, and of the same sign.

The angle between successive images of the same sign is $\frac{2\pi}{n}$. If we consider either of the conducting planes as a plane of symmetry, we shall find the positive and negative images placed symmetrically with regard to that plane, so that for every positive image there is a negative image in the same normal, and at an equal distance on the opposite side of the plane.

If we now invert this system with respect to any point, the two planes become two spheres, or a sphere and a plane intersecting at an angle $\frac{\pi}{n}$, the influencing point P being within this angle.

The successive images lié on the circle which passes through P and intersects both spheres at right angles.

To find the position of the images we may make use of the principle that a point and its image are in the same radius of the sphere, and draw successive chords of the circle beginning at P and passing through the centres of the two spheres alternately.

To find the charge which must be attributed to each image, take any point in the circle of intersection, then the charge of each image is proportional to its distance from this point, and its sign is positive or negative according as it belongs to the first or the second system.

166.] We have thus found the distribution of the images when any space bounded by a conductor consisting of two spherical surfaces meeting at an angle $\frac{\pi}{n}$, and kept at potential zero, is influenced by an electrified point.

We may by inversion deduce the case of a conductor consisting of two spherical segments meeting at a re-entering angle $\frac{\pi}{n}$, charged to potential unity and placed in free space.

For this purpose we invert the system with respect to P. The circle on which the images formerly lay now becomes a straight line through the centres of the spheres.

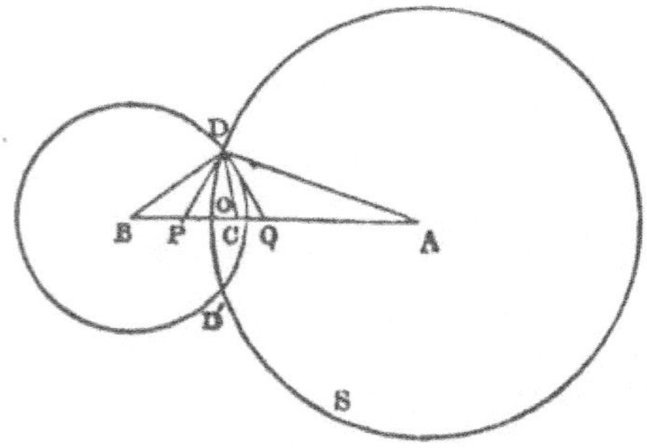

Fig. 11.

If the figure (11) represents a section through the line of centres AB, and if D, D' are the points where the circle of intersection cuts the plane of the paper, then, to find the successive images, draw DA a radius of the first circle, and draw DC, DB, &c., making angles $\frac{\pi}{n}$, $\frac{2\pi}{n}$, &c. with DA. The points C, B, &c. at which they cut the line of centres will be the positions of the positive images, and the charge of each will be represented by its distances from D. The last of these images will be at the centre of the second circle.

To find the negative images draw DP, DQ, &c., making angles $\frac{\pi}{n}$, $\frac{2\pi}{n}$, &c. with the line of centres. The intersections of these lines with the line of centres will give the positions of the negative images, and the charge of each will be represented by its distance from D.

The surface-density at any point of either sphere is the sum of the surface-densities due to the system of images. For instance, the surface-density at any point S of the sphere whose centre is A, is

$$\sigma = \frac{1}{4\pi DA}\{1+(AD^2-AB^2)\frac{DB}{BS^3}+(AD^2-AC^2)\frac{DC}{CS^3}+\&c.\},$$

where A, B, C, &c. are the positive series of images.

When S is on the circle of intersection the density is zero.

To find the total charge on each of the spherical segments, we may find the surface-integral of the induction through that segment due to each of the images.

The total charge on the segment whose centre is A due to the image at A whose charge is DA is

$$DA\,\frac{DA+OA}{2\,DA} = \tfrac{1}{2}(DA+OA),$$

where O is the centre of the circle of intersection.

In the same way the charge on the same segment due to the image at B is $\frac{1}{2}(DB+OB)$, and so on, lines such as OB measured from O to the left being reckoned negative.

Hence the total charge on the segment whose centre is A is

$$\tfrac{1}{2}(DA+DB+DC+\&c.)+\tfrac{1}{2}(OA+OB+OC+\&c.),$$
$$-\tfrac{1}{2}(DP+DQ+\&c.)-\tfrac{1}{2}(OP+OQ+\&c.).$$

167.] The method of electrical images may be applied to any space bounded by plane or spherical surfaces all of which cut one another in angles which are submultiples of two right angles.

In order that such a system of spherical surfaces may exist, every solid angle of the figure must be trihedral, and two of its angles must be right angles, and the third either a right angle or a submultiple of two right angles.

Hence the cases in which the number of images is finite are—

(1) A single spherical surface or a plane.

(2) Two planes, a sphere and a plane, or two spheres intersecting at an angle $\frac{\pi}{n}$.

(3) These two surfaces with a third, which may be either plane or spherical, cutting both orthogonally.

(4) These three surfaces with a fourth cutting the first two orthogonally and the third at an angle $\frac{\pi}{n'}$. Of these four surfaces one at least must be spherical.

We have already examined the first and second cases. In the first case we have a single image. In the second case we have $2n-1$ images arranged in two series in a circle which passes through the influencing point and is orthogonal to both surfaces.

In the third case we have, besides these images, their images with respect to the third surface, that is, $4n-1$ images in all besides the influencing point.

In the fourth case we first draw through the influencing point a circle orthogonal to the first two surfaces, and determine on it the positions and magnitudes of the n negative images and the $n-1$ positive images. Then through each of these $2n$ points, including the influencing point, we draw a circle orthogonal to the third and fourth surfaces, and determine on it two series of images, n' in each series. We shall obtain in this way, besides the influencing point, $2nn'-1$ positive and $2nn'$ negative images. These $4nn'$ points are the intersections of n circles with n' other circles, and these circles belong to the two systems of lines of curvature of a cyclide.

If each of these points is charged with the proper quantity of electricity, the surface whose potential is zero will consist of $n+n'$ spheres, forming two series of which the successive spheres of the first set intersect at angles $\frac{\pi}{n}$, and those of the second set at angles $\frac{\pi}{n'}$, while every sphere of the first set is orthogonal to every sphere of the second set.

Case of Two Spheres cutting Orthogonally. See Fig. IV at the end of this volume.

168.] Let A and B, Fig. 12, be the centres of two spheres cutting each other orthogonally in D and D', and let the straight line DD' cut the line of centres in C. Then C is the image of A with respect to the sphere B, and also the image of B with respect to the sphere whose centre is A. If $AD=\alpha$, $BD=\beta$, then $AB=\sqrt{\alpha^2+\beta^2}$, and if we place at A, B, C quantities of electricity equal to α, β, and $-\frac{\alpha\beta}{\sqrt{\alpha^2+\beta^2}}$ respectively, then both spheres will be equipotential surfaces whose potential is unity.

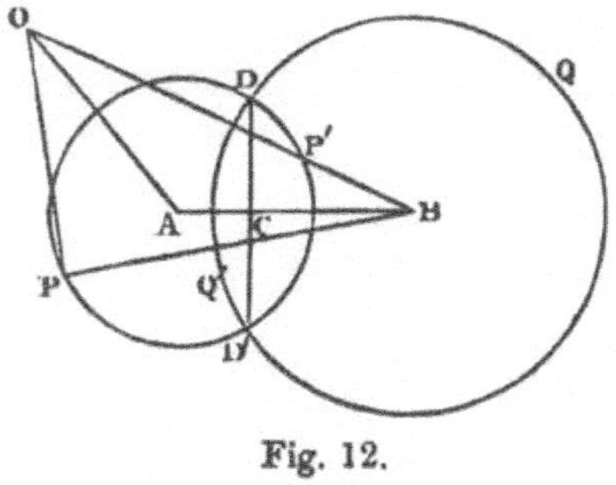

Fig. 12.

We may therefore determine from this system the distribution of electricity in the following cases:

(1) On the conductor $PDQD'$ formed of the larger segments of both spheres. Its potential is 1, and its charge is

$$\alpha+\beta-\frac{\alpha\beta}{\sqrt{\alpha^2+\beta^2}}=AD+BD-CD.$$

This quantity therefore measures the capacity of such a figure when free from the inductive action of other bodies.

The density at any point P of the sphere whose centre is A, and the density at any point Q of the sphere whose centre is B, are respectively

$$\frac{1}{4\pi\alpha}\left(1-\left(\frac{\beta}{BP}\right)^3\right) \text{ and } \frac{1}{4\pi\beta}\left(1-\left(\frac{\alpha}{AQ}\right)^3\right).$$

At the points of intersection, D, D', the density is zero.

If one of the spheres is very much larger than the other, the density at the vertex of the smaller sphere is ultimately three times that at the vertex of the larger sphere.

(2) The lens $P'DQ'D'$ formed by the two smaller segments of the spheres, charged with a quantity of electricity $= -\dfrac{\alpha\beta}{\sqrt{\alpha^2+\beta^2}}$, and acted on by points A and B, charged with quantities α and β, is also at potential unity, and the density at any point is expressed by the same formulae.

(3) The meniscus $DPD'Q'$ formed by the difference of the segments charged with a quantity α, and acted on by points B and C, charged respectively with quantities β and $\dfrac{-\alpha\beta}{\sqrt{\alpha^2+\beta^2}}$, is also in equilibrium at potential unity.

(4) The other meniscus $QDP'D'$ under the action of A and C.

We may also deduce the distribution of electricity on the following internal surfaces.

The hollow lens $P'DQ'D$ under the influence of the internal electrified point C at the centre of the circle DD'.

The hollow meniscus under the influence of a point at the centre of the concave surface.

The hollow formed of the two larger segments of both spheres under the influence of the three points A, B, C.

But, instead of working out the solutions of these cases, we shall apply the principle of electrical images to determine the density of the electricity induced at the point P of the external surface of the conductor $PDQD'$ by the action of a point at O charged with unit of electricity.

Let $OA = a, \quad OB = b, \quad OP = r, \quad BP = p,$

$AD = \alpha, \quad BD = \beta, \quad AB = \sqrt{\alpha^2 + \beta^2}.$

Invert the system with respect to a sphere of radius unity and centre O.

The two spheres will remain spheres, cutting each other orthogonally, and having their centres in the same radii with A and B. If we indicate by accented letters the quantities corresponding to the inverted system,

$$a' = \frac{a}{a^2 - \alpha^2}, \quad b' = \frac{b}{b^2 - \beta^2}, \quad \alpha' = \frac{\alpha}{a^2 - \alpha^2}, \quad \beta' = \frac{\beta}{b^2 - \beta^2},$$

$$r' = \frac{1}{r}, \quad p'^2 = \frac{\beta^2 r^2 + (b^2 - \beta^2)(p^2 - \beta^2)}{r^2 (b^2 - \beta^2)^2}.$$

If, in the inverted system, the potential of the surface is unity, then the density at the point P' is

$$\sigma' = \frac{1}{4\pi \alpha'} \left(1 - \left(\frac{\beta'}{p'}\right)^3\right).$$

If, in the original system, the density at P is σ, then

$$\frac{\sigma}{\sigma'} = \frac{1}{r^3},$$

and the potential is $\frac{1}{r}$. By placing at O a negative charge of electricity equal to unity, the potential will become zero over the surface, and the density at P will be

$$\sigma = \frac{1}{4\pi} \frac{a^2 - \alpha^2}{\alpha r^3} \left(1 - \frac{\beta^3 r^3}{(\beta^2 r^2 + (b^2 - \beta^2)(p^2 - \beta^2))^{\frac{3}{2}}}\right).$$

This gives the distribution of electricity on one of the spherical surfaces due to a charge placed at O. The distribution on the other spherical surface may be found by exchanging a and b, α and β, and putting q or AQ instead of p.

To find the total charge induced on the conductor by the electrified point at O, let us examine the inverted system.

In the inverted system we have a charge α' at A', and β' at B', and a negative charge $\frac{\alpha' \beta'}{\sqrt{\alpha'^2 + \beta'^2}}$ at a point C' in the line $A'B'$, such that $A'C' : C'B' :: \alpha'^2 : \beta'^2$.

If $OA' = a'$, $OB' = b'$, $OC' = c'$, we find

$$c'^2 = \frac{a'^2 \beta'^2 + b'^2 \alpha'^2 - \alpha'^2 \beta'^2}{\alpha'^2 + \beta'^2}.$$

Inverting this system the charges become

$$\frac{\alpha'}{a'} = \frac{\alpha}{a}, \quad \frac{\beta'}{b'} = \frac{\beta}{b};$$

and

$$-\frac{\alpha'\beta'}{\sqrt{\alpha^2+\beta^2}}\frac{1}{c'} = -\frac{\alpha\beta}{\sqrt{a^2\beta^2+b^2\alpha^2-\alpha^2\beta^2}}.$$

Hence the whole charge on the conductor due to a unit of negative electricity at O is

$$\frac{\alpha}{a}+\frac{\beta}{b}-\frac{\alpha\beta}{\sqrt{a^2\beta^2+b^2\alpha^2-\alpha^2\beta^2}}.$$

Distribution of Electricity on Three Spherical Surfaces which Intersect at Right Angles.

169.] Let the radii of the spheres be α, β, γ, then

$$BC = \sqrt{\beta^2+\gamma^2}, \quad CA = \sqrt{\gamma^2+\alpha^2}, \quad AB = \sqrt{\alpha^2+\beta^2}.$$

Let PQR, Fig. 13, be the feet of the perpendiculars from ABC on the opposite sides of the triangle, and let O be the intersection of perpendiculars.

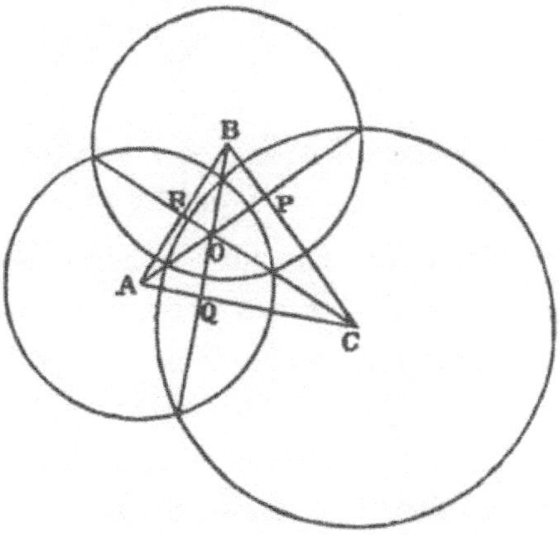

Fig. 13.

Then P is the image of B in the sphere γ, and also the image of C in the sphere β. Also O is the image of P in the sphere α.

Let charges α, β, and γ be placed at A, B, and C.

Then the charge to be placed at P is

$$-\frac{\beta\gamma}{\sqrt{\beta^2+\gamma^2}} = -\frac{1}{\sqrt{\frac{1}{\beta^2}+\frac{1}{\gamma^2}}}.$$

Also $AP = \dfrac{\sqrt{\beta^2\gamma^2+\gamma^2\alpha^2+\alpha^2\beta^2}}{\sqrt{\beta^2+\gamma^2}}$, so that the charge at O, considered as the image of P, is

$$\frac{\alpha\beta\gamma}{\sqrt{\beta^2\gamma^2+\gamma^2\alpha^2+\alpha^2\beta^2}} = \frac{1}{\sqrt{\frac{1}{\alpha^2}+\frac{1}{\beta^2}+\frac{1}{\gamma^2}}}.$$

In the same way we may find the system of images which are

electrically equivalent to four spherical surfaces at potential unity intersecting at right angles.

If the radius of the fourth sphere is δ, and if we make the charge at the centre of this sphere $= \delta$, then the charge at the intersection of the line of centres of any two spheres, say α and β, with their plane of intersection, is

$$-\frac{1}{\sqrt{\frac{1}{\alpha^2}+\frac{1}{\beta^2}}}.$$

The charge at the intersection of the plane of any three centres ABC with the perpendicular from D is

$$+\frac{1}{\sqrt{\frac{1}{\alpha^2}+\frac{1}{\beta^2}+\frac{1}{\gamma^2}}},$$

and the charge at the intersection of the four perpendiculars is

$$-\frac{1}{\sqrt{\frac{1}{\alpha^2}+\frac{1}{\beta^2}+\frac{1}{\gamma^2}+\frac{1}{\delta^2}}}.$$

System of Four Spheres Intersecting at Right Angles under the Action of an Electrified Point.

170.] Let the four spheres be A, B, C, D, and let the electrified point be O. Draw four spheres A_1, B_1, C_1, D_1, of which any one, A_1, passes through O and cuts three of the spheres, in this case B, C, and D, at right angles. Draw six spheres (ab), (ac), (ad), (bc), (bd), (cd), of which each passes through O and through the circle of intersection of two of the original spheres.

The three spheres B_1, C_1, D_1 will intersect in another point besides O. Let this point be called A', and let B', C', and D' be the intersections of C_1, D_1, A_1, of D_1, A_1, B_1, and of A_1, B_1, C_1 respectively. Any two of these spheres, A_1, B_1, will intersect one of the six (cd) in a point $(a'b')$. There will be six such points.

Any one of the spheres, A_1, will intersect three of the six (ab), (ac), (ad) in a point a'. There will be four such points. Finally, the six spheres (ab), (ac), (ad), (cd), (db), (bc), will intersect in one point S.

If we now invert the system with respect to a sphere of radius R and centre O, the four spheres A, B, C, D will be inverted into spheres, and the other ten spheres will become planes. Of the points of intersection the first four A', B', C', D' will become the centres of the spheres, and the others will correspond to the other eleven points in the preceding article. These fifteen points form the image of O in the system of four spheres.

At the point A', which is the image of O in the sphere A, we must place a charge equal to the image of O, that is, $-\frac{a}{a}$, where a is the radius of the sphere A, and a is the distance of its centre from O. In the same way we must place the proper charges at B', C', D'.

The charges for each of the other eleven points may be found from the expressions in the last article by substituting α', β', γ', δ' for α, β, γ, δ, and multiplying the result for each point by the distance of the point from O, where

$$\alpha' = -\frac{\alpha}{a^2-\alpha^2}, \quad \beta' = -\frac{\beta}{b^2-\beta^2}, \quad \gamma' = -\frac{\gamma}{c^2-\gamma^2}, \quad \delta' = -\frac{\delta}{d^2-\delta^2}.$$

[The cases discussed in Arts. 169, 170 may be dealt with as follows: Taking three coordinate planes at right angles, let us place at the system of eight points $\left(\pm\frac{1}{2\alpha},\ \pm\frac{1}{2\beta},\ \pm\frac{1}{2\gamma}\right)$ charges $\pm e$, the minus charges being at the points which have 1 or 3 negative coordinates. Then it is obvious the coordinate planes are at potential zero. Now let us invert with regard to any point and we have the case of three spheres cutting orthogonally under the influence of an electrified point. If we invert with regard to one of the electrified points, we find the solution for the case of a conductor in the form of three spheres of radii α, β, γ cutting orthogonally and freely charged.

If to the above system of electrified points we superadd their images in a sphere with its centre at the origin we see that, in addition to the three coordinate planes, the surface of the sphere forms also a part of the surface of zero potential.]

Two Spheres not Intersecting.

171.] When a space is bounded by two spherical surfaces which do not intersect, the successive images of an influencing point within this space form two infinite series, all of which lie beyond the spherical surfaces, and therefore fulfil the condition of the applicability of the method of electrical images.

Any two non-intersecting spheres may be inverted into two concentric spheres by assuming as the point of inversion either of the two common inverse points of the pair of spheres.

We shall begin, therefore, with the case of two uninsulated concentric spherical surfaces, subject to the induction of an electrified point placed between them.

Let the radius of the first be b, and that of the second be^{ϖ}, and let the distance of the influencing point from the centre be $r = be^{u}$.

Then all the successive images will be on the same radius as the influencing point.

Let Q_0, Fig. 14, be the image of P in the first sphere, P_1 that of Q_0 in the second sphere, Q_1 that of P_1 in the first sphere, and so on; then

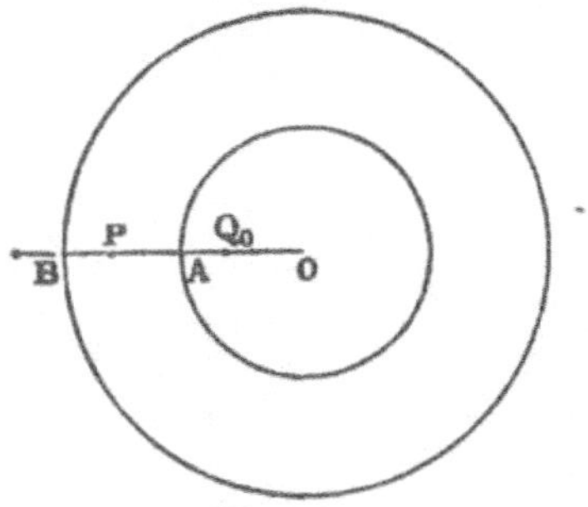

Fig. 14.

$$OP_s . OQ_s = b^2,$$

and $$OP_s . OQ_{s-1} = b^2 e^{2\varpi},$$

also $$OQ_0 = be^{-u},$$

$$OP_1 = be^{u+2\varpi},$$

$$OQ_1 = be^{-(u+2\varpi)}, \text{ \&c.}$$

Hence $$OP_s = be^{(u+2s\varpi)},$$

$$OQ_s = be^{-(u+2s\varpi)},$$

If the charge of P is denoted by P, then

$$P_s = Pe^{s\varpi}, \qquad Q_s = -Pe^{-(u+s\varpi)}.$$

Next, let Q_1' be the image of P in the second sphere, P_1' that of Q_1' in the first, &c.,

$$OQ_1' = be^{2\varpi-u}, \qquad OP_1' = be^{u-2\varpi},$$

$$OQ_2' = be^{4\varpi-u}, \qquad OP_2' = be^{u-4\varpi},$$

$$OQ_s' = be^{2s\varpi-u}, \qquad OP_s' = be^{u-2s\varpi},$$

$$Q_s' = -Pe^{s\varpi-u}, \qquad P_s' = Pe^{-s\varpi}.$$

Of these images all the P's are positive, and all the Q's negative, all the P''s and Q's belong to the first sphere, and all the P's and Q''s to the second.

The images within the first sphere form a converging series, the sum of which is

$$-P\frac{e^{\varpi-u}-1}{e^{\varpi}-1}.$$

This therefore is the quantity of electricity on the first or interior sphere. The images outside the second sphere form a diverging series, but the surface-integral of each with respect to the spherical surface is zero. The charge of electricity on the exterior spherical surface is therefore

$$P\left(\frac{e^{\varpi-u}-1}{e^{\varpi}-1}-1\right) = -P\frac{e^{\varpi}-e^{\varpi-u}}{e^{\varpi}-1}.$$

If we substitute for these expressions their values in terms of OA, OB, and OP, we find

$$\text{charge on } A = -P\frac{OA}{OP}\frac{PB}{AB},$$

$$\text{charge on } B = -P\frac{OB}{OP}\frac{AP}{AB}.$$

If we suppose the radii of the spheres to become infinite, the case becomes that of a point placed between two parallel planes A and B. In this case these expressions become

$$\text{charge on } A = -P\frac{PB}{AB},$$

$$\text{charge on } B = -P\frac{AP}{AB}.$$

172.] In order to pass from this case to that of any two spheres not intersecting each other, we begin by finding the two common inverse points O, O' through which all circles pass that are orthogonal to both spheres. Then, if we invert the system with respect to either of these points, the spheres become concentric, as in the first case.

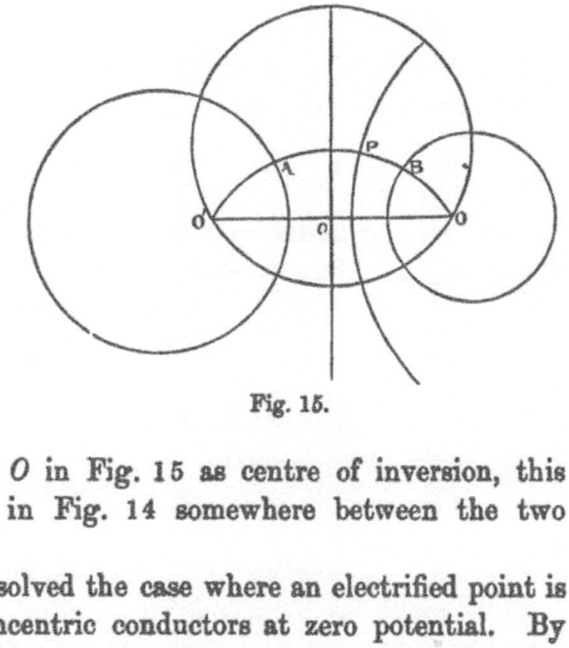

Fig. 15.

If we take the point O in Fig. 15 as centre of inversion, this point will be situated in Fig. 14 somewhere between the two spherical surfaces.

Now in Art. 171 we solved the case where an electrified point is placed between two concentric conductors at zero potential. By inversion of that case with regard to the point O we shall therefore deduce the distributions on two spherical conductors at potential zero, exterior to one another, induced by an electrified point in their neighbourhood. In Art. 173 it will be shewn how the results thus obtained may be employed in finding the distributions on two spherical charged conductors subject to their mutual influence only.

The radius $OAPB$ in Fig. 14 on which the successive images lie becomes in Fig. 15 an arc of a circle through O and O', and the ratio of $O'P$ to OP is equal to Ce^{u} where C is a numerical quantity.

If we put $\theta = \log \frac{O'P}{OP}, \quad \alpha = \log \frac{O'A}{OA}, \quad \beta = \log \frac{O'B}{OB},$

then $\beta - \alpha = \varpi, \quad u + \alpha = \theta.$

All the successive images of P will lie on the arc $OAPBO'$.

The position of the image of P in A is Q_0 where

$$\theta(Q_0) = \log \frac{O'Q}{OQ} = 2\alpha - \theta.$$

That of Q_0 in B is P_1 where

$$\theta(P_1) = \log \frac{O'P_1}{OP_1} = \theta + 2\varpi.$$

Similarly

$$\theta(P_s) = \theta + 2s\varpi, \qquad \theta(Q_s) = 2\alpha - \theta - 2s\varpi.$$

In the same way if the successive images of P in B, A, B, &c. are Q_0', P_1', Q_1', &c.,

$$\theta(Q_0') = 2\beta - \theta, \qquad \theta(P_1') = \theta - 2\varpi;$$

$$\theta(P_s') = \theta - 2s\varpi, \qquad \theta(Q_s') = 2\beta - \theta + 2s\varpi.$$

To find the charge of any image P_s we observe that in the inverted figure its charge is

$$P\sqrt{\frac{OP_s}{OP}}.$$

In the original figure we must multiply this by OP_s. Hence the charge of P_s in the dipolar figure is

$$P\sqrt{\frac{OP_s . O'P_s}{OP . O'P}}.$$

If we make $\xi = \sqrt{OP . O'P}$, and call ξ the parameter of the point P, then we may write

$$P_s = \frac{\xi_s}{\xi} P,$$

or the charge of any image is proportional to its parameter.

If we make use of the curvilinear coordinates θ and ϕ, such that

$$e^{\theta + \sqrt{-1}\phi} = \frac{x + \sqrt{-1}y - k}{x + \sqrt{-1}y + k},$$

where $2k$ is the distance OO', then

$$x = -\frac{k \sinh\theta}{\cosh\theta - \cos\phi}, \qquad y = \frac{k \sin\phi}{\cosh\theta - \cos\phi};$$

$$x^2 + (y - k\cot\phi)^2 = k^2 \operatorname{cosec}^2\phi,$$

$$(x + k\coth\theta)^2 + y^2 = k^2 \operatorname{cosech}^2\theta,$$

$$\cot\phi = \frac{x^2+y^2-k^2}{2ky}, \quad \coth\theta = -\frac{x^2+y^2+k^2}{2kx};$$

$$\xi = \frac{\sqrt{2}k}{\sqrt{\cosh\theta - \cos\phi}}*.$$

Since the charge of each image is proportional to its parameter, ξ, and is to be taken positively or negatively according as it is of the form P or Q, we find

$$P_s = \frac{P\sqrt{\cosh\theta-\cos\phi}}{\sqrt{\cosh(\theta+2s\varpi)-\cos\phi}},$$

$$Q_s = -\frac{P\sqrt{\cosh\theta-\cos\phi}}{\sqrt{\cosh(2\alpha-\theta-2s\varpi)-\cos\phi}},$$

$$P'_s = \frac{P\sqrt{\cosh\theta-\cos\phi}}{\sqrt{\cosh(\theta-2s\varpi)-\cos\phi}},$$

$$Q'_s = -\frac{P\sqrt{\cosh\theta-\cos\phi}}{\sqrt{\cosh(2\beta-\theta+2s\varpi)-\cos\phi}}.$$

We have now obtained the positions and charges of the two infinite series of images. We have next to determine the total charge on the sphere A by finding the sum of all the images within it which are of the form Q or P'. We may write this

$$P\sqrt{\cosh\theta-\cos\phi}\sum_{s=1}^{s=\infty}\frac{1}{\sqrt{\cosh(\theta-2s\varpi)-\cos\phi}},$$

$$-P\sqrt{\cosh\theta-\cos\phi}\sum_{s=0}^{s=\infty}\frac{1}{\sqrt{\cosh(2\alpha-\theta-2s\varpi)-\cos\phi}}.$$

In the same way the total induced charge on B is

$$P\sqrt{\cosh\theta-\cos\phi}\sum_{s=1}^{s=\infty}\frac{1}{\sqrt{\cosh(\theta+2s\varpi)-\cos\phi}},$$

$$-P\sqrt{\cosh\theta-\cos\phi}\sum_{s=0}^{s=\infty}\frac{1}{\sqrt{\cosh(2\beta-\theta+2s\varpi)-\cos\phi}}.$$

173.] We shall apply these results to the determination of the

* In these expressions we must remember that

$$2\cosh\theta = e^{\theta}+e^{-\theta}, \qquad 2\sinh\theta = e^{\theta}-e^{-\theta},$$

and the other functions of θ are derived from these by the same definitions as the corresponding trigonometrical functions.

The method of applying dipolar coordinates to this case was given by Thomson in *Liouville's Journal* for 1847. See Thomson's reprint of *Electrical Papers*, § 211, 212. In the text I have made use of the investigation of Prof. Betti, *Nuovo Cimento*, vol. xx, for the analytical method, but I have retained the idea of electrical images as used by Thomson in his original investigation, *Phil. Mag.*, 1853.

coefficients of capacity and induction of two spheres whose radii are a and b, and the distance between whose centres is c.

Let the sphere A be at potential unity, and the sphere B at potential zero.

Then the successive images of a charge a placed at the centre of the sphere A will be those of the actual distribution of electricity. All the images will lie on the axis between the poles and the centres of the spheres, and it will be observed that of the four systems of images determined in Art. 172, only the first and fourth exist in this case.

If we put

$$k = \frac{\sqrt{a^4+b^4+c^4-2b^2c^2-2c^2a^2-2a^2b^2}}{2c};$$

then $\sinh\alpha = -\frac{k}{a}$, $\qquad \sinh\beta = \frac{k}{b}$.

The values of θ and ϕ for the centre of the sphere A are

$$\theta = 2\alpha, \qquad \phi = 0.$$

Hence in the equations we must substitute a or $-k\frac{1}{\sinh\alpha}$ for P, 2α for θ and 0 for ϕ, remembering that P itself forms part of the charge of A. We thus find for the coefficient of capacity of A

$$q_{aa} = \quad k\sum_{s=0}^{s=\infty}\frac{1}{\sinh(s\varpi-\alpha)},$$

for the coefficient of induction of A on B or of B on A

$$q_{ab} = -k\sum_{s=1}^{s=\infty}\frac{1}{\sinh s\varpi}.$$

We may, in like manner, by supposing B at potential unity and A at potential zero, determine the value of q_{bb}. We shall find, with our present notation,

$$q_{bb} = \quad k\sum_{s=0}^{s=\infty}\frac{1}{\sinh(\beta+s\varpi)}.$$

To calculate these quantities in terms of a and b, the radii of the spheres, and of c the distance between their centres, we observe that if

$$K = \sqrt{a^4+b^4+c^4-2b^2c^2-2c^2a^2-2a^2b^2},$$

we may write

$$\sinh\alpha = -\frac{K}{2ac}, \quad \sinh\beta = \frac{K}{2bc}, \quad \sinh\varpi = \frac{K}{2ab},$$

$$\cosh\alpha = \frac{c^2+a^2-b^2}{2ca}, \quad \cosh\beta = \frac{c^2+b^2-a^2}{2cb}, \quad \cosh\varpi = \frac{c^2-a^2-b^2}{2ab},$$

and make use of

$$\sinh(\alpha+\beta) = \sinh\alpha\cosh\beta + \cosh\alpha\sinh\beta,$$
$$\cosh(\alpha+\beta) = \cosh\alpha\cosh\beta + \sinh\alpha\sinh\beta.$$

By this process or by the direct calculation of the successive images as shewn in Sir W. Thomson's paper, we find

$$q_{aa} = a + \frac{a^2 b}{c^2-b^2} + \frac{a^3 b^2}{(c^2-b^2+ac)(c^2-b^2-ac)} + \&c.,$$

$$q_{ab} = -\frac{ab}{c} - \frac{a^2 b^2}{c(c^2-a^2-b^2)} - \frac{a^3 b^3}{c(c^2-a^2-b^2+ab)(c^2-a^2-b^2-ab)} - \&c.$$

$$q_{bb} = b + \frac{ab^2}{c^2-a^2} + \frac{a^2 b^3}{(c^2-a^2+bc)(c^2-a^2-bc)} + \&c.$$

174.] We have then the following equations to determine the charges E_a and E_b of the two spheres when electrified to potentials V_a and V_b respectively,

$$E_a = V_a q_{aa} + V_b q_{ab},$$
$$E_b = V_a q_{ab} + V_b q_{bb}.$$

If we put $$q_{aa} q_{bb} - q_{ab}^2 = D = \frac{1}{D'},$$

and $$p_{aa} = q_{bb} D', \qquad p_{ab} = -q_{ab} D', \qquad p_{bb} = q_{aa} D',$$

whence $$p_{aa} p_{bb} - p_{ab}^2 = D';$$

then the equations to determine the potentials in terms of the charges are

$$V_a = p_{aa} E_a + p_{ab} E_b,$$
$$V_b = p_{ab} E_a + p_{bb} E_b,$$

and p_{aa}, p_{ab}, and p_{bb} are the coefficients of potential.

The total energy of the system is, by Art. 85,

$$Q = \tfrac{1}{2}(E_a V_a + E_b V_b),$$
$$= \tfrac{1}{2}(V_a^2 q_{aa} + 2 V_a V_b q_{ab} + V_b^2 q_{bb}),$$
$$= \tfrac{1}{2}(E_a^2 p_{aa} + 2 E_a E_b p_{ab} + E_b^2 p_{bb}).$$

The repulsion between the spheres is therefore, by Arts. 92, 93,

$$F = \tfrac{1}{2}\left\{V_a^2 \frac{dq_{aa}}{dc} + 2V_a V_b \frac{dq_{ab}}{dc} + V_b^2 \frac{dq_{bb}}{dc}\right\},$$
$$= -\tfrac{1}{2}\left\{E_a^2 \frac{dp_{aa}}{dc} + 2E_a E_b \frac{dp_{ab}}{dc} + E_b^2 \frac{dp_{bb}}{dc}\right\},$$

where c is the distance between the centres of the spheres.

Of these two expressions for the repulsion, the first, which expresses it in terms of the potentials of the spheres and the variations of the coefficients of capacity and induction, is the most convenient for calculation.

We have therefore to differentiate the q's with respect to c. These quantities are expressed as functions of k, a, β, and ϖ, and must be differentiated on the supposition that a and b are constant. From the equations

$$k = -a \sinh a = b \sinh\beta = -c \frac{\sinh a \sinh\beta}{\sinh\varpi},$$

$$\frac{dk}{dc} = \frac{\cosh a \cosh\beta}{\sinh\varpi},$$

we find

$$\frac{da}{dc} = \frac{\sinh a \cosh\beta}{k \sinh\varpi},$$

$$\frac{d\beta}{dc} = \frac{\cosh a \sinh\beta}{k \sinh\varpi},$$

$$\frac{d\varpi}{dc} = \frac{1}{k};$$

whence we find

$$\frac{dq_{aa}}{dc} = \frac{\cosh a \cosh\beta}{\sinh\varpi}\frac{q_{aa}}{k} - \sum_{s=0}^{s=\infty} \frac{(sc + b\cosh\beta)\cosh(s\varpi - a)}{c(\sinh(s\varpi - a))^2},$$

$$\frac{dq_{ab}}{dc} = \frac{\cosh a \cosh\beta}{\sinh\varpi}\frac{q_{ab}}{k} + \sum_{s=1}^{s=\infty} \frac{s\cosh s\varpi}{(\sinh s\varpi)^2},$$

$$\frac{dq_{bb}}{dc} = \frac{\cosh a \cosh\beta}{\sinh\varpi}\frac{q_{bb}}{k} - \sum_{s=0}^{s=\infty} \frac{(sc - a\cosh a)\cosh(\beta + s\varpi)}{c(\sinh(\beta + s\varpi))^2}.$$

Sir William Thomsom has calculated the force between two spheres of equal radius separated by any distance less than the diameter of one of them. For greater distances it is not necessary to use more than two or three of the successive images.

The series for the differential coefficients of the q's with respect to c are easily obtained by direct differention.

$$\frac{dq_{aa}}{dc} = -\frac{2a^2bc}{(c^2 - b^2)^2} - \frac{2a^3b^2c(2c^2 - 2b^2 - a^2)}{(c^2 - b^2 + ac)^2(c^2 - b^2 - ac)^2} - \&c.,$$

$$\frac{dq_{ab}}{dc} = \frac{ab}{c^2} + \frac{a^2b^2(3c^2 - a^2 - b^2)}{c^2(c^2 - a^2 - b^2)}$$

$$+ \frac{a^3b^3\{(5c^2 - a^2 - b^2)(c^2 - a^2 - b^2) - a^2b^2\}}{c^2(c^2 - a^2 - b^2 + ab)^2(c^2 - a^2 - b^2 - ab)^2} - \&c.,$$

$$\frac{dq_{bb}}{dc} = -\frac{2ab^2c}{(c^2 - a^2)^2} - \frac{2a^2b^3c(2c^2 - 2a^2 - b^2)}{(c^2 - a^2 + bc)^2(c^2 - a^2 - bc)^2} - \&c.$$

Distribution of Electricity on Two Spheres in Contact.

175.] If we suppose the two spheres at potential unity and not influenced by any other point, then, if we invert the system with respect to the point of contact, we shall have two parallel planes, distant $\frac{1}{2a}$ and $\frac{1}{2b}$ from the point of inversion, and electrified by the action of a unit of electricity at that point.

There will be a series of positive images, each equal to unity, at distances $s\left(\frac{1}{a}+\frac{1}{b}\right)$ from the origin, where s may have any integer value from $-\infty$ to $+\infty$.

There will also be a series of negative images each equal to -1, the distances of which from the origin, reckoned in the direction of a, are $\frac{1}{a}+s\left(\frac{1}{a}+\frac{1}{b}\right)$.

When this system is inverted back again into the form of the two spheres in contact, we have a corresponding series of negative images, the distances of which from the point of contact are of the form $\frac{1}{s\left(\frac{1}{a}+\frac{1}{b}\right)}$, where s is positive for the sphere A and negative for the sphere B. The charge of each image, when the potential of the spheres is unity, is numerically equal to its distance from the point of contact, and is always negative.

There will also be a series of positive images whose distances from the point of contact measured in the direction of the centre of a, are of the form $\frac{1}{\frac{1}{a}+s\left(\frac{1}{a}+\frac{1}{b}\right)}$.

When s is zero, or a positive integer, the image is in the sphere A.

When s is a negative integer the image is in the sphere B.

The charge of each image is measured by its distance from the origin and is always positive.

The total charge of the sphere A is therefore

$$E_a=\sum_{s=0}^{s=\infty}\frac{1}{\frac{1}{a}+s\left(\frac{1}{a}+\frac{1}{b}\right)}-\frac{ab}{a+b}\sum_{s=1}^{s=\infty}\frac{1}{s}.$$

Each of these series is infinite, but if we combine them in the form

$$E_a = \sum_{s=1}^{s=\infty} \frac{a^2 b}{s(a+b)\{s(a+b)-a\}}$$

the series becomes converging.

In the same way we find for the charge of the sphere B,

$$E_b = \sum_{s=1}^{s=\infty} \frac{ab}{s(a+b)-b} - \frac{ab}{a+b} \sum_{s=-1}^{s=-\infty} \frac{1}{s},$$

$$= \sum_{s=1}^{s=\infty} \frac{ab^2}{s(a+b)\{s(a+b)-b\}}.$$

The expression for E_a is obviously equal to

$$\frac{ab}{a+b} \int_0^1 \frac{\theta^{\frac{b}{a+b}-1} - 1}{1-\theta} d\theta,$$

in which form the result in this case was given by Poisson.

It may also be shewn (Legendre *Traité des Fonctions Elliptiques*, ii, 438) that the above series for E_a is equal to

$$a - \left\{\gamma + \Psi\left(\frac{b}{a+b}\right)\right\} \frac{ab}{a+b},$$

where $\gamma = \cdot 57712\ldots$, and $\Psi(x) = \frac{d}{dx} \log \Gamma(1+x)$.

The values of Ψ have been tabulated by Gauss (*Werke*, Band iii, pp. 161–162.)

If we denote for an instant $b \div (a+b)$ by x, we find for the difference of the charges E_a and E_b,

$$-\frac{d}{dx} \log \Gamma(x) \Gamma(1-x) \times \frac{ab}{a+b},$$

$$= \frac{d}{dx} \log \sin \pi x \times \frac{ab}{a+b},$$

$$= \frac{\pi ab}{a+b} \cot \frac{\pi b}{a+b}.$$

When the spheres are equal the charge of each for potential unity is

$$E_a = a \sum_{s=1}^{s=\infty} \frac{1}{2s(2s-1)},$$

$$= a\left(1 - \tfrac{1}{2} + \tfrac{1}{3} - \tfrac{1}{4} + \&c.\right),$$

$$= a \log_e 2 = \cdot 69314718\, a.$$

When the sphere A is very small compared with the sphere B the charge on A is

$$E_a = \frac{a^2}{b} \sum_{s=1}^{s=\infty} \frac{1}{s^2} \text{ approximately};$$

$$\text{or} \quad E_a = \frac{\pi^2}{6} \frac{a^2}{b}.$$

The charge on B is nearly the same as if A were removed, or

$$E_b = b.$$

The mean density on each sphere is found by dividing the charge by the surface. In this way we get

$$\sigma_a = \frac{E_a}{4\pi a^2} = \frac{\pi}{24 b},$$

$$\sigma_b = \frac{E_b}{4\pi b^2} = \frac{1}{4\pi b},$$

$$\sigma_a = \frac{\pi^2}{6}\sigma_b.$$

Hence, if a very small sphere is made to touch a very large one, the mean density on the small sphere is equal to that on the large sphere multiplied by $\frac{\pi^2}{6}$, or 1.644936.

Application of Electrical Inversion to the case of a Spherical Bowl.

176.] One of the most remarkable illustrations of the power of Sir W. Thomson's method of Electrical Images is furnished by his investigation of the distribution of electricity on a portion of a spherical surface bounded by a small circle. The results of this investigation, without proof, were communicated to M. Liouville and published in his *Journal* in 1847. The complete investigation is given in the reprint of Thomson's *Electrical Papers*, Article XV. I am not aware that a solution of the problem of the distribution of electricity on a finite portion of any curved surface has been given by any other mathematician.

As I wish to explain the method rather than to verify the calculation, I shall not enter at length into either the geometry or the integration, but refer my readers to Thomson's work.

Distribution of Electricity on an Ellipsoid.

177.] It is shewn by a well-known method * that the attraction of a shell bounded by two similar and similarly situated and concentric ellipsoids is such that there is no resultant attraction on any point within the shell. If we suppose the thickness of the shell to diminish indefinitely while its density increases, we ultimately arrive at the conception of a surface-density varying as the perpendicular from the centre on the tangent plane, and since the resultant attraction of this superficial distribution on any

* Thomson and Tait's *Natural Philosophy*, § 520, or Art. 150 of this book.

point within the ellipsoid is zero, electricity, if so distributed on the surface, will be in equilibrium.

Hence, the surface-density at any point of an ellipsoid undisturbed by external influence varies as the distance of the tangent plane from the centre.

Distribution of Electricity on a Disk.

By making two of the axes of the ellipsoid equal, and making the third vanish, we arrive at the case of a circular disk, and at an expression for the surface-density at any point P of such a disk when electrified to the potential V and left undisturbed by external influence. If σ be the surface-density on one side of the disk, and if KPL be a chord drawn through the point P, then

$$\sigma = \frac{V}{2\pi^2\sqrt{KP.PL}}.$$

Application of the Principle of Electric Inversion.

178.] Take any point Q as the centre of inversion, and let R be the radius of the sphere of inversion. Then the plane of the disk becomes a spherical surface passing through Q, and the disk itself becomes a portion of the spherical surface bounded by a circle. We shall call this portion of the surface the *bowl*.

If S' is the disk electrified to potential V' and free from external influence, then its electrical image S will be a spherical segment at potential zero, and electrified by the influence of a quantity $V'R$ of electricity placed at Q.

We have therefore by the process of inversion obtained the solution of the problem of the distribution of electricity on a bowl or a plane disk when under the influence of an electrified point in the surface of the sphere or plane produced.

Influence of an Electrified Point placed on the unoccupied part of the Spherical Surface.

The form of the solution, as deduced by the principles already given and by the geometry of inversion, is as follows:

If C is the central point or pole of the spherical bowl S, and if a is the distance from C to any point in the edge of the segment, then, if a quantity q of electricity is placed at a point Q in the surface of the sphere produced, and if the bowl S is maintained at potential zero, the density σ at any point P of the bowl will be

$$\sigma = \frac{1}{2\pi^2}\frac{q}{QP^2}\sqrt{\frac{CQ^2-a^2}{a^2-CP^2}},$$

CQ, CP, and QP being the straight lines joining the points, C, Q, and P.

It is remarkable that this expression is independent of the radius of the spherical surface of which the bowl is a part. It is therefore applicable without alteration to the case of a plane disk.

Influence of any Number of Electrified Points.

Now let us consider the sphere as divided into two parts, one of which, the spherical segment on which we have determined the electric distribution, we shall call the *bowl*, and the other the remainder, or unoccupied part of the sphere on which the influencing point Q is placed.

If any number of influencing points are placed on the remainder of the sphere, the electricity induced by these on any point of the bowl may be obtained by the summation of the densities induced by each separately.

179.] Let the whole of the remaining surface of the sphere be uniformly electrified, the surface-density being ρ, then the density at any point of the bowl may be obtained by ordinary integration over the surface thus electrified.

We shall thus obtain the solution of the case in which the bowl is at potential zero, and electrified by the influence of the remaining portion of the spherical surface rigidly electrified with density ρ.

Now let the whole system be insulated and placed within a sphere of diameter f, and let this sphere be uniformly and rigidly electrified so that its surface-density is ρ'.

There will be no resultant force within this sphere, and therefore the distribution of electricity on the bowl will be unaltered, but the potential of all points within the sphere will be increased by a quantity V where $$V = 2\pi\rho' f.$$

Hence the potential at every point of the bowl will now be V.

Now let us suppose that this sphere is concentric with the sphere of which the bowl forms a part, and that its radius exceeds that of the latter sphere by an infinitely small quantity.

We have now the case of the bowl maintained at potential V and influenced by the remainder of the sphere rigidly electrified with superficial density $\rho+\rho'$.

180.] We have now only to suppose $\rho+\rho' = 0$, and we get the case of the bowl maintained at potential V and free from external influence.

If σ is the density on either surface of the bowl at a given point when the bowl is at potential zero, and is influenced by the rest of the sphere electrified to density ρ, then, when the bowl is maintained at potential V, we must increase the density on the outside of the bowl by ρ', the density on the supposed enveloping sphere.

The result of this investigation is that if f is the diameter of the sphere, a the chord of the radius of the bowl, and r the chord of the distance of P from the pole of the bowl, then the surface-density σ on the *inside* of the bowl is

$$\sigma = \frac{V}{2\pi^2 f}\left\{\sqrt{\frac{f^2-a^2}{a^2-r^2}} - \tan^{-1}\sqrt{\frac{f^2-a^2}{a^2-r^2}}\right\},$$

and the surface-density on the outside of the bowl at the same point is

$$\sigma + \frac{V}{2\pi f}.$$

In the calculation of this result no operation is employed more abstruse than ordinary integration over part of a spherical surface. To complete the theory of the electrification of a spherical bowl we only require the geometry of the inversion of spherical surfaces.

181.] Let it be required to find the surface-density induced at any point of the bowl by a quantity q of electricity placed at a point Q, not now in the spherical surface produced.

Invert the bowl with respect to Q, the radius of the sphere of inversion being R. The bowl S will be inverted into its image S', and the point P will have P' for its image. We have now to determine the density σ' at P' when the bowl S' is maintained at potential V', such that $q = V'R$, and is not influenced by any external force.

The density σ at the point P of the original bowl is then

$$\sigma = -\frac{\sigma' R^3}{QP^3},$$

this bowl being at potential zero, and influenced by a quantity q of electricity placed at Q.

The result of this process is as follows:

Let the figure represent a section through the centre, O, of the sphere, the pole, C, of the bowl, and the influencing point Q. D is a point which corresponds in the inverted figure to the unoccupied pole of the rim of the bowl, and may be found by the following construction.

Draw through Q the chords EQE' and FQF', then if we suppose the radius of the sphere of inversion to be a mean proportional between the segments into which a chord is divided at Q, $E'F'$ will be the image of EF. Bisect the arc $F'CE'$ in D', so that $F'D' = D'E'$, and draw $D'QD$ to meet the sphere in D. D is the point required. Also through O, the centre of the sphere, and Q draw $HOQH'$ meeting the sphere in H and H'. Then if P be any point in the bowl, the surface-density at P on the side which is separated from Q by the completed spherical surface, induced by a quantity q of electricity at Q, will be

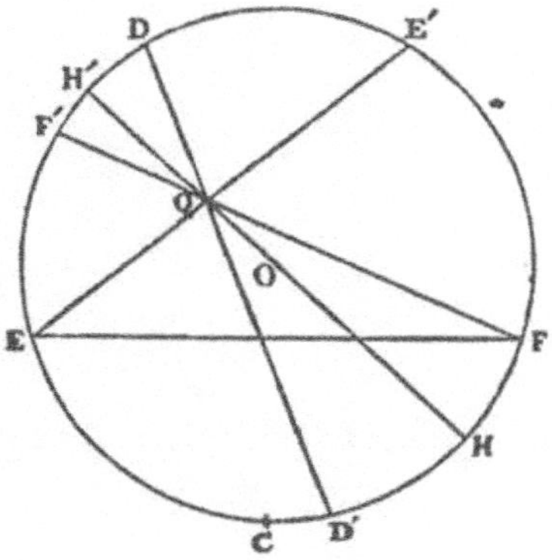

Fig. 16.

$$\sigma = \frac{q}{2\pi^2} \frac{QH.QH'}{HH'.PQ^3} \left\{ \frac{PQ}{DQ} \left(\frac{CD^2 - a^2}{a^2 - CP^2} \right)^{\frac{1}{2}} - \tan^{-1} \left[\frac{PQ}{DQ} \left(\frac{CD^2 - a^2}{a^2 - CP^2} \right)^{\frac{1}{2}} \right] \right\},$$

where a denotes the chord drawn from C, the pole of the bowl, to the rim of the bowl.

On the side next to Q the surface-density is

$$\sigma + \frac{q}{2\pi} \frac{QH.QH'}{HH'.PQ^3}.$$

CHAPTER XII.

THEORY OF CONJUGATE FUNCTIONS IN TWO DIMENSIONS.

182.] THE number of independent cases in which the problem of electrical equilibrium has been solved is very small. The method of spherical harmonics has been employed for spherical conductors, and the methods of electrical images and of inversion are still more powerful in the cases to which they can be applied. The case of surfaces of the second degree is the only one, as far as I know, in which both the equipotential surfaces and the lines of force are known when the lines of force are not plane curves.

But there is an important class of problems in the theory of electrical equilibrium, and in that of the conduction of currents, in which we have to consider space of two dimensions only.

For instance, if throughout the part of the electric field under consideration, and for a considerable distance beyond it, the surfaces of all the conductors are generated by the motion of straight lines parallel to the axis of z, and if the part of the field where this ceases to be the case is so far from the part considered that the electrical action of the distant part on the field may be neglected, then the electricity will be uniformly distributed along each generating line, and if we consider a part of the field bounded by two planes perpendicular to the axis of z and at distance unity, the potential and the distributions of electricity will be functions of x and y only.

If $\rho\, dx\, dy$ denotes the quantity of electricity in an element whose base is $dx\, dy$ and height unity, and $\sigma\, ds$ the quantity on an element of area whose base is the linear element ds and height unity, then the equation of Poisson may be written

$$\frac{d^2 V}{dx^2} + \frac{d^2 V}{dy^2} + 4\pi\rho = 0.$$

When there is no free electricity, this is reduced to the equation of Laplace;

$$\frac{d^2V}{dx^2}+\frac{d^2V}{dy^2}=0.$$

The general problem of electric equilibrium may be stated as follows:—

A continuous space of two dimensions, bounded by closed curves C_1, C_2, &c being given, to find the form of a function, V, such that at these boundaries its value may be V_1, V_2, &c. respectively, being constant for each boundary, and that within this space V may be everywhere finite, continuous, and single valued, and may satisfy Laplace's equation.

I am not aware that any perfectly general solution of even this question has been given, but the method of transformation given in Art. 190 is applicable to this case, and is much more powerful than any known method applicable to three dimensions.

The method depends on the properties of conjugate functions of two variables.

Definition of Conjugate Functions.

183.] Two quantities α and β are said to be conjugate functions of x and y, if $\alpha+\sqrt{-1}\,\beta$ is a function of $x+\sqrt{-1}\,y$.

It follows from this definition that

$$\frac{d\alpha}{dx}=\frac{d\beta}{dy}, \quad \text{and} \quad \frac{d\alpha}{dy}+\frac{d\beta}{dx}=0; \tag{1}$$

$$\frac{d^2\alpha}{dx^2}+\frac{d^2\alpha}{dy^2}=0, \quad \frac{d^2\beta}{dx^2}+\frac{d^2\beta}{dy^2}=0. \tag{2}$$

Hence both functions satisfy Laplace's equation. Also

$$\frac{d\alpha}{dx}\frac{d\beta}{dy}-\frac{d\alpha}{dy}\frac{d\beta}{dx}=\overline{\frac{d\alpha}{dx}}\Big|^2+\overline{\frac{d\alpha}{dy}}\Big|^2=\overline{\frac{d\beta}{dx}}\Big|^2+\overline{\frac{d\beta}{dy}}\Big|^2=R^2. \tag{3}$$

If x and y are rectangular coordinates, and if ds_1 is the intercept of the curve ($\beta=$ constant) between the curves α and $\alpha+d\alpha$, and ds_2 the intercept of α between the curves β and $\beta+d\beta$, then

$$\frac{ds_1}{d\alpha}=\frac{ds_2}{d\beta}=\frac{1}{R}, \tag{4}$$

and the curves intersect at right angles.

If we suppose the potential $V=V_0+k\alpha$, where k is some constant, then V will satisfy Laplace's equation, and the curves (α) will be equipotential curves. The curves (β) will be lines of force, and the surface-integral of R over unit-length of a cylindrical surface whose projection on the plane of xy is the curve AB will be $k(\beta_B - \beta_A)$, where β_A and β_B are the values of β at the extremities of the curve.

If one series of curves corresponding to values of α in arithmetical progression be drawn on the plane, and another series corresponding to a series of values of β having the same common difference, then the two series of curves will everywhere intersect at right angles, and, if the common difference is small enough, the elements into which the plane is divided will be ultimately little squares, whose sides, in different parts of the field, are in different directions and of different magnitudes, being inversely proportional to R.

If two or more of the equipotential lines (α) are closed curves enclosing a continuous space between them, we may take these for the surfaces of conductors at potentials $(V_0 + k\alpha_1)$, $(V_0 + k\alpha_2)$, &c. respectively. The quantity of electricity upon any one of these between the lines of force β_1 and β_2 will be $\frac{k}{4\pi}(\beta_2 - \beta_1)$.

The number of equipotential lines between two conductors will therefore indicate their difference of potential, and the number of lines of force which emerge from a conductor will indicate the quantity of electricity upon it.

We must next state some of the most important theorems relating to conjugate functions, and in proving them we may use either the equations (1), containing the differential coefficients, or the original definition, which makes use of imaginary symbols.

184.] THEOREM I. *If x' and y' are conjugate functions with respect to x and y, and if x'' and y'' are also conjugate functions with respect to x and y, then the functions $x' + x''$ and $y' + y''$ will be conjugate functions with respect to x and y.*

For
$$\frac{dx'}{dx} = \frac{dy'}{dy}, \text{ and } \frac{dx''}{dx} = \frac{dy''}{dy};$$

therefore
$$\frac{d(x'+x'')}{dx} = \frac{d(y'+y'')}{dy}.$$

Also
$$\frac{dx'}{dy} = -\frac{dy'}{dx}, \text{ and } \frac{dx''}{dy} = -\frac{dy''}{dx};$$

therefore
$$\frac{d(x'+x'')}{dy} = -\frac{d(y'+y'')}{dx};$$

or $x + x''$ and $y' + y''$ are conjugate with respect to x and y.

Graphic Representation of a Function which is the Sum of Two Given Functions.

Let a function (α) of x and y be graphically represented by a series of curves in the plane of xy, each of these curves corresponding to a value of α which belongs to a series of such values increasing by a common difference, δ.

Let any other function, β, of x and y be represented in the same way by a series of curves corresponding to a series of values of β having the same common difference as those of α.

Then to represent the function $\alpha+\beta$ in the same way, we must draw a series of curves through the intersections of the two former series, from the intersection of the curves (α) and (β) to that of the curves $(\alpha+\delta)$ and $(\beta-\delta)$, then through the intersection of $(\alpha+2\delta)$ and $(\beta-2\delta)$, and so on. At each of these points the function will have the same value, namely $\alpha+\beta$. The next curve must be drawn through the points of intersection of (α) and $(\beta+\delta)$, of $(\alpha+\delta)$ and (β), of $(\alpha+2\delta)$ and $(\beta-\delta)$, and so on. The function belonging to this curve will be $\alpha+\beta+\delta$.

In this way, when the series of curves (α) and the series (β) are drawn, the series $(\alpha+\beta)$ may be constructed. These three series of curves may be drawn on separate pieces of transparent paper, and when the first and second have been properly superposed, the third may be drawn.

The combination of conjugate functions by addition in this way enables us to draw figures of many interesting cases with very little trouble when we know how to draw the simpler cases of which they are compounded. We have, however, a far more powerful method of transformation of solutions, depending on the following theorem.

185.] THEOREM II. *If x'' and y'' are conjugate functions with respect to the variables x' and y', and if x' and y' are conjugate functions with respect to x and y, then x'' and y'' will be conjugate functions with respect to x and y.*

For
$$\frac{dx''}{dx} = \frac{dx''}{dx'}\frac{dx'}{dx} + \frac{dx''}{dy'}\frac{dy'}{dx},$$
$$= \frac{dy''}{dy'}\frac{dy'}{dy} + \frac{dy''}{dx'}\frac{dx'}{dy},$$
$$= \frac{dy''}{dy};$$

$$\text{and} \quad \frac{dx''}{dy} = \frac{dx''}{dx'}\frac{dx'}{dy} + \frac{dx''}{dy'}\frac{dy'}{dy},$$
$$= -\frac{dy''}{dy'}\frac{dy'}{dx} - \frac{dy''}{dx'}\frac{dx'}{dx},$$
$$= -\frac{dy''}{dx};$$

and these are the conditions that x'' and y'' should be conjugate functions of x and y.

This may also be shewn from the original definition of conjugate functions. For $x'' + \sqrt{-1}\,y''$ is a function of $x' + \sqrt{-1}\,y'$, and $x' + \sqrt{-1}\,y'$ is a function of $x + \sqrt{-1}\,y$. Hence, $x'' + \sqrt{-1}\,y''$ is a function of $x + \sqrt{-1}\,y$.

In the same way we may shew that if x' and y' are conjugate functions of x and y, then x and y are conjugate functions of x' and y'.

This theorem may be interpreted graphically as follows:—

Let x', y' be taken as rectangular coordinates, and let the curves corresponding to values of x'' and of y'' taken in regular arithmetical series be drawn on paper. A double system of curves will thus be drawn cutting the paper into little squares. Let the paper be also ruled with horizontal and vertical lines at equal intervals, and let these lines be marked with the corresponding values of x' and y'.

Next, let another piece of paper be taken in which x and y are made rectangular coordinates and a double system of curves x', y' is drawn, each curve being marked with the corresponding value of x' or y'. This system of curvilinear coordinates will correspond, point for point, to the rectilinear system of coordinates x', y' on the first piece of paper.

Hence, if we take any number of points on the curve x'' on the first paper, and note the values of x' and y' at these points, and mark the corresponding points on the second paper, we shall find a number of points on the transformed curve x''. If we do the same for all the curves x'', y'' on the first paper, we shall obtain on the second paper a double series of curves x'', y'' of a different form, but having the same property of cutting the paper into little squares.

186.] THEOREM III. *If V is any function of x' and y', and if x' and y' are conjugate functions of x and y, then*

$$\iint\left(\frac{d^2V}{dx^2}+\frac{d^2V}{dy^2}\right)dx\,dy=\iint\left(\frac{d^2V}{dx'^2}+\frac{d^2V}{dy'^2}\right)dx'\,dy',$$

the integration being between the same limits.

For
$$\frac{dV}{dx}=\frac{dV}{dx'}\frac{dx'}{dx}+\frac{dV}{dy'}\frac{dy'}{dx},$$

$$\frac{d^2V}{dx^2}=\frac{d^2V}{dx'^2}\left(\frac{dx'}{dx}\right)^2+2\frac{d^2V}{dx'dy'}\frac{dx'}{dx}\frac{dy'}{dx}+\frac{d^2V}{dy'^2}\left.\overline{\frac{dy'}{dx}}\right|^2$$
$$+\frac{dV}{dx'}\frac{d^2x'}{dx^2}+\frac{dV}{dy'}\frac{d^2y'}{dx^2};$$

and
$$\frac{d^2V}{dy^2}=\frac{d^2V}{dx'^2}\left.\overline{\frac{dx'}{dy}}\right|^2+2\frac{d^2V}{dx'dy'}\frac{dx'}{dy}\frac{dy'}{dy}+\frac{d^2V}{dy'^2}\left.\overline{\frac{dy'}{dy}}\right|^2$$
$$+\frac{dV}{dx'}\frac{d^2x'}{dy^2}+\frac{dV}{dy'}\frac{d^2y'}{dy^2}.$$

Adding the last two equations, and remembering the conditions of conjugate functions (1), we find

$$\frac{d^2V}{dx^2}+\frac{d^2V}{dy^2}=\frac{d^2V}{dx'^2}\left(\left.\overline{\frac{dx'}{dx}}\right|^2+\left.\overline{\frac{dx'}{dy}}\right|^2\right)+\frac{d^2V}{dy'^2}\left(\left.\overline{\frac{dy'}{dx}}\right|^2+\left.\overline{\frac{dy'}{dy}}\right|^2\right),$$
$$=\left(\frac{d^2V}{dx'^2}+\frac{d^2V}{dy'^2}\right)\left(\frac{dx'}{dx}\frac{dy'}{dy}-\frac{dx'}{dy}\frac{dy'}{dx}\right).$$

Hence

$$\iint\left(\frac{d^2V}{dx^2}+\frac{d^2V}{dy^2}\right)dx\,dy=\iint\left(\frac{d^2V}{dx'^2}+\frac{d^2V}{dy'^2}\right)\left(\frac{dx'}{dx}\frac{dy'}{dy}-\frac{dx'}{dy}\frac{dy'}{dx}\right)dx\,dy,$$
$$=\iint\left(\frac{d^2V}{dx'^2}+\frac{d^2V}{dy'^2}\right)dx'\,dy'.$$

If V is a potential, then, by Poisson's equation

$$\frac{d^2V}{dx^2}+\frac{d^2V}{dy^2}+4\pi\rho=0,$$

and we may write the result

$$\iint\rho\,dx\,dy=\iint\rho'\,dx'\,dy',$$

or the quantity of electricity in corresponding portions of two systems is the same if the coordinates of one system are conjugate functions of those of the other.

Additional Theorems on Conjugate Functions.

187.] THEOREM IV. *If x_1 and y_1, and also x_2 and y_2, are conjugate functions of x and y, then, if*

$$X = x_1 x_2 - y_1 y_2, \quad \text{and} \quad Y = x_1 y_2 + x_2 y_1,$$

X and Y will be conjugate functions of x and y.

For $$X + \sqrt{-1}\,Y = (x_1 + \sqrt{-1}\,y_1)(x_2 + \sqrt{-1}\,y_2).$$

THEOREM V. *If ϕ be a solution of the equation*

$$\frac{d^2\phi}{dx^2} + \frac{d^2\phi}{dy^2} = 0,$$

and if $$2R = \log\left(\overline{\frac{d\phi}{dx}}\Big|^2 + \overline{\frac{d\phi}{dy}}\Big|^2\right), \quad \text{and} \quad \Theta = -\tan^{-1}\frac{\dfrac{d\phi}{dx}}{\dfrac{d\phi}{dy}},$$

R and Θ will be conjugate functions of x and y.

For R and Θ are conjugate functions of $\frac{d\phi}{dx}$ and $\frac{d\phi}{dy}$, and these are conjugate functions of x and y.

EXAMPLE I.—*Inversion.*

188.] As an example of the general method of transformation let us take the case of inversion in two dimensions.

If O is a fixed point in a plane, and OA a fixed direction, and if $r = OP = ae^\rho$, and $\theta = AOP$, and if x, y are the rectangular coordinates of P with respect to O,

$$\left.\begin{aligned} \rho &= \log\frac{1}{a}\sqrt{x^2+y^2}, & \theta &= \tan^{-1}\frac{y}{x}, \\ x &= ae^\rho\cos\theta, & y &= ae^\rho\sin\theta, \end{aligned}\right\} \tag{5}$$

ρ and θ are conjugate functions of x and y.

If $\rho' = n\rho$ and $\theta' = n\theta$, ρ' and θ' will be conjugate functions of ρ and θ. In the case in which $n = -1$ we have

$$r' = \frac{a^2}{r}, \quad \text{and} \quad \theta' = -\theta, \tag{6}$$

which is the case of ordinary inversion combined with turning the figure 180° round OA.

Inversion in Two Dimensions.

In this case if r and r' represent the distances of corresponding points from O, e and e' the total electrification of a body, S and S' superficial elements, V and V' solid elements, σ and σ' surface-densities, ρ and ρ' volume densities, ϕ and ϕ' corresponding potentials,

$$\left.\begin{array}{l}\frac{r'}{r}=\frac{S'}{S}=\frac{a^2}{r^2}=\frac{r'^2}{a^2}, \quad \frac{V'}{V}=\frac{a^4}{r^4}=\frac{r'^4}{a^4},\\ \frac{e'}{e}=1, \quad \frac{\sigma'}{\sigma}=\frac{r^2}{a^2}=\frac{a^2}{r'^2}, \quad \frac{\rho'}{\rho}=\frac{r^4}{a^4}=\frac{a^4}{r'^4},\\ \frac{\phi'}{\phi}=1.\end{array}\right\} \qquad (7)$$

EXAMPLE II.—*Electric Images in Two Dimensions.*

189.] Let A be the centre of a circle of radius $AQ = b$, and let E be a charge at A, then the potential at any point P is

$$\phi = 2\,E \log \frac{b}{AP}; \qquad (8)$$

and if the circle is a section of a hollow conducting cylinder, the surface-density at any point Q is $-\frac{E}{2\pi b}$.

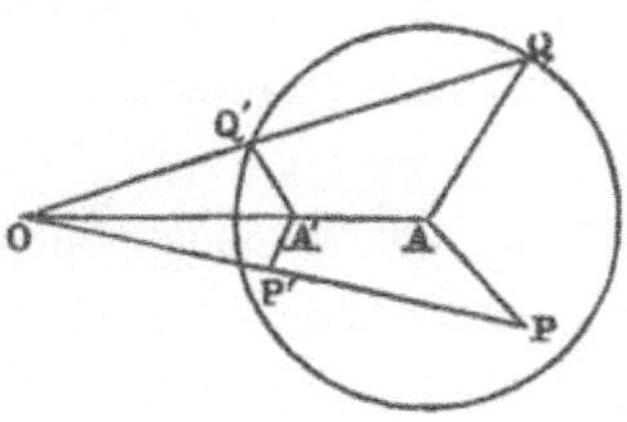

Fig. 17.

Invert the system with respect to a point O, making

$$AO = mb, \quad \text{and} \quad a^2 = (m^2-1)\,b^2;$$

then we have a charge at A' equal to that at A, where $AA' = \frac{b}{m}$.

The density at Q' is

$$-\frac{E}{2\pi b}\,\frac{b^2-\overline{AA'}|^2}{A'Q'^2},$$

and the potential at any point P' within the circle is

$$\begin{aligned}\phi' = \phi &= 2\,E\,(\log b - \log AP),\\ &= 2\,E\,(\log OP' - \log A'P' - \log m).\end{aligned} \qquad (9)$$

This is equivalent to a combination of a charge E at A', and a charge $-E$ at O, which is the image of A', with respect to the circle. The imaginary charge at O is equal and opposite to that at A'.

If the point P' is defined by its polar coordinates referred to the centre of the circle, and if we put

$$\rho = \log r - \log b, \quad \text{and} \quad \rho_0 = \log AA' - \log b,$$

then $\quad AP' = b e^{\rho}, \quad AA' = b e^{\rho_0}, \quad AO = b e^{-\rho_0}; \qquad (10)$

and the potential at the point (ρ, θ) is

$$\begin{aligned}\phi = E \log\,&(e^{-2\rho_0} - 2\,e^{-\rho_0} e^{\rho} \cos\theta + e^{2\rho})\\ &- E \log\,(e^{2\rho_0} - 2\,e^{\rho_0} e^{\rho} \cos\theta + e^{2\rho}) + 2E\rho_0.\end{aligned} \qquad (11)$$

This is the potential at the point (ρ, θ) due to a charge E, placed at the point $(\rho_0, 0)$, with the condition that when $\rho = 0$, $\phi = 0$.

In this case ρ and θ are the conjugate functions in equations (5): ρ is the logarithm of the ratio of the radius vector of a point to the radius of the circle, and θ is an angle.

The centre is the only singular point in this system of coordinates, and the line-integral of $\int \frac{d\theta}{ds}\, ds$ round a closed curve is zero or 2π, according as the closed curve excludes or includes the centre.

EXAMPLE III.—*Neumann's Transformation of this Case**.

190.] Now let α and β be any conjugate functions of x and y, such that the curves (α) are equipotential curves, and the curves (β) are lines of force due to a system consisting of a charge of half a unit at the origin, and an electrified system disposed in any manner at a certain distance from the origin.

Let us suppose that the curve for which the potential is α_0 is a closed curve, such that no part of the electrified system except the half-unit at the origin lies within this curve.

Then all the curves (α) between this curve and the origin will be closed curves surrounding the origin, and all the curves (β) will meet in the origin, and will cut the curves (α) orthogonally.

The coordinates of any point within the curve (α_0) will be determined by the values of α and β at that point, and if the point travels round one of the curves (α) in the positive direction, the value of β will increase by 2π for each complete circuit.

If we now suppose the curve (α_0) to be the section of the inner surface of a hollow cylinder of any form maintained at potential zero under the influence of a charge of linear density E on a line of which the origin is the projection, then we may leave the external electrified system out of consideration, and we have for the potential at any point (α) within the curve

$$\phi = 2E(\alpha - \alpha_0), \tag{12}$$

and for the quantity of electricity on any part of the curve α_0 between the points corresponding to β_1 and β_2,

$$Q = \frac{1}{2\pi} E(\beta_1 - \beta_2). \tag{13}$$

If in this way, or in any other, we have determined the distribution of potential for the case of a given curve of section when the charge is placed at a given point taken as origin, we may pass to the case in which the charge is placed at any other point by an application of the general method of transformation.

* See Crelle's *Journal*, 1861.

Let the values of α and β for the point at which the charge is placed be α_1 and β_1, then substituting in equation (11) $\alpha-\alpha_0$ for ρ, and $\beta-\beta_1$ for θ, we find for the potential at any point whose coordinates are α and β,

$$\phi = E\log\left(1-2e^{\alpha+\alpha_1-2\alpha_0}\cos(\beta-\beta_1)+e^{2(\alpha+\alpha_1-2\alpha_0)}\right)$$
$$-E\log\left(1-2e^{\alpha-\alpha_1}\cos(\beta-\beta_1)+e^{2(\alpha-\alpha_1)}\right)-2E(\alpha_1-\alpha_0). \qquad (14)$$

This expression for the potential becomes zero when $\alpha=\alpha_0$, and is finite and continuous within the curve α_0 except at the point (α_1, β_1), at which point the second term becomes infinite, and in its immediate neighbourhood is ultimately equal to $-2E\log r'$, where r' is the distance from that point.

We have therefore obtained the means of deducing the solution of Green's problem for a charge at any point within a closed curve when the solution for a charge at any other point is known.

The charge induced upon an element of the curve α_0 between the points β and $\beta+d\beta$ by a charge E placed at the point (α_1, β_1) is, with the notation of Art. 183,

$$-\frac{1}{4\pi}\frac{d\phi}{ds_1}ds_2,$$

where ds_1 is measured inwards and α is to be put equal to α_0 after differentiation.

This becomes, by (4) of Art. 183,

$$\frac{1}{4\pi}\frac{d\phi}{d\alpha}d\beta, \quad (\alpha=\alpha_0);$$

i.e.
$$-\frac{E}{2\pi}\frac{1-e^{2(\alpha_1-\alpha_0)}}{1-2e^{(\alpha_1-\alpha_0)}\cos(\beta-\beta_1)+e^{2(\alpha_1-\alpha_0)}}d\beta. \qquad (15)$$

From this expression we may find the potential at any point (α_1, β_1) within the closed curve, when the value of the potential at every point of the closed curve is given as a function of β, and there is no electrification within the closed curve.

For, by Art. 86, the part of the potential at (α_1, β_1), due to the maintenance of the portion $d\beta$ of the closed curve at the potential V is nV, where n is the charge induced on $d\beta$ by unit of electrification at (α_1, β_1). Hence, if V is the potential at a point on the closed curve defined as a function of β, and ϕ the potential at the point (α_1, β_1) within the closed curve, there being no electrification within the curve,

$$\phi = \frac{1}{2\pi}\int_0^{2\pi}\frac{(1-e^{2(\alpha_1-\alpha_0)})V\,d\beta}{1-2e^{(\alpha_1-\alpha_0)}\cos(\beta-\beta_1)+e^{2(\alpha_1-\alpha_0)}}. \qquad (16)$$

EXAMPLE IV.—*Distribution of Electricity near an Edge of a Conductor formed by Two Plane Faces.*

191.] In the case of an infinite plane face of a conductor charged with electricity to the surface-density σ_0, we find for the potential at a distance y from the plane

$$V = C - 4\pi\sigma_0 y,$$

where C is the value of the potential of the conductor itself.

Assume a straight line in the plane as a polar axis, and transform into polar coordinates, and we find for the potential

$$V = C - 4\pi\sigma_0 a e^{\rho} \sin\theta,$$

and for the quantity of electricity on a parallelogram of breadth unity, and length ae^{ρ} measured from the axis

$$E = \sigma_0 a e^{\rho}.$$

Now let us make $\rho = n\rho'$ and $\theta = n\theta'$, then, since ρ' and θ' are conjugate to ρ and θ, the equations

$$V = C - 4\pi\sigma_0 a e^{n\rho'} \sin n\theta'$$

and $$E = \sigma_0 a e^{n\rho'}$$

express a possible distribution of electricity and of potential.

If we write r for $ae^{\rho'}$, r will be the distance from the axis; we may also put θ instead of θ' for the angle. We shall have

$$V = C - 4\pi\sigma_0 \frac{r^n}{a^{n-1}} \sin n\theta,$$

$$E = \sigma_0 \frac{r^n}{a^{n-1}}.$$

V will be equal to C whenever $n\theta = \pi$ or a multiple of π.

Let the edge be a salient angle of the conductor, the inclination of the faces being a, then the angle of the dielectric is $2\pi - a$, so that when $\theta = 2\pi - a$ the point is in the other face of the conductor. We must therefore make

$$n(2\pi - a) = \pi, \quad \text{or} \quad n = \frac{\pi}{2\pi - a}.$$

Then $$V = C - 4\pi\sigma_0 a \left(\frac{r}{a}\right)^{\frac{\pi}{2\pi - a}} \sin\frac{\pi\theta}{2\pi - a},$$

$$E = \sigma_0 a \left(\frac{r}{a}\right)^{\frac{\pi}{2\pi - a}}.$$

The surface-density σ at any distance r from the edge is

$$\sigma = \frac{dE}{dr} = \frac{\pi}{2\pi - a}\sigma_0 \left(\frac{r}{a}\right)^{\frac{a-\pi}{2\pi - a}}.$$

When the angle is a salient one α is less than π, and the surface-density varies according to some inverse power of the distance from the edge, so that at the edge itself the density becomes infinite, although the whole charge reckoned from the edge to any finite distance from it is always finite.

Thus, when $\alpha = 0$ the edge is infinitely sharp, like the edge of a mathematical plane. In this case the density varies inversely as the square root of the distance from the edge.

When $\alpha = \frac{\pi}{3}$ the edge is like that of an equilateral prism, and the density varies inversely as the $\frac{2}{5}$ power of the distance.

When $\alpha = \frac{\pi}{2}$ the edge is a right angle, and the density is inversely as the cube root of the distance.

When $\alpha = \frac{2\pi}{3}$ the edge is like that of a regular hexagonal prism, and the density is inversely as the fourth root of the distance.

When $\alpha = \pi$ the edge is obliterated, and the density is constant.

When $\alpha = \frac{4}{3}\pi$ the edge is like that in the inside of the hexagonal prism, and the density is *directly* as the square root of the distance from the edge.

When $\alpha = \frac{3}{2}\pi$ the edge is a re-entrant right angle, and the density is directly as the distance from the edge.

When $\alpha = \frac{5}{3}\pi$ the edge is a re-entrant angle of 60°, and the density is directly as the square of the distance from the edge.

In reality, in all cases in which the density becomes infinite at any point, there is a discharge of electricity into the dielectric at that point, as is explained in Art. 55.

Example V.—*Ellipses and Hyperbolas.* Fig. X.

192.] We have seen that if

$$x_1 = e^{\phi}\cos\psi, \qquad y_1 = e^{\phi}\sin\psi, \tag{1}$$

x and y will be conjugate functions of ϕ and ψ.

Also, if
$$x_2 = e^{-\phi}\cos\psi, \qquad y_2 = -e^{-\phi}\sin\psi, \tag{2}$$

x_2 and y_2 will be conjugate functions. Hence, if

$$2x = x_1 + x_2 = (e^{\phi} + e^{-\phi})\cos\psi, \quad 2y = y_1 + y_2 = (e^{\phi} - e^{-\phi})\sin\psi, \tag{3}$$

x and y will also be conjugate functions of ϕ and ψ.

In this case the points for which ϕ is constant lie in the ellipse whose axes are $e^{\phi} + e^{-\phi}$ and $e^{\phi} - e^{-\phi}$.

The points for which ψ is constant lie in the hyperbola whose axes are $2\cos\psi$ and $2\sin\psi$.

On the axis of x, between $x=-1$ and $x=+1$,

$$\phi = 0, \qquad \psi = \cos^{-1}x. \tag{4}$$

On the axis of x, beyond these limits on either side, we have

$$\begin{array}{lll} x > \ \ 1, & \psi = 0, & \phi = \log\left(x+\sqrt{x^2-1}\right), \\ x < -1, & \psi = \pi, & \phi = \log\left(\sqrt{x^2-1}-x\right). \end{array} \tag{5}$$

Hence, if ϕ is the potential function, and ψ the function of flow, we have the case of electricity flowing from the positive to the negative side of the axis of x through the space between the points -1 and $+1$, the parts of the axis beyond these limits being impervious to electricity.

Since, in this case, the axis of y is a line of flow, we may suppose it also impervious to electricity.

We may also consider the ellipses to be sections of the equipotential surfaces due to an indefinitely long flat conductor of breadth 2, charged with half a unit of electricity per unit of length.

If we make ψ the potential function, and ϕ the function of flow, the case becomes that of an infinite plane from which a strip of breadth 2 has been cut away and the plane on one side charged to potential π while the other remains at zero.

These cases may be considered as particular cases of the quadric surfaces treated of in Chapter X. The forms of the curves are given in Fig. X.

Example VI.—Fig. XI.

193.] Let us next consider x' and y' as functions of x and y, where

$$x' = b\log\sqrt{x^2+y^2}, \qquad y' = b\tan^{-1}\frac{y}{x}, \tag{6}$$

x' and y' will be also conjugate functions of ϕ and ψ.

The curves resulting from the transformation of Fig. X with respect to these new coordinates are given in Fig. XI.

If x' and y' are rectangular coordinates, then the properties of the axis of x in the first figure will belong to a series of lines parallel to x' in the second figure for which $y' = bn'\pi$, where n' is any integer.

The positive values of x' on these lines will correspond to values of x greater than unity, for which, as we have already seen,

$$\psi = n\pi, \qquad \phi = \log\left(x+\sqrt{x^2-1}\right) = \log\left(e^{\frac{x'}{b}}+\sqrt{e^{\frac{2x'}{b}}-1}\right). \tag{7}$$

The negative values of x' on the same lines will correspond to values of x less than unity, for which, as we have seen,

$$\phi = 0, \qquad \psi = \cos^{-1} x = \cos^{-1} e^{\frac{x'}{b}}. \tag{8}$$

The properties of the axis of y in the first figure will belong to a series of lines in the second figure parallel to x', for which

$$y' = b\pi (n' + \tfrac{1}{2}). \tag{9}$$

The value of ψ along these lines is $\psi = \pi(n + \frac{1}{2})$ for all points both positive and negative, and

$$\phi = \log(y + \sqrt{y^2+1}) = \log\left(e^{\frac{x'}{b}} + \sqrt{e^{\frac{2x'}{b}} + 1}\right). \tag{10}$$

[The curves for which ϕ and ψ are constant may be traced directly from the equations

$$x' = \tfrac{1}{2} b \log \tfrac{1}{4}(e^{2\phi} + e^{-2\phi} + 2\cos 2\psi),$$

$$y' = b \tan^{-1}\left(\frac{e^{\phi} - e^{-\phi}}{e^{\phi} + e^{-\phi}} \tan\psi\right).$$

As the figure repeats itself for intervals of πb in the values of y' it will be sufficient to trace the lines for one such interval.

Now there will be two cases, according as ϕ or ψ changes sign with y'. Let us suppose that ϕ so changes sign. Then any curve for which ψ is constant will be symmetrical about the axis of x', cutting that axis orthogonally at some point on its negative side. If we begin with this point for which $\phi = 0$, and gradually increase ϕ, the curve will bend round from being at first orthogonal to being, for large values of ϕ, at length parallel to the axis of x'. The positive side of the axis of x' is one of the system, viz. ψ is there zero, and when $y' = \pm \frac{1}{2}\pi b$, $\psi = \frac{1}{2}\pi$. The lines for which ψ has constant values ranging from 0 to $\frac{1}{2}\pi$ form therefore a system of curves embracing the positive side of the axes of x'.

The curves for which ϕ has constant values cut the system ψ orthogonally, the values of ϕ ranging from $+\infty$ to $-\infty$. For any one of the curves ϕ drawn above the axis of x the value of ϕ is positive, along the negative side of the axis of x' the value is zero, and for any curve below the axis of x' the value is negative.

We have seen that the system ψ is symmetrical about the axis of x; let PQR be any curve cutting that system orthogonally and terminating in P and R in the lines $y' = \pm \frac{1}{2}\pi b$, the point Q being in the axis of x'. Then the curve PQR is symmetrical about the axis of x', but if c be the value of ϕ along PQ, the value of ϕ along QR will be $-c$. This discontinuity in the value of ϕ will be accounted for by an electrical distribution in the case which will be discussed in Art. 195.

If we next suppose that ψ and not ϕ changes sign with y', the values of ϕ will range from 0 to ∞. When $\phi = 0$ we have the negative side of the axis of x', and when $\phi = \infty$ we have a line at an infinite distance perpendicular to the axis of x'. Along any line PQR between these two the value of ϕ is constant throughout its entire length and positive.

Any value ψ now experiences an abrupt change at the point where the curve along which it is constant crosses the negative side of the axis of x', the sign of ψ changing there. The significance of this discontinuity will appear in Art. 197.

The lines we have shewn how to trace are drawn in Fig. XI if we limit ourselves to two-thirds of that diagram, cutting off the uppermost third.]

194.] If we consider ϕ as the potential function, and ψ as the function of flow, we may consider the case to be that of an indefinitely long strip of metal of breadth πb with a non-conducting division extending from the origin indefinitely in the positive direction, and thus dividing the positive part of the strip into two separate channels. We may suppose this division to be a narrow slit in the sheet of metal.

If a current of electricity is made to flow along one of these divisions and back again along the other, the entrance and exit of the current being at an indefinite distance on the positive side of the origin, the distribution of potential and of current will be given by the functions ϕ and ψ respectively.

If, on the other hand, we make ψ the potential, and ϕ the function of flow, then the case will be that of a current in the general direction of y', flowing through a sheet in which a number of non-conducting divisions are placed parallel to x', extending from the axis of y' to an indefinite distance in the negative direction.

195.] We may also apply the results to two important cases in statical electricity.

(1) Let a conductor in the form of a plane sheet, bounded by a straight edge but otherwise unlimited, be placed in the plane of xz on the positive side of the origin, and let two infinite conducting planes be placed parallel to it and at distances $\frac{1}{2}\pi b$ on either side. Then, if ψ is the potential function, its value is 0 for the middle conductor and $\frac{1}{2}\pi$ for the two planes.

Let us consider the quantity of electricity on a part of the middle conductor, extending to a distance 1 in the direction of z, and from the origin to $x' = a$.

The electricity on the part of this strip extending from x_1' to x_2' is $\frac{1}{4\pi}(\phi_2 - \phi_1)$.

Hence from the origin to $x' = a$ the amount is

$$E = \frac{1}{4\pi}\log\left(e^{\frac{a}{b}} + \sqrt{e^{\frac{2a}{b}} - 1}\right). \tag{11}$$

If a is large compared with b, this becomes

$$E = \frac{1}{4\pi}\log 2\, e^{\frac{a}{b}},$$

$$= \frac{a + b\log_e 2}{4\pi b}. \tag{12}$$

Hence the quantity of electricity on the plane bounded by the straight edge is greater than it would have been if the electricity had been uniformly distributed over it with the same density that it has at a distance from the boundary, and it is equal to the quantity of electricity having the same uniform surface-density, but extending to a breadth equal to $b\log_e 2$ beyond the actual boundary of the plate.

This imaginary uniform distribution is indicated by the dotted straight lines in Fig. XI. The vertical lines represent lines of force, and the horizontal lines equipotential surfaces, on the hypothesis that the density is uniform over both planes, produced to infinity in all directions.

196.] Electrical condensers are sometimes formed of a plate placed midway between two parallel plates extending considerably beyond the intermediate one on all sides. If the radius of curvature of the boundary of the intermediate plate is great compared with the distance between the plates, we may treat the boundary as approximately a straight line, and calculate the capacity of the condenser by supposing the intermediate plate to have its area extended by a strip of uniform breadth round its boundary, and assuming the surface-density on the extended plate the same as it is in the parts not near the boundary.

Thus, if S be the actual area of the plate, L its circumference and B the distance between the large plates, we have

$$b = \frac{1}{\pi}B, \tag{13}$$

and the breadth of the additional strip is

$$a = \frac{\log_e 2}{\pi} \cdot B, \tag{14}$$

so that the extended area is

$$S' = S + \frac{\log_e 2}{\pi} BL. \tag{15}$$

The capacity of the middle plate is

$$\frac{1}{2\pi}\frac{S'}{B} = \frac{1}{2\pi}\left\{\frac{S}{B} + L\frac{1}{\pi}\log_e 2\right\}. \tag{16}$$

Correction for the Thickness of the Plate.

Since the middle plate is generally of a thickness which cannot be neglected in comparison with the distance between the plates, we may obtain a better representation of the facts of the case by supposing the section of the intermediate plate to correspond with the curve $\psi = \psi'$.

The plate will be of nearly uniform thickness, $\beta = 2b\psi'$, at a distance from the boundary, but will be rounded near the edge.

The position of the actual edge of the plate is found by putting $y' = 0$, whence

$$x' = b \log_e \cos\psi'. \tag{17}$$

The value of ϕ at this edge is 0, and at a point for which $x' = a$ it is

$$\frac{a + b\log_e 2}{b}.$$

Hence, altogether, the quantity of electricity on the plate is the same as if a strip of breadth

$$\frac{B}{\pi}\left(\log_e 2 + \log_e \cos\frac{\pi\beta}{2B}\right),$$

$$\text{i.e.}\quad \frac{B}{\pi}\log_e\left(2\cos\frac{\pi\beta}{2B}\right), \tag{18}$$

had been added to the plate, the density being assumed to be everywhere the same as it is at a distance from the boundary.

Density near the Edge.

The surface-density at any point of the plate is

$$\frac{1}{4\pi}\frac{d\phi}{dx'} = \frac{1}{4\pi b}\frac{e^{\frac{x'}{b}}}{\sqrt{e^{\frac{2x'}{b}} - 1}}$$

$$= \frac{1}{4\pi b}\left(1 + \tfrac{1}{2}e^{-\frac{2x'}{b}} + \tfrac{3}{8}e^{-\frac{4x'}{b}} - \&c.\right). \tag{19}$$

The quantity within brackets rapidly approaches unity as x' increases, so that at a distance from the boundary equal to n times the breadth of the strip a, the actual density is greater than the normal density by about $\frac{1}{2^{2n+1}}$ of the normal density.

In like manner we may calculate the density on the infinite planes

$$= \frac{1}{4\pi b} \frac{e^{\frac{x'}{b}}}{\sqrt{e^{\frac{2x'}{b}}+1}}. \tag{20}$$

When $x' = 0$, the density is $2^{-\frac{1}{2}}$ of the normal density.

At n times the breadth of the strip on the positive side, the density is less than the normal density by about $\frac{1}{2^{n+1}}$.

At n times the breadth of the strip on the negative side, the density is about $\frac{1}{2^n}$ of the normal density.

These results indicate the degree of accuracy to be expected in applying this method to plates of limited extent, or in which irregularities may exist not very far from the boundary. The same distribution would exist in the case of an infinite series of similar plates at equal distances, the potentials of these plates being alternately $+V$ and $-V$. In this case we must take the distance between the plates equal to B.

197.] (2) The second case we shall consider is that of an infinite series of planes parallel to xz at distances $B = \pi b$, and all cut off by the plane of yz, so that they extend only on the negative side of this plane. If we make ϕ the potential function, we may regard these planes as conductors at potential zero.

Let us consider the curves for which ϕ is constant.

When $y' = n\pi b$, that is, in the prolongation of each of the planes, we have

$$x' = b \log \tfrac{1}{2}(e^{\phi} + e^{-\phi}) \tag{21}$$

when $y' = (n + \frac{1}{2}) b\pi$, that is, in the intermediate positions

$$x' = b \log \tfrac{1}{2}(e^{\phi} - e^{-\phi}). \tag{22}$$

Hence, when ϕ is large, the curve for which ϕ is constant is an undulating line whose mean distance from the axis of y' is approximately

$$a = b(\phi - \log_e 2), \tag{23}$$

and the amplitude of the undulations on either side of this line is

$$\tfrac{1}{2} b \log \frac{e^{\phi} + e^{-\phi}}{e^{\phi} - e^{-\phi}}. \tag{24}$$

When ϕ is large this becomes $be^{-2\phi}$, so that the curve approaches to the form of a straight line parallel to the axis of y' at a distance a from that axis on the positive side.

If we suppose a plane for which $x' = a$, kept at a constant potential while the system of parallel planes is kept at a different potential, then, since $b\phi = a + b\log_e 2$, the surface-density of the electricity induced on the plane is equal to that which would have been induced on it by a plane parallel to itself at a potential equal to that of the series of planes, but at a distance greater than that of the edges of the planes by $b\log_e 2$.

If B is the distance between two of the planes of the series, $B = \pi b$, so that the additional distance is

$$a = B\frac{\log_e 2}{\pi}. \tag{25}$$

198.] Let us next consider the space included between two of the equipotential surfaces, one of which consists of a series of parallel waves, while the other corresponds to a large value of ϕ, and may be considered as approximately plane.

If D is the depth of these undulations from the crest to the trough of each wave, then we find for the corresponding value of ϕ,

$$\phi = \tfrac{1}{2}\log\frac{e^{\frac{D}{b}}+1}{e^{\frac{D}{b}}-1}. \tag{26}$$

The value of x' at the crest of the wave is

$$b\log\tfrac{1}{2}(e^{\phi}+e^{-\phi}). \tag{27}$$

*Hence, if A is the distance from the crests of the waves to the opposite plane, the capacity of the system composed of the plane surface and the undulated surface is the same as that of two planes at a distance $A+a'$, where

$$a' = \frac{B}{\pi}\log_e\frac{2}{1+e^{-\pi\frac{D}{B}}}. \tag{28}$$

* Let Φ be the potential of the plane, ϕ of the undulating surface. The quantity of electricity on the plane per unit area is $1+4\pi b$. Hence the capacity

$$= 1 + 4\pi b(\Phi-\phi),$$
$$= 1 + 4\pi(A+a'), \text{ suppose.}$$

Then $A + a' = b(\Phi-\phi)$.

But $A + b\log\tfrac{1}{2}(e^{\phi}+e^{-\phi}) = b(\Phi - \log 2)$;

$$\therefore\ a' = -b\phi + b(\log 2 + \log\tfrac{1}{2}(e^{\phi}+e^{-\phi}))$$
$$= b\log(1+e^{-2\phi})$$
$$= b\log\frac{2}{1+e^{-\frac{D}{b}}}, \text{ by (26).}$$

199.] If a single groove of this form be made in a conductor having the rest of its surface plane, and if the other conductor is a plane surface at a distance A, the capacity of the one conductor with respect to the other will be diminished. The amount of this diminution will be less than the $\frac{1}{n}$th part of the diminution due to n such grooves side by side, for in the latter case the average electrical force between the conductors will be less than in the former case, so that the induction on the surface of each groove will be diminished on account of the neighbouring grooves.

If L is the length, B the breadth, and D the depth of the groove, the capacity of a portion of the opposite plane whose area is S will be

$$\frac{S-LB}{4\pi A}+\frac{LB}{4\pi(A+a')}=\frac{S}{4\pi A}-\frac{LB}{4\pi A}\frac{a'}{A+a'}. \tag{29}$$

If A is large compared with B or a', the correction becomes by (28)

$$\frac{L}{4\pi^2}\frac{B^2}{A^2}\log_e\frac{2}{1+e^{-\pi\frac{D}{B}}}, \tag{30}$$

and for a slit of infinite depth, putting $D=\infty$, the correction is

$$\frac{L}{4\pi^2}\frac{B^2}{A^2}\log_e 2. \tag{31}$$

To find the surface-density on the series of parallel plates we must find $\sigma=\frac{1}{4\pi}\frac{d\psi}{dx'}$ when $\phi=0$. We find

$$\sigma=\frac{1}{4\pi b}\frac{1}{\sqrt{e^{-2\frac{x}{b}}-1}}. \tag{32}$$

The average density on the plane plate at distance A from the edges of the series of plates is $\bar{\sigma}=\frac{1}{4\pi b}$. Hence, at a distance from the edge of one of the plates equal to na the surface-density is $\frac{1}{\sqrt{2^{2n}-1}}$ of this average density.

200.] Let us next attempt to deduce from these results the distribution of electricity in the figure formed by rotating the plane of the figure about the axis $y'=-R$. In this case, Poisson's equation will assume the form

$$\frac{d^2V}{dx'^2}+\frac{d^2V}{dy'^2}+\frac{1}{R+y'}\frac{dV}{dy'}+4\pi\rho=0. \tag{33}$$

Let us assume $V=\phi$, the function given in Art. 193, and determine the value of ρ from this equation. We know that the first two terms disappear, and therefore

$$\rho = -\frac{1}{4\pi}\frac{1}{R+y'}\frac{d\phi}{dy'}. \tag{34}$$

If we suppose that, in addition to the surface-density already investigated, there is a distribution of electricity in space according to the law just stated, the distribution of potential will be represented by the curves in Fig. XI.

Now from this figure it is manifest that $\frac{d\phi}{dy'}$ is generally very small except near the boundaries of the plates, so that the new distribution may be approximately represented by what actually exists, namely a certain superficial distribution near the edges of the plates.

If therefore we integrate $\iint \rho\, dx'\, dy'$ between the limits $y'=0$ and $y'=\frac{\pi}{2}b$, and from $x'=-\infty$ to $x=+\infty$, we shall find the whole additional charge on one side of the plates due to the curvature.

Since $\frac{d\phi}{dy'} = -\frac{d\psi}{dx'}$, we have

$$\int_{-\infty}^{\infty}\rho\, dx' = \int_{-\infty}^{\infty}\frac{1}{4\pi}\frac{1}{R+y}\frac{d\psi}{dx'}dx'$$

$$= \frac{1}{4\pi}\frac{1}{R+y'}(\psi_{\infty}-\psi_{-\infty})$$

$$= \frac{1}{8}\frac{1}{R+y'}\left(2\frac{y'}{B}-1\right). \tag{35}$$

Integrating with respect to y', we find

$$\int_0^{\frac{B}{2}}\int_{-\infty}^{\infty}\rho\, dx'\, dy' = \frac{1}{8}-\frac{1}{8}\frac{2R+B}{B}\log\frac{2R+B}{2R} \tag{36}$$

$$= -\frac{1}{32}\frac{B}{R}+\frac{1}{192}\frac{B^2}{R^2}+\&c. \tag{37}$$

This is half the total quantity of electricity which we must suppose distributed in space near the edge of one of the cylindric plates per unit of circumference. Since it is only close to the edge of the plate that the density is sensible, we may suppose it all condensed on the surface of the plate without altering sensibly its action on the opposed plane surface, and in calculating the attraction between that surface and the cylindric surface we may suppose this electricity to belong to the cylindric surface.

If there had been no curvature the superficial charge on the positive surface of the plate per unit of length would have been

$$-\int_{-\infty}^{0} \frac{1}{4\pi} \frac{d\phi}{dy'} dx' = \frac{1}{4\pi} (\psi_0 - \psi_{-\infty}) = -\frac{1}{8}.$$

Hence, if we add to it the whole of the above distribution, this charge must be multiplied by the factor $\left(1 + \frac{1}{2}\frac{B}{R}\right)$ to get the total charge on the positive side.

*In the case of a disk of radius R placed midway between two infinite parallel plates at a distance B, we find for the capacity of the disk

$$\frac{R^2}{B} + 2\frac{\log_e 2}{\pi} R + \frac{1}{2} B. \qquad (38)$$

* [In Art. 200, in estimating the total space distribution we might perhaps more correctly take for it the integral $\iint \rho\, 2\pi (R + y')\, dx'dy'$, which gives, per unit circumference of the edge of radius R, $-\frac{1}{32}\frac{B}{R}$, thus leading to the same correction as in the text.

The case of the disk may be treated in like manner as follows:

Let the figure of Art. 195 revolve round a line perpendicular to the plates and at a distance $+R$ from the edge of the middle one. That edge will therefore envelope a circle, which will be the edge of the disk. As in Art. 200, we begin with Poisson's equation, which in this case will be

$$\frac{d^2V}{dy'^2} + \frac{d^2V}{dx'^2} - \frac{1}{R - x'}\frac{dV}{dx'} + 4\pi\rho = 0.$$

We now assume that $V = \psi$, the potential function of Art. 195. We must therefore suppose electricity to exist in the region between the plates whose volume density ρ is

$$\frac{1}{4\pi}\frac{1}{R - x'}\frac{d\psi}{dx'}.$$

The total amount is

$$2\int_0^{\frac{B}{2}}\int_{-\infty}^{R} \rho . 2\pi (R - x')\, dx'\, dy'.$$

Now if R is large in comparison with the distance between the plates this result will be seen, on an examination of the potential lines in Fig. XI, to be sensibly the same as

$$\int_0^{\frac{B}{2}}\int_{-\infty}^{\infty} \frac{d\psi}{dx'} dx'dy'; \quad \text{that is, } -\tfrac{1}{8}\pi B.$$

The total surface distribution if we include both sides of the disk is

$$2\int_0^R \left(-\frac{1}{4\pi}\frac{d\psi}{dy'}\right)_{y'=0} 2\pi (R - x')\, dx'$$

$$= -\int_0^R (R - x') \left(\frac{d\phi}{dx'}\right)_{y'=0} dx'$$

$$= -\int_0^R \phi_{y'=0}\, dx'$$

$$= -\frac{R^2\pi}{2B} - B\log_e 2 + \frac{B}{\pi}(\log_e 2 - \tfrac{1}{2}\pi + 1).$$

If, therefore, the volume distribution between the plates be supposed to be concentrated on the disk the expression for the capacity, the difference of the potentials of the plates and disk being $\frac{\pi}{2}$, becomes

$$\frac{R^2}{B} + 2\frac{\log_e 2}{\pi} R + \tfrac{1}{4} B - \frac{2}{\pi^2}(\log_e 2 - \tfrac{1}{2}\pi + 1) B,$$

a result differing from that in the text by $\frac{B}{4}$ nearly.]

Theory of Thomson's Guard-ring.

201.] In some of Sir W. Thomson's electrometers, a large plane surface is kept at one potential, and at a distance a from this surface is placed a plane disk of radius R surrounded by a large plane plate called a Guard-ring with a circular aperture of radius R' concentric with the disk. This disk and plate are kept at potential zero.

The interval between the disk and the guard-plate may be regarded as a circular groove of infinite depth, and of breadth $R'-R$, which we denote by B.

The charge on the disk due to unit potential of the large disk, supposing the density uniform, would be $\frac{R^2}{4A}$.

The charge on one side of a straight groove of breadth B and length $L = 2\pi R$, and of infinite depth, may be estimated by the number of lines of force emanating from the large disk and falling upon the side of the groove. Referring to Art. 197 and footnote we see that the charge will therefore be

$$\tfrac{1}{2}LB \times \frac{1}{4\pi b},$$

$$\text{i.e.}\quad \tfrac{1}{4}\frac{RB}{A+a'},$$

since in this case $\Phi = 1$, $\phi = 0$, and therefore $b = A + a'$.

But since the groove is not straight, but has a radius of curvature R, this must be multiplied by the factor $\left(1 + \tfrac{1}{2}\frac{B}{R}\right)$.

The whole charge on the disk is therefore

$$\frac{R^2}{4A} + \tfrac{1}{4}\frac{RB}{A+a'}\left(1 + \frac{B}{2R}\right) \tag{39}$$

$$= \frac{R^2+R'^2}{8A} - \frac{R'^2-R^2}{8A}\cdot\frac{a'}{A+a'}. \tag{40}$$

The value of a' cannot be greater than

$$\frac{B\log 2}{\pi}, = 0.22B \text{ nearly.}$$

If B is small compared with either A or R this expression will give a sufficiently good approximation to the charge on the disk due to unity of difference of potential. The ratio of A to R may have any value, but the radii of the large disk and of the guard-ring must exceed R by several multiples of A.

Example VII.—Fig. XII.

202.] Helmholtz, in his memoir on discontinuous fluid motion *, has pointed out the application of several formulae in which the coordinates are expressed as functions of the potential and its conjugate function.

One of these may be applied to the case of an electrified plate of finite size placed parallel to an infinite plane surface connected with the earth.

Since $x_1 = A\phi$ and $y_1 = A\psi$,

and also $x_2 = Ae^{\phi}\cos\psi$ and $y_2 = Ae^{\phi}\sin\psi$,

are conjugate functions of ϕ and ψ, the functions formed by adding x_1 to x_2 and y_1 to y_2 will be also conjugate. Hence, if

$$x = A\phi + Ae^{\phi}\cos\psi,$$

$$y = A\psi + Ae^{\phi}\sin\psi.$$

then x and y will be conjugate with respect to ϕ and ψ, and ϕ and ψ will be conjugate with respect to x and y.

Now let x and y be rectangular coordinates, and let $k\psi$ be the potential, then $k\phi$ will be conjugate to $k\psi$, k being any constant.

Let us put $\psi = \pi$, then $y = A\pi$, $x = A(\phi - e^{\phi})$.

If ϕ varies from $-\infty$ to 0, and then from 0 to $+\infty$, x varies from $-\infty$ to $-A$ and from $-A$ to $-\infty$. Hence the equipotential surface, for which $\psi = \pi$, is a plane parallel to x at a distance $b = \pi A$ from the origin, and extending from $-\infty$ to $x = -A$.

Let us consider a portion of this plane, extending from

$$x = -(A+a) \text{ to } x = -A \text{ and from } z = 0 \text{ to } z = c,$$

let us suppose its distance from the plane of xz to be $y = b = A\pi$, and its potential to be $V = k\psi = k\pi$.

The charge of electricity on the portion of the plane considered is found by ascertaining the values of ϕ at its extremities.

We have therefore to determine ϕ from the equation

$$x = -(A+a) = A(\phi - e^{\phi}),$$

ϕ will have a negative value ϕ_1 and a positive value ϕ_2; at the edge of the plane, where $x = -A$, $\phi = 0$.

Hence the charge on the one side is $-ck\phi_1 \div 4\pi$, and that on the other side is $ck\phi_2 \div 4\pi$.

* *Königl. Akad. der Wissenschaften*, zu Berlin, April 23, 1868.

Both these charges are positive and their sum is

$$\frac{ck(\phi_2-\phi_1)}{4\pi}.$$

If we suppose that a is large compared with A,

$$\phi_1 = -\frac{a}{A} - 1 + e^{-\frac{a}{A}-1+e^{-\frac{a}{A}-1+\&c.}}$$

$$\phi_2 = \log\left\{\frac{a}{A} + 1 + \log\left(\frac{a}{A} + 1 + \&c.\right)\right\}.$$

If we neglect the exponential terms in ϕ_1 we shall find that the charge on the negative surface exceeds that which it would have if the superficial density had been uniform and equal to that at a distance from the boundary, by a quantity equal to the charge on a strip of breadth $A = \frac{b}{\pi}$ with the uniform superficial density.

The total capacity of the part of the plane considered is

$$C = \frac{c}{4\pi^2}(\phi_2-\phi_1).$$

The total charge is CV, and the attraction towards the infinite plane, whose equation is $y = 0$ and potential $\psi = 0$, is

$$-\tfrac{1}{2}V^2\frac{dC}{db} = V^2\frac{ac}{8\pi b^2}\left(1 + \frac{\frac{A}{a}}{1+\frac{a}{A}\log\frac{a}{A}} + e^{-\frac{a}{A}} + \&c.\right.$$

$$= \frac{V^2 c}{8\pi b^2}\left\{a + \frac{b}{\pi} - \frac{b^2}{\pi^2 a}\log\frac{a}{A} + \&c.\right\}.$$

The equipotential lines and lines of force are given in Fig. XII.

EXAMPLE VIII. *Theory of a Grating of Parallel Wires.* Fig. XIII.

203.] In many electrical instruments a wire grating is used to prevent certain parts of the apparatus from being electrified by induction. We know that if a conductor be entirely surrounded by a metallic vessel at the same potential with itself, no electricity can be induced on the surface of the conductor by any electrified body outside the vessel. The conductor, however, when completely surrounded by metal, cannot be seen, and therefore, in certain cases, an aperture is left which is covered with a grating of fine wire. Let us investigate the effect of this grating in diminishing the effect of electrical induction. We shall suppose the grating to consist of a series of parallel wires in one plane and at equal intervals, the diameter of the wires being small compared with the distance between them, while the nearest portions of the electrified bodies on the one side and of the protected conductor on the other are at distances from the plane of the screen, which are considerable compared with the distance between consecutive wires.

204.] The potential at a distance r' from the axis of a straight wire of infinite length charged with a quantity of electricity λ per unit of length is

$$V = -2\lambda \log r' + C. \tag{1}$$

We may express this in terms of polar coordinates referred to an axis whose distance from the wire is unity, in which case we must make

$$r'^2 = 1 - 2r\cos\theta + r^2, \tag{2}$$

and if we suppose that the axis of reference is also charged with the linear density λ', we find

$$V = -\lambda \log(1 - 2r\cos\theta + r^2) - 2\lambda' \log r + C. \tag{3}$$

If we now make

$$r = e^{2\pi\frac{y}{a}}, \qquad \theta = \frac{2\pi x}{a}, \tag{4}$$

then, by the theory of conjugate functions,

$$V = -\lambda \log\left(1 - 2e^{\frac{2\pi y}{a}} \cos\frac{2\pi x}{a} + e^{\frac{4\pi y}{a}}\right) - 2\lambda' \log e^{\frac{2\pi y}{a}} + C, \tag{5}$$

where x and y are rectangular coordinates, will be the value of the potential due to an infinite series of fine wires parallel to z in the plane of xz, and passing through points in the axis of x for which x is a multiple of a.

Each of these wires is charged with a linear density λ.

The term involving λ' indicates an electrification, producing a constant force $\frac{4\pi\lambda'}{a}$ in the direction of y.

The forms of the equipotential surfaces and lines of force when $\lambda' = 0$ are given in Fig. XIII. The equipotential surfaces near the wires are nearly cylinders, so that we may consider the solution approximately true, even when the wires are cylinders of a diameter which is finite but small compared with the distance between them.

The equipotential surfaces at a distance from the wires become more and more nearly planes parallel to that of the grating.

If in the equation we make $y = b_1$, a quantity large compared with a, we find approximately,

$$V_1 = -\frac{4\pi b_1}{a}(\lambda + \lambda') + C \text{ nearly}. \tag{6}$$

If we next make $y = -b_2$, where b_2 is a positive quantity large compared with a, we find approximately,

$$V_2 = \frac{4\pi b_2}{a}\lambda' + C \text{ nearly.} \tag{7}$$

If c is the radius of the wires of the grating, c being small compared with a, we may find the potential of the grating itself by supposing that the surface of the wire coincides with the equipotential surface which cuts the plane of xz at a distance c from the axis of z. To find the potential of the grating we therefore put $x = c$, and $y = 0$, whence

$$V = -2\lambda \log 2 \sin\frac{\pi c}{a} + C. \tag{8}$$

205.] We have now obtained expressions representing the electrical state of a system consisting of a grating of wires whose diameter is small compared with the distance between them, and two plane conducting surfaces, one on each side of the grating, and at distances which are great compared with the distance between the wires.

The surface-density σ_1 on the first plane is got from the equation (6)

$$4\pi\sigma_1 = \frac{dV_1}{db_1} = -\frac{4\pi}{a}(\lambda + \lambda'). \tag{9}$$

That on the second plane σ_2 from the equation (7)

$$4\pi\sigma_2 = \frac{dV_2}{db_2} = \frac{4\pi}{a}\lambda'. \tag{10}$$

If we now write

$$\alpha = -\frac{a}{2\pi}\log_e\left(2\sin\frac{\pi c}{a}\right), \tag{11}$$

and eliminate λ and λ' from the equations (6), (7), (8), (9), (10), we find

$$4\pi\sigma_1\left(b_1 + b_2 + \frac{b_1 b_2}{\alpha}\right) = V_1\left(1 + \frac{b_2}{\alpha}\right) - V_2 - V\frac{b_2}{\alpha}, \tag{12}$$

$$4\pi\sigma_2\left(b_1 + b_2 + \frac{b_1 b_2}{\alpha}\right) = -V_1 + V_2\left(1 + \frac{b_1}{\alpha}\right) - V\frac{b_1}{\alpha}. \tag{13}$$

When the wires are infinitely thin, α becomes infinite, and the terms in which it is the denominator disappear, so that the case is reduced to that of two parallel planes without a grating interposed.

If the grating is in metallic communication with one of the planes, say the first, $V = V_1$, and the right-hand side of the equation for σ_1 becomes $V_1 - V_2$. Hence the density σ_1 induced on the first plane when the grating is interposed is to that which would have been induced on it if the grating were removed, the second plane being maintained at the same potential, as 1 to $1 + \dfrac{b_1 b_2}{\alpha(b_1 + b_2)}$.

We should have found the same value for the effect of the grating in diminishing the electrical influence of the first surface on the second, if we had supposed the grating connected with the second surface. This is evident since b_1 and b_2 enter into the expression in the same way. It is also a direct result of the theorem of Art. 88.

The induction of the one electrified plane on the other through the grating is the same as if the grating were removed, and the distance between the planes increased from $b_1 + b_2$ to

$$b_1 + b_2 + \frac{b_1 b_2}{\alpha}.$$

If the two planes are kept at potential zero, and the grating electrified to a given potential, the quantity of electricity on the grating will be to that which would be induced on a plane of equal area placed in the same position as

$$b_1 b_2 : b_1 b_2 + \alpha (b_1 + b_2).$$

This investigation is approximate only when b_1 and b_2 are large compared with α, and when α is large compared with c. The quantity α is a line which may be of any magnitude. It becomes infinite when c is indefinitely diminished.

If we suppose $c = \frac{1}{2} a$ there will be no apertures between the wires of the grating, and therefore there will be no induction through it. We ought therefore to have for this case $\alpha = 0$. The formula (11), however, gives in this case

$$\alpha = -\frac{a}{2\pi} \log_e 2, \qquad = -0{\cdot}11 a,$$

which is evidently erroneous, as the induction can never be altered in sign by means of the grating. It is easy, however, to proceed to a higher degree of approximation in the case of a grating of cylindrical wires. I shall merely indicate the steps of this process.

Method of Approximation.

206.] Since the wires are cylindrical, and since the distribution of electricity on each is symmetrical with respect to the diameter parallel to y, the proper expansion of the potential is of the form

$$V = C_0 \log r + \Sigma C_i r^i \cos i\theta, \tag{14}$$

where r is the distance from the axis of one of the wires, and θ the angle between r and y, and, since the wire is a conductor, when r is made equal to the radius V must be constant, and therefore the coefficient of each of the multiple cosines of θ must vanish.

For the sake of conciseness let us assume new coordinates ξ, η, &c. such that

$$a\xi = 2\pi x, \quad a\eta = 2\pi y, \quad a\rho = 2\pi r, \quad a\beta = 2\pi b, \text{ \&c.,} \tag{15}$$

and let

$$F_\beta = \log\left(\epsilon^{\eta+\beta} + \epsilon^{-(\eta+\beta)} - 2\cos\xi\right). \tag{16}$$

Then if we make

$$V = A_0 F + A_1 \frac{dF}{d\eta} + A_2 \frac{d^2F}{d\eta^2} + \text{\&c.} \tag{17}$$

by giving proper values to the coefficients A we may express any potential which is a function of η and $\cos\xi$, and does not become infinite except when $\eta + \beta = 0$ and $\cos\xi = 1$.

When $\beta = 0$ the expansion of F in terms of ρ and θ is

$$F_0 = 2\log\rho + \tfrac{1}{12}\rho^2\cos 2\theta - \tfrac{1}{1440}\rho^4\cos 4\theta + \text{\&c.} \tag{18}$$

For finite values of β the expansion of F is

$$F_\beta = \beta + 2\log(1-e^{-\beta}) + \frac{1+e^{-\beta}}{1-e^{-\beta}}\rho\cos\theta - \frac{e^{-\beta}}{(1-e^{-\beta})^2}\rho^2\cos 2\theta + \text{\&c.} \tag{19}$$

In the case of the grating with two conducting planes whose equations are $\eta = \beta_1$ and $\eta = -\beta_2$, that of the plane of the grating being $\eta = 0$, there will be two infinite series of images of the grating. The first series will consist of the grating itself together with an infinite series of images on both sides, equal and similarly electrified. The axes of these imaginary cylinders lie in planes whose equations are of the form

$$\eta = \pm 2n(\beta_1 + \beta_2), \tag{20}$$

n being an integer.

The second series will consist of an infinite series of images for which the coefficients A_0, A_2, A_4, &c. are equal and opposite to the same quantities in the grating itself, while A_1, A_3, &c. are equal and of the same sign. The axes of these images are in planes whose equations are of the form

$$\eta = 2\beta_2 \pm 2m(\beta_1 + \beta_2), \tag{21}$$

m being an integer.

The potential due to any finite series of such images will depend on whether the number of images is odd or even. Hence the potential due to an infinite series is indeterminate, but if we add to it the function $B\eta + C$, the conditions of the problem will be sufficient to determine the electrical distribution.

We may first determine V_1 and V_2, the potentials of the two conducting planes, in terms of the coefficients A_0, A_1, &c., and of B and C. We must then determine σ_1 and σ_2, the surface-density at any point of these planes. The mean values of σ_1 and σ_2 are given by the equations

$$4\pi\sigma_1 = \frac{2\pi}{a}(A_0 - B), \quad 4\pi\sigma_2 = \frac{2\pi}{a}(A_0 + B.) \tag{22}$$

We must then expand the potentials due to the grating itself and to all the images in terms of ρ and cosines of multiples of θ, adding to the result $$B\rho\cos\theta + C.$$

The terms independent of θ then give V the potential of the grating, and the coefficient of the cosine of each multiple of θ equated to zero gives an equation between the indeterminate coefficients.

In this way as many equations may be found as are sufficient to eliminate all these coefficients and to leave two equations to determine σ_1 and σ_2 in terms of V_1, V_2, and V.

These equations will be of the form

$$\begin{aligned} V_1 - V &= 4\pi\sigma_1(b_2 + \alpha - \gamma) + 4\pi\sigma_2(\alpha + \gamma), \\ V_2 - V &= 4\pi\sigma_1(\alpha + \gamma) + 4\pi\sigma_2(b_2 + \alpha - \gamma). \end{aligned} \tag{23}$$

The quantity of electricity induced on one of the planes protected by the grating, the other plane being at a given difference of potential, will be the same as if the plates had been at a distance

$$\frac{(\alpha-\gamma)(b_1+b_2) + b_1 b_2 - 4\alpha\gamma}{\alpha+\gamma} \text{ instead of } b_1 + b_2.$$

The values of α and γ are approximately as follows,

$$\alpha = \frac{a}{2\pi}\left\{\log\frac{a}{2\pi c} - \frac{5}{3}\cdot\frac{\pi^4 c^4}{15a^4 + \pi^4 c^4} \right.$$
$$\left. + 2e^{-4\pi\frac{b_1+b_2}{a}}\left(1 + e^{-4\pi\frac{b_1}{a}} + e^{-4\pi\frac{b_2}{a}} + \&c.\right) + \&c.\right\}, \tag{24}$$

$$\gamma = \frac{3\pi a c^2}{3a^2 + \pi^2 c^2}\left(\frac{e^{-4\pi\frac{b_1}{a}}}{1 - e^{-4\pi\frac{b_1}{a}}} - \frac{e^{-4\pi\frac{b_2}{a}}}{1 - e^{-4\pi\frac{b_2}{a}}}\right) + \&c. \tag{25}$$

CHAPTER XIII.

ELECTROSTATIC INSTRUMENTS.

On Electrostatic Instruments.

THE instruments which we have to consider at present may be divided into the following classes:

(1) Electrical machines for the production and augmentation of electrification.

(2) Multipliers, for increasing electrification in a known ratio.

(3) Electrometers, for the measurement of electric potentials and charges.

(4) Accumulators, for holding large electrical charges.

Electrical Machines.

207.] In the common electrical machine a plate or cylinder of glass is made to revolve so as to rub against a surface of leather, on which is spread an amalgam of zinc and mercury. The surface of the glass becomes electrified positively and that of the rubber negatively. As the electrified surface of the glass moves away from the negative electrification of the rubber it acquires a high positive potential. It then comes opposite to a set of sharp metal points in connexion with the conductor of the machine. The positive electrification of the glass induces a negative electrification of the points, which is the more intense the sharper the points and the nearer they are to the glass.

When the machine works properly there is a discharge through the air between the glass and the points, the glass loses part of its positive charge, which is transferred to the points and so to the insulated prime conductor of the machine, and to any other body with which it is in electric communication.

The portion of the glass which is advancing towards the rubber has thus a smaller positive charge than that which is leaving it at the same time, so that the rubber, and the conductors in communication with it, become negatively electrified.

The highly positive surface of the glass where it leaves the rubber is more attracted by the negative charge of the rubber than the partially discharged surface which is advancing towards the rubber. The electrical forces therefore act as a resistance to the force employed in turning the machine. The work done in turning the machine is therefore greater than that spent in overcoming ordinary friction and other resistances, and the excess is employed in producing a state of electrification whose energy is equivalent to this excess.

The work done in overcoming friction is at once converted into heat in the bodies rubbed together. The electrical energy may be also converted either into mechanical energy or into heat.

If the machine does not store up mechanical energy, all the energy will be converted into heat, and the only difference between the heat due to friction and that due to electrical action is that the former is generated at the rubbing surfaces while the latter may be generated in conductors at a distance *.

We have seen that the electrical charge on the surface of the glass is attracted by the rubber. If this attraction were sufficiently intense there would be a discharge between the glass and the rubber, instead of between the glass and the collecting points. To prevent this, flaps of silk are attached to the rubber. These become negatively electrified and adhere to the glass, and so diminish the potential near the rubber.

The potential therefore increases more gradually as the glass moves away from the rubber, and therefore at any one point there is less attraction of the charge on the glass towards the rubber, and consequently less danger of direct discharge to the rubber.

In some electrical machines the moving part is of ebonite instead of glass, and the rubbers of wool or fur. The rubber is then electrified positively and the prime conductor negatively.

The Electrophorus of Volta.

208.] The electrophorus consists of a plate of resin or of ebonite backed with metal, and a plate of metal of the same size. An insulating handle can be screwed to the back of either of these plates. The ebonite plate has a metal pin which connects the metal

* It is probable that in many cases where dynamical energy is converted into heat by friction, part of the energy may be first transformed into electrical energy and then converted into heat as the electrical energy is spent in maintaining currents of short circuit close to the rubbing surfaces. See Sir W. Thomson, 'On the Electrodynamic Qualities of Metals.' *Phil. Trans.*, 1856, p. 650.

plate with the metal back of the ebonite plate when the two plates are in contact.

The ebonite plate is electrified negatively by rubbing it with wool or cat's skin. The metal plate is then brought near the ebonite by means of the insulating handle. No direct discharge passes between the ebonite and the metal plate, but the potential of the metal plate is rendered negative by induction, so that when it comes within a certain distance of the metal pin a spark passes, and if the metal plate be now carried to a distance it is found to have a positive charge which may be communicated to a conductor. The metal at the back of the ebonite plate is found to have a negative charge equal and opposite to the charge of the metal plate.

In using the instrument to charge a condenser or accumulator one of the plates is laid on a conductor in communication with the earth, and the other is first laid on it, then removed and applied to the electrode of the condenser, then laid on the fixed plate and the process repeated. If the ebonite plate is fixed the condenser will be charged positively. If the metal plate is fixed the condenser will be charged negatively.

The work done by the hand in separating the plates is always greater than the work done by the electrical attraction during the approach of the plates, so that the operation of charging the condenser involves the expenditure of work. Part of this work is accounted for by the energy of the charged condenser, part is spent in producing the noise and heat of the sparks, and the rest in overcoming other resistances to the motion.

On Machines producing Electrification by Mechanical Work.

209.] In the ordinary frictional electrical machine the work done in overcoming friction is far greater than that done in increasing the electrification. Hence any arrangement by which the electrification may be produced entirely by mechanical work against the electrical forces is of scientific importance if not of practical value. The first machine of this kind seems to have been Nicholson's Revolving Doubler, described in the *Philosophical Transactions* for 1788 as 'an instrument which by the turning of a Winch produces the two states of Electricity without friction or communication with the Earth.'

210.] It was by means of the revolving doubler that Volta succeeded in developing from the electrification of the pile an electrification capable of affecting his electrometer. Instruments on the same principle have been invented independently by Mr. C. F. Varley* and Sir W. Thomson.

These instruments consist essentially of insulated conductors of various forms, some fixed and others moveable. The moveable conductors are called Carriers, and the fixed ones may be called Inductors, Receivers, and Regenerators. The inductors and receivers are so formed that when the carriers arrive at certain points in their revolution they are almost completely surrounded by a conducting body. As the inductors and receivers cannot completely surround the carrier and at the same time allow it to move freely in and out without a complicated arrangement of moveable pieces, the instrument is not theoretically perfect without a pair of regenerators, which store up the small amount of electricity which the carriers retain when they emerge from the receivers.

For the present, however, we may suppose the inductors and receivers to surround the carrier completely when it is within them, in which case the theory is much simplified.

We shall suppose the machine to consist of two inductors A and C, and of two receivers B and D, with two carriers F and G.

Suppose the inductor A to be positively electrified so that its potential is A, and that the carrier F is within it and is at potential F. Then, if Q is the coefficient of induction (taken positive) between A and F, the quantity of electricity on the carrier will be $Q(F-A)$.

If the carrier, while within the inductor, is put in connexion with the earth, then $F=0$, and the charge on the carrier will be $-QA$, a negative quantity. Let the carrier be carried round till it is within the receiver B, and let it then come in contact with a spring so as to be in electrical connexion with B. It will then, as was shewn in Art. 32, become completely discharged, and will communicate its whole negative charge to the receiver B.

The carrier will next enter the inductor C, which we shall suppose charged negatively. While within C it is put in connexion with the earth and thus acquires a positive charge, which it carries off and communicates to the receiver D, and so on.

In this way, if the potentials of the inductors remain always constant, the receivers B and D receive successive charges, which are the same for every revolution of the carrier, and thus every revolution produces an equal increment of electricity in the receivers.

* Specification of Patent, Jan. 27, 1860, No. 206.

But by putting the inductor A in communication with the receiver D, and the inductor C with the receiver B, the potentials of the inductors will be continually increased, and the quantity of electricity communicated to the receivers in each revolution will continually increase.

For instance, let the potential of A and D be U, and that of B and C, V, then, since the potential of the carrier is zero when it is within A, being in contact with earth, its charge is $z = -QU$. The carrier enters B with this charge and communicates it to B. If the capacity of B and C is B, their potential will be changed from V to $V - \frac{Q}{B} U$.

If the other carrier has at the same time carried a charge $-Q'V$ from C to D, it will change the potential of A and D from U to $U - \frac{Q'}{A} V$, if Q' is the coefficient of induction between the carrier and C, and A the capacity of A and D. If, therefore, U_n and V_n be the potentials of the two inductors after n half revolutions, and U_{n+1} and V_{n+1} after $n+1$ half revolutions,

$$U_{n+1} = U_n - \frac{Q'}{A} V_n,$$

$$V_{n+1} = V_n - \frac{Q}{B} U_n.$$

If we write $p^2 = \frac{Q}{B}$ and $q^2 = \frac{Q'}{A}$, we find

$$pU_{n+1} + qV_{n+1} = (pU_n + qV_n)(1 - pq) = (pU_0 + qV_0)(1 - pq)^{n+1},$$

$$pU_{n+1} - qV_{n+1} = (pU_n - qV_n)(1 + pq) = (pU_0 - qV_0)(1 + pq)^{n+1}.$$

Hence

$$U_n = U_0\left((1 - pq)^n + (1 + pq)^n\right) + \frac{q}{p} V_0\left((1 - pq)^n - (1 + pq)^n\right),$$

$$V_n = \frac{p}{q} U_0\left((1 - pq)^n - (1 + pq)^n\right) + V_0\left((1 - pq)^n + (1 + pq)^n\right).$$

It appears from these equations that the quantity $pU + qV$ continually diminishes, so that whatever be the initial state of electrification the receivers are ultimately oppositely electrified, so that the potentials of A and B are in the ratio of p to $-q$.

On the other hand, the quantity $pU - qV$ continually increases, so that, however little pU may exceed or fall short of qV at first, the difference will be increased in a geometrical ratio in each revolution till the electromotive forces become so great that the insulation of the apparatus is overcome.

Instruments of this kind may be used for various purposes.

For producing a copious supply of electricity at a high potential, as is done by means of Mr. Varley's large machine.

For adjusting the charge of a condenser, as in the case of Thomson's electrometer, the charge of which can be increased or diminished by a few turns of a very small machine of this kind, which is then called a Replenisher.

For multiplying small differences of potential. The inductors may be charged at first to an exceedingly small potential, as, for instance, that due to a thermo-electric pair, then, by turning the machine, the difference of potentials may be continually multiplied till it becomes capable of measurement by an ordinary electrometer. By determining by experiment the ratio of increase of this difference due to each turn of the machine, the original electromotive force with which the inductors were charged may be deduced from the number of turns and the final electrification.

In most of these instruments the carriers are made to revolve about an axis and to come into the proper positions with respect to the inductors by turning an axle. The connexions are made by means of springs so placed that the carriers come in contact with them at the proper instants.

211.] Sir W. Thomson *, however, has constructed a machine for multiplying electrical charges in which the carriers are drops of water falling out of the inside of an inductor into an insulated receiver. The receiver is thus continually supplied with electricity of opposite sign to that of the inductor. If the inductor is electrified positively, the receiver will receive a continually increasing charge of negative electricity.

The water is made to escape from the receiver by means of a funnel, the nozzle of which is almost surrounded by the metal of the receiver. The drops falling from this nozzle are therefore nearly free from electrification. Another inductor and receiver of the same construction are arranged so that the inductor of the one system is in connexion with the receiver of the other. The rate of increase of charge of the receivers is thus no longer constant, but increases in a geometrical progression with the time, the charges of the two receivers being of opposite signs. This increase goes on till the falling drops are so diverted from their course by

* *Proc. R. S.*, June 20, 1867.

the electrical action that they fall outside of the receiver or even strike the inductor.

In this instrument the energy of the electrification is drawn from that of the falling drops.

212.] Several other electrical machines have been constructed in which the principle of electric induction is employed. Of these the most remarkable is that of Holtz, in which the carrier is a glass plate varnished with gum-lac and the inductors are pieces of pasteboard. Sparks are prevented from passing between the parts of the apparatus by means of two glass plates, one on each side of the revolving carrier plate. This machine is found to be very effective, and not to be much affected by the state of the atmosphere. The principle is the same as in the revolving doubler and the instruments developed out of the same idea, but as the carrier is an insulating plate and the inductors are imperfect conductors, the complete explanation of the action is more difficult than in the case where the carriers are good conductors of known form and are charged and discharged at definite points.

213.] In the electrical machines already described sparks occur whenever the carrier comes in contact with a conductor at a different potential from its own.

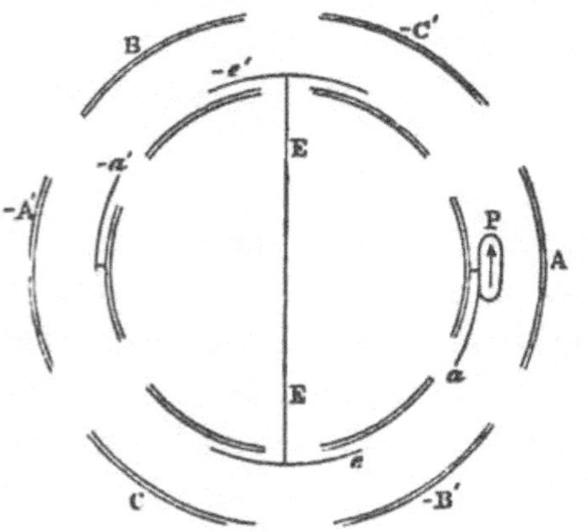

Fig. 18.

Now we have shewn that whenever this occurs there is a loss of energy, and therefore the whole work employed in turning the machine is not converted into electrification in an available form, but part is spent in producing the heat and noise of electric sparks.

I have therefore thought it desirable to shew how an electrical machine may be constructed which is not subject to this loss of efficiency. I do not propose it as a useful form of machine, but as an example of the method by which the contrivance called in heat-engines a regenerator may be applied to an electrical machine to prevent loss of work.

In the figure let A, B, C, A', B', C' represent hollow fixed conductors, so arranged that the carrier P passes in succession within each of them. Of these A, A' and B, B' nearly surround the carrier when it is at the middle point of its passage, but C, C' do not cover it so much.

We shall suppose A, B, C to be connected with a Leyden jar of great capacity at potential V, and A', B', C' to be connected with another jar at potential $-V'$.

P is one of the carriers moving in a circle from A to C', &c., and touching in its course certain springs, of which a and a' are connected with A and A' respectively, and e, e' are connected with the earth.

Let us suppose that when the carrier P is in the middle of A the coefficient of induction between P and A is $-A$. The capacity of P in this position is greater than A, since it is not completely surrounded by the receiver A. Let it be $A+a$.

Then if the potential of P is U, and that of A, V, the charge on P will be $(A+a)\,U-AV$.

Now let P be in contact with the spring a when in the middle of the receiver A, then the potential of P is V, the same as that of A, and its charge is therefore aV.

If P now leaves the spring a it carries with it the charge aV. As P leaves A its potential diminishes, and it diminishes still more when it comes within the influence of C', which is negatively electrified.

If when P comes within C' its coefficient of induction on C' is $-C'$, and its capacity is $C'+c'$, then, if U is the potential of P the charge on P is

$$(C'+c')\,U+C'V'=aV.$$

If
$$C'V'=aV,$$

then at this point U the potential of P will be reduced to zero.

Let P at this point come in contact with the spring e' which is connected with the earth. Since the potential of P is equal to that of the spring there will be no spark at contact.

This conductor C', by which the carrier is enabled to be connected to earth without a spark, answers to the contrivance called a regenerator in heat-engines. We shall therefore call it a Regenerator.

Now let P move on, still in contact with the earth-spring e', till it comes into the middle of the inductor B, the potential of which is V. If $-B$ is the coefficient of induction between P and B at this point, then, since $U=0$ the charge on P will be $-BV$.

When P moves away from the earth-spring it carries this charge with it. As it moves out of the positive inductor B towards the negative receiver A' its potential will be increasingly negative. At the middle of A', if it retained its charge, its potential would be

$$-\frac{A'V'+BV}{A'+a'},$$

and if BV is greater than $a'V'$ its numerical value will be greater than that of V'. Hence there is some point before P reaches the middle of A' where its potential is $-V'$. At this point let it come in contact with the negative receiver-spring a'. There will be no spark since the two bodies are at the same potential. Let P move on to the middle of A', still in contact with the spring, and therefore at the same potential with A'. During this motion it communicates a negative charge to A'. At the middle of A' it leaves the spring and carries away a charge $-a'V'$ towards the positive regenerator C, where its potential is reduced to zero and it touches the earth-spring e. It then slides along the earth-spring into the negative inductor B', during which motion it acquires a positive charge $B'V'$ which it finally communicates to the positive receiver A, and the cycle of operations is repeated.

During this cycle the positive receiver has lost a charge aV and gained a charge $B'V'$. Hence the total gain of positive electricity is

$$B'V'-aV.$$

Similarly the total gain of negative electricity is $BV-a'V'$.

By making the inductors so as to be as close to the surface of the carrier as is consistent with insulation, B and B' may be made large, and by making the receivers so as nearly to surround the carrier when it is within them, a and a' may be made very small, and then the charges of both the Leyden jars will be increased in every revolution.

The conditions to be fulfilled by the regenerators are

$$C'V'=aV, \quad \text{and} \quad CV=a'V'.$$

Since a and a' are small the regenerators do not require to be either large or very close to the carriers.

On Electrometers and Electroscopes.

214.] An electrometer is an instrument by means of which electrical charges or electrical potentials may be measured. Instruments by means of which the existence of electric charges or of differences of potential may be indicated, but which are not capable of affording numerical measures, are called Electroscopes.

An electroscope if sufficiently sensitive may be used in electrical measurements, provided we can make the measurement depend on the absence of electrification. For instance, if we have two charged bodies A and B we may use the method described in Chapter I to determine which body has the greater charge. Let the body A be carried by an insulating support into the interior of an insulated closed vessel C. Let C be connected to earth and again insulated. There will then be no external electrification on C. Now let A be removed, and B introduced into the interior of C, and the electrification of C tested by an electroscope. If the charge of B is equal to that of A there will be no electrification, but if it is greater or less there will be electrification of the same kind as that of B, or the opposite kind.

Methods of this kind, in which the thing to be observed is the non-existence of some phenomenon, are called *null* or *zero* methods. They require only an instrument capable of detecting the existence of the phenomenon.

In another class of instruments for the registration of phenomena the instruments may be depended upon to give always the same indication for the same value of the quantity to be registered, but the readings of the scale of the instrument are not proportional to the values of the quantity, and the relation between these readings and the corresponding value is unknown, except that the one is some continuous function of the other. Several electrometers depending on the mutual repulsion of parts of the instrument which are similarly electrified are of this class. The use of such instruments is to register phenomena, not to measure them. Instead of the true values of the quantity to be measured, a series of numbers is obtained, which may be used afterwards to determine these values when the scale of the instrument has been properly investigated and tabulated.

In a still higher class of instruments the scale readings are proportional to the quantity to be measured, so that all that is required for the complete measurement of the quantity is a knowledge of the coefficient by which the scale readings must be multiplied to obtain the true value of the quantity.

Instruments so constructed that they contain within themselves the means of independently determining the true values of quantities are called Absolute Instruments.

Coulomb's Torsion Balance.

215.] A great number of the experiments by which Coulomb

established the fundamental laws of electricity were made by measuring the force between two small spheres charged with electricity, one of which was fixed while the other was held in equilibrium by two forces, the electrical action between the spheres, and the torsional elasticity of a glass fibre or metal wire. See Art. 38.

The balance of torsion consists of a horizontal arm of gum-lac, suspended by a fine wire or glass fibre, and carrying at one end a little sphere of elder pith, smoothly gilt. The suspension wire is fastened above to the vertical axis of an arm which can be moved round a horizontal graduated circle, so as to twist the upper end of the wire about its own axis any number of degrees.

The whole of this apparatus is enclosed in a case. Another little sphere is so mounted on an insulating stem that it can be charged and introduced into the case through a hole, and brought so that its centre coincides with a definite point in the horizontal circle described by the suspended sphere. The position of the suspended sphere is ascertained by means of a graduated circle engraved on the cylindrical glass case of the instrument.

Now suppose both spheres charged, and the suspended sphere in equilibrium in a known position such that the torsion-arm makes an angle θ with the radius through the centre of the fixed sphere. The distance of the centres is then $2a \sin \frac{1}{2}\theta$, where a is the radius of the torsion-arm, and if F is the force between the spheres the moment of this force about the axis of torsion is $Fa \cos \frac{1}{2}\theta$.

Let both spheres be completely discharged, and let the torsion-arm now be in equilibrium at an angle ϕ with the radius through the fixed sphere.

Then the angle through which the electrical force twisted the torsion-arm must have been $\theta - \phi$, and if M is the moment of the torsional elasticity of the fibre, we shall have the equation

$$Fa \cos \tfrac{1}{2}\theta = M(\theta - \phi).$$

Hence, if we can ascertain M, we can determine F, the actual force between the spheres at the distance $2a \sin \frac{1}{2}\theta$.

To find M, the moment of torsion, let I be the moment of inertia of the torsion-arm, and T the time of a double vibration of the arm under the action of the torsional elasticity, then

$$M = 4\pi^2 \frac{I}{T^2}.$$

In all electrometers it is of the greatest importance to know what force we are measuring. The force acting on the suspended sphere is due partly to the direct action of the fixed sphere, but partly also to the electrification, if any, of the sides of the case.

If the case is made of glass it is impossible to determine the electrification of its surface otherwise than by very difficult measurements at every point. If, however, either the case is made of metal, or if a metallic case which almost completely encloses the apparatus is placed as a screen between the spheres and the glass case, the electrification of the inside of the metal screen will depend entirely on that of the spheres, and the electrification of the glass case will have no influence on the spheres. In this way we may avoid any indefiniteness due to the action of the case.

To illustrate this by an example in which we can calculate all the effects, let us suppose that the case is a sphere of radius b, that the centre of motion of the torsion-arm coincides with the centre of the sphere and that its radius is a; that the charges on the two spheres are E_1 and E, and that the angle between their positions is θ; that the fixed sphere is at a distance a_1 from the centre, and that r is the distance between the two small spheres.

Neglecting for the present the effect of induction on the distribution of electricity on the small spheres, the force between them will be a repulsion

$$= \frac{EE_1}{r^2},$$

and the moment of this force round a vertical axis through the centre will be

$$\frac{EE_1 aa_1 \sin\theta}{r^3}.$$

The image of E_1 due to the spherical surface of the case is a point in the same radius at a distance $\frac{b^2}{a_1}$ with a charge $-E_1\frac{b}{a_1}$, and the moment of the attraction between E and this image about the axis of suspension is

$$EE_1\frac{b}{a_1}\frac{a\frac{b^2}{a_1}\sin\theta}{\left\{a^2-2\frac{ab^2}{a_1}\cos\theta+\frac{b^4}{a_1^{\,2}}\right\}^{\frac{3}{2}}}$$

$$= EE_1\frac{aa_1\sin\theta}{b^3\left\{1-2\frac{aa_1}{b^2}\cos\theta+\frac{a^2a_1^2}{b^4}\right\}^{\frac{3}{2}}}.$$

If b, the radius of the spherical case, is large compared with a

and a_1, the distances of the spheres from the centre, we may neglect the second and third terms of the factor in the denominator. The whole moment tending to turn the torsion-arm may then be written

$$EE_1 aa_1 \sin\theta \left\{\frac{1}{r^3} - \frac{1}{b^3}\right\} = M(\theta - \phi).$$

Electrometers for the Measurement of Potentials.

216.] In all electrometers the moveable part is a body charged with electricity, and its potential is different from that of certain of the fixed parts round it. When, as in Coulomb's method, an insulated body having a certain charge is used, it is the charge which is the direct object of measurement. We may, however, connect the balls of Coulomb's electrometer, by means of fine wires, with different conductors. The charges of the balls will then depend on the values of the potentials of these conductors and on the potential of the case of the instrument. The charge on each ball will be approximately equal to its radius multiplied by the excess of its potential over that of the case of the instrument, provided the radii of the balls are small compared with their distances from each other and from the sides or opening of the case.

Coulomb's form of apparatus, however, is not well adapted for measurements of this kind, owing to the smallness of the force between spheres at the proper distances when the difference of potentials is small. A more convenient form is that of the Attracted Disk Electrometer. The first electrometers on this principle were constructed by Sir W. Snow Harris *. They have since been brought to great perfection, both in theory and construction, by Sir W. Thomson †.

When two disks at different potentials are brought face to face with a small interval between them there will be a nearly uniform electrification on the opposite faces and very little electrification on the backs of the disks, provided there are no other conductors or electrified bodies in the neighbourhood. The charge on the positive disk will be approximately proportional to its area, and to the difference of potentials of the disks, and inversely as the distance between them. Hence, by making the areas of the disks large

* *Phil. Trans.* 1834.

† See an excellent report on Electrometers by Sir W. Thomson. *Report of the British Association*, Dundee, 1867.

and the distance between them small, a small difference of potential may give rise to a measurable force of attraction.

The mathematical theory of the distribution of electricity over two disks thus arranged is given at Art. 202, but since it is impossible to make the case of the apparatus so large that we may suppose the disks insulated in an infinite space, the indications of the instrument in this form are not easily interpreted numerically.

217.] The addition of the guard-ring to the attracted disk is one of the chief improvements which Sir W. Thomson has made on the apparatus.

Instead of suspending the whole of one of the disks and determining the force acting upon it, a central portion of the disk is separated from the rest to form the attracted disk, and the outer ring forming the remainder of the disk is fixed. In this way the force is measured only on that part of the disk where it is most regular, and the want of uniformity of the electrification near the

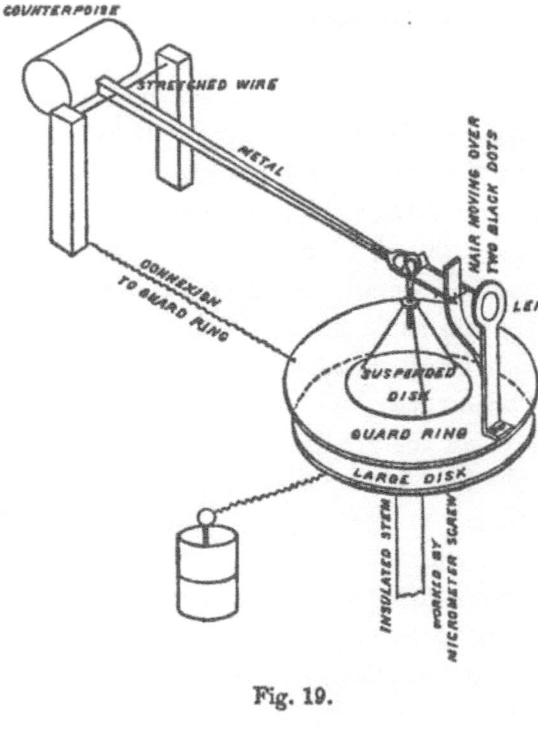

Fig. 19.

edge is of no importance, as it occurs on the guard-ring and not on the suspended part of the disk.

Besides this, by connecting the guard-ring with a metal case surrounding the back of the attracted disk and all its suspending apparatus, the electrification of the back of the disk is rendered impossible, for it is part of the inner surface of a closed hollow conductor all at the same potential.

Thomson's Absolute Electrometer therefore consists essentially of two parallel plates at different potentials, one of which is made so that a certain area, no part of which is near the edge of the plate, is moveable under the action of electric force. To fix our ideas we may suppose the attracted disk and guard-ring uppermost. The fixed disk is horizontal, and is mounted on an insulating stem which has a measurable vertical motion given to it by means of a micrometer screw. The guard-ring is at least as large as the fixed disk; its lower surface is truly plane and parallel to the fixed disk. A delicate balance is erected on the guard-ring to which is suspended a light moveable disk which almost fills the circular aperture in the guard-ring without rubbing against its sides. The lower surface of the suspended disk must be truly plane, and we must have the means of knowing when its plane coincides with that of the lower surface of the guard-ring, so as to form a single plane interrupted only by the narrow interval between the disk and its guard-ring.

For this purpose the lower disk is screwed up till it is in contact with the guard-ring, and the suspended disk is allowed to rest upon the lower disk, so that its lower surface is in the same plane as that of the guard-ring. Its position with respect to the guard-ring is then ascertained by means of a system of fiducial marks. Sir W. Thomson generally uses for this purpose a black hair attached to the moveable part. This hair moves up or down just in front of two black dots on a white enamelled ground and is viewed along with these dots by means of a plano convex lens with the plane side next the eye. If the hair as seen through the lens appears straight and bisects the interval between the black dots it is said to be in its *sighted position*, and indicates that the suspended disk with which it moves is in its proper position as regards height. The horizontality of the suspended disk may be tested by comparing the reflexion of part of any object from its upper surface with that of the remainder of the same object from the upper surface of the guard-ring.

The balance is then arranged so that when a known weight is placed on the centre of the suspended disk it is in equilibrium in its sighted position, the whole apparatus being freed from electrification by putting every part in metallic communication. A metal case is placed over the guard-ring so as to enclose the balance and suspended disk, sufficient apertures being left to see the fiducial marks.

The guard-ring, case, and suspended disk are all in metallic communication with each other, but are insulated from the other parts of the apparatus.

Now let it be required to measure the difference of potentials of two conductors. The conductors are put in communication with the upper and lower disks respectively by means of wires, the weight is taken off the suspended disk, and the lower disk is moved up by means of the micrometer screw till the electrical attraction brings the suspended disk down to its sighted position. We then know that the attraction between the disks is equal to the weight which brought the disk to its sighted position.

If W be the numerical value of the weight, and g the force of gravity, the force is Wg, and if A is the area of the suspended disk, D the distance between the disks, and V the difference of the potentials of the disks *,

$$Wg = \frac{V^2 A}{8\pi D^2}, \quad \text{or} \quad V = D\sqrt{\frac{8\pi g W}{A}}.$$

* Let us denote the radius of the suspended disk by R, and that of the aperture of the guard-ring by R', then the breadth of the annular interval between the disk and the ring will be $B = R' - R$.

If the distance between the suspended disk and the large fixed disk is D, and the difference of potentials between these disks is V, then, by the investigation in Art. 201, the quantity of electricity on the suspended disk will be

$$Q = V\left\{\frac{R^2+R'^2}{8D} - \frac{R'^2-R^2}{8D}\frac{a}{D+a}\right\},$$

where $a = B\dfrac{\log_e 2}{\pi}$, or $a = 0.220635\,(R'-R)$.

If the surface of the guard-ring is not exactly in the plane of the surface of the suspended disk, let us suppose that the distance between the fixed disk and the guard-ring is not D but $D+z = D'$, then it appears from the investigation in Art. 225 that there will be an additional charge of electricity near the edge of the disk on account of its height z above the general surface of the guard-ring. The whole charge in this case is therefore, approximately,

$$Q = V\left\{\frac{R^2+R'^2}{8D} - \frac{R'^2-R^2}{8D}\frac{a}{D+a} + \frac{R+R'}{D}(D'-D)\log_e\frac{4\pi(R+R')}{D'-D}\right\},$$

and in the expression for the attraction we must substitute for A, the area of the disk, the corrected quantity

$$A = \tfrac{1}{2}\pi\left\{R^2+R'^2-(R'^2-R^2)\frac{a}{D+a}+8(R+R')(D'-D)\log_e\frac{4\pi(R+R')}{D'-D}\right\},$$

where R = radius of suspended disk,
R' = radius of aperture in the guard-ring,
D = distance between fixed and suspended disks,
D' = distance between fixed disk and guard-ring,
a = $0.220635\,(R'-R)$.

When a is small compared with D we may neglect the second term, and when $D'-D$ is small we may neglect the last term.

If the suspended disk is circular, of radius R, and if the radius of the aperture of the guard-ring is R', then

$$A = \tfrac{1}{2}\pi(R^2 + R'^2), \text{ and } V = 4D\sqrt{\frac{gW}{R^2+R'^2}}.$$

218.] Since there is always some uncertainty in determining the micrometer reading corresponding to $D = 0$, and since any error in the position of the suspended disk is most important when D is small, Sir W. Thomson prefers to make all his measurements depend on differences of the electromotive force V. Thus, if V and V' are two potentials, and D and D' the corresponding distances,

$$V - V' = (D - D')\sqrt{\frac{8\pi gW}{A}}.$$

For instance, in order to measure the electromotive force of a galvanic battery, two electrometers are used.

By means of a condenser, kept charged if necessary by a replenisher, the lower disk of the principal electrometer is maintained at a constant potential. This is tested by connecting the lower disk of the principal electrometer with the lower disk of a secondary electrometer, the suspended disk of which is connected with the earth. The distance between the disks of the secondary electrometer and the force required to bring the suspended disk to its sighted position being constant, if we raise the potential of the condenser till the secondary electrometer is in its sighted position, we know that the potential of the lower disk of the principal electrometer exceeds that of the earth by a constant quantity which we may call V.

If we now connect the positive electrode of the battery to earth, and connect the suspended disk of the principal electrometer to the negative electrode, the difference of potentials between the disks will be $V + v$, if v is the electromotive force of the battery. Let D be the reading of the micrometer in this case, and let D' be the reading when the suspended disk is connected with earth, then

$$v = (D - D')\sqrt{\frac{8\pi gW}{A}}.$$

In this way a small electromotive force v may be measured by the electrometer with the disks at conveniently measurable distances. When the distance is too small a small change of absolute distance makes a great change in the force, since the force varies inversely as the square of the distance, so that any error in the absolute distance introduces a large error in the result unless the distance is large compared with the limits of error of the micrometer screw.

The effect of small irregularities of form in the surfaces of the disks and of the interval between them diminish according to the inverse cube and higher inverse powers of the distance, and whatever be the form of a corrugated surface, the eminences of which just reach a plane surface, the electrical effect at any distance which is considerable compared to the breadth of the corrugations, is the same as that of a plane at a certain small distance behind the plane of the tops of the eminences. See Arts. 197, 198.

By means of the auxiliary electrification, tested by the auxiliary electrometer, a proper interval between the disks is secured.

The auxiliary electrometer may be of a simpler construction, in which there is no provision for the determination of the force of attraction in absolute measure, since all that is wanted is to secure a constant electrification. Such an electrometer may be called a *gauge* electrometer.

This method of using an auxiliary electrification besides the electrification to be measured is called the Heterostatic method of electrometry, in opposition to the Idiostatic method in which the whole effect is produced by the electrification to be measured.

In several forms of the attracted disk electrometer, the attracted disk is placed at one end of an arm which is supported by being attached to a platinum wire passing through its centre of gravity and kept stretched by means of a spring. The other end of the arm carries the hair which is brought to a sighted position by altering the distance between the disks, and so adjusting the force of the electric attraction to a constant value. In these electrometers this force is not in general determined in absolute measure, but is known to be constant, provided the torsional elasticity of the platinum wire does not change.

The whole apparatus is placed in a Leyden jar, of which the inner surface is charged and connected with the attracted disk and guard-ring. The other disk is worked by a micrometer screw and is connected first with the earth and then with the conductor whose potential is to be measured. The difference of readings multiplied by a constant to be determined for each electrometer gives the potential required.

219.] The electrometers already described are not self-acting, but require for each observation an adjustment of a micrometer screw, or some other movement which must be made by the observer. They are therefore not fitted to act as self-registering instruments, which must of themselves move into the proper position. This condition is fulfilled by Thomson's Quadrant Electrometer.

The electrical principle on which this instrument is founded may be thus explained :—

A and *B* are two fixed conductors which may be at the same or at different potentials. *C* is a moveable conductor at a high potential, which is so placed that part of it is opposite to the surface of *A* and part opposite to that of *B*, and that the proportions of these parts are altered as *C* moves.

For this purpose it is most convenient to make *C* moveable about an axis, and make the opposed surfaces of *A*, of *B*, and of *C* portions of surfaces of revolution about the same axis.

In this way the distance between the surface of *C* and the opposed surfaces of *A* or of *B* remains always the same, and the motion of *C* in the positive direction simply increases the area opposed to *B* and diminishes the area opposed to *A*.

If the potentials of *A* and *B* are equal there will be no force urging *C* from *A* to *B*, but if the potential of *C* differs from that of *B* more than from that of *A*, then *C* will tend to move so as to increase the area of its surface opposed to *B*.

By a suitable arrangement of the apparatus this force may be made nearly constant for different positions of *C* within certain limits, so that if *C* is suspended by a torsion fibre, its deflexions will be nearly proportional to the difference of potentials between *A* and *B* multiplied by the difference of the potential of *C* from the mean of those of *A* and *B*.

C is maintained at a high potential by means of a condenser provided with a replenisher and tested by a gauge electrometer, and *A* and *B* are connected with the two conductors the difference of whose potentials is to be measured. The higher the potential of *C* the more sensitive is the instrument. This electrification of *C*, being independent of the electrification to be measured, places this electrometer in the heterostatic class.

We may apply to this electrometer the general theory of systems of conductors given in Arts. 93, 127.

Let *A*, *B*, *C* denote the potentials of the three conductors respectively. Let *a*, *b*, *c* be their respective capacities, *p* the coefficient of induction between *B* and *C*, *q* that between *C* and *A*, and *r* that between A and B. All these coefficients will in general vary with the position of C, and if C is so arranged that the extremities of A and B are not near those of C as long as the motion of C is confined within certain limits, we may ascertain the form of these coefficients. If θ represents the deflexion of C from A towards B, then the part of the surface of A opposed to C will diminish as θ increases. Hence if A is kept at potential 1 while B and C are kept at potential 0, the charge on A will be $a = a_0 - \alpha\theta$, where a_0 and α are constants, and a is the capacity of A.

If A and B are symmetrical, the capacity of B is $b = b_0 + \alpha\theta$.

The capacity of C is not altered by the motion, for the only effect of the motion is to bring a different part of C opposite to the interval between A and B. Hence $c = c_0$.

The quantity of electricity induced on C when B is raised to potential unity is $p = p_0 - \alpha\theta$.

The coefficient of induction between A and C is $q = q_0 + \alpha\theta$.

The coefficient of induction between A and B is not altered by the motion of C, but remains $r = r_0$.

Hence the electrical energy of the system is

$$W = \tfrac{1}{2}A^2a + \tfrac{1}{2}B^2b + \tfrac{1}{2}C^2c + BCp + CAq + ABr,$$

and if Θ is the moment of the force tending to increase θ,

$$\Theta = \frac{dW}{d\theta}, \text{ } A, B, C \text{ being supposed constant,}$$

$$= \tfrac{1}{2}A^2\frac{da}{d\theta} + \tfrac{1}{2}B^2\frac{db}{d\theta} + \tfrac{1}{2}C^2\frac{dc}{d\theta} + BC\frac{dp}{d\theta} + CA\frac{dq}{d\theta} + AB\frac{dr}{d\theta},$$

$$= -\tfrac{1}{2}A^2\alpha + \tfrac{1}{2}B^2\alpha - BC\alpha + CA\alpha;$$

$$\text{or} \quad \Theta = \alpha(A-B)\left(C - \tfrac{1}{2}(A+B)\right).$$

In the present form of Thomson's Quadrant Electrometer the conductors A and B are in the form of a cylindrical box completely divided into four quadrants, separately insulated, but joined by wires so that two opposite quadrants are connected with A and the two others with B.

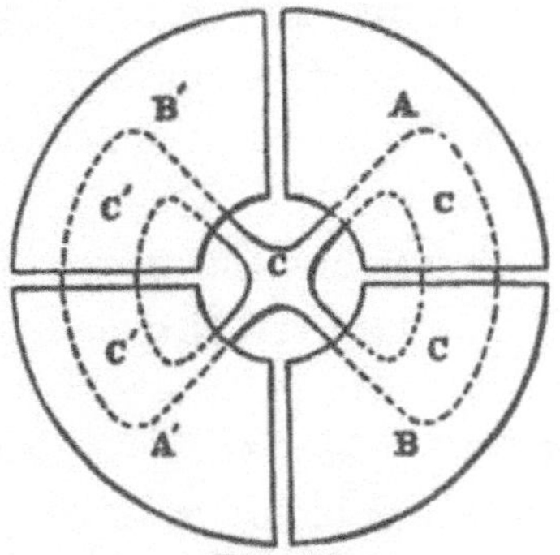

Fig. 20.

The conductor C is suspended so as to be capable of turning about a vertical axis, and may consist of two opposite flat quadrantal arcs supported by their radii at their extremities. In the position of equilibrium these quadrants should be partly within A and partly within B, and the supporting radii should be near the middle of the quadrants of the hollow base, so that the divisions of the box and the extremities and supports of C may be as far from each other as possible.

The conductor C is kept permanently at a high potential by being connected with the inner coating of the Leyden jar which forms the case of the instrument. B and A are connected, the first with the earth, and the other with the body whose potential is to be measured.

If the potential of this body is zero, and if the instrument be in adjustment, there ought to be no force tending to make C move, but if the potential of A is of the same sign as that of C, then C will tend to move from A to B with a nearly uniform force, and the suspension apparatus will be twisted till an equal force is called into play and produces equilibrium. Within certain limits the deflexions of C will be proportional to the product

$$(A-B)(C-\tfrac{1}{2}(A+B)).$$

By increasing the potential of C the sensibility of the instrument may be increased, and for small values of $\frac{1}{2}(A+B)$ the deflexions will be nearly proportional to $(A-B)\,C$.

On the Measurement of Electric Potential.

220.] In order to determine large differences of potential in absolute measure we may employ the attracted disk electrometer, and compare the attraction with the effect of a weight. If at the same time we measure the difference of potential of the same conductors by means of the quadrant electrometer, we shall ascertain the absolute value of certain readings of the scale of the quadrant electrometer, and in this way we may deduce the value of the scale readings of the quadrant electrometer in terms of the potential of the suspended part, and the moment of torsion of the suspension apparatus.

To ascertain the potential of a charged conductor of finite size we may connect the conductor with one electrode of the electrometer, while the other is connected to earth or to a body of constant potential. The electrometer reading will give the potential of the conductor after the division of its electricity between it and the part of the electrometer with which it is put in contact. If K denote the capacity of the conductor, and K' that of this part of the electrometer, and if V, V' denote the potentials of these bodies before making contact, then their common potential after making contact will be

$$\overline{V} = \frac{KV + K'V'}{K + K'}.$$

Hence the original potential of the conductor was

$$V = \overline{V} + \frac{K'}{K}(\overline{V} - V').$$

If the conductor is not large compared with the electrometer, K' will be comparable with K, and unless we can ascertain the values of K and K' the second term of the expression will have a doubtful value. But if we can make the potential of the electrode of the electrometer very nearly equal to that of the body before making contact, then the uncertainty of the values of K and K' will be of little consequence.

If we know the value of the potential of the body approximately, we may charge the electrode by means of a 'replenisher' or otherwise to this approximate potential, and the next experiment will give a closer approximation. In this way we may measure the potential of a conductor whose capacity is small compared with that of the electrometer.

To Measure the Potential at any Point in the Air.

221.] *First Method.* Place a sphere, whose radius is small compared with the distance of electrified conductors, with its centre at the given point. Connect it by means of a fine wire with the earth, then insulate it, and carry it to an electrometer and ascertain the total charge on the sphere.

Then, if V be the potential at the given point, and a the radius of the sphere, the charge on the sphere will be $-Va = Q$, and if V' be the potential of the sphere as measured by an electrometer when placed in a room whose walls are connected with the earth, then

$$Q = V'a,$$

whence

$$V + V' = 0,$$

or the potential of the air at the point where the centre of the sphere was placed is equal but of opposite sign to the potential of the sphere after being connected to earth, then insulated, and brought into a room.

This method has been employed by M. Delmann of Creuznach in

measuring the potential at a certain height above the earth's surface.

Second Method. We have supposed the sphere placed at the given point and first connected to earth, and then insulated, and carried into a space surrounded with conducting matter at potential zero.

Now let us suppose a fine insulated wire carried from the electrode of the electrometer to the place where the potential is to be measured. Let the sphere be first discharged completely. This may be done by putting it into the inside of a vessel of the same metal which nearly surrounds it and making it touch the vessel. Now let the sphere thus discharged be carried to the end of the wire and made to touch it. Since the sphere is not electrified it will be at the potential of the air at the place. If the electrode wire is at the same potential it will not be affected by the contact, but if the electrode is at a different potential it will by contact with the sphere be made nearer to that of the air than it was before. By a succession of such operations, the sphere being alternately discharged and made to touch the electrode, the potential of the electrode of the electrometer will continually approach that of the air at the given point.

222.] To measure the potential of a conductor without touching it, we may measure the potential of the air at any point in the neighbourhood of the conductor, and calculate that of the conductor from the result. If there be a hollow nearly surrounded by the conductor, then the potential at any point of the air in this hollow will be very nearly that of the conductor.

In this way it has been ascertained by Sir W. Thomson that if two hollow conductors, one of copper and the other of zinc, are in metallic contact, then the potential of the air in the hollow surrounded by zinc is positive with reference to that of the air in the hollow surrounded by copper.

Third Method. If by any means we can cause a succession of small bodies to detach themselves from the end of the electrode, the potential of the electrode will approximate to that of the surrounding air. This may be done by causing shot, filings, sand, or water to drop out of a funnel or pipe connected with the electrode. The point at which the potential is measured is that at which the stream ceases to be continuous and breaks into separate parts or drops.

Another convenient method is to fasten a slow match to the

electrode. The potential is very soon made equal to that of the air at the burning end of the match. Even a fine metallic point is sufficient to create a discharge by means of the particles of the air when the difference of potentials is considerable, but if we wish to reduce this difference to zero, we must use one of the methods stated above.

If we only wish to ascertain the sign of the difference of the potentials at two places, and not its numerical value, we may cause drops or filings to be discharged at one of the places from a nozzle connected with the other place, and catch the drops or filings in an insulated vessel. Each drop as it falls is charged with a certain amount of electricity, and it is completely discharged into the vessel. The charge of the vessel therefore is continually accumulating, and after a sufficient number of drops have fallen, the charge of the vessel may be tested by the roughest methods. The sign of the charge is positive if the potential of the nozzle is positive relatively to that of the surrounding air.

MEASUREMENT OF SURFACE-DENSITY OF ELECTRIFICATION.

Theory of the Proof Plane.

223.] In testing the results of the mathematical theory of the distribution of electricity on the surface of conductors, it is necessary to be able to measure the surface-density at different points of the conductor. For this purpose Coulomb employed a small disk of gilt paper fastened to an insulating stem of gum-lac. He applied this disk to various points of the conductor by placing it so as to coincide as nearly as possible with the surface of the conductor. He then removed it by means of the insulating stem, and measured the charge of the disk by means of his electrometer.

Since the surface of the disk, when applied to the conductor, nearly coincided with that of the conductor, he concluded that the surface-density on the outer surface of the disk was nearly equal to that on the surface of the conductor at that place, and that the charge on the disk when removed was nearly equal to that on an area of the surface of the conductor equal to that of one side of the disk. This disk, when employed in this way, is called Coulomb's Proof Plane.

As objections have been raised to Coulomb's use of the proof plane, I shall make some remarks on the theory of the experiment.

This experiment consists in bringing a small conducting body into contact with the surface of the conductor at the point where the density is to be measured, and then removing the body and determining its charge.

We have first to shew that the charge on the small body when in contact with the conductor is proportional to the surface-density which existed at the point of contact before the small body was placed there.

We shall suppose that all the dimensions of the small body, and especially its dimension in the direction of the normal at the point of contact, are small compared with either of the radii of curvature of the conductor at the point of contact. Hence the variation of the resultant force due to the conductor supposed rigidly electrified within the space occupied by the small body may be neglected, and we may treat the surface of the conductor near the small body as a plane surface.

Now the charge which the small body will take by contact with a plane surface will be proportional to the resultant force normal to the surface, that is, to the surface-density. We shall ascertain the amount of the charge for particular forms of the body.

We have next to shew that when the small body is removed no spark will pass between it and the conductor, so that it will carry its charge with it. This is evident, because when the bodies are in contact their potentials are the same, and therefore the density on the parts nearest to the point of contact is extremely small. When the small body is removed to a very short distance from the conductor, which we shall suppose to be electrified positively, then the electrification at the point nearest to the small body is no longer zero but positive, but, since the charge of the small body is positive, the positive electrification close to the small body will be less than at other neighbouring points of the surface. Now the passage of a spark depends in general on the magnitude of the resultant force, and this on the surface-density. Hence, since we suppose that the conductor is not so highly electrified as to be discharging electricity from the other parts of its surface, it will not discharge a spark to the small body from a part of its surface which we have shewn to have a smaller surface-density.

224.] We shall now consider various forms of the small body.

Suppose it to be a small hemisphere applied to the conductor so as to touch it at the centre of its flat side.

Let the conductor be a large sphere, and let us modify the form of the hemisphere so that its surface is a little more than a hemisphere, and meets the surface of the sphere at right angles. Then we have a case of which we have already obtained the exact solution. See Art. 167.

If A and B be the centres of the two spheres cutting each other at right angles, DD' a diameter of the circle of intersection, and C the centre of that circle, then if V is the potential of a conductor whose outer surface coincides with that of the two spheres, the quantity of electricity on the exposed surface of the sphere A is

$$\tfrac{1}{2}V(AD+BD+AC-CD-BC),$$

and that on the exposed surface of the sphere B is

$$\tfrac{1}{2}V(AD+BD+BC-CD-AC),$$

the total charge being the sum of these, or

$$V(AD+BD-CD).$$

If α and β are the radii of the spheres, then, when α is large compared with β, the charge on B is to that on A in the ratio of

$$\frac{3}{4}\frac{\beta^2}{\alpha^2}\left(1+\frac{1}{3}\frac{\beta}{\alpha}+\frac{1}{6}\frac{\beta^2}{\alpha^2}+\&c.\right) \text{ to } 1.$$

Now let σ be the uniform surface-density on A when B is removed, then the charge on A is

$$4\pi\alpha^2\sigma,$$

and therefore the charge on B is

$$3\pi\beta^2\sigma\left(1+\frac{1}{3}\frac{\beta}{\alpha}+\&c.\right),$$

or, when β is very small compared with α, the charge on the hemisphere B is equal to three times that due to a surface-density σ extending over an area equal to that of the circular base of the hemisphere.

It appears from Art. 175 that if a small sphere is made to touch an electrified body, and is then removed to a distance from it, the mean surface-density on the sphere is to the surface-density of the body at the point of contact as π^2 is to 6, or as 1.645 to 1.

225.] The most convenient form for the proof plane is that of a circular disk. We shall therefore shew how the charge on a circular disk laid on an electrified surface is to be measured.

For this purpose we shall construct a value of the potential function so that one of the equipotential surfaces resembles a circular flattened protuberance whose general form is somewhat like that of a disk lying on a plane.

Let σ be the surface-density of a plane, which we shall suppose to be that of xy.

The potential due to this electrification will be

$$V = -4\pi\sigma z.$$

Now let two disks of radius a be rigidly electrified with surface-densities $-\sigma'$ and $+\sigma'$. Let the first of these be placed on the plane of xy with its centre at the origin, and the second parallel to it at the very small distance c.

Then it may be shewn, as we shall see in the theory of magnetism, that the potential of the two disks at any point is $\omega\sigma' c$, where ω is the solid angle subtended by the edge of either disk at the point. Hence the potential of the whole system will be

$$V = -4\pi\sigma z + \sigma' c\omega.$$

The forms of the equipotential surfaces and lines of induction are given on the left-hand side of Fig. XX, at the end of Vol. II.

Let us trace the form of the surface for which $V = 0$. This surface is indicated by the dotted line.

Putting the distance of any point from the axis of $z = r$, then, when r is much less than a, and z is small, we find

$$\omega = 2\pi - 2\pi\frac{z}{a} + \&c.$$

Hence, for values of r considerably less than a, the equation of the zero equipotential surface is

$$0 = -4\pi\sigma z + 2\pi\sigma' c - 2\pi\sigma'\frac{zc}{a} + \&c.;$$

$$\text{or} \quad z_0 = \frac{\sigma' c}{2\sigma + \sigma'\frac{c}{a}}.$$

Hence this equipotential surface near the axis is nearly flat.

Outside the disk, where r is greater than a, ω is zero when z is zero, so that the plane of xy is part of the equipotential surface.

To find where these two parts of the surface meet, let us find at what point of this plane $\frac{dV}{dz} = 0$.

When r is very nearly equal to a, the solid angle ω becomes approximately a lune of the sphere of unit radius whose angle is $\tan^{-1}\{z \div (r-a)\}$, that is, ω is $2\tan^{-1}\{z \div (r-a)\}$, so that

$$\frac{dV}{dz} = -4\pi\sigma + \frac{2\sigma' c}{r-a}.$$

Hence, when

$$\frac{dV}{dz} = 0, \qquad r_0 = a + \frac{\sigma' c}{2\pi\sigma} = a + \frac{z_0}{\pi}, \text{ nearly.}$$

The equipotential surface $V=0$ is therefore composed of a disk-like figure of radius r_0, and nearly uniform thickness z_0, and of the part of the infinite plane of xy which lies beyond this figure.

The surface-integral over the whole disk gives the charge of electricity on it. It may be found, as in the theory of a circular current in Part IV, Art. 704, to be

$$Q = 4\pi a\sigma' c\{\log\frac{8a}{r_0-a}-2\}+\pi\sigma r_0^2.$$

The charge on an equal area of the plane surface is $\pi\sigma r_0^2$, hence the charge on the disk exceeds that on an equal area of the plane in the ratio of

$$1+8\frac{z}{r}\log\frac{8\pi r}{z} \text{ to unity,}$$

where z is the thickness and r the radius of the disk, z being supposed small compared with r.

On Electric Accumulators and the Measurement of Capacity.

226.] An Accumulator or Condenser is an apparatus consisting of two conducting surfaces separated by an insulating dielectric medium.

A Leyden jar is an accumulator in which an inside coating of tinfoil is separated from the outside coating by the glass of which the jar is made. The original Leyden phial was a glass vessel containing water which was separated by the glass from the hand which held it.

The outer surface of any insulated conductor may be considered as one of the surfaces of an accumulator, the other being the earth or the walls of the room in which it is placed, and the intervening air being the dielectric medium.

The capacity of an accumulator is measured by the quantity of electricity with which the inner surface must be charged to make the difference between the potentials of the surfaces unity.

Since every electrical potential is the sum of a number of parts found by dividing each electrical element by its distance from a point, the ratio of a quantity of electricity to a potential must have the dimensions of a line. Hence electrostatic capacity is a linear quantity, or we may measure it in feet or metres without ambiguity.

In electrical researches accumulators are used for two principal purposes, for receiving and retaining large quantities of electricity in as small a compass as possible, and for measuring definite quantities of electricity by means of the potential to which they raise the accumulator.

For the retention of electrical charges nothing has been devised more perfect than the Leyden jar. The principal part of the loss arises from the electricity creeping along the damp uncoated surface of the glass from the one coating to the other. This may be checked in a great degree by artificially drying the air within the jar, and by varnishing the surface of the glass where it is exposed to the atmosphere. In Sir W. Thomson's electroscopes there is a very small percentage of loss from day to day, and I believe that none of this loss can be traced to direct conduction either through air or through glass when the glass is good, but that it arises chiefly from superficial conduction along the various insulating stems and glass surfaces of the instrument.

In fact, the same electrician has communicated a charge to sulphuric acid in a large bulb with a long neck, and has then hermetically sealed the neck by fusing it, so that the charge was completely surrounded by glass, and after some years the charge was found still to be retained.

It is only, however, when cold, that glass insulates in this way, for the charge escapes at once if the glass is heated to a temperature below 100°C.

When it is desired to obtain great capacity in small compass, accumulators in which the dielectric is sheet caoutchouc, mica, or paper impregnated with paraffin are convenient.

227.] For accumulators of the second class, intended for the measurement of quantities of electricity, all solid dielectrics must be employed with great caution on account of the property which they possess called Electric Absorption.

The only safe dielectric for such accumulators is air, which has this inconvenience, that if any dust or dirt gets into the narrow space between the opposed surfaces, which ought to be occupied only by air, it not only alters the thickness of the stratum of air, but may establish a connexion between the opposed surfaces, in which case the accumulator will not hold a charge.

To determine in absolute measure, that is to say in feet or metres, the capacity of an accumulator, we must either first ascertain its form and size, and then solve the problem of the distribution of electricity on its opposed surfaces, or we must compare its capacity with that of another accumulator, for which this problem has been solved.

As the problem is a very difficult one, it is best to begin with an accumulator constructed of a form for which the solution is known.

Thus the capacity of an insulated sphere in an unlimited space is known to be measured by the radius of the sphere.

A sphere suspended in a room was actually used by MM. Kohlrausch and Weber, as an absolute standard with which they compared the capacity of other accumulators.

The capacity, however, of a sphere of moderate size is so small when compared with the capacities of the accumulators in common use that the sphere is not a convenient standard measure.

Its capacity might be greatly increased by surrounding the sphere with a hollow concentric spherical surface of somewhat greater radius. The capacity of the inner surface is then a fourth proportional to the thickness of the stratum of air and the radii of the two surfaces.

Sir W. Thomson has employed this arrangement as a standard of capacity, but the difficulties of working the surfaces truly spherical, of making them truly concentric, and of measuring their distance and their radii with sufficient accuracy, are considerable.

We are therefore led to prefer for an absolute measure of capacity a form in which the opposed surfaces are parallel planes.

The accuracy of the surface of the planes can be easily tested, and their distance can be measured by a micrometer screw, and may be made capable of continuous variation, which is a most important property of a measuring instrument.

The only difficulty remaining arises from the fact that the planes must necessarily be bounded, and that the distribution of electricity near the boundaries of the planes has not been rigidly calculated. It is true that if we make them equal circular disks, whose radius is large compared with the distance between them, we may treat the edges of the disks as if they were straight lines, and calculate the distribution of electricity by the method due to Helmholtz, and described in Art. 202. But it will be noticed that in this case part of the electricity is distributed on the back of each disk, and that in the calculation it has been supposed that there are no conductors in the neighbourhood, which is not and cannot be the case in a small instrument.

228.] We therefore prefer the following arrangement, due to Sir W. Thomson, which we may call the Guard-ring arrangement, by means of which the quantity of electricity on an insulated disk may be exactly determined in terms of its potential.

The Guard-ring Accumulator.

Bb is a cylindrical vessel of conducting material of which the outer surface of the upper face is accurately plane. This upper surface consists of two parts, a disk A, and a broad ring BB surrounding the disk, separated from it by a very small interval all round, just sufficient to prevent sparks passing. The upper surface of the disk is accurately in the same plane with that of the guard-ring. The disk is supported by pillars of insulating material GG. C is a metal disk, the under surface of which is accurately plane and parallel to BB. The disk C is considerably larger than A. Its distance from A is adjusted and measured by means of a micrometer screw, which is not given in the figure.

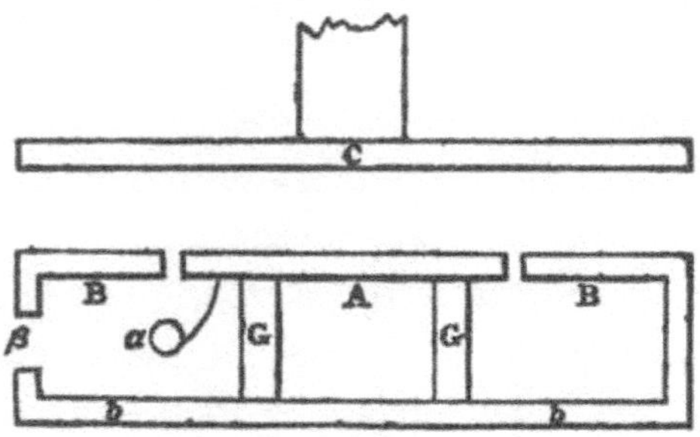

Fig. 21.

This accumulator is used as a measuring instrument as follows:—

Suppose C to be at potential zero, and the disk A and vessel Bb both at potential V. Then there will be no electrification on the back of the disk because the vessel is nearly closed and is all at the same potential. There will be very little electrification on the edges of the disk because BB is at the same potential with the disk. On the face of the disk the electrification will be nearly uniform, and therefore the whole charge on the disk will be almost exactly represented by its area multiplied by the surface-density on a plane, as given in Art. 124.

In fact, we learn from the investigation in Art. 201 that the charge on the disk is

$$V\left\{\frac{R^2+R'^2}{8A}-\frac{R'^2-R^2}{8A}\frac{a}{A+a}\right\},$$

where R is the radius of the disk, R' that of the hole in the guard-ring, A the distance between A and C, and a a quantity which cannot exceed $(R'-R)\dfrac{\log_e 2}{\pi}$.

If the interval between the disk and the guard-ring is small compared with the distance between A and C, the second term will be very small, and the charge on the disk will be nearly

$$V\frac{R^2+R'^2}{8A}.$$

Now let the vessel Bb be put in connexion with the earth. The charge on the disk A will no longer be uniformly distributed, but it will remain the same in quantity, and if we now discharge A we shall obtain a quantity of electricity, the value of which we know in terms of V, the original difference of potentials and the measurable quantities R, R' and A.

On the Comparison of the Capacity of Accumulators.

229.] The form of accumulator which is best fitted to have its capacity determined in absolute measure from the form and dimensions of its parts is not generally the most suitable for electrical experiments. It is desirable that the measures of capacity in actual use should be accumulators having only two conducting surfaces, one of which is as nearly as possible surrounded by the other. The guard-ring accumulator, on the other hand, has three independent conducting portions which must be charged and discharged in a certain order. Hence it is desirable to be able to compare the capacities of two accumulators by an electrical process, so as to test accumulators which may afterwards serve as secondary standards.

I shall first shew how to test the equality of the capacity of two guard-ring accumulators.

Let A be the disk, B the guard-ring with the rest of the conducting vessel attached to it, and C the large disk of one of these accumulators, and let A', B', and C' be the corresponding parts of the other.

If either of these accumulators is of the more simple kind, having only two conductors, we have only to suppress B or B', and to suppose A to be the inner and C the outer conducting surface, C, in this case being understood to surround A.

Let the following connexions be made.

Let B be kept always connected with C', and B' with C, that is, let each guard-ring be connected with the large disk of the other condenser.

(1) Let A be connected with B and C' and with J, the electrode of a Leyden jar, and let A' be connected with B' and C and with the earth.

(2) Let A, B, and C' be insulated from J.

(3) Let A be insulated from B and C', and A' from B' and C.

(4) Let B and C' be connected with B' and C and with the earth.

(5) Let A be connected with A'.

(6) Let A and A' be connected with an electroscope E.

We may express these connexions as follows:—

$$(1)\quad 0=C=B'=A' \quad | \quad A=B=C'=J.$$
$$(2)\quad 0=C=B'=A' \quad | \quad A=B=C' \mid J.$$
$$(3)\quad 0=C=B' \mid A' \quad | \quad A \mid B=C'.$$
$$(4)\quad 0=C=B' \mid A' \quad | \quad A \mid B=C'=0.$$
$$(5)\quad 0=C=B' \mid A' \quad = \quad A \mid B=C'=0.$$
$$(6)\quad 0=C=B' \mid A'=E=A \mid B=C'=0.$$

Here the sign of equality expresses electrical connexion, and the vertical stroke expresses insulation.

In (1) the two accumulators are charged oppositely, so that A is positive and A' negative, the charges on A and A' being uniformly distributed on the upper surface opposed to the large disk of each accumulator.

In (2) the jar is removed, and in (3) the charges on A and A' are insulated.

In (4) the guard-rings are connected with the large disks, so that the charges on A and A', though unaltered in magnitude, are now distributed over their whole surface.

In (5) A is connected with A'. If the charges are equal and of opposite signs, the electrification will be entirely destroyed, and in (6) this is tested by means of the electroscope E.

The electroscope E will indicate positive or negative electrification according as A or A' has the greater capacity.

By means of a key of proper construction, the whole of these operations can be performed in due succession in a very small fraction of a second, and the capacities adjusted till no electrification can be detected by the electroscope, and in this way the capacity of an accumulator may be adjusted to be equal to that of any other, or to the sum of the capacities of several accumulators, so that a system of accumulators may be formed, each of which has its capacity determined in absolute measure, i.e. in feet or in metres, while at the same time it is of the construction most suitable for electrical experiments.

This method of comparison will probably be found useful in determining the specific capacity for electrostatic induction of different dielectrics in the form of plates or disks. If a disk of the dielectric is interposed between A and C, the disk being considerably larger than A, then the capacity of the accumulator will be altered and made equal to that of the same accumulator when A and C are nearer together. If the accumulator with the dielectric plate, and with A and C at distance x, is of the same capacity as the same accumulator without the dielectric, and with A and C at distance x', then, if a is the thickness of the plate, and K its specific dielectric inductive capacity referred to air as a standard,

$$K = \frac{a}{a + x' - x}.$$

The combination of three cylinders, described in Art. 127, has been employed by Sir W. Thomson as an accumulator whose capacity may be increased or diminished by measurable quantities.

The experiments of MM. Gibson and Barclay with this apparatus are described in the *Proceedings of the Royal Society*, Feb. 2, 1871, and *Phil. Trans.*, 1871, p. 573. They found the specific inductive capacity of paraffin to be 1.975, that of air being unity.

PART II.

ELECTROKINEMATICS.

CHAPTER I.

THE ELECTRIC CURRENT.

230.] We have seen, in Art. 45, that when a conductor is in electrical equilibrium the potential at every point of the conductor must be the same.

If two conductors A and B are charged with electricity so that the potential of A is higher than that of B, then, if they are put in communication by means of a metallic wire C touching both of them, part of the charge of A will be transferred to B, and the potentials of A and B will become in a very short time equalized.

231.] During this process certain phenomena are observed in the wire C, which are called the phenomena of the electric conflict or current.

The first of these phenomena is the transference of positive electrification from A to B and of negative electrification from B to A. This transference may be also effected in a slower manner by bringing a small insulated body into contact with A and B alternately. By this process, which we may call electrical convection, successive small portions of the electrification of each body are transferred to the other. In either case a certain quantity of electricity, or of the state of electrification, passes from one place to another along a certain path in the space between the bodies.

Whatever therefore may be our opinion of the nature of electricity, we must admit that the process which we have described constitutes a current of electricity. This current may be described as a current of positive electricity from A to B, or a current of negative electricity from B to A, or as a combination of these two currents.

According to Fechner's and Weber's theory it is a combination of a current of positive electricity with an exactly equal current of negative electricity in the opposite direction through the same substance. It is necessary to remember this exceedingly artificial hypothesis regarding the constitution of the current in order to understand the statement of some of Weber's most valuable experimental results.

If, as in Art. 36, we suppose P units of positive electricity transferred from A to B, and N units of negative electricity transferred from B to A in unit of time, then, according to Weber's theory, $P = N$, and P or N is to be taken as the numerical measure of the current.

We, on the contrary, make no assumption as to the relation between P and N, but attend only to the result of the current, namely, the transference of $P+N$ of positive electrification from A to B, and we shall consider $P+N$ the true measure of the current. The current, therefore, which Weber would call 1 we shall call 2.

On Steady Currents.

232.] In the case of the current between two insulated conductors at different potentials the operation is soon brought to an end by the equalization of the potentials of the two bodies, and the current is therefore essentially a Transient current.

But there are methods by which the difference of potentials of the conductors may be maintained constant, in which case the current will continue to flow with uniform strength as a Steady Current.

The Voltaic Battery.

The most convenient method of producing a steady current is by means of the Voltaic Battery.

For the sake of distinctness we shall describe Daniell's Constant Battery:—

A solution of sulphate of zinc is placed in a cell of porous earthenware, and this cell is placed in a vessel containing a saturated solution of sulphate of copper. A piece of zinc is dipped into the sulphate of zinc, and a piece of copper is dipped into the sulphate of copper. Wires are soldered to the zinc and to the copper above the surface of the liquid. This combination is called a cell or element of Daniell's battery. See Art. 272.

233.] If the cell is insulated by being placed on a non-conducting stand, and if the wire connected with the copper is put in contact with an insulated conductor *A*, and the wire connected with the zinc is put in contact with *B*, another insulated conductor of the same metal as *A*, then it may be shewn by means of a delicate electrometer that the potential of *A* exceeds that of *B* by a certain quantity. This difference of potentials is called the Electromotive Force of the Daniell's Cell.

If *A* and *B* are now disconnected from the cell and put in communication by means of a wire, a transient current passes through the wire from *A* to *B*, and the potentials of *A* and *B* become equal. *A* and *B* may then be charged again by the cell, and the process repeated as long as the cell will work. But if *A* and *B* be connected by means of the wire *C*, and at the same time connected with the battery as before, then the cell will maintain a constant current through *C*, and also a constant difference of potentials between *A* and *B*. This difference will not, as we shall see, be equal to the whole electromotive force of the cell, for part of this force is spent in maintaining the current through the cell itself.

A number of cells placed in series so that the zinc of the first cell is connected by metal with the copper of the second, and so on, is called a Voltaic Battery. The electromotive force of such a battery is the sum of the electromotive forces of the cells of which it is composed. If the battery is insulated it may be charged with electricity as a whole, but the potential of the copper end will always exceed that of the zinc end by the electromotive force of the battery, whatever the absolute value of either of these potentials may be. The cells of the battery may be of very various construction, containing different chemical substances and different metals, provided they are such that chemical action does not go on when no current passes.

234.] Let us now consider a voltaic battery with its ends insulated from each other. The copper end will be positively or vitreously electrified, and the zinc end will be negatively or resinously electrified.

Let the two ends of the battery be now connected by means of a wire. An electric current will commence, and will in a very short time attain a constant value. It is then said to be a Steady Current.

Properties of the Current.

235.] The current forms a closed circuit in the direction from copper to zinc through the wires, and from zinc to copper through the solutions.

If the circuit be broken by cutting any of the wires which connect the copper of one cell with the zinc of the next in order, the current will be stopped, and the potential of the end of the wire in connexion with the copper will be found to exceed that of the end of the wire in connexion with the zinc by a constant quantity, namely, the total electromotive force of the circuit.

Electrolytic Action of the Current.

236.] As long as the circuit is broken no chemical action goes on in the cells, but as soon as the circuit is completed, zinc is dissolved from the zinc in each of the Daniell's cells, and copper is deposited on the copper.

The quantity of sulphate of zinc increases, and the quantity of sulphate of copper diminishes unless more is constantly supplied.

The quantity of zinc dissolved and also that of copper deposited is the same in each of the Daniell's cells throughout the circuit, whatever the size of the plates of the cell, and if any of the cells be of a different construction, the amount of chemical action in it bears a constant proportion to the action in the Daniell's cell. For instance, if one of the cells consists of two platinum plates dipped into sulphuric acid diluted with water, oxygen will be given off at the surface of the plate where the current enters the liquid, namely, the plate in metallic connexion with the copper of Daniell's cell, and hydrogen at the surface of the plate where the current leaves the liquid, namely, the plate connected with the zinc of Daniell's cell.

The volume of the hydrogen is exactly twice the volume of the oxygen given off in the same time, and the weight of the oxygen is exactly eight times the weight of the hydrogen.

In every cell of the circuit the weight of each substance dissolved, deposited, or decomposed is equal to a certain quantity called the electrochemical equivalent of that substance, multiplied by the strength of the current and by the time during which it has been flowing.

For the experiments which established this principle, see the seventh and eighth series of Faraday's *Experimental Researches;*

and for an investigation of the apparent exceptions to the rule, see Miller's *Chemical Physics* and Wiedemann's *Galvanismus.*

237.] Substances which are decomposed in this way are called Electrolytes. The process is called Electrolysis. The places where the current enters and leaves the electrolyte are called Electrodes. Of these the electrode by which the current enters is called the Anode, and that by which it leaves the electrolyte is called the Cathode. The components into which the electrolyte is resolved are called Ions: that which appears at the anode is called the Anion, and that which appears at the cathode is called the Cation.

Of these terms, which were, I believe, invented by Faraday with the help of Dr. Whewell, the first three, namely, electrode, electrolysis, and electrolyte have been generally adopted, and the mode of conduction of the current in which this kind of decomposition and transfer of the components takes place is called Electrolytic Conduction.

If a homogeneous electrolyte is placed in a tube of variable section, and if the electrodes are placed at the ends of this tube, it is found that when the current passes, the anion appears at the anode and the cation at the cathode, the quantities of these ions being electrochemically equivalent, and such as to be together equivalent to a certain quantity of the electrolyte. In the other parts of the tube, whether the section be large or small, uniform or varying, the composition of the electrolyte remains unaltered. Hence the amount of electrolysis which takes place across every section of the tube is the same. Where the section is small the action must therefore be more intense than where the section is large, but the total amount of each ion which crosses any complete section of the electrolyte in a given time is the same for all sections.

The strength of the current may therefore be measured by the amount of electrolysis in a given time. An instrument by which the quantity of the electrolytic products can be readily measured is called a Voltameter.

The strength of the current, as thus measured, is the same at every part of the circuit, and the total quantity of the electrolytic products in the voltameter after any given time is proportional to the amount of electricity which passes any section in the same time.

238.] If we introduce a voltameter at one part of the circuit of a voltaic battery, and break the circuit at another part, we may suppose the measurement of the current to be conducted thus.

Let the ends of the broken circuit be A and B, and let A be the anode and B the cathode. Let an insulated ball be made to touch A and B alternately, it will carry from A to B a certain measurable quantity of electricity at each journey. This quantity may be measured by an electrometer, or it may be calculated by multiplying the electromotive force of the circuit by the electrostatic capacity of the ball. Electricity is thus carried from A to B on the insulated ball by a process which may be called Convection. At the same time electrolysis goes on in the voltameter and in the cells of the battery, and the amount of electrolysis in each cell may be compared with the amount of electricity carried across by the insulated ball. The quantity of a substance which is electrolysed by one unit of electricity is called an Electrochemical equivalent of that substance.

This experiment would be an extremely tedious and troublesome one if conducted in this way with a ball of ordinary magnitude and a manageable battery, for an enormous number of journeys would have to be made before an appreciable quantity of the electrolyte was decomposed. The experiment must therefore be considered as a mere illustration, the actual measurements of electrochemical equivalents being conducted in a different way. But the experiment may be considered as an illustration of the process of electrolysis itself, for if we regard electrolytic conduction as a species of convection in which an electrochemical equivalent of the anion travels with negative electricity in the direction of the anode, while an equivalent of the cation travels with positive electricity in the direction of the cathode, the whole amount of transfer of electricity being one unit, we shall have an idea of the process of electrolysis, which, so far as I know, is not inconsistent with known facts, though, on account of our ignorance of the nature of electricity and of chemical compounds, it may be a very imperfect representation of what really takes place.

Magnetic Action of the Current.

239.] Oersted discovered that a magnet placed near a straight electric current tends to place itself at right angles to the plane passing through the magnet and the current. See Art. 475.

If a man were to place his body in the line of the current so that the current from copper through the wire to zinc should flow from his head to his feet, and if he were to direct his face towards the centre of the magnet, then that end of the magnet which tends to point to the north would, when the current flows, tend to point towards the man's right hand.

The nature and laws of this electromagnetic action will be discussed when we come to the fourth part of this treatise. What we are concerned with at present is the fact that the electric current has a magnetic action which is exerted outside the current, and by which its existence can be ascertained and its intensity measured without breaking the circuit or introducing anything into the current itself.

The amount of the magnetic action has been ascertained to be strictly proportional to the strength of the current as measured by the products of electrolysis in the voltameter, and to be quite independent of the nature of the conductor in which the current is flowing, whether it be a metal or an electrolyte.

240.] An instrument which indicates the strength of an electric current by its magnetic effects is called a Galvanometer.

Galvanometers in general consist of one or more coils of silk-covered wire within which a magnet is suspended with its axis horizontal. When a current is passed through the wire the magnet tends to set itself with its axis perpendicular to the plane of the coils. If we suppose the plane of the coils to be placed parallel to the plane of the earth's equator, and the current to flow round the coil from east to west in the direction of the apparent motion of the sun, then the magnet within will tend to set itself with its magnetization in the same direction as that of the earth considered as a great magnet, the north pole of the earth being similar to that end of the compass needle which points south.

The galvanometer is the most convenient instrument for measuring the strength of electric currents. We shall therefore assume the possibility of constructing such an instrument in studying the laws of these currents, reserving the discussion of the principles of the instrument for our fourth part. When therefore we say that an electric current is of a certain strength we suppose that the measurement is effected by the galvanometer.

CHAPTER II.

CONDUCTION AND RESISTANCE.

241.] If by means of an electrometer we determine the electric potential at different points of a circuit in which a constant electric current is maintained, we shall find that in any portion of the circuit consisting of a single metal of uniform temperature throughout, the potential at any point exceeds that at any other point farther on in the direction of the current by a quantity depending on the strength of the current and on the nature and dimensions of the intervening portion of the circuit. The difference of the potentials at the extremities of this portion of the circuit is called the External electromotive force acting on it. If the portion of the circuit under consideration is not homogeneous, but contains transitions from one substance to another, from metals to electrolytes, or from hotter to colder parts, there may be, besides the external electromotive force, Internal electromotive forces which must be taken into account.

The relations between Electromotive Force, Current, and Resistance were first investigated by Dr. G. S. Ohm, in a work published in 1827, entitled *Die Galvanische Kette Mathematisch Bearbeitet*, translated in Taylor's *Scientific Memoirs*. The result of these investigations in the case of homogeneous conductors is commonly called 'Ohm's Law.'

Ohm's Law.

The electromotive force acting between the extremities of any part of a circuit is the product of the strength of the current and the resistance of that part of the circuit.

Here a new term is introduced, the Resistance of a conductor, which is defined to be the ratio of the electromotive force to the strength of the current which it produces. The introduction of this term would have been of no scientific value unless Ohm had shewn, as he did experimentally, that it corresponds to a real physical quantity, that is, that it has a definite value which is altered only when the nature of the conductor is altered.

In the first place, then, the resistance of a conductor is independent of the strength of the current flowing through it.

In the second place the resistance is independent of the electric potential at which the conductor is maintained, and of the density of the distribution of electricity on the surface of the conductor.

It depends entirely on the nature of the material of which the conductor is composed, the state of aggregation of its parts, and its temperature.

The resistance of a conductor may be measured to within one ten thousandth or even one hundred thousandth part of its value, and so many conductors have been tested that our assurance of the truth of Ohm's Law is now very high. In the sixth chapter we shall trace its applications and consequences.

Generation of Heat by the Current.

242.] We have seen that when an electromotive force causes a current to flow through a conductor, electricity is transferred from a place of higher to a place of lower potential. If the transfer had been made by convection, that is, by carrying successive charges on a ball from the one place to the other, work would have been done by the electrical forces on the ball, and this might have been turned to account. It is actually turned to account in a partial manner in those dry pile circuits where the electrodes have the form of bells, and the carrier ball is made to swing like a pendulum between the two bells and strike them alternately. In this way the electrical action is made to keep up the swinging of the pendulum and to propagate the sound of the bells to a distance. In the case of the conducting wire we have the same transfer of electricity from a place of high to a place of low potential without any external work being done. The principle of the Conservation of Energy therefore leads us to look for internal work in the conductor. In an electrolyte this internal work consists partly of the separation of its components. In other conductors it is entirely converted into heat.

The energy converted into heat is in this case the product of the electromotive force into the quantity of electricity which passes. But the electromotive force is the product of the current into the resistance, and the quantity of electricity is the product of the current into the time. Hence the quantity of heat multiplied by the mechanical equivalent of unit of heat is equal to the square of the strength of the current multiplied into the resistance and into the time.

The heat developed by electric currents in overcoming the resistance of conductors has been determined by Dr. Joule, who first established that the heat produced in a given time is proportional to the square of the current, and afterwards by careful absolute measurements of all the quantities concerned, verified the equation

$$JH = C^2 Rt,$$

where J is Joule's dynamical equivalent of heat, H the number of units of heat, C the strength of the current, R the resistance of the conductor, and t the time during which the current flows. These relations between electromotive force, work, and heat, were first fully explained by Sir W. Thomson in a paper on the application of the principle of mechanical effect to the measurement of electromotive forces*.

243.] The analogy between the theory of the conduction of electricity and that of the conduction of heat is at first sight almost complete. If we take two systems geometrically similar, and such that the conductivity for heat at any part of the first is proportional to the conductivity for electricity at the corresponding part of the second, and if we also make the temperature at any part of the first proportional to the electric potential at the corresponding point of the second, then the flow of heat across any area of the first will be proportional to the flow of electricity across the corresponding area of the second.

Thus, in the illustration we have given, in which flow of electricity corresponds to flow of heat, and electric potential to temperature, electricity tends to flow from places of high to places of low potential, exactly as heat tends to flow from places of high to places of low temperature.

244.] The theory of potential and that of temperature may therefore be made to illustrate one another; there is, however, one remarkable difference between the phenomena of electricity and those of heat.

Suspend a conducting body within a closed conducting vessel by a silk thread, and charge the vessel with electricity. The potential

* *Phil. Mag.*, Dec. 1851.

of the vessel and of all within it will be instantly raised, but however long and however powerfully the vessel be electrified, and whether the body within be allowed to come in contact with the vessel or not, no signs of electrification will appear within the vessel, nor will the body within shew any electrical effect when taken out.

But if the vessel is raised to a high temperature, the body within will rise to the same temperature, but only after a considerable time, and if it is then taken out it will be found hot, and will remain so till it has continued to emit heat for some time.

The difference between the phenomena consists in the fact that bodies are capable of absorbing and emitting heat, whereas they have no corresponding property with respect to electricity. A body cannot be made hot without a certain amount of heat being supplied to it, depending on the mass and specific heat of the body, but the electric potential of a body may be raised to any extent in the way already described without communicating any electricity to the body.

245.] Again, suppose a body first heated and then placed inside the closed vessel. The outside of the vessel will be at first at the temperature of surrounding bodies, but it will soon get hot, and will remain hot till the heat of the interior body has escaped.

It is impossible to perform a corresponding electrical experiment. It is impossible so to electrify a body, and so to place it in a hollow vessel, that the outside of the vessel shall at first shew no signs of electrification but shall afterwards become electrified. It was for some phenomenon of this kind that Faraday sought in vain under the name of an absolute charge of electricity.

Heat may be hidden in the interior of a body so as to have no external action, but it is impossible to isolate a quantity of electricity so as to prevent it from being constantly in inductive relation with an equal quantity of electricity of the opposite kind.

There is nothing therefore among electric phenomena which corresponds to the capacity of a body for heat. This follows at once from the doctrine which is asserted in this treatise, that electricity obeys the same condition of continuity as an incompressible fluid. It is therefore impossible to give a bodily charge of electricity to any substance by forcing an additional quantity of electricity into it. See Arts. 61, 111, 329, 334.

CHAPTER III.

ELECTROMOTIVE FORCE BETWEEN BODIES IN CONTACT.

The Potentials of Different Substances in Contact.

246.] IF we define the potential of a hollow conducting vessel as the potential of the air inside the vessel, we may ascertain this potential by means of an electrometer as described in Part I, Art. 222.

If we now take two hollow vessels of different metals, say copper and zinc, and put them in metallic contact with each other, and then test the potential of the air inside each vessel, the potential of the air inside the zinc vessel will be positive as compared with that inside the copper vessel. The difference of potentials depends on the nature of the surface of the insides of the vessels, being greatest when the zinc is bright and when the copper is coated with oxide.

It appears from this that when two different metals are in contact there is in general an electromotive force acting from the one to the other, so as to make the potential of the one exceed that of the other by a certain quantity. This is Volta's theory of Contact Electricity.

If we take a certain metal, say copper, as the standard, then if the potential of iron in contact with copper at the zero potential is I, and that of zinc in contact with copper at zero is Z, then the potential of zinc in contact with iron at zero will be $Z-I$.

It appears from this result, which is true of any three metals, that the differences of potential of any two metals at the same temperature in contact is equal to the difference of their potentials when in contact with a third metal, so that if a circuit be formed of any number of metals at the same temperature there will be electrical equilibrium as soon as they have acquired their proper potentials, and there will be no current kept up in the circuit.

247.] If, however, the circuit consist of two metals and an electrolyte, the electrolyte, according to Volta's theory, tends to reduce the potentials of the metals in contact with it to equality, so that the electromotive force at the metallic junction is no longer balanced, and a continuous current is kept up. The energy of this current is supplied by the chemical action which takes place between the electrolyte and the metals.

248.] The electric effect may, however, be produced without chemical action if by any other means we can produce an equalization of the potentials of two metals in contact. Thus, in an experiment due to Sir W. Thomson *, a copper funnel is placed in contact with a vertical zinc cylinder, so that when copper filings are allowed to pass through the funnel, they separate from each other and from the funnel near the middle of the zinc cylinder, and then fall into an insulated receiver placed below. The receiver is then found to be charged negatively, and the charge increases as the filings continue to pour into it. At the same time the zinc cylinder with the copper funnel in it becomes charged more and more positively.

If now the zinc cylinder were connected with the receiver by a wire, there would be a positive current in the wire from the cylinder to the receiver. The stream of copper filings, each filing charged negatively by induction, constitutes a negative current from the funnel to the receiver, or, in other words, a positive current from the receiver to the copper funnel. The positive current, therefore, passes through the air (by the filings) from zinc to copper, and through the metallic junction from copper to zinc, just as in the ordinary voltaic arrangement, but in this case the force which keeps up the current is not chemical action but gravity, which causes the filings to fall, in spite of the electrical attraction between the positively charged funnel and the negatively charged filings.

249.] A remarkable confirmation of the theory of contact electricity is supplied by the discovery of Peltier, that, when a current of electricity crosses the junction of two metals, the junction is heated when the current is in one direction, and cooled when it is in the other direction. It must be remembered that a current in its passage through a metal always produces heat, because it meets with resistance, so that the cooling effect on the whole conductor must always be less than the heating effect. We must therefore distinguish between the generation of heat in each metal,

* *North British Review*, 1864, p. 353; and *Proc. R. S.*, June 20, 1867.

due to ordinary resistance, and the generation or absorption of heat at the junction of two metals. We shall call the first the frictional generation of heat by the current, and, as we have seen, it is proportional to the square of the current, and is the same whether the current be in the positive or the negative direction. The second we may call the Peltier effect, which changes its sign with that of the current.

The total heat generated in a portion of a compound conductor consisting of two metals may be expressed by

$$H = \frac{R}{J} C^2 t - \Pi C t,$$

where H is the quantity of heat, J the mechanical equivalent of unit of heat, R the resistance of the conductor, C the current, and t the time; Π being the coefficient of the Peltier effect, that is, the heat absorbed at the junction by unit of current in unit of time.

Now the heat generated is mechanically equivalent to the work done against electrical forces in the conductor, that is, it is equal to the product of the current into the electromotive force producing it. Hence, if E is the external electromotive force which causes the current to flow through the conductor,

$$JH = CEt = RC^2 t - J\Pi C t,$$

whence $$E = RC - J\Pi.$$

It appears from this equation that the external electromotive force required to drive the current through the compound conductor is less than that due to its resistance alone by the electromotive force $J\Pi$. Hence $J\Pi$ represents the electromotive contact force at the junction acting in the positive direction.

This application, due to Sir W. Thomson *, of the dynamical theory of heat to the determination of a local electromotive force is of great scientific importance, since the ordinary method of connecting two points of the compound conductor with the electrodes of a galvanometer or electroscope by wires would be useless, owing to the contact forces at the junctions of the wires with the materials of the compound conductor. In the thermal method, on the other hand, we know that the only source of energy is the current of electricity, and that no work is done by the current in a certain portion of the circuit except in heating that portion of the conductor. If, therefore, we can measure the amount of the

* *Proc. R. S. Edin.*, Dec. 15, 1851; and *Trans. R. S. Edin.*, 1854.

current and the amount of heat produced or absorbed, we can determine the electromotive force required to urge the current through that portion of the conductor, and this measurement is entirely independent of the effect of contact forces in other parts of the circuit.

The electromotive force at the junction of two metals, as determined by this method, does not account for Volta's electromotive force as described in Art. 246. The latter is in general far greater than that of this Article, and is sometimes of opposite sign. Hence the assumption that the potential of a metal is to be measured by that of the air in contact with it must be erroneous, and the greater part of Volta's electromotive force must be sought for, not at the junction of the two metals, but at one or both of the surfaces which separate the metals from the air or other medium which forms the third element of the circuit.

250.] The discovery by Seebeck of thermoelectric currents in circuits of different metals with their junctions at different temperatures, shews that these contact forces do not always balance each other in a complete circuit. It is manifest, however, that in a complete circuit of different metals at uniform temperature the contact forces must balance each other. For if this were not the case there would be a current formed in the circuit, and this current might be employed to work a machine or to generate heat in the circuit, that is, to do work, while at the same time there is no expenditure of energy, as the circuit is all at the same temperature, and no chemical or other change takes place. Hence, if the Peltier effect at the junction of two metals a and b be represented by Π_{ab} when the current flows from a to b, then for a circuit of two metals at the same temperature we must have

$$\Pi_{ab}+\Pi_{ba} = 0,$$

and for a circuit of three metals a, b, c, we must have

$$\Pi_{bc}+\Pi_{ca}+\Pi_{ab} = 0.$$

It follows from this equation that the three Peltier effects are not independent, but that one of them can be deduced from the other two. For instance, if we suppose c to be a standard metal, and if we write $P_a = J\Pi_{ac}$ and $P_b = J\Pi_{bc}$, then

$$J\Pi_{ab} = P_a - P_b.$$

The quantity P_a is a function of the temperature, and depends on the nature of the metal a.

251.] It has also been shewn by Magnus that if a circuit is

formed of a single metal no current will be formed in it, however the section of the conductor and the temperature may vary in different parts.

Since in this case there is conduction of heat and consequent dissipation of energy, we cannot, as in the former case, consider this result as self-evident. The electromotive force, for instance, between two portions of a circuit might have depended on whether the current was passing from a thick portion of the conductor to a thin one, or the reverse, as well as on its passing rapidly or slowly from a hot portion to a cold one, or the reverse, and this would have made a current possible in an unequally heated circuit of one metal.

Hence, by the same reasoning as in the case of Peltier's phenomenon, we find that if the passage of a current through a conductor of one metal produces any thermal effect which is reversed when the current is reversed, this can only take place when the current flows from places of high to places of low temperature, or the reverse, and if the heat generated in a conductor of one metal in flowing from a place where the temperature is x to a place where it is y, is H, then

$$JH = RC^2t - S_{xy}Ct,$$

and the electromotive force tending to maintain the current will be S_{xy}.

If x, y, z be the temperatures at three points of a homogeneous circuit, we must have

$$S_{yz} + S_{zx} + S_{xy} = 0,$$

according to the result of Magnus. Hence, if we suppose z to be the zero temperature, and if we put

$$Q_x = S_{xz} \quad \text{and} \quad Q_y = S_{yz},$$

we find
$$S_{xy} = Q_x - Q_y,$$

where Q_x is a function of the temperature x, the form of the function depending on the nature of the metal.

If we now consider a circuit of two metals a and b in which the temperature is x where the current passes from a to b, and y where it passes from b to a, the electromotive force will be

$$F = P_{ax} - P_{bx} + Q_{bx} - Q_{by} + P_{by} - P_{ay} + Q_{ay} - Q_{ax},$$

where P_{ax} signifies the value of P for the metal a at the temperature x, or

$$F = P_{ax} - Q_{ax} - (P_{ay} - Q_{ay}) - (P_{bx} - Q_{bx}) + P_{by} - Q_{by}.$$

Since in unequally heated circuits of different metals there are in general thermoelectric currents, it follows that P and Q are in general different for the same metal and same temperature.

252.] The existence of the quantity Q was first demonstrated by Sir W. Thomson, in the memoir we have referred to, as a deduction from the phenomenon of thermoelectric inversion discovered by Cumming *, who found that the order of certain metals in the thermoelectric scale is different at high and at low temperatures, so that for a certain temperature two metals may be neutral to each other. Thus, in a circuit of copper and iron if one junction be kept at the ordinary temperature while the temperature of the other is raised, a current sets from copper to iron through the hot junction, and the electromotive force continues to increase till the hot junction has reached a temperature T, which, according to Thomson, is about 284°C. When the temperature of the hot junction is raised still further the electromotive force is reduced, and at last, if the temperature be raised high enough, the current is reversed. The reversal of the current may be obtained more easily by raising the temperature of the colder junction. If the temperature of both junctions is above T the current sets from iron to copper through the hotter junction, that is, in the reverse direction to that observed when both junctions are below T.

Hence, if one of the junctions is at the neutral temperature T and the other is either hotter or colder, the current will set from copper to iron through the junction at the neutral temperature.

253.] From this fact Thomson reasoned as follows :—

Suppose the other junction at a temperature lower than T. The current may be made to work an engine or to generate heat in a wire, and this expenditure of energy must be kept up by the transformation of heat into electric energy, that is to say, heat must disappear somewhere in the circuit. Now at the temperature T iron and copper are neutral to each other, so that no reversible thermal effect is produced at the hot junction, and at the cold junction there is, by Peltier's principle, an evolution of heat by the current. Hence the only place where the heat can disappear is in the copper or iron portions of the circuit, so that either a current in iron from hot to cold must cool the iron, or a current in copper from cold to hot must cool the copper, or both these effects may take place. By an elaborate series of ingenious experiments Thomson succeeded in detecting the reversible thermal action of the current in passing between parts of different temperatures,

* *Cambridge Transactions*, 1823.

and he found that the current produced opposite effects in copper and in iron *.

When a stream of a material fluid passes along a tube from a hot part to a cold part it heats the tube, and when it passes from cold to hot it cools the tube, and these effects depend on the specific capacity for heat of the fluid. If we supposed electricity, whether positive or negative, to be a material fluid, we might measure its specific heat by the thermal effect on an unequally heated conductor. Now Thomson's experiments shew that positive electricity in copper and negative electricity in iron carry heat with them from hot to cold. Hence, if we supposed either positive or negative electricity to be a fluid, capable of being heated and cooled, and of communicating heat to other bodies, we should find the supposition contradicted by iron for positive electricity and by copper for negative electricity, so that we should have to abandon both hypotheses.

This scientific prediction of the reversible effect of an electric current upon an unequally heated conductor of one metal is another instructive example of the application of the theory of Conservation of Energy to indicate new directions of scientific research. Thomson has also applied the Second Law of Thermodynamics to indicate relations between the quantities which we have denoted by P and Q, and has investigated the possible thermoelectric properties of bodies whose structure is different in different directions. He has also investigated experimentally the conditions under which these properties are developed by pressure, magnetization, &c.

254.] Professor Tait † has recently investigated the electromotive force of thermoelectric circuits of different metals, having their junctions at different temperatures. He finds that the electromotive force of a circuit may be expressed very accurately by the formula

$$E = a(t_1 - t_2)\left[t_0 - \tfrac{1}{2}(t_1 + t_2)\right],$$

where t_1 is the absolute temperature of the hot junction, t_2 that of the cold junction, and t_0 the temperature at which the two metals are neutral to each other. The factor a is a coefficient depending on the nature of the two metals composing the circuit. This law has been verified through considerable ranges of temperature by Professor Tait and his students, and he hopes to make the thermoelectric circuit available as a thermometric instrument in his

* 'On the Electrodynamic Qualities of Metals.' *Phil. Trans.*, 1856.

† *Proc. R. S. Edin.*, Session 1870-71, p. 308, also Dec. 18, 1871.

experiments on the conduction of heat, and in other cases in which the mercurial thermometer is not convenient or has not a sufficient range.

According to Tait's theory, the quantity which Thomson calls the specific heat of electricity is proportional to the absolute temperature in each pure metal, though its magnitude and even its sign vary in different metals. From this he has deduced by thermodynamic principles the following results. Let $k_a t$, $k_b t$, $k_c t$ be the specific heats of electricity in three metals a, b, c, and let T_{bc}, T_{ca}, T_{ab} be the temperatures at which pairs of these metals are neutral to each other, then the equations

$$(k_b - k_c)\,T_{bc} + (k_c - k_a)\,T_{ca} + (k_a - k_b)\,T_{ab} = 0,$$
$$J\Pi_{ab} = (k_a - k_b)\,t\,(T_{ab} - t),$$
$$E_{ab} = (k_a - k_b)\,(t_1 - t_2)\,[T_{ab} - \tfrac{1}{2}(t_1 + t_2)]$$

express the relation of the neutral temperatures, the value of the Peltier effect, and the electromotive force of a thermoelectric circuit.

CHAPTER IV.

ELECTROLYSIS.

Electrolytic Conduction.

255.] I HAVE already stated that when an electric current in any part of its circuit passes through certain compound substances called Electrolytes, the passage of the current is accompanied by a certain chemical process called Electrolysis, in which the substance is resolved into two components called Ions, of which one, called the Anion, or the electronegative component, appears at the Anode, or place where the current enters the electrolyte, and the other, called the Cation, appears at the Cathode, or the place where the current leaves the electrolyte.

The complete investigation of Electrolysis belongs quite as much to Chemistry as to Electricity. We shall consider it from an electrical point of view, without discussing its application to the theory of the constitution of chemical compounds.

Of all electrical phenomena electrolysis appears the most likely to furnish us with a real insight into the true nature of the electric current, because we find currents of ordinary matter and currents of electricity forming essential parts of the same phenomenon.

It is probably for this very reason that, in the present imperfectly formed state of our ideas about electricity, the theories of electrolysis are so unsatisfactory.

The fundamental law of electrolysis, which was established by Faraday, and confirmed by the experiments of Beetz, Hittorf, and others down to the present time, is as follows:—

The number of electrochemical equivalents of an electrolyte which are decomposed by the passage of an electric current during a given time is equal to the number of units of electricity which are transferred by the current in the same time.

The electrochemical equivalent of a substance is that quantity of the substance which is electrolysed by a unit current passing through the substance for a unit of time, or, in other words, by the passage of a unit of electricity. When the unit of electricity is defined in absolute measure the absolute value of the electrochemical equivalent of each substance can be determined in grains or in grammes.

The electrochemical equivalents of different substances are proportional to their ordinary chemical equivalents. The ordinary chemical equivalents, however, are the mere numerical ratios in which the substances combine, whereas the electrochemical equivalents are quantities of matter of a determinate magnitude, depending on the definition of the unit of electricity.

Every electrolyte consists of two components, which, during the electrolysis, appear where the current enters and leaves the electrolyte, and nowhere else. Hence, if we conceive a surface described within the substance of the electrolyte, the amount of electrolysis which takes place through this surface, as measured by the electrochemical equivalents of the components transferred across it in opposite directions, will be proportional to the total electric current through the surface.

The actual transfer of the ions through the substance of the electrolyte in opposite directions is therefore part of the phenomenon of the conduction of an electric current through an electrolyte. At every point of the electrolyte through which an electric current is passing there are also two opposite material currents of the anion and the cation, which have the same lines of flow with the electric current, and are proportional to it in magnitude.

It is therefore extremely natural to suppose that the currents of the ions are convection currents of electricity, and, in particular, that every molecule of the cation is charged with a certain fixed quantity of positive electricity, which is the same for the molecules of all cations, and that every molecule of the anion is charged with an equal quantity of negative electricity.

The opposite motion of the ions through the electrolyte would then be a complete physical representation of the electric current. We may compare this motion of the ions with the motion of gases and liquids through each other during the process of diffusion, there being this difference between the two processes, that, in diffusion, the different substances are only mixed together and the mixture is not homogeneous, whereas in electrolysis they are chemically combined and the electrolyte is homogeneous. In diffusion the determining cause of the motion of a substance in a given direction is a diminution of the quantity of that substance per unit of volume in that direction, whereas in electrolysis the motion of each ion is due to the electromotive force acting on the charged molecules.

256.] Clausius*, who has bestowed much study on the theory of the molecular agitation of bodies, supposes that the molecules of all bodies are in a state of constant agitation, but that in solid bodies each molecule never passes beyond a certain distance from its original position, whereas in fluids a molecule, after moving a certain distance from its original position, is just as likely to move still farther from it as to move back again. Hence the molecules of a fluid apparently at rest are continually changing their positions, and passing irregularly from one part of the fluid to another. In a compound fluid he supposes that not only the compound molecules travel about in this way, but that, in the collisions which occur between the compound molecules, the molecules of which they are composed are often separated and change partners, so that the same individual atom is at one time associated with one atom of the opposite kind, and at another time with another. This process Clausius supposes to go on in the liquid at all times, but when an electromotive force acts on the liquid the motions of the molecules, which before were indifferently in all directions, are now influenced by the electromotive force, so that the positively charged molecules have a greater tendency towards the cathode than towards the anode, and the negatively charged molecules have a greater tendency to move in the opposite direction. Hence the molecules of the cation will during their intervals of freedom struggle towards the cathode, but will continually be checked in their course by pairing for a time with molecules of the anion, which are also struggling through the crowd, but in the opposite direction.

257.] This theory of Clausius enables us to understand how it is, that whereas the actual decomposition of an electrolyte requires an electromotive force of finite magnitude, the conduction of the current in the electrolyte obeys the law of Ohm, so that every electromotive force within the electrolyte, even the feeblest, produces a current of proportionate magnitude.

According to the theory of Clausius, the decomposition and recomposition of the electrolyte is continually going on even when there is no current, and the very feeblest electromotive force is

* Pogg. *Ann.* bd. ci. s. 338 (1857).

sufficient to give this process a certain degree of direction, and so to produce the currents of the ions and the electric current, which is part of the same phenomenon. Within the electrolyte, however, the ions are never set free in finite quantity, and it is this liberation of the ions which requires a finite electromotive force. At the electrodes the ions accumulate, for the successive portions of the ions, as they arrive at the electrodes, instead of finding molecules of the opposite ion ready to combine with them, are forced into company with molecules of their own kind, with which they cannot combine. The electromotive force required to produce this effect is of finite magnitude, and forms an opposing electromotive force which produces a reversed current when other electromotive forces are removed. When this reversed electromotive force, owing to the accumulation of the ions at the electrode, is observed, the electrodes are said to be Polarized.

258.] One of the best methods of determining whether a body is or is not an electrolyte is to place it between platinum electrodes and to pass a current through it for some time, and then, disengaging the electrodes from the voltaic battery, and connecting them with a galvanometer, to observe whether a reverse current, due to polarization of the electrodes, passes through the galvanometer. Such a current, being due to accumulation of different substances on the two electrodes, is a proof that the substance has been electrolytically decomposed by the original current from the battery. This method can often be applied where it is difficult, by direct chemical methods, to detect the presence of the products of decomposition at the electrodes. See Art. 271.

259.] So far as we have gone the theory of electrolysis appears very satisfactory. It explains the electric current, the nature of which we do not understand, by means of the currents of the material components of the electrolyte, the motion of which, though not visible to the eye, is easily demonstrated. It gives a clear explanation, as Faraday has shewn, why an electrolyte which conducts in the liquid state is a non-conductor when solidified, for unless the molecules can pass from one part to another no electrolytic conduction can take place, so that the substance must be in a liquid state, either by fusion or by solution, in order to be a conductor.

But if we go on, and assume that the molecules of the ions within the electrolyte are actually charged with certain definite quantities of electricity, positive and negative, so that the electrolytic current is simply a current of convection, we find that this tempting hypothesis leads us into very difficult ground.

In the first place, we must assume that in every electrolyte each molecule of the cation, as it is liberated at the cathode, communicates to the cathode a charge of positive electricity, the amount of which is the same for every molecule, not only of that cation but of all other cations. In the same way each molecule of the anion when liberated, communicates to the anode a charge of negative electricity, the numerical magnitude of which is the same as that of the positive charge due to a molecule of a cation, but with sign reversed.

If, instead of a single molecule, we consider an assemblage of molecules, constituting an electrochemical equivalent of the ion, then the total charge of all the molecules is, as we have seen, one unit of electricity, positive or negative.

260.] We do not as yet know how many molecules there are in an electrochemical equivalent of any substance, but the molecular theory of chemistry, which is corroborated by many physical considerations, supposes that the number of molecules in an electrochemical equivalent is the same for all substances. We may therefore, in molecular speculations, assume that the number of molecules in an electrochemical equivalent is N, a number unknown at present, but which we may hereafter find means to determine *.

Each molecule, therefore, on being liberated from the state of combination, parts with a charge whose magnitude is $\frac{1}{N}$, and is positive for the cation and negative for the anion. This definite quantity of electricity we shall call the molecular charge. If it were known it would be the most natural unit of electricity.

Hitherto we have only increased the precision of our ideas by exercising our imagination in tracing the electrification of molecules and the discharge of that electrification.

The liberation of the ions and the passage of positive electricity from the anode and into the cathode are simultaneous facts. The ions, when liberated, are not charged with electricity, hence, when they are in combination, they have the molecular charges as above described.

The electrification of a molecule, however, though easily spoken of, is not so easily conceived.

We know that if two metals are brought into contact at any

* See note to Art. 5.

point, the rest of their surfaces will be electrified, and if the metals are in the form of two plates separated by a narrow interval of air, the charge on each plate may become of considerable magnitude. Something like this may be supposed to occur when the two components of an electrolyte are in combination. Each pair of molecules may be supposed to touch at one point, and to have the rest of their surface charged with electricity due to the electromotive force of contact.

But to explain the phenomenon, we ought to shew why the charge thus produced on each molecule is of a fixed amount, and why, when a molecule of chlorine is combined with a molecule of zinc, the molecular charges are the same as when a molecule of chlorine is combined with a molecule of copper, although the electromotive force between chlorine and zinc is much greater than that between chlorine and copper. If the charging of the molecules is the effect of the electromotive force of contact, why should electromotive forces of different intensities produce exactly equal charges?

Suppose, however, that we leap over this difficulty by simply asserting the fact of the constant value of the molecular charge, and that we call this constant molecular charge, for convenience in description, *one molecule of electricity.*

This phrase, gross as it is, and out of harmony with the rest of this treatise, will enable us at least to state clearly what is known about electrolysis, and to appreciate the outstanding difficulties.

Every electrolyte must be considered as a binary compound of its anion and its cation. The anion or the cation or both may be compound bodies, so that a molecule of the anion or the cation may be formed by a number of molecules of simple bodies. A molecule of the anion and a molecule of the cation combined together form one molecule of the electrolyte.

In order to act as an anion in an electrolyte, the molecule which so acts must be charged with what we have called one molecule of negative electricity, and in order to act as a cation the molecule must be charged with one molecule of positive electricity.

These charges are connected with the molecules only when they are combined as anion and cation in the electrolyte.

When the molecules are electrolysed, they part with their charges to the electrodes, and appear as unelectrified bodies when set free from combination.

If the same molecule is capable of acting as a cation in one electrolyte and as an anion in another, and also of entering into compound bodies which are not electrolytes, then we must suppose that it receives a positive charge of electricity when it acts as a cation, a negative charge when it acts as an anion, and that it is without charge when it is not in an electrolyte.

Iodine, for instance, acts as an anion in the iodides of the metals and in hydriodic acid, but is said to act as a cation in the bromide of iodine.

This theory of molecular charges may serve as a method by which we may remember a good many facts about electrolysis. It is extremely improbable that when we come to understand the true nature of electrolysis we shall retain in any form the theory of molecular charges, for then we shall have obtained a secure basis on which to form a true theory of electric currents, and so become independent of these provisional theories.

261.] One of the most important steps in our knowledge of electrolysis has been the recognition of the secondary chemical processes which arise from the evolution of the ions at the electrodes.

In many cases the substances which are found at the electrodes are not the actual ions of the electrolysis, but the products of the action of these ions on the electrolyte.

Thus, when a solution of sulphate of soda is electrolysed by a current which also passes through dilute sulphuric acid, equal quantities of oxygen are given off at the anodes, and equal quantities of hydrogen at the cathodes, both in the sulphate of soda and in the dilute acid.

But if the electrolysis is conducted in suitable vessels, such as U-shaped tubes or vessels with a porous diaphragm, so that the substance surrounding each electrode can be examined separately, it is found that at the anode of the sulphate of soda there is an equivalent of sulphuric acid as well as an equivalent of oxygen, and at the cathode there is an equivalent of soda as well as two equivalents of hydrogen.

It would at first sight seem as if, according to the old theory of the constitution of salts, the sulphate of soda were electrolysed into its constituents sulphuric acid and soda, while the water of the solution is electrolysed at the same time into oxygen and hydrogen. But this explanation would involve the admission that the same current which passing through dilute sulphuric acid electrolyses one equivalent of water, when it passes through solution of sulphate of soda electrolyses one equivalent of the salt as well as one equivalent of the water, and this would be contrary to the law of electrochemical equivalents.

But if we suppose that the components of sulphate of soda are not SO_3 and NaO but SO_4 and Na,—not sulphuric acid and soda but sulphion and sodium—then the sulphion travels to the anode and is set free, but being unable to exist in a free state it breaks up into sulphuric acid and oxygen, one equivalent of each. At the same time the sodium is set free at the cathode, and there decomposes the water of the solution, forming one equivalent of soda and two of hydrogen.

In the dilute sulphuric acid the gases collected at the electrodes are the constituents of water, namely one volume of oxygen and two volumes of hydrogen. There is also an increase of sulphuric acid at the anode, but its amount is not equal to an equivalent.

It is doubtful whether pure water is an electrolyte or not. The greater the purity of the water, the greater the resistance to electrolytic conduction. The minutest traces of foreign matter are sufficient to produce a great diminution of the electrical resistance of water. The electric resistance of water as determined by different observers has values so different that we cannot consider it as a determined quantity. The purer the water the greater its resistance, and if we could obtain really pure water it is doubtful whether it would conduct at all.

As long as water was considered an electrolyte, and was, indeed, taken as the type of electrolytes, there was a strong reason for maintaining that it is a binary compound, and that two volumes of hydrogen are chemically equivalent to one volume of oxygen. If, however, we admit that water is not an electrolyte, we are free to suppose that equal volumes of oxygen and of hydrogen are chemically equivalent.

The dynamical theory of gases leads us to suppose that in perfect gases equal volumes always contain an equal number of molecules, and that the principal part of the specific heat, that, namely, which depends on the motion of agitation of the molecules among each other, is the same for equal numbers of molecules of all gases. Hence we are led to prefer a chemical system in which equal volumes of oxygen and of hydrogen are regarded as equivalent, and in which water is regarded as a compound of two equivalents of hydrogen and one of oxygen, and therefore probably not capable of direct electrolysis.

While electrolysis fully establishes the close relationship between electrical phenomena and those of chemical combination, the fact that every chemical compound is not an electrolyte shews that chemical combination is a process of a higher order of complexity than any purely electrical phenomenon. Thus the combinations of the metals with each other, though they are good conductors, and their components stand at different points of the scale of electrification by contact, are not, even when in a fluid state, decomposed by the current. Most of the combinations of the substances which act as anions are not conductors, and therefore are not electrolytes. Besides these we have many compounds, containing the same components as electrolytes, but not in equivalent proportions, and these are also non-conductors, and therefore not electrolytes.

On the Conservation of Energy in Electrolysis.

262.] Consider any voltaic circuit consisting partly of a battery, partly of a wire, and partly of an electrolytic cell.

During the passage of unit of electricity through any section of the circuit, one electrochemical equivalent of each of the substances in the cells, whether voltaic or electrolytic, is electrolysed.

The amount of mechanical energy equivalent to any given chemical process can be ascertained by converting the whole energy due to the process into heat, and then expressing the heat in dynamical measure by multiplying the number of thermal units by Joule's mechanical equivalent of heat.

Where this direct method is not applicable, if we can estimate the heat given out by the substances taken first in the state before the process and then in the state after the process during their reduction to a final state, which is the same in both cases, then the thermal equivalent of the process is the difference of the two quantities of heat.

In the case in which the chemical action maintains a voltaic circuit, Joule found that the heat developed in the voltaic cells is less than that due to the chemical process within the cell, and that the remainder of the heat is developed in the connecting wire, or, when there is an electromagnetic engine in the circuit, part of the heat may be accounted for by the mechanical work of the engine.

For instance, if the electrodes of the voltaic cell are first connected by a short thick wire, and afterwards by a long thin wire, the heat developed in the cell for each grain of zinc dissolved is greater in the first case than in the second, but the heat developed in the wire is greater in the second case than in the first. The sum of the heat developed in the cell and in the wire for each grain of zinc dissolved is the same in both cases. This has been established by Joule by direct experiment.

The ratio of the heat generated in the cell to that generated in the wire is that of the resistance of the cell to that of the wire, so that if the wire were made of sufficient resistance nearly the whole of the heat would be generated in the wire, and if it were made of sufficient conducting power nearly the whole of the heat would be generated in the cell.

Let the wire be made so as to have great resistance, then the heat generated in it is equal in dynamical measure to the product of the quantity of electricity which is transmitted, multiplied by the electromotive force under which it is made to pass through the wire.

263.] Now during the time in which an electrochemical equivalent of the substance in the cell undergoes the chemical process which gives rise to the current, one unit of electricity passes through the wire. Hence, the heat developed by the passage of one unit of electricity is in this case measured by the electromotive force. But this heat is that which one electrochemical equivalent of the substance generates, whether in the cell or in the wire, while undergoing the given chemical process.

Hence the following important theorem, first proved by Thomson (*Phil. Mag.*, Dec. 1851):—

'The electromotive force of an electrochemical apparatus is in absolute measure equal to the mechanical equivalent of the chemical action on one electrochemical equivalent of the substance.'

The thermal equivalents of many chemical actions have been determined by Andrews, Hess, Favre and Silbermann, &c., and from these their mechanical equivalents can be deduced by multiplication by the mechanical equivalent of heat.

This theorem not only enables us to calculate from purely thermal data the electromotive forces of different voltaic arrangements, and the electromotive forces required to effect electrolysis in different cases, but affords the means of actually measuring chemical affinity.

It has long been known that chemical affinity, or the tendency which exists towards the going on of a certain chemical change, is stronger in some cases than in others, but no proper measure of this tendency could be made till it was shewn that this tendency in certain cases is exactly equivalent to a certain electromotive force, and can therefore be measured according to the very same principles used in the measurement of electromotive forces.

Chemical affinity being therefore, in certain cases, reduced to the form of a measurable quantity, the whole theory of chemical processes, of the rate at which they go on, of the displacement of one substance by another, &c., becomes much more intelligible than when chemical affinity was regarded as a quality *sui generis*, and irreducible to numerical measurement.

When the volume of the products of electrolysis is greater than that of the electrolyte, work is done during the electrolysis in overcoming the pressure. If the volume of an electrochemical equivalent of the electrolyte is increased by a volume v when electrolysed under a pressure p, then the work done during the passage of a unit of electricity in overcoming pressure is vp, and the electromotive force required for electrolysis must include a part equal to vp, which is spent in performing this mechanical work.

If the products of electrolysis are gases which, like oxygen and hydrogen, are much rarer than the electrolyte, and fulfil Boyle's law very exactly, vp will be very nearly constant for the same temperature, and the electromotive force required for electrolysis will not depend in any sensible degree on the pressure. Hence it has been found impossible to check the electrolytic decomposition of dilute sulphuric acid by confining the decomposed gases in a small space.

When the products of electrolysis are liquid or solid the quantity vp will increase as the pressure increases, so that if v is positive an increase of pressure will increase the electromotive force required for electrolysis.

In the same way, any other kind of work done during electrolysis will have an effect on the value of the electromotive force, as, for instance, if a vertical current passes between two zinc electrodes in a solution of sulphate of zinc a greater electromotive force will be required when the current in the solution flows upwards than when it flows downwards, for, in the first case, it carries zinc from the lower to the upper electrode, and in the second from the upper to the lower. The electromotive force required for this purpose is less than the millionth part of that of a Daniell's cell per foot.

CHAPTER V.

ELECTROLYTIC POLARIZATION.

264.] WHEN an electric current is passed through an electrolyte bounded by metal electrodes, the accumulation of the ions at the electrodes produces the phenomenon called Polarization, which consists in an electromotive force acting in the opposite direction to the current, and producing an apparent increase of the resistance.

When a continuous current is employed, the resistance appears to increase rapidly from the commencement of the current, and at last reaches a value nearly constant. If the form of the vessel in which the electrolyte is contained is changed, the resistance is altered in the same way as a similar change of form of a metallic conductor would alter its resistance, but an additional apparent resistance, depending on the nature of the electrodes, has always to be added to the true resistance of the electrolyte.

265.] These phenomena have led some to suppose that there is a finite electromotive force required for a current to pass through an electrolyte. It has been shewn, however, by the researches of Lenz, Neumann, Beetz, Wiedemann *, Paalzow †, and recently by those of MM. F. Kohlrausch and W. A. Nippoldt ‡, that the conduction in the electrolyte itself obeys Ohm's Law with the same precision as in metallic conductors, and that the apparent resistance at the bounding surface of the electrolyte and the electrodes is entirely due to polarization.

266.] The phenomenon called polarization manifests itself in the case of a continuous current by a diminution in the current, indicating a force opposed to the current. Resistance is also perceived as a force opposed to the current, but we can distinguish

* *Galvanismus*, bd. i.

† *Berlin Monatsbericht*, July, 1868.

‡ Pogg. *Ann*. bd. cxxxviii. s. 286 (October, 1869).

between the two phenomena by instantaneously removing or reversing the electromotive force.

The resisting force is always opposite in direction to the current, and the external electromotive force required to overcome it is proportional to the strength of the current, and changes its direction when the direction of the current is changed. If the external electromotive force becomes zero the current simply stops.

The electromotive force due to polarization, on the other hand, is in a fixed direction, opposed to the current which produced it. If the electromotive force which produced the current is removed, the polarization produces a current in the opposite direction.

The difference between the two phenomena may be compared with the difference between forcing a current of water through a long capillary tube, and forcing water through a tube of moderate length up into a cistern. In the first case if we remove the pressure which produces the flow the current will simply stop. In the second case, if we remove the pressure the water will begin to flow down again from the cistern.

To make the mechanical illustration more complete, we have only to suppose that the cistern is of moderate depth, so that when a certain amount of water is raised into it, it begins to overflow. This will represent the fact that the total electromotive force due to polarization has a maximum limit.

267.] The cause of polarization appears to be the existence at the electrodes of the products of the electrolytic decomposition of the fluid between them. The surfaces of the electrodes are thus rendered electrically different, and an electromotive force between them is called into action, the direction of which is opposite to that of the current which caused the polarization.

The ions, which by their presence at the electrodes produce the phenomena of polarization, are not in a perfectly free state, but are in a condition in which they adhere to the surface of the electrodes with considerable force.

The electromotive force due to polarization depends upon the density with which the electrode is covered with the ion, but it is not proportional to this density, for the electromotive force does not increase so rapidly as this density.

This deposit of the ion is constantly tending to become free, and either to diffuse into the liquid, to escape as a gas, or to be precipitated as a solid.

The rate of this dissipation of the polarization is exceedingly small for slight degrees of polarization, and exceedingly rapid near the limiting value of polarization.

268.] We have seen, Art. 262, that the electromotive force acting in any electrolytic process is numerically equal to the mechanical equivalent of the result of that process on one electrochemical equivalent of the substance. If the process involves a diminution of the intrinsic energy of the substances which take part in it, as in the voltaic cell, then the electromotive force is in the direction of the current. If the process involves an increase of the intrinsic energy of the substances, as in the case of the electrolytic cell, the electromotive force is in the direction opposite to that of the current, and this electromotive force is called polarization.

In the case of a steady current in which electrolysis goes on continuously, and the ions are separated in a free state at the electrodes, we have only by a suitable process to measure the intrinsic energy of the separated ions, and compare it with that of the electrolyte in order to calculate the electromotive force required for the electrolysis. This will give the maximum polarization.

But during the first instants of the process of electrolysis the ions when deposited at the electrodes are not in a free state, and their intrinsic energy is less than their energy in a free state, though greater than their energy when combined in the electrolyte. In fact, the ion in contact with the electrode is in a state which when the deposit is very thin may be compared with that of chemical combination with the electrode, but as the deposit increases in density, the succeeding portions are no longer so intimately combined with the electrode, but simply adhere to it, and at last the deposit, if gaseous, escapes in bubbles, if liquid, diffuses through the electrolyte, and if solid, forms a precipitate.

In studying polarization we have therefore to consider

(1) The superficial density of the deposit, which we may call σ. This quantity σ represents the number of electrochemical equivalents of the ion deposited on unit of area. Since each electrochemical equivalent deposited corresponds to one unit of electricity transmitted by the current, we may consider σ as representing either a surface-density of matter or a surface-density of electricity.

(2) The electromotive force of polarization, which we may call p. This quantity p is the difference between the electric potentials of the two electrodes when the current through the electrolyte

is so feeble that the proper resistance of the electrolyte makes no sensible difference between these potentials.

The electromotive force p at any instant is numerically equal to the mechanical equivalent of the electrolytic process going on at that instant which corresponds to one electrochemical equivalent of the electrolyte. This electrolytic process, it must be remembered, consists in the deposit of the ions on the electrodes, and the state in which they are deposited depends on the actual state of the surface of the electrodes, which may be modified by previous deposits.

Hence the electromotive force at any instant depends on the previous history of the electrode. It is, speaking very roughly, a function of σ, the density of the deposit, such that $p = 0$ when $\sigma = 0$, but p approaches a limiting value much sooner than σ does. The statement, however, that p is a function of σ cannot be considered accurate. It would be more correct to say that p is a function of the chemical state of the superficial layer of the deposit, and that this state depends on the density of the deposit according to some law involving the time.

269.] (3) The third thing we must take into account is the dissipation of the polarization. The polarization when left to itself diminishes at a rate depending partly on the intensity of the polarization or the density of the deposit, and partly on the nature of the surrounding medium, and the chemical, mechanical, or thermal action to which the surface of the electrode is exposed.

If we determine a time T such that at the rate at which the deposit is dissipated, the whole deposit would be removed in the time T, we may call T the modulus of the time of dissipation. When the density of the deposit is very small, T is very large, and may be reckoned by days or months. When the density of the deposit approaches its limiting value T diminishes very rapidly, and is probably a minute fraction of a second. In fact, the rate of dissipation increases so rapidly that when the strength of the current is maintained constant, the separated gas, instead of contributing to increase the density of the deposit, escapes in bubbles as fast as it is formed.

270.] There is therefore a great difference between the state of polarization of the electrodes of an electrolytic cell when the polarization is feeble, and when it is at its maximum value. For instance, if a number of electrolytic cells of dilute sulphuric acid with platinum electrodes are arranged in series, and if a small electromotive force, such as that of one Daniell's cell, be made to act on the circuit, the electromotive force will produce a current of exceedingly short duration, for after a very short time the electromotive force arising from the polarization of the cell will balance that of the Daniell's cell.

The dissipation will be very small in the case of so feeble a state of polarization, and it will take place by a very slow absorption of the gases and diffusion through the liquid. The rate of this dissipation is indicated by the exceedingly feeble current which still continues to flow without any visible separation of gases.

If we neglect this dissipation for the short time during which the state of polarization is set up, and if we call Q the total quantity of electricity which is transmitted by the current during this time, then if A is the area of one of the electrodes, and σ the density of the deposit, supposed uniform,

$$Q = A\sigma.$$

If we now disconnect the electrodes of the electrolytic apparatus from the Daniell's cell, and connect them with a galvanometer capable of measuring the whole discharge through it, a quantity of electricity nearly equal to Q will be discharged as the polarization disappears.

271.] Hence we may compare the action of this apparatus, which is a form of Ritter's Secondary Pile, with that of a Leyden jar.

Both the secondary pile and the Leyden jar are capable of being charged with a certain amount of electricity, and of being afterwards discharged. During the discharge a quantity of electricity nearly equal to the charge passes in the opposite direction. The difference between the charge and the discharge arises partly from dissipation, a process which in the case of small charges is very slow, but which, when the charge exceeds a certain limit, becomes exceedingly rapid. Another part of the difference between the charge and the discharge arises from the fact that after the electrodes have been connected for a time sufficient to produce an apparently complete discharge, so that the current has completely disappeared, if we separate the electrodes for a time, and afterwards connect them, we obtain a second discharge in the same direction as the original discharge. This is called the residual discharge, and is a phenomenon of the Leyden jar as well as of the secondary pile.

The secondary pile may therefore be compared in several respects to a Leyden jar. There are, however, certain important differences. The charge of a Leyden jar is very exactly proportional to the electromotive force of the charge, that is, to the difference of potentials of the two surfaces, and the charge corresponding to unit of electromotive force is called the capacity of the jar, a constant quantity. The corresponding quantity, which may be called the capacity of the secondary pile, increases when the electromotive force increases.

The capacity of the jar depends on the area of the opposed surfaces, on the distance between them, and on the nature of the substance between them, but not on the nature of the metallic surfaces themselves. The capacity of the secondary pile depends on the area of the surfaces of the electrodes, but not on the distance between them, and it depends on the nature of the surface of the electrodes, as well as on that of the fluid between them. The maximum difference of the potentials of the electrodes in each element of a secondary pile is very small compared with the maximum difference of the potentials of those of a charged Leyden jar, so that in order to obtain much electromotive force a pile of many elements must be used.

On the other hand, the superficial density of the charge in the secondary pile is immensely greater than the utmost superficial density of the charge which can be accumulated on the surfaces of a Leyden jar, insomuch that Mr. C. F. Varley *, in describing the construction of a condenser of great capacity, recommends a series of gold or platinum plates immersed in dilute acid as preferable in point of cheapness to induction plates of tinfoil separated by insulating material.

The form in which the energy of a Leyden jar is stored up is the state of constraint of the dielectric between the conducting surfaces, a state which I have already described under the name of electric polarization, pointing out those phenomena attending this state which are at present known, and indicating the imperfect state of our knowledge of what really takes place. See Arts. 62, 111.

The form in which the energy of the secondary pile is stored up is the chemical condition of the material stratum at the surface of the electrodes, consisting of the ions of the electrolyte and the substance of the electrodes in a relation varying from chemical combination to superficial condensation, mechanical adherence, or simple juxtaposition.

The seat of this energy is close to the surfaces of the electrodes,

* Specification of C. F. Varley, 'Electric Telegraphs, &c.,' Jan. 1860.

and not throughout the substance of the electrolyte, and the form in which it exists may be called electrolytic polarization.

After studying the secondary pile in connexion with the Leyden jar, the student should again compare the voltaic battery with some form of the electrical machine, such as that described in Art. 211.

Mr. Varley has lately * found that the capacity of one square inch is from 175 to 542 microfarads and upwards for platinum plates in dilute sulphuric acid, and that the capacity increases with the electromotive force, being about 175 for 0.02 of a Daniell's cell, and 542 for 1.6 Daniell's cells.

But the comparison between the Leyden jar and the secondary pile may be carried still farther, as in the following experiment, due to Buff †. It is only when the glass of the jar is cold that it is capable of retaining a charge. At a temperature below 100°C the glass becomes a conductor. If a test-tube containing mercury is placed in a vessel of mercury, and if a pair of electrodes are connected, one with the inner and the other with the outer portion of mercury, the arrangement constitutes a Leyden jar which will hold a charge at ordinary temperatures. If the electrodes are connected with those of a voltaic battery, no current will pass as long as the glass is cold, but if the apparatus is gradually heated a current will begin to pass, and will increase rapidly in intensity as the temperature rises, though the glass remains apparently as hard as ever.

This current is manifestly electrolytic, for if the electrodes are disconnected from the battery, and connected with a galvanometer, a considerable reverse current passes, due to polarization of the surfaces of the glass.

If, while the battery is in action the apparatus is cooled, the current is stopped by the cold glass as before, but the polarization of the surfaces remains. The mercury may be removed, the surfaces may be washed with nitric acid and with water, and fresh mercury introduced. If the apparatus is then heated, the current of polarization appears as soon as the glass is sufficiently warm to conduct it.

We may therefore regard glass at 100°C, though apparently a solid body, as an electrolyte, and there is considerable reason to believe that in most instances in which a dielectric has a slight degree of conductivity the conduction is electrolytic. The

* *Proc. R. S.* Jan. 12, 1871.

† *Annalen der Chemie und Pharmacie*, bd. xc. 257 (1854).

existence of polarization may be regarded as conclusive evidence of electrolysis, and if the conductivity of a substance increases as the temperature rises, we have good grounds for suspecting that it is electrolytic.

On Constant Voltaic Elements.

272.] When a series of experiments is made with a voltaic battery in which polarization occurs, the polarization diminishes during the time the current is not flowing, so that when it begins to flow again the current is stronger than after it has flowed for some time. If, on the other hand, the resistance of the circuit is diminished by allowing the current to flow through a short shunt, then, when the current is again made to flow through the ordinary circuit, it is at first weaker than its normal strength on account of the great polarization produced by the use of the short circuit.

To get rid of these irregularities in the current, which are exceedingly troublesome in experiments involving exact measurements, it is necessary to get rid of the polarization, or at least to reduce it as much as possible.

It does not appear that there is much polarization at the surface of the zinc plate when immersed in a solution of sulphate of zinc or in dilute sulphuric acid. The principal seat of polarization is at the surface of the negative metal. When the fluid in which the negative metal is immersed is dilute sulphuric acid, it is seen to become covered with bubbles of hydrogen gas, arising from the electrolytic decomposition of the fluid. Of course these bubbles, by preventing the fluid from touching the metal, diminish the surface of contact and increase the resistance of the circuit. But besides the visible bubbles it is certain that there is a thin coating of hydrogen, probably not in a free state, adhering to the metal, and as we have seen that this coating is able to produce an electromotive force in the reverse direction, it must necessarily diminish the electromotive force of the battery.

Various plans have been adopted to get rid of this coating of hydrogen. It may be diminished to some extent by mechanical means, such as stirring the liquid, or rubbing the surface of the negative plate. In Smee's battery the negative plates are vertical, and covered with finely divided platinum from which the bubbles of hydrogen easily escape, and in their ascent produce a current of liquid which helps to brush off other bubbles as they are formed.

A far more efficacious method, however, is to employ chemical means. These are of two kinds. In the batteries of Grove and Bunsen the negative plate is immersed in a fluid rich in oxygen, and the hydrogen, instead of forming a coating on the plate, combines with this substance. In Grove's battery the plate is of platinum immersed in strong nitric acid. In Bunsen's first battery it is of carbon in the same acid. Chromic acid is also used for the same purpose, and has the advantage of being free from the acid fumes produced by the reduction of nitric acid.

A different mode of getting rid of the hydrogen is by using copper as the negative metal, and covering the surface with a coat of oxide. This, however, rapidly disappears when it is used as the negative electrode. To renew it Joule has proposed to make the copper plates in the form of disks, half immersed in the liquid, and to rotate them slowly, so that the air may act on the parts exposed to it in turn.

The other method is by using as the liquid an electrolyte, the cation of which is a metal highly negative to zinc.

In Daniell's battery a copper plate is immersed in a saturated solution of sulphate of copper. When the current flows through the solution from the zinc to the copper no hydrogen appears on the copper plate, but copper is deposited on it. When the solution is saturated, and the current is not too strong, the copper appears to act as a true cation, the anion SO_4 travelling towards the zinc.

When these conditions are not fulfilled hydrogen is evolved at the cathode, but immediately acts on the solution, throwing down copper, and uniting with SO_4 to form oil of vitriol. When this is the case, the sulphate of copper next the copper plate is replaced by oil of vitriol, the liquid becomes colourless, and polarization by hydrogen gas again takes place. The copper deposited in this way is of a looser and more friable structure than that deposited by true electrolysis.

To ensure that the liquid in contact with the copper shall be saturated with sulphate of copper, crystals of this substance must be placed in the liquid close to the copper, so that when the solution is made weak by the deposition of the copper, more of the crystals may be dissolved.

We have seen that it is necessary that the liquid next the copper should be saturated with sulphate of copper. It is still more necessary that the liquid in which the zinc is immersed should be free from sulphate of copper. If any of this salt makes its way to the surface of the zinc it is reduced, and copper is deposited on the zinc. The zinc, copper, and fluid then form a little circuit in which rapid electrolytic action goes on, and the zinc is eaten away by an action which contributes nothing to the useful effect of the battery.

To prevent this, the zinc is immersed either in dilute sulphuric acid or in a solution of sulphate of zinc, and to prevent the solution of sulphate of copper from mixing with this liquid, the two liquids are separated by a division consisting of bladder or porous earthenware, which allows electrolysis to take place through it, but effectually prevents mixture of the fluids by visible currents.

In some batteries sawdust is used to prevent currents. The experiments of Graham, however, shew that the process of diffusion goes on nearly as rapidly when two liquids are separated by a division of this kind as when they are in direct contact, provided there are no visible currents, and it is probable that if a septum is employed which diminishes the diffusion, it will increase in exactly the same ratio the resistance of the element, because electrolytic conduction is a process the mathematical laws of which have the same form as those of diffusion, and whatever interferes with one must interfere equally with the other. The only difference is that diffusion is always going on, whereas the current flows only when the battery is in action.

In all forms of Daniell's battery the final result is that the sulphate of copper finds its way to the zinc and spoils the battery. To retard this result indefinitely, Sir W. Thomson * has constructed Daniell's battery in the following form.

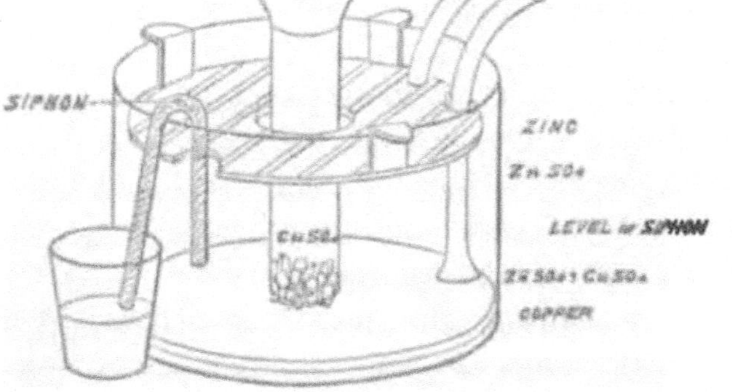

Fig. 22.

In each cell the copper plate is placed horizontally at the bottom

* *Proc. R. S.*, Jan. 19, 1871.

and a saturated solution of sulphate of zinc is poured over it. The zinc is in the form of a grating and is placed horizontally near the surface of the solution. A glass tube is placed vertically in the solution with its lower end just above the surface of the copper plate. Crystals of sulphate of copper are dropped down this tube, and, dissolving in the liquid, form a solution of greater density than that of sulphate of zinc alone, so that it cannot get to the zinc except by diffusion. To retard this process of diffusion, a siphon, consisting of a glass tube stuffed with cotton wick, is placed with one extremity midway between the zinc and copper, and the other in a vessel outside the cell, so that the liquid is very slowly drawn off near the middle of its depth. To supply its place, water, or a weak solution of sulphate of zinc, is added above when required. In this way the greater part of the sulphate of copper rising through the liquid by diffusion is drawn off by the siphon before it reaches the zinc, and the zinc is surrounded by liquid nearly free from sulphate of copper, and having a very slow downward motion in the cell, which still further retards the upward motion of the sulphate of copper. During the action of the battery copper is deposited on the copper plate, and SO_4 travels slowly through the liquid to the zinc with which it combines, forming sulphate of zinc. Thus the liquid at the bottom becomes less dense by the deposition of the copper, and the liquid at the top becomes more dense by the addition of the zinc. To prevent this action from changing the order of density of the strata, and so producing instability and visible currents in the vessel, care must be taken to keep the tube well supplied with crystals of sulphate of copper, and to feed the cell above with a solution of sulphate of zinc sufficiently dilute to be lighter than any other stratum of the liquid in the cell.

Daniell's battery is by no means the most powerful in common use. The electromotive force of Grove's cell is 192,000,000, of Daniell's 107,900,000 and that of Bunsen's 188,000,000.

The resistance of Daniell's cell is in general greater than that of Grove's or Bunsen's of the same size.

These defects, however, are more than counterbalanced in all cases where exact measurements are required, by the fact that Daniell's cell exceeds every other known arrangement in constancy of electromotive force. It has also the advantage of continuing in working order for a long time, and of emitting no gas.

CHAPTER VI.

LINEAR ELECTRIC CURRENTS.

On Systems of Linear Conductors.

273.] ANY conductor may be treated as a linear conductor if it is arranged so that the current must always pass in the same manner between two portions of its surface which are called its electrodes. For instance, a mass of metal of any form the surface of which is entirely covered with insulating material except at two places, at which the exposed surface of the conductor is in metallic contact with electrodes formed of a perfectly conducting material, may be treated as a linear conductor. For if the current be made to enter at one of these electrodes and escape at the other the lines of flow will be determinate, and the relation between electromotive force, current and resistance will be expressed by Ohm's Law, for the current in every part of the mass will be a linear function of E. But if there be more possible electrodes than two, the conductor may have more than one independent current through it, and these may not be conjugate to each other. See Art. 282.

Ohm's Law.

274.] Let E be the electromotive force in a linear conductor from the electrode A_1 to the electrode A_2. (See Art. 69.) Let C be the strength of the electric current along the conductor, that is to say, let C units of electricity pass across every section in the direction $A_1 A_2$ in unit of time, and let R be the resistance of the conductor, then the expression of Ohm's Law is

$$E = CR. \tag{1}$$

Linear Conductors arranged in Series.

275.] Let A_1, A_2 be the electrodes of the first conductor and let the second conductor be placed with one of its electrodes in contact with A_2, so that the second conductor has for its electrodes A_2, A_3. The electrodes of the third conductor may be denoted by A_3 and A_4.

Let the electromotive forces along these conductors be denoted by E_{12}, E_{23}, E_{34}, and so on for the other conductors.

Let the resistances of the conductors be

$$R_{12},\quad R_{23},\quad R_{34},\ \&c.$$

Then, since the conductors are arranged in series so that the same current C flows through each, we have by Ohm's Law,

$$E_{12} = CR_{12},\quad E_{23} = CR_{23},\quad E_{34} = CR_{34}. \tag{2}$$

If E is the resultant electromotive force, and R the resultant resistance of the system, we must have by Ohm's Law,

$$E = CR. \tag{3}$$

Now $$E = E_{12}+E_{23}+E_{34}, \tag{4}$$

the sum of the separate electromotive forces,

$$= C(R_{12}+R_{23}+R_{34}) \text{ by equations (2).}$$

Comparing this result with (3), we find

$$R = R_{12}+R_{23}+R_{34}. \tag{5}$$

Or, *the resistance of a series of conductors is the sum of the resistances of the conductors taken separately.*

Potential at any Point of the Series.

Let A and C be the electrodes of the series, B a point between them, a, c, and b the potentials of these points respectively. Let R_1 be the resistance of the part from A to B, R_2 that of the part from B to C, and R that of the whole from A to C, then, since

$$a-b = R_1 C,\quad b-c = R_2 C,\quad \text{and}\quad a-c = RC,$$

the potential at B is

$$b = \frac{R_2 a + R_1 c}{R}, \tag{6}$$

which determines the potential at B when the potentials at A and C are given.

Resistance of a Multiple Conductor.

276.] Let a number of conductors ABZ, ACZ, ADZ be arranged side by side with their extremities in contact with the same two points A and Z. They are then said to be arranged in multiple arc.

Let the resistances of these conductors be R_1, R_2, R_3 respectively, and the currents C_1, C_2, C_3, and let the resistance of the multiple conductor be R, and the total current C. Then, since the potentials at A and Z are the same for all the conductors, they have the same difference, which we may call E. We then have

$$E = C_1 R_1 = C_2 R_2 = C_3 R_3 = CR,$$

but

$$C = C_1 + C_2 + C_3,$$

whence

$$\frac{1}{R} = \frac{1}{R_1} + \frac{1}{R_2} + \frac{1}{R_3}. \tag{7}$$

Or, *the reciprocal of the resistance of a multiple conductor is the sum of the reciprocals of the component conductors.*

If we call the reciprocal of the resistance of a conductor the conductivity of the conductor, then we may say that *the conductivity of a multiple conductor is the sum of the conductivities of the component conductors.*

Current in any Branch of a Multiple Conductor.

From the equations of the preceding article, it appears that if C_1 is the current in any branch of the multiple conductor, and R_1 the resistance of that branch,

$$C_1 = C\frac{R}{R_1}, \tag{8}$$

where C is the total current, and R is the resistance of the multiple conductor as previously determined.

Longitudinal Resistance of Conductors of Uniform Section.

277.] Let the resistance of a cube of a given material to a current parallel to one of its edges be ρ, the side of the cube being unit of length, ρ is called the 'specific resistance of that material for unit of volume.'

Consider next a prismatic conductor of the same material whose length is l, and whose section is unity. This is equivalent to l cubes arranged in series. The resistance of the conductor is therefore $l\rho$.

Finally, consider a conductor of length l and uniform section s. This is equivalent to s conductors similar to the last arranged in multiple arc. The resistance of this conductor is therefore

$$R = \frac{l\rho}{s}.$$

When we know the resistance of a uniform wire we can determine the specific resistance of the material of which it is made if we can measure its length and its section.

The sectional area of small wires is most accurately determined by calculation from the length, weight, and specific gravity of the specimen. The determination of the specific gravity is sometimes inconvenient, and in such cases the resistance of a wire of unit length and unit mass is used as the 'specific resistance per unit of weight.'

If r is this resistance, l the length, and m the mass of a wire, then

$$R = \frac{l^2 r}{m}.$$

On the Dimensions of the Quantities involved in these Equations.

278.] The resistance of a conductor is the ratio of the electromotive force acting on it to the current produced. The conductivity of the conductor is the reciprocal of this quantity, or in other words, the ratio of the current to the electromotive force producing it.

Now we know that in the electrostatic system of measurement the ratio of a quantity of electricity to the potential of the conductor on which it is spread is the capacity of the conductor, and is measured by a line. If the conductor is a sphere placed in an unlimited field, this line is the radius of the sphere. The ratio of a quantity of electricity to an electromotive force is therefore a line, but the ratio of a quantity of electricity to a current is the time during which the current flows to transmit that quantity. Hence the ratio of a current to an electromotive force is that of a line to a time, or in other words, it is a velocity.

The fact that the conductivity of a conductor is expressed in the electrostatic system of measurement by a velocity may be verified by supposing a sphere of radius r charged to potential V, and then connected with the earth by the given conductor. Let the sphere contract, so that as the electricity escapes through the conductor the potential of the sphere is always kept equal to V. Then the charge on the sphere is rV at any instant, and the current is $\frac{d}{dt}(rV)$, but, since V is constant, the current is $\frac{dr}{dt}V$, and the electromotive force through the conductor is V.

The conductivity of the conductor is the ratio of the current to the electromotive force, or $\frac{dr}{dt}$, that is, the *velocity* with which the radius of the sphere must diminish in order to maintain the potential constant when the charge is allowed to pass to earth through the conductor.

In the electrostatic system, therefore, the conductivity of a conductor is a velocity, and of the dimensions $[LT^{-1}]$.

The resistance of the conductor is therefore of the dimensions $[L^{-1}T]$.

The specific resistance per unit of volume is of the dimension of $[T]$, and the specific conductivity per unit of volume is of the dimension of $[T^{-1}]$.

The numerical magnitude of these coefficients depends only on the unit of time, which is the same in different countries.

The specific resistance per unit of weight is of the dimensions $[L^{-3}MT]$.

279.] We shall afterwards find that in the electromagnetic system of measurement the resistance of a conductor is expressed by a velocity, so that in this system the dimensions of the resistance of a conductor are $[LT^{-1}]$.

The conductivity of the conductor is of course the reciprocal of this.

The specific resistance per unit of volume in this system is of the dimensions $[L^2T^{-1}]$, and the specific resistance per unit of weight is of the dimensions $[L^{-1}T^{-1}M]$.

On Linear Systems of Conductors in general.

280.] The most general case of a linear system is that of n points, $A_1, A_2, \dots A_n$, connected together in pairs by $\frac{1}{2}n(n-1)$ linear conductors. Let the conductivity (or reciprocal of the resistance) of that conductor which connects any pair of points, say A_p and A_q, be called K_{pq}, and let the current from A_p to A_q be C_{pq}. Let P_p and P_q be the electric potentials at the points A_p and A_q respectively, and let the internal electromotive force, if there be any, along the conductor from A_p to A_q be E_{pq}.

The current from A_p to A_q is, by Ohm's Law,

$$C_{pq} = K_{pq}(P_p - P_q + E_{pq}). \tag{1}$$

Among these quantities we have the following sets of relations:

The conductivity of a conductor is the same in either direction, or

$$K_{pq} = K_{qp}. \tag{2}$$

The electromotive force and the current are directed quantities, so that

$$E_{pq} = -E_{qp}, \quad \text{and} \quad C_{pq} = -C_{qp}. \tag{3}$$

Let $P_1, P_2, \dots P_n$ be the potentials at $A_1, A_2, \dots A_n$ respectively, and let $Q_1, Q_2, \dots Q_n$ be the quantities of electricity which enter the system in unit of time at each of these points respectively. These are necessarily subject to the condition of 'continuity'

$$Q_1 + Q_2 \dots + Q_n = 0, \tag{4}$$

since electricity can neither be indefinitely accumulated nor produced within the system.

The condition of 'continuity' at any point A_p is

$$Q_p = C_{p1} + C_{p2} + \text{\&c.} + C_{pn}. \tag{5}$$

Substituting the values of the currents in terms of equation (1), this becomes

$$Q_p = (K_{p1} + K_{p2} + \text{\&c.} + K_{pn}) P_p - (K_{p1} P_1 + K_{p2} P_2 + \text{\&c.} + K_{pn} P_n) \\ + (K_{p1} E_{p1} + \text{\&c.} + K_{pn} E_{pn}). \tag{6}$$

The symbol K_{pp} does not occur in this equation. Let us therefore give it the value

$$K_{pp} = -(K_{p1} + K_{p2} + \text{\&c.} + K_{pn}); \tag{7}$$

that is, let K_{pp} be a quantity equal and opposite to the sum of all the conductivities of the conductors which meet in A_p. We may then write the condition of continuity for the point A_p,

$$K_{p1} P_1 + K_{p2} P_2 + \text{\&c.} + K_{pp} P_p + \text{\&c.} + K_{pn} P_n \\ = K_{p1} E_{p1} + \text{\&c.} + K_{pn} E_{pn} - Q_p. \tag{8}$$

By substituting 1, 2, &c. n for p in this equation we shall obtain n equations of the same kind from which to determine the n potentials P_1, P_2, &c., P_n.

Since, however, if we add the system of equations (8) the result is identically zero by (3), (4) and (7), there will be only $n-1$ independent equations. These will be sufficient to determine the differences of the potentials of the points, but not to determine the absolute potential of any. This, however, is not required to calculate the currents in the system.

If we denote by D the determinant

$$\begin{vmatrix} K_{11}, & K_{12}, & \dots\dots K_{1(n-1)}, \\ K_{21}, & K_{22}, & \dots\dots K_{2(n-1)}, \\ - - - & - - - & - - - - \\ K_{(n-1)1}, & K_{(n-1)2}, & \dots\dots K_{(n-1)(n-1)}; \end{vmatrix} \tag{9}$$

and by D_{pq}, the minor of K_{pq}, we find for the value of $P_p - P_n$,

$$(P_p - P_n) D = (K_{12} E_{12} + \text{\&c.} - Q_1) D_{p1} + (K_{21} E_{21} + \text{\&c.} - Q_2) D_{p2} + \text{\&c.} \\ + (K_{q1} E_{q1} + \text{\&c.} + K_{qn} E_{qn} - Q_q) D_{pq} + \text{\&c.} \tag{10}$$

In the same way the excess of the potential of any other point, say A_q, over that of A_n may be determined. We may then determine the current between A_p and A_q from equation (1), and so solve the problem completely.

281.] We shall now demonstrate a reciprocal property of any two conductors of the system, answering to the reciprocal property we have already demonstrated for statical electricity in Art. 88.

The coefficient of Q_q in the expression for P_p is $\frac{D_{pq}}{D}$. That of Q_p in the expression for P_q is $\frac{D_{qp}}{D}$.

Now D_{pq} differs from D_{qp} only by the substitution of the symbols such as K_{qp} for K_{pq}. But, by equation (2), these two symbols are equal, since the conductivity of a conductor is the same both ways. Hence

$$D_{pq} = D_{qp}. \qquad (11)$$

It follows from this that the part of the potential at A_p arising from the introduction of a unit current at A_q is equal to the part of the potential at A_q arising from the introduction of a unit current at A_p.

We may deduce from this a proposition of a more practical form.

Let A, B, C, D be any four points of the system, and let the effect of a current Q, made to enter the system at A and leave it at B, be to make the potential at C exceed that at D by P. Then, if an equal current Q be made to enter the system at C and leave it at D, the potential at A will exceed that at B by the same quantity P.

If an electromotive force E be introduced, acting in the conductor from A to B, and if this causes a current C from X to Y, then the same electromotive force E introduced into the conductor from X to Y will cause an equal current C from A to B.

The electromotive force E may be that of a voltaic battery introduced between the points named, care being taken that the resistance of the conductor is the same before and after the introduction of the battery.

282 *a*.] If an electromotive force E_{pq} act along the conductor $A_p A_q$, the current produced along another conductor of the system $A_r A_s$ is easily found to be

$$K_{rs} K_{pq} E_{pq} (D_{rp} + D_{sq} - D_{rq} - D_{sp}) \div D.$$

There will be no current if

$$D_{rp} + D_{sq} - D_{rq} - D_{sp} = 0. \qquad (12)$$

But, by (11), the same equation holds if, when the electromotive force acts along $A_r A_s$, there is no current in $A_p A_q$. On account of this reciprocal relation the two conductors referred to are said to be *conjugate*.

The theory of conjugate conductors has been investigated by

Kirchhoff, who has stated the conditions of a linear system in the following manner, in which the consideration of the potential is avoided.

(1) (Condition of 'continuity.') At any point of the system the sum of all the currents which flow towards that point is zero.

(2) In any complete circuit formed by the conductors the sum of the electromotive forces taken round the circuit is equal to the sum of the products of the current in each conductor multiplied by the resistance of that conductor.

We obtain this result by adding equations of the form (1) for the complete circuit, when the potentials necessarily disappear.

*282 *b*.] If the conducting wires form a simple network and if we suppose that a current circulates round each mesh, then the actual current in the wire which forms a thread of each of two neighbouring meshes will be the difference between the two currents circulating in the two meshes, the currents being reckoned positive when they circulate in a direction opposite to the motion of the hands of a watch. It is easy to establish in this case the following proposition:—Let x be the current, E the electromotive force, and R the total resistance in any mesh; let also $y, z, \ldots$ be currents circulating in neighbouring meshes which have threads in common with that in which x circulates, the resistances of those parts being $s, t, \ldots$; then

$$Rx - sy - tz - \&c. = E.$$

To illustrate the use of this rule we will take the arrangement known as Wheatstone's Bridge, adopting the figure and notation of Art. 347. We have then the three following equations representing the application of the rule in the case of the three circuits OBC, OCA, OAB in which the currents x, y, z respectively circulate, viz.

$$\begin{aligned} (a+\beta+\gamma)x \quad & -\gamma y & -\beta z &= E, \\ -\gamma\, x + (b+\gamma+a)y & & -a\, z &= 0, \\ -\beta\, x \quad & -a y & +(c+a+\beta)z &= 0. \end{aligned}$$

From these equations we may now determine the value of $x-y$ the galvanometer current in the branch OA, but the reader is referred to Art. 347 et seq. where this and other questions connected with Wheatstone's Bridge are discussed.

Heat Generated in the System.

283.] The mechanical equivalent of the quantity of heat generated

* [Extracted from notes of Professor Maxwell's lectures by Mr. J. A. Fleming, B.A., St. John's College.]

in a conductor whose resistance is R by a current C in unit of time is, by Art. 242,
$$JH = RC^2. \tag{13}$$

We have therefore to determine the sum of such quantities as RC^2 for all the conductors of the system.

For the conductor from A_p to A_q the conductivity is K_{pq}, and the resistance R_{pq}, where
$$K_{pq} \,.\, R_{pq} = 1. \tag{14}$$

The current in this conductor is, according to Ohm's Law,
$$C_{pq} = K_{pq}(P_p - P_q). \tag{15}$$

We shall suppose, however, that the value of the current is not that given by Ohm's Law, but X_{pq}, where
$$X_{pq} = C_{pq} + Y_{pq}. \tag{16}$$

To determine the heat generated in the system we have to find the sum of all the quantities of the form
$$R_{pq} X^2_{pq},$$
or
$$JH = \Sigma \{R_{pq} C^2_{pq} + 2R_{pq} C_{pq} Y_{pq} + R_{pq} Y^2_{pq}\}. \tag{17}$$

Giving C_{pq} its value, and remembering the relation between K_{pq} and R_{pq}, this becomes
$$\Sigma \left[(P_p - P_q)(C_{pq} + 2Y_{pq}) + R_{pq} Y^2_{pq}\right]. \tag{18}$$

Now since both C and X must satisfy the condition of continuity at A_p, we have
$$Q_p = C_{p1} + C_{p2} + \&c. + C_{pn}, \tag{19}$$
$$Q_p = X_{p1} + X_{p2} + \&c. + X_{pn}, \tag{20}$$
therefore
$$0 = Y_{p1} + Y_{p2} + \&c. + Y_{pn}. \tag{21}$$

Adding together therefore all the terms of (18), we find
$$\Sigma (R_{pq} X^2_{pq}) = \Sigma P_p Q_p + \Sigma R_{pq} Y^2_{pq}. \tag{22}$$

Now since R is always positive and Y^2 is essentially positive, the last term of this equation must be essentially positive. Hence the first term is a minimum when Y is zero in every conductor, that is, when the current in every conductor is that given by Ohm's Law.

Hence the following theorem :

284.] In any system of conductors in which there are no internal electromotive forces the heat generated by currents distributed in accordance with Ohm's Law is less than if the currents had been distributed in any other manner consistent with the actual conditions of supply and outflow of the current.

The heat actually generated when Ohm's Law is fulfilled is mechanically equivalent to $\Sigma P_p Q_p$, that is, to the sum of the products of the quantities of electricity supplied at the different external electrodes, each multiplied by the potential at which it is supplied.

CHAPTER VII.

CONDUCTION IN THREE DIMENSIONS.

Notation of Electric Currents.

285.] At any point let an element of area dS be taken normal to the axis of x, and let Q units of electricity pass across this area from the negative to the positive side in unit of time, then, if $\frac{Q}{dS}$ becomes ultimately equal to u when dS is indefinitely diminished, u is said to be the Component of the electric current in the direction of x at the given point.

In the same way we may determine v and w, the components of the current in the directions of y and z respectively.

286.] To determine the component of the current in any other direction OR through the given point O, let l, m, n be the direction-cosines of OR; then if we cut off from the axes of x, y, z portions equal to

$$\frac{r}{l}, \frac{r}{m}, \text{ and } \frac{r}{n}$$

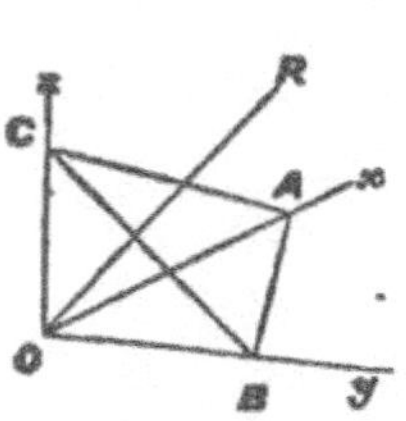

Fig. 28.

respectively at A, B and C, the triangle ABC will be normal to OR.

The area of this triangle ABC will be

$$dS = \tfrac{1}{2}\frac{r^2}{lmn},$$

and by diminishing r this area may be diminished without limit.

The quantity of electricity which leaves the tetrahedron $ABCO$ by the triangle ABC must be equal to that which enters it through the three triangles OBC, OCA, and OAB.

The area of the triangle OBC is $\tfrac{1}{2}\frac{r^2}{mn}$, and the component of the current normal to its plane is u, so that the quantity which enters through this triangle is $\frac{1}{2} r^2 \frac{u}{mn}$.

The quantities which enter through the triangles OCA and OAB respectively are

$$\frac{1}{2} r^2 \frac{v}{nl}, \quad \text{and} \quad \frac{1}{2} r^2 \frac{w}{lm}.$$

If γ is the component of the velocity in the direction OR, then the quantity which leaves the tetrahedron through ABC is

$$\frac{1}{2} r^2 \frac{\gamma}{lmn}.$$

Since this is equal to the quantity which enters through the three other triangles,

$$\frac{1}{2} \frac{r^2 \gamma}{lmn} = \frac{1}{2} r^2 \left\{ \frac{u}{mn} + \frac{v}{nl} + \frac{w}{lm} \right\};$$

multiplying by $\frac{2\,lmn}{r^2}$, we get

$$\gamma = lu + mv + nw. \tag{1}$$

If we put $\qquad u^2 + v^2 + w^2 = \Gamma^2,$

and make l', m', n' such that

$$u = l'\Gamma, \qquad v = m'\Gamma, \quad \text{and} \quad w = n'\Gamma;$$

then

$$\gamma = \Gamma(ll' + mm' + nn'). \tag{2}$$

Hence, if we define the resultant current as a vector whose magnitude is Γ, and whose direction-cosines are l', m', n', and if γ denotes the current resolved in a direction making an angle θ with that of the resultant current, then

$$\gamma = \Gamma \cos \theta; \tag{3}$$

shewing that the law of resolution of currents is the same as that of velocities, forces, and all other vectors.

287.] To determine the condition that a given surface may be a surface of flow, let

$$F(x, y, z) = \lambda \tag{4}$$

be the equation of a family of surfaces any one of which is given by making λ constant; then, if we make

$$\overline{\frac{d\lambda}{dx}}\Big|^2 + \overline{\frac{d\lambda}{dy}}\Big|^2 + \overline{\frac{d\lambda}{dz}}\Big|^2 = \frac{1}{N^2}, \tag{5}$$

the direction-cosines of the normal, reckoned in the direction in which λ increases, are

$$l = N\frac{d\lambda}{dx}, \qquad m = N\frac{d\lambda}{dy}, \qquad n = N\frac{d\lambda}{dz}. \tag{6}$$

Hence, if γ is the component of the current normal to the surface,

$$\gamma = N\left\{u\frac{d\lambda}{dx} + v\frac{d\lambda}{dy} + w\frac{d\lambda}{dz}\right\}. \tag{7}$$

If $\gamma = 0$ there will be no current through the surface, and the surface may be called a Surface of Flow, because the lines of motion are in the surface.

288.] The equation of a surface of flow is therefore

$$u\frac{d\lambda}{dx} + v\frac{d\lambda}{dy} + w\frac{d\lambda}{dz} = 0. \tag{8}$$

If this equation is true for all values of λ, all the surfaces of the family will be surfaces of flow.

289.] Let there be another family of surfaces, whose parameter is λ', then, if these are also surfaces of flow, we shall have

$$u\frac{d\lambda'}{dx} + v\frac{d\lambda'}{dy} + w\frac{d\lambda'}{dz} = 0. \tag{9}$$

If there is a third family of surfaces of flow, whose parameter is λ'', then

$$u\frac{d\lambda''}{dx} + v\frac{d\lambda''}{dy} + w\frac{d\lambda''}{dz} = 0. \tag{10}$$

If we eliminate between these three equations, u, v, and w disappear together, and we find

$$\begin{vmatrix} \frac{d\lambda}{dx}, & \frac{d\lambda}{dy}, & \frac{d\lambda}{dz} \\ \frac{d\lambda'}{dx}, & \frac{d\lambda'}{dy}, & \frac{d\lambda'}{dz} \\ \frac{d\lambda''}{dx}, & \frac{d\lambda''}{dy}, & \frac{d\lambda''}{dz} \end{vmatrix} = 0; \tag{11}$$

or $$\lambda'' = \phi(\lambda, \lambda'); \tag{12}$$

that is, λ'' is some function of λ and λ'.

290.] Now consider the four surfaces whose parameters are λ, $\lambda+\delta\lambda$, λ', and $\lambda'+\delta\lambda'$. These four surfaces enclose a quadrilateral tube, which we may call the tube $\delta\lambda.\delta\lambda'$. Since this tube is bounded by surfaces across which there is no flow, we may call it a Tube of Flow. If we take any two sections across the tube, the quantity which enters the tube at one section must be equal to the quantity which leaves it at the other, and since this quantity is therefore the same for every section of the tube, let us call it $L\delta\lambda.\delta\lambda'$ where L is a function of λ and λ', the parameters which determine the particular tube.

291.] If δS denotes the section of a tube of flow by a plane normal to x, we have by the theory of the change of the independent variables,

$$\delta\lambda\,.\,\delta\lambda' = \delta S\left(\frac{d\lambda}{dy}\frac{d\lambda'}{dz} - \frac{d\lambda}{dz}\frac{d\lambda'}{dy}\right), \tag{13}$$

and by the definition of the components of the current

$$u\,\delta S = L\,\delta\lambda\,.\,\delta\lambda'. \tag{14}$$

$$\left.\begin{aligned} \text{Hence}\qquad u &= L\left(\frac{d\lambda}{dy}\frac{d\lambda'}{dz} - \frac{d\lambda}{dz}\frac{d\lambda'}{dy}\right).\\ \text{Similarly}\qquad v &= L\left(\frac{d\lambda}{dz}\frac{d\lambda'}{dx} - \frac{d\lambda}{dx}\frac{d\lambda'}{dz}\right),\\ w &= L\left(\frac{d\lambda}{dx}\frac{d\lambda'}{dy} - \frac{d\lambda}{dy}\frac{d\lambda'}{dx}\right).\end{aligned}\right\} \tag{15}$$

292.] It is always possible when one of the functions λ or λ' is known, to determine the other so that L may be equal to unity. For instance, let us take the plane of yz, and draw upon it a series of equidistant lines parallel to y, to represent the sections of the family λ' by this plane. In other words, let the function λ' be determined by the condition that when $x = 0$ $\lambda' = z$. If we then make $L = 1$, and therefore (when $x = 0$)

$$\lambda = \int u\,dy;$$

then in the plane ($x = 0$) the amount of electricity which passes through any portion will be

$$\iint u\,dy\,dz = \iint d\lambda\,d\lambda'. \tag{16}$$

Having determined the nature of the sections of the surfaces of flow by the plane of yz, the form of the surfaces elsewhere is determined by the conditions (8) and (9). The two functions λ and λ' thus determined are sufficient to determine the current at every point by equations (15), unity being substituted for L.

On Lines of Flow.

293.] Let a series of values of λ and of λ' be chosen, the successive differences in each series being unity. The two series of surfaces defined by these values will divide space into a system of quadrilateral tubes through each of which there will be a unit current. By assuming the unit sufficiently small, the details of the current may be expressed by these tubes with any desired amount of minuteness. Then if any surface be drawn cutting the system of tubes, the quantity of the current which passes through this surface will be expressed by the *number* of tubes which cut it, since each tube carries unity of current.

The actual intersections of the surfaces may be called Lines of Flow. When the unit is taken sufficiently small, the number of lines of flow which cut a surface is approximately equal to the number of tubes of flow which cut it, so that we may consider the lines of flow as expressing not only the *direction* of the current but its *strength*, since each line of flow through a given section corresponds to a unit current.

On Current-Sheets and Current-Functions.

294.] A stratum of a conductor contained between two consecutive surfaces of flow of one system, say that of λ', is called a Current-Sheet. The tubes of flow within this sheet are determined by the function λ. If λ_A and λ_P denote the values of λ at the points A and P respectively, then the current from right to left across any line drawn on the sheet, from A to P is $\lambda_P - \lambda_A$. If AP be an element, ds, of a curve drawn on the sheet, the current which crosses this element from right to left is

$$\frac{d\lambda}{ds}\,ds.$$

This function λ, from which the distribution of the current in the sheet can be completely determined, is called the Current-Function.

Any thin sheet of metal or conducting matter bounded on both sides by air or some other non-conducting medium may be treated as a current-sheet, in which the distribution of the current may be expressed by means of a current-function. See Art. 647.

Equation of 'Continuity.'

295.] If we differentiate the three equations (15) with respect to x, y, z respectively, remembering that L is a function of λ and λ', we find

$$\frac{du}{dx} + \frac{dv}{dy} + \frac{dw}{dz} = 0. \qquad (17)$$

The corresponding equation in Hydrodynamics is called the Equation of 'Continuity.' The continuity which it expresses is the continuity of existence, that is, the fact that a material substance cannot leave one part of space and arrive at another, without going through the space between. It cannot simply vanish in the one place and appear in the other, but it must travel along a continuous path, so that if a closed surface be drawn, including the one place and excluding the other, a material substance in passing from the one place to the other must go through the closed surface. The most general form of the equation in hydrodynamics is

$$\frac{d(\rho u)}{dx} + \frac{d(\rho v)}{dy} + \frac{d(\rho w)}{dz} + \frac{d\rho}{dt} = 0; \qquad (18)$$

where ρ signifies the ratio of the quantity of the substance to the volume it occupies, that volume being in this case the differential element of volume, and (ρu), (ρv), and (ρw) signify the ratio of the quantity of the substance which crosses an element of area in unit of time to that area, these areas being normal to the axes of x, y, and z respectively. Thus understood, the equation is applicable to any material substance, solid or fluid, whether the motion be continuous or discontinuous, provided the existence of the parts of that substance is continuous. If anything, though not a substance, is subject to the condition of continuous existence in time and space, the equation will express this condition. In other parts of Physical Science, as, for instance, in the theory of electric and magnetic quantities, equations of a similar form occur. We shall call such equations 'equations of continuity' to indicate their form, though we may not attribute to these quantities the properties of matter, or even continuous existence in time and space.

The equation (17), which we have arrived at in the case of electric currents, is identical with (18) if we make $\rho = 1$, that is, if we suppose the substance homogeneous and incompressible. The equation, in the case of fluids, may also be established by either of the modes of proof given in treatises on Hydrodynamics. In one of these we trace the course and the deformation of a certain element of the fluid as it moves along. In the other, we fix our attention on an element of space, and take account of all that enters or leaves it. The former of these methods cannot be applied to electric currents, as we do not know the velocity with which the electricity passes through the body, or even whether it moves in the positive or the negative direction of the current. All that we know is the algebraical value of the quantity which crosses unit of area in unit of time, a quantity corresponding to (ρu) in the equation (18). We have no means of ascertaining the value of either of the factors ρ or u, and therefore we cannot follow a particular portion of electricity in its course through the body. The other method of investigation, in which we consider what passes through the walls of an element of volume, is applicable to electric currents, and is perhaps preferable in point of form to that which we have given, but as it may be found in any treatise on Hydrodynamics we need not repeat it here.

Quantity of Electricity which passes through a given Surface.

296.] Let Γ be the resultant current at any point of the surface. Let dS be an element of the surface, and let ϵ be the angle between Γ and the normal to the surface, then the total current through the surface will be

$$\iint \Gamma \cos \epsilon \, dS,$$

the integration being extended over the surface.

As in Art. 21, we may transform this integral into the form

$$\iint \Gamma \cos \epsilon \, dS = \iiint \left(\frac{du}{dx} + \frac{dv}{dy} + \frac{dw}{dz}\right) dx\, dy\, dz \qquad (19)$$

in the case of any closed surface, the limits of the triple integration being those included by the surface. This is the expression for the total efflux from the closed surface. Since in all cases of steady currents this must be zero whatever the limits of the integration, the quantity under the integral sign must vanish, and we obtain in this way the equation of continuity (17).

CHAPTER VIII.

RESISTANCE AND CONDUCTIVITY IN THREE DIMENSIONS.

On the most General Relations between Current and Electromotive Force.

297.] LET the components of the current at any point be u, v, w.

Let the components of the electromotive force be X, Y, Z.

The electromotive force at any point is the resultant force on a unit of positive electricity placed at that point. It may arise (1) from electrostatic action, in which case if V is the potential,

$$X = -\frac{dV}{dx}, \quad Y = -\frac{dV}{dy}, \quad Z = -\frac{dV}{dz}; \tag{1}$$

or (2) from electromagnetic induction, the laws of which we shall afterwards examine; or (3) from thermoelectric or electrochemical action at the point itself, tending to produce a current in a given direction.

We shall in general suppose that X, Y, Z represent the components of the actual electromotive force at the point, whatever be the origin of the force, but we shall occasionally examine the result of supposing it entirely due to variation of potential.

By Ohm's Law the current is proportional to the electromotive force. Hence X, Y, Z must be linear functions of u, v, w. We may therefore assume as the equations of Resistance,

$$\left.\begin{aligned} X &= R_1 u + Q_3 v + P_2 w, \\ Y &= P_3 u + R_2 v + Q_1 w, \\ Z &= Q_2 u + P_1 v + R_3 w. \end{aligned}\right\} \tag{2}$$

We may call the coefficients R the coefficients of longitudinal resistance in the directions of the axes of coordinates.

The coefficients P and Q may be called the coefficients of transverse resistance. They indicate the electromotive force in one direction required to produce a current in a different direction.

If we were at liberty to assume that a solid body may be treated as a system of linear conductors, then, from the reciprocal property (Art. 281) of any two conductors of a linear system, we might shew that the electromotive force along z required to produce a unit current parallel to y must be equal to the electromotive force along y required to produce a unit current parallel to z. This would shew that $P_1 = Q_1$, and similarly we should find $P_2 = Q_2$, and $P_3 = Q_3$. When these conditions are satisfied the system of coefficients is said to be Symmetrical. When they are not satisfied it is called a Skew system.

We have great reason to believe that in every actual case the system is symmetrical, but we shall examine some of the consequences of admitting the possibility of a skew system.

298.] The quantities u, v, w may be expressed as linear functions of X, Y, Z by a system of equations, which we may call Equations of Conductivity,

$$\left.\begin{aligned} u &= r_1 X + p_3 Y + q_2 Z, \\ v &= q_3 X + r_2 Y + p_1 Z, \\ w &= p_2 X + q_1 Y + r_3 Z; \end{aligned}\right\} \qquad (3)$$

we may call the coefficients r the coefficients of Longitudinal conductivity, and p and q those of Transverse conductivity.

The coefficients of resistance are inverse to those of conductivity. This relation may be defined as follows:

Let $[PQR]$ be the determinant of the coefficients of resistance, and $[pqr]$ that of the coefficients of conductivity, then

$$[PQR] = P_1 P_2 P_3 + Q_1 Q_2 Q_3 + R_1 R_2 R_3 - P_1 Q_1 R_1 - P_2 Q_2 R_2 - P_3 Q_3 R_3, \quad (4)$$

$$[pqr] = p_1 p_2 p_3 + q_1 q_2 q_3 + r_1 r_2 r_3 - p_1 q_1 r_1 - p_2 q_2 r_2 - p_3 q_3 r_3, \quad (5)$$

$$[PQR]\,[pqr] = 1, \quad (6)$$

$$[PQR]\, p_1 = (P_2 P_3 - Q_1 R_1), \qquad [pqr]\, P_1 = (p_2 p_3 - q_1 r_1), \quad (7)$$

&c. &c.

The other equations may be formed by altering the symbols P, Q, R, p, q, r, and the suffixes 1, 2, 3 in cyclical order.

Rate of Generation of Heat.

299.] To find the work done by the current in unit of time in overcoming resistance, and so generating heat, we multiply the components of the current by the corresponding components of the electromotive force. We thus obtain the following expressions for W, the quantity of work expended in unit of time:

$$W = Xu + Yv + Zw; \tag{8}$$

$$= R_1 u^2 + R_2 v^2 + R_3 w^2 + (P_1 + Q_1) vw + (P_2 + Q_2) wu + (P_3 + Q_3) uv; \tag{9}$$

$$= r_1 X^2 + r_2 Y^2 + r_3 Z^2 + (p_1 + q_1) YZ + (p_2 + q_2) ZX + (p_3 + q_3) XY. \tag{10}$$

By a proper choice of axes, either of the two latter equations may be deprived of the terms involving the products of u, v, w or of X, Y, Z. The system of axes, however, which reduces W to the form

$$R_1 u^2 + R_2 v^2 + R_3 w^2$$

is not in general the same as that which reduces it to the form

$$r_1 X^2 + r_2 Y^2 + r_3 Z^2.$$

It is only when the coefficients P_1, P_2, P_3 are equal respectively to Q_1, Q_2, Q_3 that the two systems of axes coincide.

If with Thomson * we write

$$\left.\begin{aligned} P &= S + T, & Q &= S - T; \\ \text{and} \quad p &= s + t, & q &= s - t; \end{aligned}\right\} \tag{11}$$

then we have

$$\left.\begin{aligned} [PQR] = {} & R_1 R_2 R_3 + 2 S_1 S_2 S_3 - S_1^2 R_1 - S_2^2 R_2 - S_3^2 R_3 \\ & + 2 (S_1 T_2 T_3 + S_2 T_3 T_1 + S_3 T_1 T_2) + R_1 T_1^2 + R_2 T_2^2 + R_3 T_3^2; \end{aligned}\right\} \tag{12}$$

and

$$\left.\begin{aligned} [PQR] r_1 &= R_2 R_3 - S_1^2 + T_1^2, \\ [PQR] s_1 &= T_2 T_3 + S_2 S_3 - R_1 S_1, \\ [PQR] t_1 &= -R_1 T_1 + S_2 T_3 + S_3 T_2. \end{aligned}\right\} \tag{13}$$

If therefore we cause S_1, S_2, S_3 to disappear, s_1 will not also disappear unless the coefficients T are zero.

Condition of Stability.

300.] Since the equilibrium of electricity is stable, the work spent in maintaining the current must always be positive. The conditions that W may be positive are that the three coefficients R_1, R_2, R_3, and the three expressions

$$\left.\begin{aligned} & 4 R_2 R_3 - (P_1 + Q_1)^2, \\ & 4 R_3 R_1 - (P_2 + Q_2)^2, \\ & 4 R_1 R_2 - (P_3 + Q_3)^2, \end{aligned}\right\} \tag{14}$$

must all be positive.

There are similar conditions for the coefficients of conductivity.

* *Trans. R. S. Edin.*, 1853-4, p. 165.

Equation of Continuity in a Homogeneous Medium.

301.] If we express the components of the electromotive force as the derivatives of the potential V, the equation of continuity

$$\frac{du}{dx}+\frac{dv}{dy}+\frac{dw}{dz}=0 \qquad (15)$$

becomes in a homogeneous medium

$$r_1\frac{d^2V}{dx^2}+r_2\frac{d^2V}{dy^2}+r_3\frac{d^2V}{dz^2}+2s_1\frac{d^2V}{dy\,dz}+2s_2\frac{d^2V}{dz\,dx}+2s_3\frac{d^2V}{dx\,dy}=0. \qquad (16)$$

If the medium is not homogeneous there will be terms arising from the variation of the coefficients of conductivity in passing from one point to another.

This equation corresponds to Laplace's equation in an isotropic medium.

302.] If we put

$$[rs]=r_1r_2r_3+2s_1s_2s_3-r_1s_1^2-r_2s_2^2-r_3s_3^2, \qquad (17)$$

and $[AB]=A_1A_2A_3+2B_1B_2B_3-A_1B_1^2-A_2B_2^2-A_3B_3^2$, (18)

where

$$\left.\begin{aligned}[rs]A_1&=r_2r_3-s_1^2,\\ [rs]B_1&=s_2s_3-r_1s_1,\\ &\cdots\cdots\end{aligned}\right\} \qquad (19)$$

and so on, the system A, B will be inverse to the system r, s, and if we make

$$A_1x^2+A_2y^2+A_3z^2+2B_1yz+2B_2zx+2B_3xy=[AB]\rho^2, \qquad (20)$$

we shall find that

$$V=\frac{c}{4\pi}\frac{1}{\rho} \qquad (21)$$

is a solution of the equation.

In the case in which the coefficients T are zero, the coefficients A and B become identical with R and S. When T exists this is not the case.

In the case therefore of electricity flowing out from a centre in an infinite, homogeneous, but not isotropic, medium, the equipotential surfaces are ellipsoids, for each of which ρ is constant. The axes of these ellipsoids are in the directions of the principal axes of conductivity, and these do not coincide with the principal axes of resistance unless the system is symmetrical.

By a transformation of this equation we may take for the axes of x, y, z the principal axes of conductivity. The coefficients of the forms s and B will then be reduced to zero, and each coefficient of the form A will be the reciprocal of the corresponding coefficient of the form r. The expression for ρ will be

$$\frac{x^2}{r_1}+\frac{y^2}{r_2}+\frac{z^2}{r_3}=\frac{\rho^2}{r_1 r_2 r_3}. \tag{22}$$

303.] The theory of the complete system of equations of resistance and of conductivity is that of linear functions of three variables, and it is exemplified in the theory of Strains *, and in other parts of physics. The most appropriate method of treating it is that by which Hamilton and Tait treat a linear and vector function of a vector. We shall not, however, expressly introduce Quaternion notation.

The coefficients T_1, T_2, T_3 may be regarded as the rectangular components of a vector T, the absolute magnitude and direction of which are fixed in the body, and independent of the direction of the axes of reference. The same is true of t_1, t_2, t_3, which are the components of another vector t.

The vectors T and t do not in general coincide in direction.

Let us now take the axis of z so as to coincide with the vector T, and transform the equations of resistance accordingly. They will then have the form

$$\left.\begin{aligned} X &= R_1 u + S_3 v + S_2 w - Tv,\\ Y &= S_3 u + R_2 v + S_1 w + Tu,\\ Z &= S_2 u + S_1 v + R_3 w. \end{aligned}\right\} \tag{23}$$

It appears from these equations that we may consider the electromotive force as the resultant of two forces, one of them depending only on the coefficients R and S, and the other depending on T alone. The part depending on R and S is related to the current in the same way that the perpendicular on the tangent plane of an ellipsoid is related to the radius vector. The other part, depending on T, is equal to the product of T into the resolved part of the current perpendicular to the axis of T, and its direction is perpendicular to T and to the current, being always in the direction in which the resolved part of the current would lie if turned 90° in the positive direction round T.

If we consider the current and T as vectors, the part of the electromotive force due to T is the vector part of the product, $T\times$ current.

The coefficient T may be called the Rotatory coefficient. We have reason to believe that it does not exist in any known sub-

* See Thomson and Tait's *Natural Philosophy*, § 154.

stance. It should be found, if anywhere, in magnets, which have a polarization in one direction, probably due to a rotational phenomenon in the substance.

304.] Assuming then that there is no rotatory coefficient, we shall shew how Thomson's Theorem given in Art. 100 may be extended to prove that the heat generated by the currents in the system in a given time is a unique minimum.

To simplify the algebraical work let the axes of coordinates be chosen so as to reduce expression (9), and therefore also in this case expression (10), to three terms; and let us consider the general characteristic equation (16) which thus reduces to

$$r_1 \frac{d^2 V}{dx^2} + r_2 \frac{d^2 V}{dy^2} + r_3 \frac{d^2 V}{dz^2} = 0. \tag{24}$$

Also, let a, b, c be three functions of x, y, z satisfying the condition

$$\frac{da}{dx} + \frac{db}{dy} + \frac{dc}{dz} = 0; \tag{25}$$

and let

$$\left. \begin{aligned} a &= -r_1 \frac{dV}{dx} + u, \\ b &= -r_2 \frac{dV}{dy} + v, \\ c &= -r_3 \frac{dV}{dz} + w. \end{aligned} \right\} \tag{26}$$

Finally, let the triple-integral

$$W = \iiint (R_1 a^2 + R_2 b^2 + R_3 c^2)\, dx\, dy\, dz \tag{27}$$

be extended over spaces bounded as in the enunciation of Art. 100; such viz. that V is constant over certain portions or else the normal component of the vector a, b, c is given, the latter condition being accompanied by the further restriction that the integral of this component over the whole bounding surface must be zero: then W will be a minimum when

$$u = 0, \qquad v = 0, \qquad w = 0.$$

For we have in this case

$$r_1 R_1 = 1, \qquad r_2 R_2 = 1, \qquad r_3 R_3 = 1;$$

and therefore, by (26),

$$\begin{aligned} W = &\iiint \left(r_1 \left|\frac{dV}{dx}\right|^2 + r_2 \left|\frac{dV}{dy}\right|^2 + r_3 \left|\frac{dV}{dz}\right|^2 \right) dx\, dy\, dz \\ &+ \iiint (R_1 u^2 + R_2 v^2 + R_3 w^2)\, dx\, dy\, dz \\ &- 2 \iiint \left(u \frac{dV}{dx} + v \frac{dV}{dy} + w \frac{dV}{dz} \right) dx\, dy\, dz. \end{aligned} \tag{28}$$

But since $$\frac{du}{dx} + \frac{dv}{dy} + \frac{dw}{dz} = 0, \tag{29}$$
the third term vanishes by virtue of the conditions at the limits.

The first term of (28) is therefore the unique minimum value of *W*.

305.] As this proposition is of great importance in the theory of electricity, it may be useful to present the following proof of the most general case in a form free from analytical operations.

Let us consider the propagation of electricity through a conductor of any form, homogeneous or heterogeneous.

Then we know that

(1) If we draw a line along the path and in the direction of the electric current, the line must pass from places of high potential to places of low potential.

(2) If the potential at every point of the system be altered in a given uniform ratio, the currents will be altered in the same ratio, according to Ohm's Law.

(3) If a certain distribution of potential gives rise to a certain distribution of currents, and a second distribution of potential gives rise to a second distribution of currents, then a third distribution in which the potential is the sum or difference of those in the first and second will give rise to a third distribution of currents, such that the total current passing through a given finite surface in the third case is the sum or difference of the currents passing through it in the first and second cases. For, by Ohm's Law, the additional current due to an alteration of potentials is independent of the original current due to the original distribution of potentials.

(4) If the potential is constant over the whole of a closed surface, and if there are no electrodes or intrinsic electromotive forces within it, then there will be no currents within the closed surface, and the potential at any point within it will be equal to that at the surface.

If there are currents within the closed surface they must either be closed curves, or they must begin and end either within the closed surface or at the surface itself.

But since the current must pass from places of high to places of low potential, it cannot flow in a closed curve.

Since there are no electrodes within the surface the current cannot begin or end within the closed surface, and since the potential at all points of the surface is the same, there can be no current along lines passing from one point of the surface to another.

Hence there are no currents within the surface, and therefore there can be no difference of potential, as such a difference would produce currents, and therefore the potential within the closed surface is everywhere the same as at the surface.

(5) If there is no electric current through any part of a closed surface, and no electrodes or intrinsic electromotive forces within the surface, there will be no currents within the surface, and the potential will be uniform.

We have seen that the currents cannot form closed curves, or begin or terminate within the surface, and since by the hypothesis they do not pass through the surface, there can be no currents, and therefore the potential is constant.

(6) If the potential is uniform over part of a closed surface, and if there is no current through the remainder of the surface, the potential within the surface will be uniform for the same reasons.

(7) If over part of the surface of a body the potential of every point is known, and if over the rest of the surface of the body the current passing through the surface at each point is known, then only one distribution of potentials at points within the body can exist.

For if there were two different values of the potential at any point within the body, let these be V_1 in the first case and V_2 in the second case, and let us imagine a third case in which the potential of every point of the body is the excess of potential in the first case over that in the second. Then on that part of the surface for which the potential is known the potential in the third case will be zero, and on that part of the surface through which the currents are known the currents in the third case will be zero, so that by (6) the potential everywhere within the surface will be zero, or there is no excess of V_1 over V_2, or the reverse. Hence there is only one possible distribution of potentials. This proposition is true whether the solid be bounded by one closed surface or by several.

On the Approximate Calculation of the Resistance of a Conductor of a given Form.

306.] The conductor here considered has its surface divided into three portions. Over one of these portions the potential is maintained at a constant value. Over a second portion the potential has a constant value different from the first. The whole of the remainder of the surface is impervious to electricity. We may suppose the conditions of the first and second portions to be fulfilled by applying to the conductor two electrodes of perfectly conducting material, and that of the remainder of the surface by coating it with perfectly non-conducting material.

Under these circumstances the current in every part of the conductor is simply proportional to the difference between the potentials of the electrodes. Calling this difference the electromotive force, the total current from the one electrode to the other is the product of the electromotive force by the conductivity of the conductor as a whole, and the resistance of the conductor is the reciprocal of the conductivity.

It is only when a conductor is approximately in the circumstances above defined that it can be said to have a definite resistance, or conductivity as a whole. A resistance coil, consisting of a thin wire terminating in large masses of copper, approximately satisfies these conditions, for the potential in the massive electrodes is nearly constant, and any differences of potential in different points of the same electrode may be neglected in comparison with the difference of the potentials of the two electrodes.

A very useful method of calculating the resistance of such conductors has been given, so far as I know, for the first time, by Lord Rayleigh, in a paper on the Theory of Resonance *.

It is founded on the following considerations.

If the specific resistance of any portion of the conductor be changed, that of the remainder being unchanged, the resistance of the whole conductor will be increased if that of the portion is increased, and diminished if that of the portion be diminished.

This principle may be regarded as self-evident, but it may easily be shewn that the value of the expression for the resistance of a system of conductors between two points selected as electrodes, increases as the resistance of each member of the system increases.

It follows from this that if a surface of any form be described in the substance of the conductor, and if we further suppose this surface to be an infinitely thin sheet of a perfectly conducting substance, the resistance of the conductor as a whole will be diminished unless the surface is one of the equipotential surfaces in the natural state of the conductor, in which case no effect will be produced by making it a perfect conductor, as it is already in electrical equilibrium.

* *Phil. Trans.*, 1871, p. 77. See Art. 102.

If therefore we draw within the conductor a series of surfaces, the first of which coincides with the first electrode, and the last with the second, while the intermediate surfaces are bounded by the non-conducting surface and do not intersect each other, and if we suppose each of these surfaces to be an infinitely thin sheet of perfectly conducting matter, we shall have obtained a system the resistance of which is certainly not greater than that of the original conductor, and is equal to it only when the surfaces we have chosen are the natural equipotential surfaces.

To calculate the resistance of the artificial system is an operation of much less difficulty than the original problem. For the resistance of the whole is the sum of the resistances of all the strata contained between the consecutive surfaces, and the resistance of each stratum can be found thus:

Let dS be an element of the surface of the stratum, ν the thickness of the stratum perpendicular to the element, ρ the specific resistance, E the difference of potential of the perfectly conducting surfaces, and dC the current through dS, then

$$dC = E\frac{1}{\rho\nu}dS, \tag{1}$$

and the whole current through the stratum is

$$C = E\iint\frac{1}{\rho\nu}dS, \tag{2}$$

the integration being extended over the whole stratum bounded by the non-conducting surface of the conductor.

Hence the conductivity of the stratum is

$$\frac{C}{E} = \iint\frac{1}{\rho\nu}dS, \tag{3}$$

and the resistance of the stratum is the reciprocal of this quantity.

If the stratum be that bounded by the two surfaces for which the function F has the values F and $F+dF$ respectively, then

$$\frac{dF}{\nu} = \nabla F = \left[\left(\frac{dF}{dx}\right)^2 + \left(\frac{dF}{dy}\right)^2 + \left(\frac{dF}{dz}\right)^2\right]^{\frac{1}{2}}, \tag{4}$$

and the resistance of the stratum is

$$\frac{dF}{\iint\frac{1}{\rho}\nabla F\,dS}. \tag{5}$$

To find the resistance of the whole artificial conductor, we have only to integrate with respect to F, and we find

$$R_1 = \int \frac{dF}{\iint \frac{1}{\rho} \nabla F dS}. \tag{6}$$

The resistance R of the conductor in its natural state is greater than the value thus obtained, unless all the surfaces we have chosen are the natural equipotential surfaces. Also, since the true value of R is the absolute maximum of the values of R_1 which can thus be obtained, a small deviation of the chosen surfaces from the true equipotential surfaces will produce an error of R which is comparatively small.

This method of determining a lower limit of the value of the resistance is evidently perfectly general, and may be applied to conductors of any form, even when ρ, the specific resistance, varies in any manner within the conductor.

The most familiar example is the ordinary method of determining the resistance of a straight wire of variable section. In this case the surfaces chosen are planes perpendicular to the axis of the wire, the strata have parallel faces, and the resistance of a stratum of section S and thickness ds is

$$dR_1 = \frac{\rho ds}{S}, \tag{7}$$

and that of the whole wire of length s is

$$R_1 = \int \frac{\rho ds}{S}, \tag{8}$$

where S is the transverse section and is a function of s.

This method in the case of wires whose section varies slowly with the length gives a result very near the truth, but it is really only a lower limit, for the true resistance is always greater than this, except in the case where the section is perfectly uniform.

307.] To find the higher limit of the resistance, let us suppose a surface drawn in the conductor to be rendered impermeable to electricity. The effect of this must be to increase the resistance of the conductor unless the surface is one of the natural surfaces of flow. By means of two systems of surfaces we can form a set of tubes which will completely regulate the flow, and the effect, if there is any, of this system of impermeable surfaces must be to increase the resistance above its natural value.

The resistance of each of the tubes may be calculated by the method already given for a fine wire, and the resistance of the whole conductor is the reciprocal of the sum of the reciprocals of the resistances of all the tubes. The resistance thus found is greater than the natural resistance, except when the tubes follow the natural lines of flow.

In the case already considered, where the conductor is in the form of an elongated solid of revolution, let us measure x along the axis, and let the radius of the section at any point be b. Let one set of impermeable surfaces be the planes through the axis for each of which ϕ is constant, and let the other set be surfaces of revolution for which

$$y^2 = \psi b^2, \tag{9}$$

where ψ is a numerical quantity between 0 and 1.

Let us consider a portion of one of the tubes bounded by the surfaces ϕ and $\phi + d\phi$, ψ and $\psi + d\psi$, x and $x + dx$.

The section of the tube taken perpendicular to the axis is

$$y\,dy\,d\phi = \tfrac{1}{2} b^2\, d\psi\, d\phi. \tag{10}$$

If θ be the angle which the tube makes with the axis

$$\tan\theta = \psi^{\frac{1}{2}} \frac{db}{dx}. \tag{11}$$

The true length of the element of the tube is $dx \sec\theta$, and its true section is

$$\tfrac{1}{2} b^2\, d\psi\, d\phi \cos\theta,$$

so that its resistance is

$$2\rho \frac{dx}{b^2\, d\psi\, d\phi} \sec^2\theta = 2\rho \frac{dx}{b^2\, d\psi\, d\phi}\left(1 + \psi \overline{\frac{db}{dx}}\bigg|^2\right). \tag{12}$$

Let
$$A = \int \frac{\rho}{b^2}\, dx, \quad \text{and} \quad B = \int \frac{\rho}{b^2}\left(\frac{db}{dx}\right)^2 dx, \tag{13}$$

the integration being extended over the whole length, x, of the conductor, then the resistance of the tube $d\psi\, d\phi$ is

$$\frac{2}{d\psi\, d\phi}(A + \psi B),$$

and its conductivity is

$$\frac{d\psi\, d\phi}{2(A + \psi B)}.$$

To find the conductivity of the whole conductor, which is the sum of the conductivities of the separate tubes, we must integrate this expression between $\phi = 0$ and $\phi = 2\pi$, and between $\psi = 0$ and $\psi = 1$. The result is

$$\frac{1}{R'} = \frac{\pi}{B} \log\left(1 + \frac{B}{A}\right), \tag{14}$$

which may be less, but cannot be greater, than the true conductivity of the conductor.

When $\frac{db}{dx}$ is always a small quantity $\frac{B}{A}$ will also be small, and we may expand the expression for the conductivity, thus

$$\frac{1}{R'} = \frac{\pi}{A}\left(1 - \tfrac{1}{2}\frac{B}{A} + \tfrac{1}{3}\frac{B^2}{A^2} - \tfrac{1}{4}\frac{B^3}{A^3} + \&c.\right). \tag{15}$$

The first term of this expression, $\frac{\pi}{A}$, is that which we should have found by the former method as the superior limit of the conductivity. Hence the true conductivity is less than the first term but greater than the whole series. The superior value of the resistance is the reciprocal of this, or

$$R' = \frac{A}{\pi}\left(1 + \tfrac{1}{2}\frac{B}{A} - \frac{1}{12}\frac{B^2}{A^2} + \frac{1}{24}\frac{B^3}{A^3} - \&c.\right). \tag{16}$$

If, besides supposing the flow to be guided by the surfaces ϕ and ψ, we had assumed that the flow through each tube is proportional to $d\psi\, d\phi$, we should have obtained as the value of the resistance under this additional constraint

$$R'' = \frac{1}{\pi}(A + \tfrac{1}{2}B), \tag{17}$$

which is evidently greater than the former value, as it ought to be, on account of the additional constraint. In Lord Rayleigh's paper this is the supposition made, and the superior limit of the resistance there given has the value (17), which is a little greater than that which we have obtained in (16).

308.] We shall now apply the same method to find the correction which must be applied to the length of a cylindrical conductor of radius a when its extremity is placed in metallic contact with a massive electrode, which we may suppose of a different metal.

For the lower limit of the resistance we shall suppose that an infinitely thin disk of perfectly conducting matter is placed between the end of the cylinder and the massive electrode, so as to bring the end of the cylinder to one and the same potential throughout. The potential within the cylinder will then be a function of its length only, and if we suppose the surface of the electrode where the cylinder meets it to be approximately plane, and all its dimensions to be large compared with the diameter of the cylinder, the distribution of potential will be that due to a conductor in the form of a disk placed in an infinite medium. See Arts. 151, 177.

If E is the difference of the potential of the disk from that of the distant parts of the electrode, C the current issuing from the surface of the disk into the electrode, and ρ' the specific resistance of the electrode; then if Q is the amount of electricity on the disk, which we assume distributed as in Art. 151, we have

$$\rho' C = \tfrac{1}{2} \cdot 4\pi Q = 2\pi \frac{aE}{\frac{\pi}{2}}, \text{ by Art. 151,}$$

$$= 4aE. \qquad (18)$$

Hence, if the length of the wire from a given point to the electrode is L, and its specific resistance ρ, the resistance from that point to any point of the electrode not near the junction is

$$R = \rho \frac{L}{\pi a^2} + \frac{\rho'}{4a},$$

and this may be written

$$R = \frac{\rho}{\pi a^2}\left(L + \frac{\rho'}{\rho}\frac{\pi a}{4}\right), \qquad (19)$$

where the second term within brackets is a quantity which must be added to the length of the cylinder or wire in calculating its resistance, and this is certainly too small a correction.

To understand the nature of the outstanding error we may observe, that whereas we have supposed the flow in the wire up to the disk to be uniform throughout the section, the flow from the disk to the electrode is not uniform, but is at any point inversely proportional to the minimum chord through that point. In the actual case the flow through the disk will not be uniform, but it will not vary so much from point to point as in this supposed case. The potential of the disk in the actual case will not be uniform, but will diminish from the middle to the edge.

309.] We shall next determine a quantity greater than the true resistance by constraining the flow through the disk to be uniform at every point. We may suppose electromotive forces introduced for this purpose acting perpendicular to the surface of the disk.

The resistance within the wire will be the same as before, but in the electrode the rate of generation of heat will be the surface-integral of the product of the flow into the potential. The rate of flow at any point is $\frac{C}{\pi a^2}$, and the potential is the same as that of an electrified surface whose surface-density is σ, where

$$2\pi\sigma = \frac{C\rho'}{\pi a^2}, \qquad (20)$$

ρ' being the specific resistance.

We have therefore to determine the potential energy of the electrification of the disk with the uniform surface-density σ.

* The potential at the edge of a disk of uniform density σ is easily found to be $4a\sigma$. The work done in adding a strip of breadth da at the circumference of the disk is $2\pi a\sigma da \,.\, 4a\sigma$, and the whole potential energy of the disk is the integral of this,

$$\text{or} \qquad P = \frac{8\pi}{3}a^3\sigma^2. \qquad (21)$$

In the case of electrical conduction the rate at which work is done in the electrode whose resistance is R' is C^2R'. But from the general equation of conduction the current across the disk per unit area is of the form

$$-\frac{1}{\rho'}\frac{dV}{d\nu}$$

$$\text{or} \qquad \frac{4\pi}{\rho'}\sigma.$$

Hence the rate at which work is done is

$$\frac{4\pi}{\rho'}P.$$

We have therefore

$$C^2R' = \frac{4\pi}{\rho'}P, \qquad (22)$$

whence, by (20) and (21),

$$R' = \frac{8\rho'}{3\pi^2 a},$$

and the correction to be added to the length of the cylinder is

$$\frac{\rho'}{\rho}\frac{8}{3\pi}a,$$

this correction being greater than the true value. The true correction to be added to the length is therefore $\frac{\rho'}{\rho}an$, where n is a number lying between $\frac{\pi}{4}$ and $\frac{8}{3\pi}$, or between 0.785 and 0.849.

† Lord Rayleigh, by a second approximation, has reduced the superior limit of n to 0.8282.

* See a Paper by Professor Cayley, *London Math. Soc. Proc.* vi. p. 47.

† *Phil. Mag.*, Nov. 1872. Lord Rayleigh subsequently obtained ·8242 as the superior limit. See *London Math. Soc. Proc.* viii. p. 74.

CHAPTER IX.

CONDUCTION THROUGH HETEROGENEOUS MEDIA.

On the Conditions to be Fulfilled at the Surface of Separation between Two Conducting Media.

310.] THERE are two conditions which the distribution of currents must fulfil in general, the condition that the potential must be continuous, and the condition of 'continuity' of the electric currents.

At the surface of separation between two media the first of these conditions requires that the potentials at two points on opposite sides of the surface, but infinitely near each other, shall be equal. The potentials are here understood to be measured by an electrometer put in connexion with the given point by means of an electrode of a given metal. If the potentials are measured by the method described in Arts. 222, 246, where the electrode terminates in a cavity of the conductor filled with air, then the potentials at contiguous points of different metals measured in this way will differ by a quantity depending on the temperature and on the nature of the two metals.

The other condition at the surface is that the current through any element of the surface is the same when measured in either medium.

Thus, if V_1 and V_2 are the potentials in the two media, then at any point in the surface of separation

$$V_1 = V_2, \tag{1}$$

and if u_1, v_1, w_1 and u_2, v_2, w_2 are the components of currents in the two media, and l, m, n the direction-cosines of the normal to the surface of separation,

$$u_1 l + v_1 m + w_1 n = u_2 l + v_2 m + w_2 n. \tag{2}$$

In the most general case the components u, v, w are linear functions of the derivatives of V, the forms of which are given in the equations

$$\left.\begin{aligned} u &= r_1 X + p_3 Y + q_2 Z, \\ v &= q_3 X + r_2 Y + p_1 Z, \\ w &= p_2 X + q_1 Y + r_3 Z, \end{aligned}\right\} \tag{3}$$

where X, Y, Z are the derivatives of V with respect to x, y, z respectively.

Let us take the case of the surface which separates a medium having these coefficients of conduction from an isotropic medium having a coefficient of conduction equal to r.

Let X', Y', Z' be the values of X, Y, Z in the isotropic medium, then we have at the surface

$$V = V', \tag{4}$$

or $$X\,dx + Y\,dy + Z\,dz = X'dx + Y'dy + Z'dz, \tag{5}$$

when $$l\,dx + m\,dy + n\,dz = 0. \tag{6}$$

This condition gives

$$X' = X + 4\pi\sigma l, \quad Y' = Y + 4\pi\sigma m, \quad Z' = Z + 4\pi\sigma n, \tag{7}$$

where σ is the surface-density.

We have also in the isotropic medium

$$u' = rX', \quad v' = rY', \quad w' = rZ', \tag{8}$$

and at the boundary the condition of flow is

$$u'l + v'm + w'n = ul + vm + wn, \tag{9}$$

or $r(lX + mY + nZ + 4\pi\sigma)$

$$= l(r_1 X + p_3 Y + q_2 Z) + m(q_3 X + r_2 Y + p_1 Z) + n(p_2 X + q_1 Y + r_3 Z), \tag{10}$$

whence

$$4\pi\sigma r = (l(r_1 - r) + mq_3 + np_2)X + (lp_3 + m(r_2 - r) + nq_1)Y + (lq_2 + mp_1 + n(r_3 - r))Z. \tag{11}$$

The quantity σ represents the surface-density of the charge on the surface of separation. In crystallized and organized substances it depends on the direction of the surface as well as on the force perpendicular to it. In isotropic substances the coefficients p and q are zero, and the coefficients r are all equal, so that

$$4\pi\sigma = \left(\frac{r_1}{r} - 1\right)(lX + mY + nZ), \tag{12}$$

where r_1 is the conductivity of the substance, r that of the external medium, and l, m, n the direction-cosines of the normal drawn towards the medium whose conductivity is r.

When both media are isotropic the conditions may be greatly

simplified, for if k is the specific resistance per unit of volume, then

$$u = -\frac{1}{k}\frac{dV}{dx}, \qquad v = -\frac{1}{k}\frac{dV}{dy}, \qquad w = -\frac{1}{k}\frac{dV}{dz}, \tag{13}$$

and if ν is the normal drawn at any point of the surface of separation from the first medium towards the second, the condition of continuity is

$$\frac{1}{k_1}\frac{dV_1}{d\nu} = \frac{1}{k_2}\frac{dV_2}{d\nu}. \tag{14}$$

If θ_1 and θ_2 are the angles which the lines of flow in the first and second media respectively make with the normal to the surface of separation, then the tangents to these lines of flow are in the same plane with the normal and on opposite sides of it, and

$$k_1 \tan\theta_1 = k_2 \tan\theta_2. \tag{15}$$

This may be called the law of refraction of lines of flow.

311.] As an example of the conditions which must be fulfilled when electricity crosses the surface of separation of two media, let us suppose the surface spherical and of radius a, the specific resistance being k_1 within and k_2 without the surface.

Let the potential, both within and without the surface, be expanded in solid harmonics, and let the part which depends on the surface harmonic S_i be

$$V_1 = (A_1 r^i + B_1 r^{-(i+1)})\, S_i, \tag{1}$$

$$V_2 = (A_2 r^i + B_2 r^{-(i+1)})\, S_i \tag{2}$$

within and without the sphere respectively.

At the surface of separation where $r = a$ we must have

$$V_1 = V_2 \quad \text{and} \quad \frac{1}{k_1}\frac{dV_1}{dr} = \frac{1}{k_2}\frac{dV_2}{dr}. \tag{3}$$

From these conditions we get the equations

$$\left.\begin{aligned} (A_1 - A_2)a^{2i+1} + B_1 - B_2 = 0, \\ \left(\frac{1}{k_1}A_1 - \frac{1}{k_2}A_2\right) i a^{2i+1} - \left(\frac{1}{k_1}B_1 - \frac{1}{k_2}B_2\right)(i+1) = 0. \end{aligned}\right\} \tag{4}$$

These equations are sufficient, when we know two of the four quantities A_1, A_2, B_1, B_2, to deduce the other two.

Let us suppose A_1 and B_1 known, then we find the following expressions for A_2 and B_2,

$$\left.\begin{aligned} A_2 &= \frac{(k_1(i+1) + k_2 i)\, A_1 + (k_1 - k_2)(i+1)\, B_1 a^{-(2i+1)}}{k_1(2i+1)}, \\ B_2 &= \frac{(k_1 - k_2)\, i A_1 a^{2i+1} + (k_1 i + k_2(i+1))\, B_1}{k_1(2i+1)}. \end{aligned}\right\} \tag{5}$$

In this way we can find the conditions which each term of the harmonic expansion of the potential must satisfy for any number of strata bounded by concentric spherical surfaces.

312.] Let us suppose the radius of the first spherical surface to be a_1, and let there be a second spherical surface of radius a_2 greater than a_1, beyond which the specific resistance is k_3. If there are no sources or sinks of electricity within these spheres there will be no infinite values of V, and we shall have $B_1 = 0$.

We then find for A_3 and B_3, the coefficients for the outer medium,

$$\left.\begin{aligned} A_3 k_1 k_2 (2i+1)^2 &= \Big[\{k_1(i+1)+k_2 i\}\{k_2(i+1)+k_3 i\} \\ &\qquad + i(i+1)(k_1-k_2)(k_2-k_3)\left(\frac{a_1}{a_2}\right)^{2i+1}\Big] A_1, \\ B_3 k_1 k_2 (2i+1)^2 &= \Big[i\{k_1(i+1)+k_2 i\}(k_2-k_3)\, a_2^{2i+1} \\ &\qquad + i(k_1-k_2)\{k_2 i + k_3(i+1)\}\, a_1^{2i+1}\Big] A_1. \end{aligned}\right\} \quad (6)$$

The value of the potential in the outer medium depends partly on the external sources of electricity, which produce currents independently of the existence of the sphere of heterogeneous matter within, and partly on the disturbance caused by the introduction of the heterogeneous sphere.

The first part must depend on solid harmonics of positive degrees only, because it cannot have infinite values within the sphere. The second part must depend on harmonics of negative degrees, because it must vanish at an infinite distance from the centre of the sphere.

Hence the potential due to the external electromotive forces must be expanded in a series of solid harmonics of positive degree. Let A_3 be the coefficient of one of these, of the form

$$A_3 S_i r^i.$$

Then we can find A_1, the corresponding coefficient for the inner sphere by equation (6), and from this deduce A_2, B_2, and B_3. Of these B_3 represents the effect on the potential in the outer medium due to the introduction of the heterogeneous spheres.

Let us now suppose $k_3 = k_1$, so that the case is that of a hollow shell for which $k = k_2$, separating an inner from an outer portion of the same medium for which $k = k_1$.

If we put

$$C = \frac{1}{(2i+1)^2 k_1 k_2 + i(i+1)(k_2-k_1)^2\left(1-\left(\frac{a_1}{a_2}\right)^{2i+1}\right)},$$

then
$$\left.\begin{aligned} A_1 &= k_1 k_2 (2i+1)^2 C A_3, \\ A_2 &= k_2 (2i+1)\big(k_1(i+1)+k_2 i\big) C A_3, \\ B_2 &= k_2 i (2i+1)(k_1-k_2)\, a_1^{2i+1} C A_3, \\ B_3 &= i (k_2-k_1)\big(k_1(i+1)+k_2 i\big)\big(a_2^{2i+1}-a_1^{2i+1}\big) C A_3. \end{aligned}\right\} \quad (7)$$

The difference between A_3 the undisturbed coefficient, and A_1 its value in the hollow within the spherical shell, is

$$A_3 - A_1 = (k_2-k_1)^2 i(i+1)\left(1-\left(\frac{a_1}{a_2}\right)^{2i+1}\right) C A_3. \quad (8)$$

Since this quantity is always positive whatever be the values of k_1 and k_2, it follows that, whether the spherical shell conducts better or worse than the rest of the medium, the electrical action in the space occupied by the shell is less than it would otherwise be. If the shell is a better conductor than the rest of the medium it tends to equalize the potential all round the inner sphere. If it is a worse conductor, it tends to prevent the electrical currents from reaching the inner sphere at all.

The case of a solid sphere may be deduced from this by making $a_1 = 0$, or it may be worked out independently.

313.] The most important term in the harmonic expansion is that in which $i = 1$, for which

$$\left.\begin{aligned} C &= \frac{1}{9k_1k_2 + 2(k_1-k_2)^2\left(1-\left(\frac{a_1}{a_2}\right)^3\right)}, \\ A_1 &= 9k_1k_2 C A_3, \qquad A_2 = 3k_2(2k_1+k_2) C A_3, \\ B_2 &= 3k_2(k_1-k_2)a_1^3 C A_3, \quad B_3 = (k_2-k_1)(2k_1+k_2)(a_2^3-a_1^3) C A_3. \end{aligned}\right\} \quad (9)$$

The case of a solid sphere of resistance k_2 may be deduced from this by making $a_1 = 0$. We then have

$$\left.\begin{aligned} A_2 &= \frac{3k_2}{k_1+2k_2} A_3, \quad B_2 = 0, \\ B_3 &= \frac{k_2-k_1}{k_1+2k_2} a_2^3 A_3. \end{aligned}\right\} \quad (10)$$

It is easy to shew from the general expressions that the value of B_3 in the case of a hollow sphere having a nucleus of resistance k_1, surrounded by a shell of resistance k_2, is the same as that of a uniform solid sphere of the radius of the outer surface, and of resistance K, where

$$K = \frac{(2k_1+k_2)a_2^3 + (k_1-k_2)a_1^3}{(2k_1+k_2)a_2^3 - 2(k_1-k_2)a_1^3} k_2. \quad (11)$$

314.] If there are n spheres of radius a_1 and resistance k_1, placed in a medium whose resistance is k_2, at such distances from each other that their effects in disturbing the course of the current may be taken as independent of each other, then if these spheres are all contained within a sphere of radius a_2, the potential at a great distance from the centre of this sphere will be of the form

$$V = \left(Ar + nB\frac{1}{r^2}\right)\cos\theta, \tag{12}$$

where the value of B is

$$B = \frac{k_1 - k_2}{2k_1 + k_2} a_1^3 A. \tag{13}$$

The ratio of the volume of the n small spheres to that of the sphere which contains them is

$$p = \frac{n a_1^3}{a_2^3}. \tag{14}$$

The value of the potential at a great distance from the sphere may therefore be written

$$V = A\left(r + p a_2^3 \frac{k_1 - k_2}{2k_1 + k_2}\frac{1}{r^2}\right)\cos\theta. \tag{15}$$

Now if the whole sphere of radius a_2 had been made of a material of specific resistance K, we should have had

$$V = A\left\{r + a_2^3 \frac{K - k_2}{2K + k_2}\frac{1}{r^2}\right\}\cos\theta. \tag{16}$$

That the one expression should be equivalent to the other,

$$K = \frac{2k_1 + k_2 + p(k_1 - k_2)}{2k_1 + k_2 - 2p(k_1 - k_2)} k_2. \tag{17}$$

This, therefore, is the specific resistance of a compound medium consisting of a substance of specific resistance k_2, in which are disseminated small spheres of specific resistance k_1, the ratio of the volume of all the small spheres to that of the whole being p. In order that the action of these spheres may not produce effects depending on their interference, their radii must be small compared with their distances, and therefore p must be a small fraction.

This result may be obtained in other ways, but that here given involves only the repetition of the result already obtained for a single sphere.

When the distance between the spheres is not great compared with their radii, and when $\frac{k_1 - k_2}{2k_1 + k_2}$ is considerable, then other terms enter into the result, which we shall not now consider. In consequence of these terms certain systems of arrangement of the spheres cause the resistance of the compound medium to be different in different directions.

Application of the Principle of Images.

315.] Let us take as an example the case of two media separated by a plane surface, and let us suppose that there is a source S of electricity at a distance a from the plane surface in the first medium, the quantity of electricity flowing from the source in unit of time being S.

If the first medium had been infinitely extended the current at any point P would have been in the direction SP, and the potential at P would have been $\frac{E}{r_1}$ where $E = \frac{Sk_1}{4\pi}$ and $r_1 = SP$.

In the actual case the conditions may be satisfied by taking a point I, the image of S in the second medium, such that IS is normal to the plane of separation and is bisected by it. Let r_2 be the distance of any point from I, then at the surface of separation

$$r_1 = r_2, \tag{1}$$

$$\frac{dr_1}{d\nu} = -\frac{dr_2}{d\nu}. \tag{2}$$

Let the potential V_1 at any point in the first medium be that due to a quantity of electricity E placed at S, together with an imaginary quantity E_2 at I, and let the potential V_2 at any point of the second medium be that due to an imaginary quantity E_1 at S, then if

$$V_1 = \frac{E}{r_1} + \frac{E_2}{r_2} \quad \text{and} \quad V_2 = \frac{E_1}{r_1}, \tag{3}$$

the superficial condition $V_1 = V_2$ gives

$$E + E_2 = E_1, \tag{4}$$

and the condition

$$\frac{1}{k_1}\frac{dV_1}{d\nu} = \frac{1}{k_2}\frac{dV_2}{d\nu} \tag{5}$$

gives

$$\frac{1}{k_1}(E - E_2) = \frac{1}{k_2}E_1, \tag{6}$$

whence

$$E_1 = \frac{2k_2}{k_1+k_2}E, \qquad E_2 = \frac{k_2-k_1}{k_1+k_2}E. \tag{7}$$

The potential in the first medium is therefore the same as would be produced in air by a charge E placed at S, and a charge E_2 at I on the electrostatic theory, and the potential in the second medium is the same as that which would be produced in air by a charge E_1 at S.

The current at any point of the first medium is the same as would have been produced by the source S together with a source $\frac{k_2-k_1}{k_1+k_2}S$ placed at I if the first medium had been infinite, and the current at any point of the second medium is the same as would have been produced by a source $\frac{2k_2S}{(k_1+k_2)}$ placed at S if the second medium had been infinite.

We have thus a complete theory of electrical images in the case of two media separated by a plane boundary. Whatever be the nature of the electromotive forces in the first medium, the potential they produce in the first medium may be found by combining their direct effect with the effect of their image.

If we suppose the second medium a perfect conductor, then $k_2 = 0$, and the image at I is equal and opposite to the source at S. This is the case of electric images, as in Thomson's theory in electrostatics.

If we suppose the second medium a perfect insulator, then $k_2 = \infty$, and the image at I is equal to the source at S and of the same sign. This is the case of images in hydrokinetics when the fluid is bounded by a rigid plane surface.

316.] The method of inversion, which is of so much use in electrostatics when the bounding surface is supposed to be that of a perfect conductor, is not applicable to the more general case of the surface separating two conductors of unequal electric resistance. The method of inversion in two dimensions is, however, applicable, as well as the more general method of transformation in two dimensions given in Art. 190 *.

Conduction through a Plate separating Two Media.

317.] Let us next consider the effect of a plate of thickness AB of a medium whose resistance is k_2, and separating two media whose resistances are k_1 and k_3, in altering the potential due to a source S in the first medium.

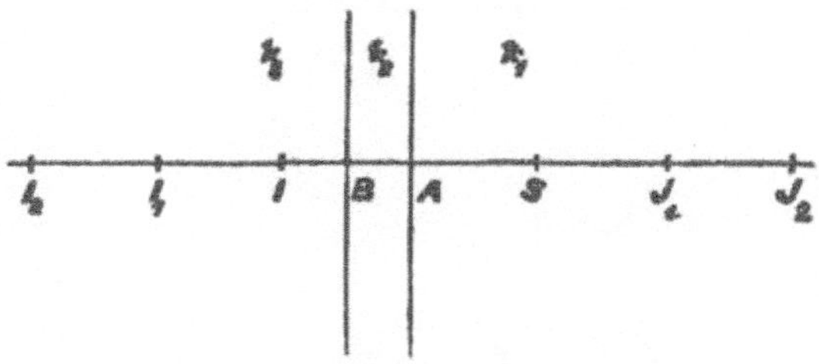

Fig. 24.

The potential will be

* See Kirchhoff, Pogg. *Ann.* lxiv. 497, and lxvii. 344; Quincke, Pogg. xcvii. 382; and Smith, *Proc. R. S. Edin.*, 1869-70, p. 79.

equal to that due to a system of charges placed in air at certain points along the normal to the plate through S.

Make

$$AI=SA, \quad BI_1=SB, \quad AJ_1=I_1A, \quad BI_2=J_1B, \quad AJ_2=I_2A, \text{ \&c.};$$

then we have two series of points at distances from each other equal to twice the thickness of the plate.

318.] The potential in the first medium at any point P is equal to

$$\frac{E}{PS}+\frac{I}{PI}+\frac{I_1}{PI_1}+\frac{I_2}{PI_2}+\text{\&c.}, \tag{8}$$

that at a point P' in the second

$$\frac{E'}{P'S}+\frac{I'}{P'I}+\frac{I_1'}{P'I_1}+\frac{I_2'}{P'I_2}+\text{\&c.}$$
$$+\frac{J_1'}{P'J_1}+\frac{J_2'}{P'J_2}+\text{\&c.}, \tag{9}$$

and that at a point P'' in the third

$$\frac{E''}{P''S}+\frac{J_1}{P''J_1}+\frac{J_2}{P''J_2}+\text{\&c.}, \tag{10}$$

where I, I', &c. represent the imaginary charges placed at the points I, &c., and the accents denote that the potential is to be taken within the plate.

Then, by the last Article, for the surface through A we have,

$$I=\frac{k_2-k_1}{k_2+k_1}E, \qquad E'=\frac{2k_2}{k_2+k_1}E. \tag{11}$$

For the surface through B we find

$$I_1'=\frac{k_3-k_2}{k_3+k_2}E', \quad E''=\frac{2k_3}{k_2+k_1}E'. \tag{12}$$

Similarly for the surface through A again,

$$J_1'=\frac{k_1-k_2}{k_1+k_2}I_1', \quad I_1=\frac{2k_1}{k_1+k_2}I_1', \tag{13}$$

and for the surface through B,

$$I_2'=\frac{k_3-k_2}{k_3+k_2}J_1', \quad J_1=\frac{2k_3}{k_3+k_2}J_1'. \tag{14}$$

If we make $\rho=\dfrac{k_1-k_2}{k_1+k_2}$ and $\rho'=\dfrac{k_3-k_2}{k_3+k_2}$,

we find for the potential in the first medium,

$$V=\frac{E}{PS}-\rho\frac{E}{PI}+(1-\rho^2)\rho'\frac{E}{PI_1}+\rho'(1-\rho^2)\rho\rho'\frac{E}{PI_2}+\text{\&c.}$$
$$+\rho'(1-\rho^2)(\rho\rho')^{n-1}\frac{E}{PI_n}. \tag{15}$$

For the potential in the third medium we find

$$V = (1+\rho')(1-\rho)E\left\{\frac{1}{PS} + \frac{\rho\rho'}{PJ_1} + \&c. + \frac{(\rho\rho')^n}{PJ_n}\right\}. \qquad (16)$$

If the first medium is the same as the third, then $k_1 = k_3$ and $\rho = \rho'$, and the potential on the other side of the plate will be

$$V = (1-\rho^2)E\left\{\frac{1}{PS} + \frac{\rho^2}{PJ_1} + \&c. + \frac{\rho^{2n}}{PJ_n}\right\}. \qquad (17)$$

If the plate is a very much better conductor than the rest of the medium, ρ is very nearly equal to 1. If the plate is a nearly perfect insulator, ρ is nearly equal to -1, and if the plate differs little in conducting power from the rest of the medium, ρ is a small quantity positive or negative.

The theory of this case was first stated by Green in his 'Theory of Magnetic Induction' (*Essay*, p. 65). His result, however, is correct only when ρ is nearly equal to 1 *. The quantity g which he uses is connected with ρ by the equations

$$g = \frac{2\rho}{3-\rho} = \frac{k_1-k_2}{k_1+2k_2}, \qquad \rho = \frac{3g}{2+g} = \frac{k_1-k_2}{k_1+k_2}.$$

If we put $\rho = \dfrac{2\pi\kappa}{1+2\pi\kappa}$, we shall have a solution of the problem of the magnetic induction excited by a magnetic pole in an infinite plate whose coefficient of magnetization is κ.

On Stratified Conductors.

319.] Let a conductor be composed of alternate strata of thickness c and c' of two substances whose coefficients of conductivity are different. Required the coefficients of resistance and conductivity of the compound conductor.

Let the plane of the strata be normal to Z. Let every symbol relating to the strata of the second kind be accented, and let every symbol relating to the compound conductor be marked with a bar thus, $\overline{X}$. Then

$$\overline{X} = X = X', \qquad (c+c')\bar{u} = cu + c'u',$$
$$\overline{Y} = Y = Y', \qquad (c+c')\bar{v} = cv + c'v';$$
$$(c+c')\overline{Z} = cZ + c'Z', \qquad \bar{w} = w = w'.$$

We must first determine u, u', v, v', Z and Z' in terms of $\overline{X}$, $\overline{Y}$ and $\bar{w}$ from the equations of resistance, Art. 297, or those

* See Sir W. Thomson's 'Note on Induced Magnetism in a Plate,' *Camb. and Dub. Math. Journ.*, Nov. 1845, or *Reprint*, art. ix. § 156.

of conductivity, Art. 298. If we put D for the determinant of the coefficients of resistance, we find

$$u r_3 D = R_2 \overline{X} - Q_3 \overline{Y} + \overline{w} q_2 D,$$
$$v r_3 D = R_1 \overline{Y} - P_3 \overline{X} + \overline{w} p_1 D,$$
$$Z r_3 \quad = -p_2 \overline{X} - q_1 \overline{Y} + \overline{w}.$$

Similar equations with the symbols accented give the values of u', v' and Z'. Having found $\overline{u}$, $\overline{v}$ and $\overline{w}$ in terms of $\overline{X}$, $\overline{Y}$ and $\overline{Z}$, we may write down the equations of conductivity of the stratified conductor. If we make $h = \frac{c}{r_3}$ and $h' = \frac{c'}{r_3'}$, we find

$$\overline{p}_1 = \frac{h p_1 + h' p_1'}{h + h'}, \qquad \overline{q}_1 = \frac{h q_1 + h' q_1'}{h + h'},$$

$$\overline{p}_2 = \frac{h p_2 + h' p_2'}{h + h'}, \qquad \overline{q}_2 = \frac{h q_2 + h' q_2'}{h + h'},$$

$$\overline{p}_3 = \frac{c p_3 + c' p_3'}{c + c'} - \frac{h h' (q_1 - q_1')(q_2 - q_2')}{(h + h')(c + c')},$$

$$\overline{q}_3 = \frac{c q_3 + c' q_3'}{c + c'} - \frac{h h' (p_1 - p_1')(p_2 - p_2')}{(h + h')(c + c')},$$

$$\overline{r}_1 = \frac{c r_1 + c' r_1'}{c + c'} - \frac{h h' (p_2 - p_2')(q_2 - q_2')}{(h + h')(c + c')},$$

$$\overline{r}_2 = \frac{c r_2 + c' r_2'}{c + c'} - \frac{h h' (p_1 - p_1')(q_1 - q_1')}{(h + h')(c + c')},$$

$$\overline{r}_3 = \frac{c + c'}{h + h'}.$$

320.] If neither of the two substances of which the strata are formed has the rotatory property of Art. 303, the value of any P or p will be equal to that of its corresponding Q or q. From this it follows that in the stratified conductor also

$$\overline{p}_1 = \overline{q}_1, \qquad \overline{p}_2 = \overline{q}_2, \qquad \overline{p}_3 = \overline{q}_3,$$

or there is no rotatory property developed by stratification, unless it exists in the materials.

321.] If we now suppose that there is no rotatory property, and also that the axes of x, y and z are the principal axes, then the p and q coefficients vanish, and

$$\overline{r}_1 = \frac{c r_1 + c r_1'}{c + c'}, \qquad \overline{r}_2 = \frac{c r_2 + c' r_2'}{c + c'}, \qquad \overline{r}_3 = \frac{c + c'}{\frac{c}{r_3} + \frac{c'}{r_3'}}.$$

If we begin with both substances isotropic, but of different conductivities, then the result of stratification will be to make the resistance greatest in the direction of a normal to the strata, and the resistance in all directions in the plane of the strata will be equal.

822.] Take an isotropic substance of conductivity r, cut it into exceedingly thin slices of thickness a, and place them alternately with slices of a substance whose conductivity is s, and thickness $k_1 a$.

Let these slices be normal to x. Then cut this compound conductor into thicker slices, of thickness b, normal to y, and alternate these with slices whose conductivity is s and thickness $k_2 b$.

Lastly, cut the new conductor into still thicker slices, of thickness c, normal to z, and alternate them with slices whose conductivity is s and thickness $k_3 c$.

The result of the three operations will be to cut the substance whose conductivity is r into rectangular parallelepipeds whose dimensions are a, b and c, where b is exceedingly small compared with c, and a is exceedingly small compared with b, and to embed these parallelepipeds in the substance whose conductivity is s, so that they are separated from each other $k_1 a$ in the direction of x, $k_2 b$ in that of y, and $k_3 c$ in that of z. The conductivities of the conductor so formed in the directions of x, y, and z are to be found by three applications in order of the results of Art. 321. We thereby obtain

$$r_1 = \frac{\{1+k_1(1+k_2)(1+k_3)\}r+(k_2+k_3+k_2k_3)s}{(1+k_2)(1+k_3)(k_1 r+s)}s,$$

$$r_2 = \frac{(1+k_2+k_2k_3)r+(k_1+k_3+k_1k_2+k_1k_3+k_1k_2k_3)s}{(1+k_3)\{k_2 r+(1+k_1+k_1k_2)s\}}s,$$

$$r_3 = \frac{(1+k_3)(r+(k_1+k_2+k_1k_2)s)}{k_3 r+(1+k_1+k_2+k_2k_3+k_3k_1+k_1k_2+k_1k_2k_3)s}s.$$

The accuracy of this investigation depends upon the three dimensions of the parallelepipeds being of different orders of magnitude, so that we may neglect the conditions to be fulfilled at their edges and angles. If we make k_1, k_2 and k_3 each unity, then

$$r_1 = \frac{5r+3s}{4r+4s}s, \qquad r_2 = \frac{3r+5s}{2r+6s}s, \qquad r_3 = \frac{2r+6s}{r+7s}s.$$

If $r = 0$, that is, if the medium of which the parallelepipeds are made is a perfect insulator, then

$$r_1 = \tfrac{3}{4}s, \qquad r_2 = \tfrac{5}{6}s, \qquad r_3 = \tfrac{6}{7}s.$$

If $r = \infty$, that is, if the parallelepipeds are perfect conductors,

$$r_1 = \tfrac{5}{4}s, \qquad r_2 = \tfrac{3}{2}s, \qquad r_3 = 2s.$$

In every case, provided $k_1 = k_2 = k_3$, it may be shewn that r_1, r_2 and r_3 are in ascending order of magnitude, so that the greatest conductivity is in the direction of the longest dimensions of the parallelepipeds, and the greatest resistance in the direction of their shortest dimensions.

323.] In a rectangular parallelepiped of a conducting solid, let there be a conducting channel made from one angle to the opposite, the channel being a wire covered with insulating material, and let the lateral dimensions of the channel be so small that the conductivity of the solid is not affected except on account of the current conveyed along the wire.

Let the dimensions of the parallelepiped in the directions of the coordinate axes be a, b, c, and let the conductivity of the channel, extending from the origin to the point (abc), be $abcK$.

The electromotive force acting between the extremities of the channel is

$$aX+bY+cZ,$$

and if C' be the current along the channel

$$C' = Kabc(aX+bY+cZ).$$

The current across the face bc of the parallelepiped is bcu, and this is made up of that due to the conductivity of the solid and of that due to the conductivity of the channel, or

$$bcu = bc(r_1 X+p_3 Y+q_2 Z)+Kabc(aX+bY+cZ),$$

or
$$u = (r_1+Ka^2)X+(p_3+Kab)Y+(q_2+Kca)Z.$$

In the same way we may find the values of v and w. The coefficients of conductivity as altered by the effect of the channel will be

$$\begin{array}{lll} r_1+Ka^2, & r_2+Kb^2, & r_3+Kc^2, \\ p_1+Kbc, & p_2+Kca, & p_3+Kab, \\ q_1+Kbc, & q_2+Kca, & q_3+Kab. \end{array}$$

In these expressions, the additions to the values of p_1, &c., due to the effect of the channel, are equal to the additions to the values of q_1, &c. Hence the values of p_1 and q_1 cannot be rendered unequal by the introduction of linear channels into every element of volume of the solid, and therefore the rotatory property of Art. 303, if it does not exist previously in a solid, cannot be introduced by such means.

324.] *To construct a framework of linear conductors which shall have any given coefficients of conductivity forming a symmetrical system.*

Let the space be divided into equal small cubes, of which let the figure represent one. Let the coordinates of the points O, L, M, N, and their potentials be as follows :—

	x	y	z	*Potential*
O	0	0	0	$X+Y+Z$
L	0	1	1	X
M	1	0	1	Y
N	1	1	0	Z

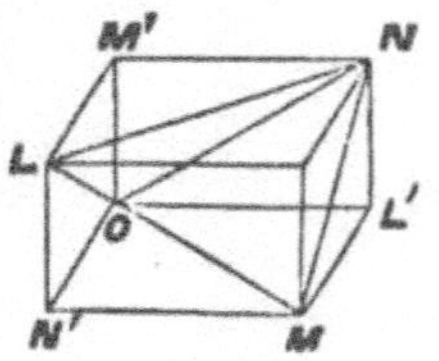

Fig. 25.

Let these four points be connected by six conductors,

$$OL, \quad OM, \quad ON, \quad MN, \quad NL, \quad LM,$$

of which the conductivities are respectively

$$A, \quad B, \quad C, \quad P, \quad Q, \quad R.$$

The electromotive forces along these conductors will be

$$Y+Z, \quad Z+X, \quad X+Y, \quad Y-Z, \quad Z-X, \quad X-Y,$$

and the currents

$$A(Y+Z), \; B(Z+X), \; C(X+Y), \; P(Y-Z), \; Q(Z-X), \; R(X-Y).$$

Of these currents, those which convey electricity in the positive direction of x are those along LM, LN, OM and ON, and the quantity conveyed is

$$u = (B+C+Q+R)X+(C-R)Y \qquad +(B-Q)Z.$$

Similarly

$$v = (C-R)X \qquad +(C+A+R+P)Y+(A-P)Z;$$

$$w = (B-Q)X \qquad +(A-P)Y \qquad +(A+B+P+Q)Z;$$

whence we find by comparison with the equations of conduction, Art. 298,

$$4A = r_2+r_3-r_1+2p_1, \qquad 4P = r_2+r_3-r_1-2p_1,$$

$$4B = r_3+r_1-r_2+2p_2, \qquad 4Q = r_3+r_1-r_2-2p_2,$$

$$4C = r_1+r_2-r_3+2p_3, \qquad 4R = r_1+r_2-r_3-2p_3.$$

CHAPTER X.

CONDUCTION IN DIELECTRICS.

325.] We have seen that when electromotive force acts on a dielectric medium it produces in it a state which we have called electric polarization, and which we have described as consisting of electric displacement within the medium in a direction which, in isotropic media, coincides with that of the electromotive force, combined with a superficial charge on every element of volume into which we may suppose the dielectric divided, which is negative on the side towards which the force acts, and positive on the side from which it acts.

When electromotive force acts on a conducting medium it also produces what is called an electric current.

Now dielectric media, with very few, if any, exceptions, are also more or less imperfect conductors, and many media which are not good insulators exhibit phenomena of dielectric induction. Hence we are led to study the state of a medium in which induction and conduction are going on at the same time.

For simplicity we shall suppose the medium isotropic at every point, but not necessarily homogeneous at different points. In this case, the equation of Poisson becomes, by Art. 83,

$$\frac{d}{dx}\left(K\frac{dV}{dx}\right)+\frac{d}{dy}\left(K\frac{dV}{dy}\right)+\frac{d}{dz}\left(K\frac{dV}{dz}\right)+4\pi\rho=0, \qquad (1)$$

where K is the 'specific inductive capacity.'

The 'equation of continuity' of electric currents becomes

$$\frac{d}{dx}\left(\frac{1}{r}\frac{dV}{dx}\right)+\frac{d}{dy}\left(\frac{1}{r}\frac{dV}{dy}\right)+\frac{d}{dz}\left(\frac{1}{r}\frac{dV}{dz}\right)-\frac{d\rho}{dt}=0, \qquad (2)$$

where r is the specific resistance referred to unit of volume.

When K or r is discontinuous, these equations must be transformed into those appropriate to surfaces of discontinuity.

In a strictly homogeneous medium r and K are both constant, so that we find

$$\frac{d^2V}{dx^2}+\frac{d^2V}{dy^2}+\frac{d^2V}{dz^2}=-4\pi\frac{\rho}{K}=r\frac{d\rho}{dt}, \tag{3}$$

whence
$$\rho = Ce^{-\frac{4\pi}{Kr}t}; \tag{4}$$

or, if we put
$$T=\frac{Kr}{4\pi}, \qquad \rho = Ce^{-\frac{t}{T}}. \tag{5}$$

This result shews that under the action of any external electric forces on a homogeneous medium, the interior of which is originally charged in any manner with electricity, the internal charges will die away at a rate which does not depend on the external forces, so that at length there will be no charge of electricity within the medium, after which no external forces can either produce or maintain a charge in any internal portion of the medium, provided the relation between electromotive force, electric polarization and conduction remains the same. When disruptive discharge occurs these relations cease to be true, and internal charge may be produced.

On Conduction through a Condenser.

326.] Let C be the capacity of a condenser, R its resistance, and E the electromotive force which acts on it, that is, the difference of potentials of the surfaces of the metallic electrodes.

Then the quantity of electricity on the side from which the electromotive force acts will be CE, and the current through the substance of the condenser in the direction of the electromotive force will be $\frac{E}{R}$.

If the electrification is supposed to be produced by an electromotive force E acting in a circuit of which the condenser forms part, and if $\frac{dQ}{dt}$ represents the current in that circuit, then

$$\frac{dQ}{dt}=\frac{E}{R}+C\frac{dE}{dt}. \tag{6}$$

Let a battery of electromotive force E_0 and resistance r_1 be introduced into this circuit, then

$$\frac{dQ}{dt}=\frac{E_0-E}{r_1}=\frac{E}{R}+C\frac{dE}{dt}. \tag{7}$$

Hence, at any time t_1,

$$E\,(=E_1)=E_0\frac{R}{R+r_1}\left(1-e^{-\frac{t_1}{T_1}}\right) \text{ where } T_1=\frac{CRr_1}{R+r_1}. \tag{8}$$

Next, let the circuit r_1 be broken for a time t_2,

$$E(=E_2)=E_1 e^{-\frac{t_2}{T_2}} \text{ where } T_2 = CR. \tag{9}$$

Finally, let the surfaces of the condenser be connected by means of a wire whose resistance is r_3 for a time t_3,

$$E(=E_3)=E_2 e^{-\frac{t_3}{T_3}} \text{ where } T_3 = \frac{CRr_3}{R+r_3}. \tag{10}$$

If Q_3 is the total discharge through this wire in the time t_3,

$$Q_3 = E_0 \frac{CR^2}{(R+r_1)(R+r_3)}\left(1-e^{-\frac{t_1}{T_1}}\right) e^{-\frac{t_2}{T_2}}\left(1-e^{-\frac{t_3}{T_3}}\right). \tag{11}$$

In this way we may find the discharge through a wire which is made to connect the surfaces of a condenser after being charged for a time t_1, and then insulated for a time t_2. If the time of charging is sufficient, as it generally is, to develope the whole charge, and if the time of discharge is sufficient for a complete discharge, the discharge is

$$Q_3 = E_0 \frac{CR^2}{(R+r_1)(R+r_3)} e^{-\frac{t_2}{CR}}. \tag{12}$$

327.] In a condenser of this kind, first charged in any way, next discharged through a wire of small resistance, and then insulated, no new electrification will appear. In most actual condensers, however, we find that after discharge and insulation a new charge is gradually developed, of the same kind as the original charge, but inferior in intensity. This is called the residual charge. To account for it we must admit that the constitution of the dielectric medium is different from that which we have just described. We shall find, however, that a medium formed of a conglomeration of small pieces of different simple media would possess this property.

Theory of a Composite Dielectric.

328.] We shall suppose, for the sake of simplicity, that the dielectric consists of a number of plane strata of different materials and of area unity, and that the electric forces act in the direction of the normal to the strata.

Let a_1, a_2, &c. be the thicknesses of the different strata.

X_1, X_2, &c. the resultant electrical forces within the strata.

p_1, p_2, &c. the currents due to conduction through the strata.

f_1, f_2, &c. the electric displacements.

u_1, u_2, &c. the total currents, due partly to conduction and partly to variation of displacement.

r_1, r_2, &c. the specific resistances referred to unit of volume.

K_1, K_2, &c. the specific inductive capacities.

k_1, k_2, &c. the reciprocals of the specific inductive capacities.

E the electromotive force due to a voltaic battery, placed in the part of the circuit leading from the last stratum towards the first, which we shall suppose good conductors.

Q the total quantity of electricity which has passed through this part of the circuit up to the time t.

R_0 the resistance of the battery with its connecting wires.

σ_{12} the surface-density of electricity on the surface which separates the first and second strata.

Then in the first stratum we have, by Ohm's Law,

$$X_1 = r_1 p_1. \tag{1}$$

By the theory of electrical displacement,

$$X_1 = 4\pi k_1 f_1. \tag{2}$$

By the definition of the total current,

$$u_1 = p_1 + \frac{df_1}{dt}, \tag{3}$$

with similar equations for the other strata, in each of which the quantities have the suffix belonging to that stratum.

To determine the surface-density on any stratum, we have an equation of the form

$$\sigma_{12} = f_2 - f_1, \tag{4}$$

and to determine its variation we have

$$\frac{d\sigma_{12}}{dt} = p_1 - p_2. \tag{5}$$

By differentiating (4) with respect to t, and equating the result to (5), we obtain

$$p_1 + \frac{df_1}{dt} = p_2 + \frac{df_2}{dt} = u, \text{ say}, \tag{6}$$

or, by taking account of (3),

$$u_1 = u_2 = \&\text{c.} = u. \tag{7}$$

That is, the total current u is the same in all the strata, and is equal to the current through the wire and battery.

We have also, in virtue of equations (1) and (2),

$$u = \frac{1}{r_1} X_1 + \frac{1}{4\pi k_1}\frac{dX_1}{dt}, \tag{8}$$

from which we may find X_1 by the inverse operation on u,

$$X_1 = \left(\frac{1}{r_1} + \frac{1}{4\pi k_1}\frac{d}{dt}\right)^{-1} u. \tag{9}$$

The total electromotive force E is

$$E = a_1 X_1 + a_2 X_2 + \&c., \tag{10}$$

or $$E = \left\{ a_1 \left(\frac{1}{r_1} + \frac{1}{4\pi k_1} \frac{d}{dt} \right)^{-1} + a_2 \left(\frac{1}{r_2} + \frac{1}{4\pi k_2} \frac{d}{dt} \right)^{-1} + \&c. \right\} u, \tag{11}$$

an equation between E, the external electromotive force, and u, the external current.

If the ratio of r to k is the same in all the strata, the equation reduces itself to

$$E + \frac{r}{4\pi k} \frac{dE}{dt} = (a_1 r_1 + a_2 r_2 + \&c.)\, u, \tag{12}$$

which is the case we have already examined, and in which, as we found, no phenomenon of residual charge can take place.

If there are n substances having different ratios of r to k, the general equation (11), when cleared of inverse operations, will be a linear differential equation, of the nth order with respect to E and of the $(n-1)$th order with respect to u, t being the independent variable.

From the form of the equation it is evident that the order of the different strata is indifferent, so that if there are several strata of the same substance we may suppose them united into one without altering the phenomena.

329.] Let us now suppose that at first f_1, f_2, &c. are all zero, and that an electromotive force E is suddenly made to act, and let us find its instantaneous effect.

Integrating (8) with respect to t, we find

$$Q = \int u\, dt = \frac{1}{r_1} \int X_1\, dt + \frac{1}{4\pi k_1} X_1 + \text{const.} \tag{13}$$

Now, since X_1 is always in this case finite, $\int X_1\, dt$ must be insensible when t is insensible, and therefore, since X_1 is originally zero, the instantaneous effect will be

$$X_1 = 4\pi k_1 Q. \tag{14}$$

Hence, by equation (10),

$$E = 4\pi (k_1 a_1 + k_2 a_2 + \&c.)\, Q, \tag{15}$$

and if C be the electric capacity of the system as measured in this instantaneous way,

$$C = \frac{Q}{E} = \frac{1}{4\pi (k_1 a_1 + k_2 a_2 + \&c.)}. \tag{16}$$

This is the same result that we should have obtained if we had neglected the conductivity of the strata.

Let us next suppose that the electromotive force E is continued uniform for an indefinitely long time, or till a uniform current of conduction equal to p is established through the system.

We have then $X_1 = r_1 p$, etc., and therefore by (10),

$$E = (r_1 a_1 + r_2 a_2 + \&c.)p. \tag{17}$$

If R be the total resistance of the system,

$$R = \frac{E}{p} = r_1 a_1 + r_2 a_2 + \&c. \tag{18}$$

In this state we have by (2),

$$f_1 = \frac{r_1}{4\pi k_1} p,$$

so that

$$\sigma_{12} = \left(\frac{r_2}{4\pi k_2} - \frac{r_1}{4\pi k_1}\right)p. \tag{19}$$

If we now suddenly connect the extreme strata by means of a conductor of small resistance, E will be suddenly changed from its original value E_0 to zero, and a quantity Q of electricity will pass through the conductor.

To determine Q we observe that if X_1' be the new value of X_1, then by (13),

$$X_1' = X_1 + 4\pi k_1 Q. \tag{20}$$

Hence, by (10), putting $E = 0$,

$$0 = a_1 X_1 + \&c. + 4\pi(a_1 k_1 + a_2 k_2 + \&c.)\, Q, \tag{21}$$

or

$$0 = E_0 + \frac{1}{C} Q. \tag{22}$$

Hence $Q = -CE_0$ where C is the capacity, as given by equation (16). The instantaneous discharge is therefore equal to the instantaneous charge.

Let us next suppose the connexion broken immediately after this discharge. We shall then have $u = 0$, so that by equation (8),

$$X_1 = X' e^{-\frac{4\pi k_1}{r_1} t}, \tag{23}$$

where X' is the initial value after the discharge.

Hence, at any time t,

$$X_1 = E_0 \left\{ \frac{r_1}{R} - 4\pi k_1 C \right\} e^{-\frac{4\pi k_1}{r_1} t}.$$

The value of E at any time is therefore

$$= E_0 \left\{ \left(\frac{a_1 r_1}{R} - 4\pi a_1 k_1 C\right) e^{-\frac{4\pi k_1}{r_1} t} + \left(\frac{a_2 r_2}{R} - 4\pi a_2 k_2 C\right) e^{-\frac{4\pi k_2}{r_2} t} + \&c. \right\}, \tag{24}$$

and the instantaneous discharge after any time t is EC. This is called the residual discharge.

If the ratio of r to k is the same for all the strata, the value of E will be reduced to zero. If, however, this ratio is not the same, let the terms be arranged according to the values of this ratio in descending order of magnitude.

The sum of all the coefficients is evidently zero, so that when $t = 0$, $E = 0$. The coefficients are also in descending order of magnitude, and so are the exponential terms when t is positive. Hence, when t is positive, E will be positive, so that the residual discharge is always of the same sign as the primary discharge.

When t is indefinitely great all the terms disappear unless any of the strata are perfect insulators, in which case r_1 is infinite for that stratum, and R is infinite for the whole system, and the final value of E is not zero but

$$E = E_0(1 - 4\pi a_1 k_1 C). \tag{25}$$

Hence, when some, but not all, of the strata are perfect insulators, a residual discharge may be permanently preserved in the system.

330.] We shall next determine the total discharge through a wire of resistance R_0 kept permanently in connexion with the extreme strata of the system, supposing the system first charged by means of a long-continued application of the electromotive force E.

At any instant we have

$$E = a_1 r_1 p_1 + a_2 r_2 p_2 + \&c. + R_0 u = 0, \tag{26}$$

and also, by (3),
$$u = p_1 + \frac{df_1}{dt}. \tag{27}$$

Hence
$$(R + R_0)\, u = a_1 r_1 \frac{df_1}{dt} + a_2 r_2 \frac{df_2}{dt} + \&c. \tag{28}$$

Integrating with respect to t in order to find Q, we get

$$(R + R_0)\, Q = a_1 r_1 (f_1' - f_1) + a_2 r_2 (f_2' - f_2) + \&c., \tag{29}$$

where f_1 is the initial, and f_1' the final value of f_1.

In this case $f_1' = 0$, and by (2) and (20) $f_1 = E_0\left(\frac{r_1}{4\pi k_1 R} - C\right)$.

Hence
$$(R + R_0)\, Q = \frac{E_0}{4\pi R}\left(\frac{a_1 r_1^2}{k_1} + \frac{a_2 r_2^2}{k_2} + \&c.\right) - E_0 CR, \tag{30}$$

$$= -\frac{CE_0}{R}\,\Sigma\Sigma\left[a_1 a_2 k_1 k_2 \left(\frac{r_1}{k_1} - \frac{r_2}{k_2}\right)^2\right], \tag{31}$$

where the summation is extended to all quantities of this form belonging to every pair of strata.

It appears from this that Q is always negative, that is to say, in the opposite direction to that of the current employed in charging the system.

This investigation shews that a dielectric composed of strata of different kinds may exhibit the phenomena known as electric absorption and residual discharge, although none of the substances of which it is made exhibit these phenomena when alone. An investigation of the cases in which the materials are arranged otherwise than in strata would lead to similar results, though the calculations would be more complicated, so that we may conclude that the phenomena of electric absorption may be expected in the case of substances composed of parts of different kinds, even though these individual parts should be microscopically small.

It by no means follows that every substance which exhibits this phenomenon is so composed, for it may indicate a new kind of electric polarization of which a homogeneous substance may be capable, and this in some cases may perhaps resemble electrochemical polarization much more than dielectric polarization.

The object of the investigation is merely to point out the true mathematical character of the so-called electric absorption, and to shew how fundamentally it differs from the phenomena of heat which seem at first sight analogous.

331.] If we take a thick plate of any substance and heat it on one side, so as to produce a flow of heat through it, and if we then suddenly cool the heated side to the same temperature as the other, and leave the plate to itself, the heated side of the plate will again become hotter than the other by conduction from within.

Now an electrical phenomenon exactly analogous to this can be produced, and actually occurs in telegraph cables, but its mathematical laws, though exactly agreeing with those of heat, differ entirely from those of the stratified condenser.

In the case of heat there is true absorption of the heat into the substance with the result of making it hot. To produce a truly analogous phenomenon in electricity is impossible, but we may imitate it in the following way in the form of a lecture-room experiment.

Let A_1, A_2, &c. be the inner conducting surfaces of a series of condensers, of which B_0, B_1, B_2, &c. are the outer surfaces.

Let A_1, A_2, &c. be connected in series by connexions of resistance R, and let a current be passed along this series from left to right.

Let us first suppose the plates B_0, B_1, B_2, each insulated and free from charge. Then the total quantity of electricity on each of the plates B must remain zero, and since the electricity on the plates A is in each case equal and opposite to that of the opposed

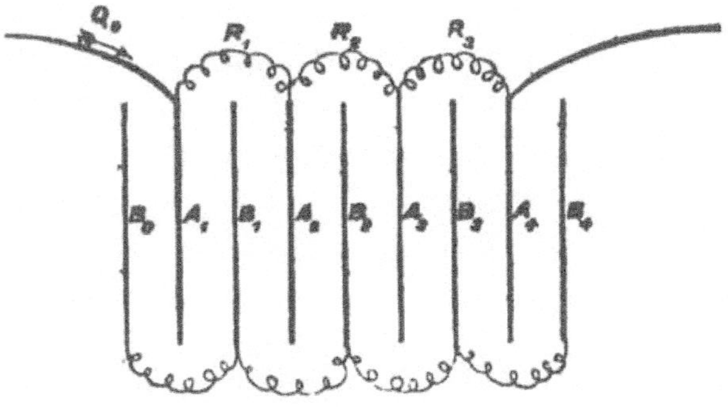

Fig. 26.

surface they will not be electrified, and no alteration of the current will be observed.

But let the plates B be all connected together, or let each be connected with the earth. Then, since the potential of A_1 is positive, while that of the plates B is zero, A_1 will be positively electrified and B_1 negatively.

If P_1, P_2, &c. are the potentials of the plates A_1, A_2, &c., and C the capacity of each, and if we suppose that a quantity of electricity equal to Q_0 passes through the wire on the left, Q_1 through the connexion R_1, and so on, then the quantity which exists on the plate A_1 is $Q_0 - Q_1$, and we have

$$Q_0 - Q_1 = C_1 P_1.$$

Similarly

$$Q_1 - Q_2 = C_2 P_2,$$

and so on.

But by Ohm's Law we have

$$P_1 - P_2 = R_1 \frac{dQ_1}{dt},$$

$$P_2 - P_3 = R_2 \frac{dQ_2}{dt}.$$

If we suppose the values of C the same for each plate, and those of R the same for each wire, we shall have a series of equations of the form

$$Q_0 - 2\,Q_1 + Q_2 = RC\frac{dQ_1}{dt},$$

$$Q_1 - 2\,Q_2 + Q_3 = RC\frac{dQ_2}{dt}.$$

If there are n quantities of electricity to be determined, and if either the total electromotive force, or some other equivalent conditions be given, the differential equation for determining any one of them will be linear and of the nth order.

By an apparatus arranged in this way, Mr. Varley succeeded in imitating the electrical action of a cable 12,000 miles long.

When an electromotive force is made to act along the wire on the left hand, the electricity which flows into the system is at first principally occupied in charging the different condensers beginning with A_1, and only a very small fraction of the current appears at the right hand till a considerable time has elapsed. If galvanometers be placed in circuit at R_1, R_2, &c. they will be affected by the current one after another, the interval between the times of equal indications being greater as we proceed to the right.

332.] In the case of a telegraph cable the conducting wire is separated from conductors outside by a cylindrical sheath of gutta-percha, or other insulating material. Each portion of the cable thus becomes a condenser, the outer surface of which is always at potential zero. Hence, in a given portion of the cable, the quantity of free electricity at the surface of the conducting wire is equal to the product of the potential into the capacity of the portion of the cable considered as a condenser.

If a_1, a_2 are the outer and inner radii of the insulating sheath, and if K is its specific dielectric capacity, the capacity of unit of length of the cable is, by Art. 126,

$$c = \frac{K}{2\log\frac{a_1}{a_2}}. \qquad (1)$$

Let v be the potential at any point of the wire, which we may consider as the same at every part of the same section.

Let Q be the total quantity of electricity which has passed through that section since the beginning of the current. Then the quantity which at the time t exists between sections at x and at $x+\delta x$, is

$$Q - \left(Q + \frac{dQ}{dx}\delta x\right), \quad \text{or} \quad -\frac{dQ}{dx}\delta x,$$

and this is, by what we have said, equal to $cv\delta x$.

Hence $$cv = -\frac{dQ}{dx}. \tag{2}$$

Again, the electromotive force at any section is $-\frac{dv}{dx}$, and by Ohm's Law,
$$-\frac{dv}{dx} = k\frac{dQ}{dt}, \tag{3}$$
where k is the resistance of unit of length of the conductor, and $\frac{dQ}{dt}$ is the strength of the current. Eliminating Q between (2) and (3), we find
$$ck\frac{dv}{dt} = \frac{d^2v}{dx^2}. \tag{4}$$

This is the partial differential equation which must be solved in order to obtain the potential at any instant at any point of the cable. It is identical with that which Fourier gives to determine the temperature at any point of a stratum through which heat is flowing in a direction normal to the stratum. In the case of heat c represents the capacity of unit of volume, or what Fourier denotes by CD, and k represents the reciprocal of the conductivity.

If the sheath is not a perfect insulator, and if k_1 is the resistance of unit of length of the sheath to conduction through it in a radial direction, then if ρ_1 is the specific resistance of the insulating material, it is easy to shew that
$$k_1 = \frac{1}{2\pi}\rho_1 \log_e \frac{a_1}{a_2}. \tag{5}$$

The equation (2) will no longer be true, for the electricity is expended not only in charging the wire to the extent represented by cv, but in escaping at a rate represented by $\frac{v}{k_1}$. Hence the rate of expenditure of electricity will be
$$-\frac{d^2Q}{dx\,dt} = c\frac{dv}{dt} + \frac{1}{k_1}v, \tag{6}$$
whence, by comparison with (3), we get
$$ck\frac{dv}{dt} = \frac{d^2v}{dx^2} - \frac{k}{k_1}v, \tag{7}$$
and this is the equation of conduction of heat in a rod or ring as given by Fourier*.

333.] If we had supposed that a body when raised to a high potential becomes electrified throughout its substance as if electricity were compressed into it, we should have arrived at equations of this very form. It is remarkable that Ohm himself,

* *Théorie de la Chaleur*, Art. 105.

misled by the analogy between electricity and heat, entertained an opinion of this kind, and was thus, by means of an erroneous opinion, led to employ the equations of Fourier to express the true laws of conduction of electricity through a long wire, long before the real reason of the appropriateness of these equations had been suspected.

Mechanical Illustration of the Properties of a Dielectric.

334.] Five tubes of equal sectional area A, B, C, D and P are arranged in circuit as in the figure. A, B, C and D are vertical and equal, and P is horizontal.

The lower halves of A, B, C, D are filled with mercury, their upper halves and the horizontal tube P are filled with water.

A tube with a stopcock Q connects the lower part of A and B with that of C and D, and a piston P is made to slide in the horizontal tube.

Let us begin by supposing that the level of the mercury in the four tubes is the same, and that it is indicated by A_0, B_0, C_0, D_0, that the piston is at P_0, and that the stopcock Q is shut.

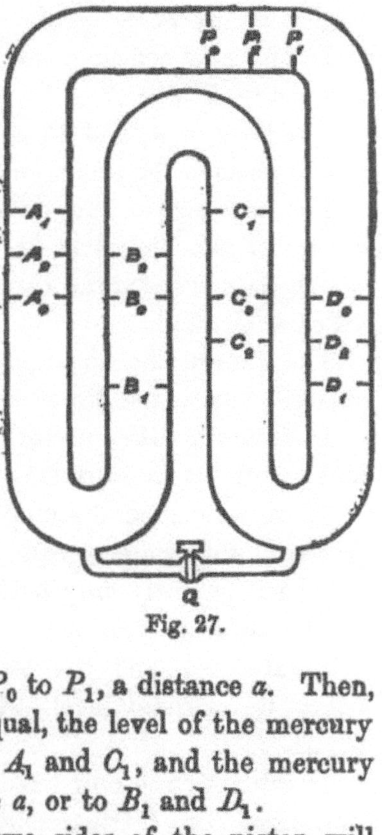

Fig. 27.

Now let the piston be moved from P_0 to P_1, a distance a. Then, since the sections of all the tubes are equal, the level of the mercury in A and C will rise a distance a, or to A_1 and C_1, and the mercury in B and D will sink an equal distance a, or to B_1 and D_1.

The difference of pressure on the two sides of the piston will be represented by $4a$.

This arrangement may serve to represent the state of a dielectric acted on by an electromotive force $4a$.

The excess of water in the tube D may be taken to represent a positive charge of electricity on one side of the dielectric, and the excess of mercury in the tube A may represent the negative charge on the other side. The excess of pressure in the tube P on the side of the piston next D will then represent the excess of potential on the positive side of the dielectric.

If the piston is free to move it will move back to P_0 and be in equilibrium there. This represents the complete discharge of the dielectric.

During the discharge there is a reversed motion of the liquids throughout the whole tube, and this represents that change of electric displacement which we have supposed to take place in a dielectric.

I have supposed every part of the system of tubes filled with incompressible liquids, in order to represent the property of all electric displacement that there is no real accumulation of electricity at any place.

Let us now consider the effect of opening the stopcock Q while the piston P is at P_1.

The level of A_1 and D_1 will remain unchanged, but that of B and C will become the same, and will coincide with B_0 and C_0.

The opening of the stopcock Q corresponds to the existence of a part of the dielectric which has a slight conducting power, but which does not extend through the whole dielectric so as to form an open channel.

The charges on the opposite sides of the dielectric remain insulated, but their difference of potential diminishes.

In fact, the difference of pressure on the two sides of the piston sinks from $4a$ to $2a$ during the passage of the fluid through Q.

If we now shut the stopcock Q and allow the piston P to move freely, it will come to equilibrium at a point P_2, and the discharge will be apparently only half of the charge.

The level of the mercury in A and B will be $\frac{1}{2}a$ above its original level, and the level in the tubes C and D will be $\frac{1}{2}a$ below its original level. This is indicated by the levels A_2, B_2, C_2, D_2.

If the piston is now fixed and the stopcock opened, mercury will flow from B to C till the level in the two tubes is again at B_0 and C_0. There will then be a difference of pressure $= a$ on the two sides of the piston P. If the stopcock is then closed and the piston P left free to move, it will again come to equilibrium at a point P_3, half way between P_2 and P_0. This corresponds to the residual charge which is observed when a charged dielectric is first discharged and then left to itself. It gradually recovers part of its charge, and if this is again discharged a third charge is formed, the successive charges diminishing in quantity. In the case of the illustrative experiment each charge is half of the preceding, and the discharges, which are $\frac{1}{2}$, $\frac{1}{4}$, &c. of the original charge, form a series whose sum is equal to the original charge.

If, instead of opening and closing the stopcock, we had allowed it to remain nearly, but not quite, closed during the whole experiment, we should have had a case resembling that of the electrification of a dielectric which is a perfect insulator and yet exhibits the phenomenon called 'electric absorption.'

To represent the case in which there is true conduction through the dielectric we must either make the piston leaky, or we must establish a communication between the top of the tube *A* and the top of the tube *D*.

In this way we may construct a mechanical illustration of the properties of a dielectric of any kind, in which the two electricities are represented by two real fluids, and the electric potential is represented by fluid pressure. Charge and discharge are represented by the motion of the piston *P*, and electromotive force by the resultant force on the piston.

CHAPTER XI.

THE MEASUREMENT OF ELECTRIC RESISTANCE.

335.] In the present state of electrical science, the determination of the electric resistance of a conductor may be considered as the cardinal operation in electricity, in the same sense that the determination of weight is the cardinal operation in chemistry.

The reason of this is that the determination in absolute measure of other electrical magnitudes, such as quantities of electricity, electromotive forces, currents, &c., requires in each case a complicated series of operations, involving generally observations of time, measurements of distances, and determinations of moments of inertia, and these operations, or at least some of them, must be repeated for every new determination, because it is impossible to preserve a unit of electricity, or of electromotive force, or of current, in an unchangeable state, so as to be available for direct comparison.

But when the electric resistance of a properly shaped conductor of a properly chosen material has been once determined, it is found that it always remains the same for the same temperature, so that the conductor may be used as a standard of resistance, with which that of other conductors can be compared, and the comparison of two resistances is an operation which admits of extrême accuracy.

When the unit of electrical resistance has been fixed on, material copies of this unit, in the form of 'Resistance Coils,' are prepared for the use of electricians, so that in every part of the world electrical resistances may be expressed in terms of the same unit. These unit resistance coils are at present the only examples of material electric standards which can be preserved, copied, and used for the purpose of measurement. Measures of electrical capacity, which are also of great importance, are still defective, on account of the disturbing influence of electric absorption.

336.] The unit of resistance may be an entirely arbitrary one, as in the case of Jacobi's Etalon, which was a certain copper wire of 22.4932 grammes weight, 7.61975 metres length, and 0.667

millimetres diameter. Copies of this have been made by Leyser of Leipsig, and are to be found in different places.

According to another method the unit may be defined as the resistance of a portion of a definite substance of definite dimensions. Thus, Siemens' unit is defined as the resistance of a column of mercury of one metre long, and one square millimetre section, at the temperature 0°C.

337.] Finally, the unit may be defined with reference to the electrostatic or the electromagnetic system of units. In practice the electromagnetic system is used in all telegraphic operations, and therefore the only systematic units actually in use are those of this system.

In the electromagnetic system, as we shall shew at the proper place, a resistance is a quantity homogeneous with a velocity, and may therefore be expressed as a velocity. See Art. 628.

338.] The first actual measurements on this system were made by Weber, who employed as his unit one millimetre per second. Sir W. Thomson afterwards used one foot per second as a unit, but a large number of electricians have now agreed to use the unit of the British Association, which professes to represent a resistance which, expressed as a velocity, is ten millions of metres per second. The magnitude of this unit is more convenient than that of Weber's unit, which is too small. It is sometimes referred to as the B.A. unit, but in order to connect it with the name of the discoverer of the laws of resistance, it is called the Ohm.

339.] To recollect its value in absolute measure it is useful to know that ten millions of metres is professedly the distance from the pole to the equator, measured along the meridian of Paris. A body, therefore, which in one second travels along a meridian from the pole to the equator would have a velocity which, on the electromagnetic system, is professedly represented by an Ohm.

I say professedly, because, if more accurate researches should prove that the Ohm, as constructed from the British Association's material standards, is not really represented by this velocity, electricians would not alter their standards, but would apply a correction. In the same way the metre is professedly one ten-millionth of a certain quadrantal arc, but though this is found not to be exactly true, the length of the metre has not been altered, but the dimensions of the earth are expressed by a less simple number.

According to the system of the British Association, the absolute value of the unit is *originally chosen* so as to represent as nearly as possible a quantity derived from the electromagnetic absolute system.

340.] When a material unit representing this abstract quantity has been made, other standards are constructed by copying this unit, a process capable of extreme accuracy—of much greater accuracy than, for instance, the copying of foot-rules from a standard foot.

These copies, made of the most permanent materials, are distributed over all parts of the world, so that it is not likely that any difficulty will be found in obtaining copies of them if the original standards should be lost.

But such units as that of Siemens can without very great labour be reconstructed with considerable accuracy, so that as the relation of the Ohm to Siemens unit is known, the Ohm can be reproduced even without having a standard to copy, though the labour is much greater and the accuracy much less than by the method of copying.

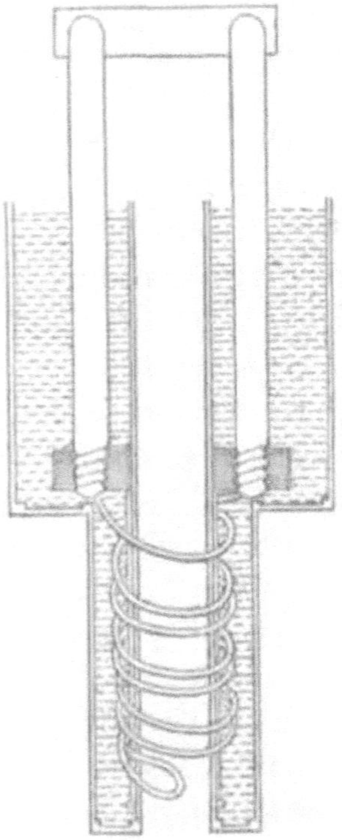

Fig. 28.

Finally, the Ohm may be reproduced by the electromagnetic method by which it was originally determined. This method, which is considerably more laborious than the determination of a foot from the seconds pendulum, is probably inferior in accuracy to that last mentioned. On the other hand, the determination of the electromagnetic unit in terms of the Ohm with an amount of accuracy corresponding to the progress of electrical science, is a most important physical research and well worthy of being repeated.

The actual resistance coils constructed to represent the Ohm were made of an alloy of two parts of silver and one of platinum in the form of wires from .5 millimetres to .8 millimetres diameter, and from one to two metres in length. These wires were soldered to stout copper electrodes. The wire itself was covered with two layers of silk, imbedded in solid paraffin, and enclosed in a thin brass case, so that it can be easily brought to a temperature at which its resistance is accurately one Ohm. This temperature is marked on the insulating support of the coil. (See Fig. 28.)

On the Forms of Resistance Coils.

341.] A Resistance Coil is a conductor capable of being easily placed in the voltaic circuit, so as to introduce into the circuit a known resistance.

The electrodes or ends of the coil must be such that no appreciable error may arise from the mode of making the connexions. For resistances of considerable magnitude it is sufficient that the electrodes should be made of stout copper wire or rod well amalgamated with mercury at the ends, and that the ends should be made to press on flat amalgamated copper surfaces placed in mercury cups.

For very great resistances it is sufficient that the electrodes should be thick pieces of brass, and that the connexions should be made by inserting a wedge of brass or copper into the interval between them. This method is found very convenient.

The resistance coil itself consists of a wire well covered with silk, the ends of which are soldered permanently to the electrodes.

The coil must be so arranged that its temperature may be easily observed. For this purpose the wire is coiled on a tube and covered with another tube, so that it may be placed in a vessel of water, and that the water may have access to the inside and the outside of the coil.

To avoid the electromagnetic effects of the current in the coil the wire is first doubled back on itself and then coiled on the tube, so that at every part of the coil there are equal and opposite currents in the adjacent parts of the wire.

When it is desired to keep two coils at the same temperature the wires are sometimes placed side by side and coiled up together. This method is especially useful when it is more important to secure equality of resistance than to know the absolute value of the resistance, as in the case of the equal arms of Wheatstone's Bridge, (Art. 347).

When measurements of resistance were first attempted, a resistance coil, consisting of an uncovered wire coiled in a spiral groove round a cylinder of insulating material, was much used. It was called a Rheostat. The accuracy with which it was found possible to compare resistances was soon found to be inconsistent with the use of any instrument in which the contacts are not more perfect than can be obtained in the rheostat. The rheostat, however, is still used for adjusting the resistance where accurate measurement is not required.

Resistance coils are generally made of those metals whose resistance is greatest and which vary least with temperature. German silver fulfils these conditions very well, but some specimens are found to change their properties during the lapse of years. Hence, for standard coils, several pure metals, and also an alloy of platinum and silver, have been employed, and the relative resistance of these during several years has been found constant up to the limits of modern accuracy.

342.] For very great resistances, such as several millions of Ohms, the wire must be either very long or very thin, and the construction of the coil is expensive and difficult. Hence tellurium and selenium have been proposed as materials for constructing standards of great resistance. A very ingenious and easy method of construction has been lately proposed by Phillips *. On a piece of ebonite or ground glass a fine pencil-line is drawn. The ends of this filament of plumbago are connected to metallic electrodes, and the whole is then covered with insulating varnish. If it should be found that the resistance of such a pencil-line remains constant, this will be the best method of obtaining a resistance of several millions of Ohms.

343.] There are various arrangements by which resistance coils may be easily introduced into a circuit.

For instance, a series of coils of which the resistances are 1, 2, 4, 8, 16, &c., arranged according to the powers of 2, may be placed in a box in series.

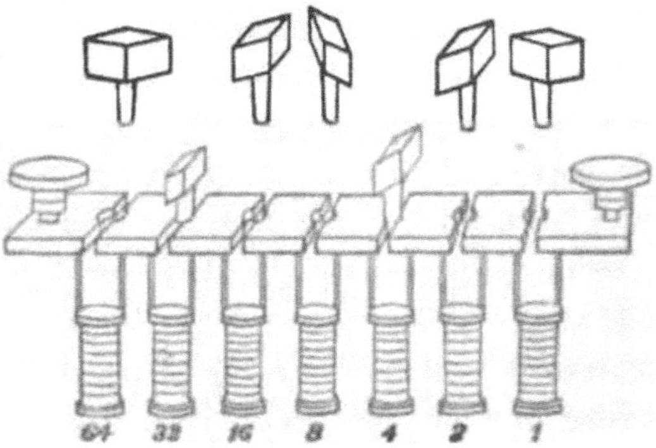

Fig. 29.

The electrodes consist of stout brass plates, so arranged on the outside of the box that by inserting a brass plug or wedge between

* *Phil. Mag.*, July, 1870.

two of them as a shunt, the resistance of the corresponding coil may be put out of the circuit. This arrangement was introduced by Siemens.

Each interval between the electrodes is marked with the resistance of the corresponding coil, so that if we wish to make the resistance box equal to 107 we express 107 in the binary scale as $64+32+8+2+1$ or 1101011. We then take the plugs out of the holes corresponding to 64, 32, 8, 2 and 1, and leave the plugs in 16 and 4.

This method, founded on the binary scale, is that in which the smallest number of separate coils is needed, and it is also that which can be most readily tested. For if we have another coil equal to 1 we can test the equality of 1 and 1′, then that of $1+1'$ and 2, then that of $1+1'+2$ and 4, and so on.

The only disadvantage of the arrangement is that it requires a familiarity with the binary scale of notation, which is not generally possessed by those accustomed to express every number in the decimal scale.

344.] A box of resistance coils may be arranged in a different way for the purpose of measuring conductivities instead of resistances.

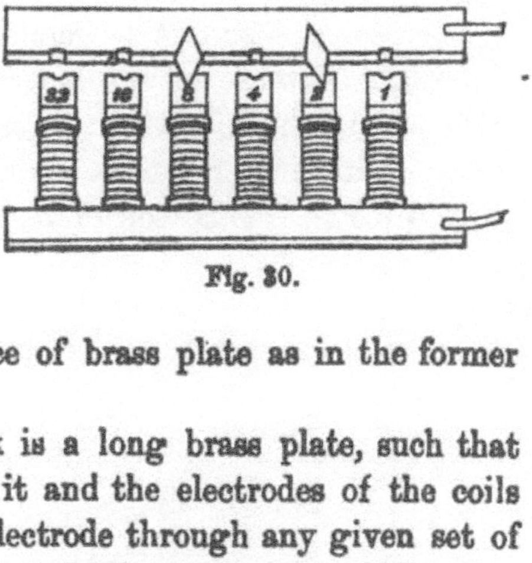

Fig. 30.

The coils are placed so that one end of each is connected with a long thick piece of metal which forms one electrode of the box, and the other end is connected with a stout piece of brass plate as in the former case.

The other electrode of the box is a long brass plate, such that by inserting brass plugs between it and the electrodes of the coils it may be connected to the first electrode through any given set of coils. The conductivity of the box is then the sum of the conductivities of the coils.

In the figure, in which the resistances of the coils are 1, 2, 4, &c., and the plugs are inserted at 2 and 8, the conductivity of the box is $\frac{1}{2}+\frac{1}{8}=\frac{5}{8}$, and the resistance of the box is therefore $\frac{8}{5}$ or 1.6.

This method of combining resistance coils for the measurement of fractional resistances was introduced by Sir W. Thomson under the name of the method of multiple arcs. See Art. 276.

On the Comparison of Resistances.

345.] If E is the electromotive force of a battery, and R the resistance of the battery and its connexions, including the galvanometer used in measuring the current, and if the strength of the current is I when the battery connexions are closed, and I_1, I_2 when additional resistances r_1, r_2 are introduced into the circuit, then, by Ohm's Law,

$$E = IR = I_1(R+r_1) = I_2(R+r_2).$$

Eliminating E, the electromotive force of the battery, and R the resistance of the battery and its connexions, we get Ohm's formula

$$\frac{r_1}{r_2} = \frac{(I-I_1)\,I_2}{(I-I_2)\,I_1}.$$

This method requires a measurement of the ratios of I, I_1 and I_2, and this implies a galvanometer graduated for absolute measurements.

If the resistances r_1 and r_2 are equal, then I_1 and I_2 are equal, and we can test the equality of currents by a galvanometer which is not capable of determining their ratios.

But this is rather to be taken as an example of a faulty method than as a practical method of determining resistance. The electromotive force E cannot be maintained rigorously constant, and the internal resistance of the battery is also exceedingly variable, so that any methods in which these are assumed to be even for a short time constant are not to be depended on.

346.] The comparison of resistances can be made with extreme

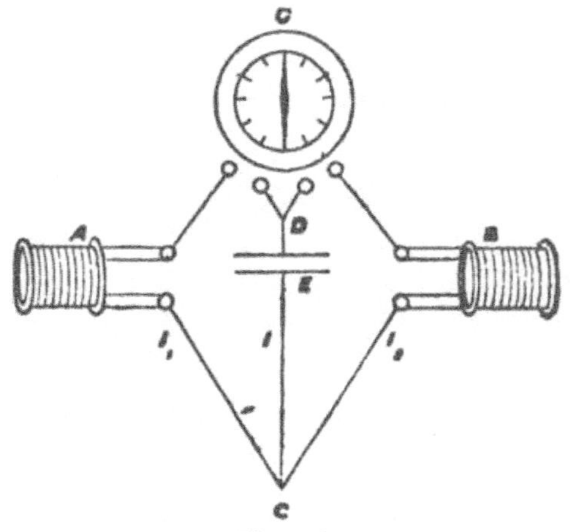

Fig. 31.

accuracy by either of two methods, in which the result is independent of variations of R and E.

The first of these methods depends on the use of the differential galvanometer, an instrument in which there are two coils, the currents in which are independent of each other, so that when the currents are made to flow in opposite directions they act in opposite directions on the needle, and when the ratio of these currents is that of m to n they have no resultant effect on the galvanometer needle.

Let I_1, I_2 be the currents through the two coils of the galvanometer, then the deflexion of the needle may be written

$$\delta = m I_1 - n I_2.$$

Now let the battery current I be divided between the coils of the galvanometer, and let resistances A and B be introduced into the first and second coils respectively. Let the remainder of the resistance of the coils and their connexions be a and β respectively, and let the resistance of the battery and its connexions between C and D be r, and its electromotive force E.

Then we find, by Ohm's Law, for the difference of potentials between C and D,

$$I_1(A+a) = I_2(B+\beta) = E - Ir,$$

and since
$$I_1 + I_2 = I,$$

$$I_1 = E\frac{B+\beta}{D}, \qquad I_2 = E\frac{A+a}{D}, \qquad I = E\frac{A+a+B+\beta}{D},$$

where
$$D = (A+a)(B+\beta) + r(A+a+B+\beta).$$

The deflexion of the galvanometer needle is therefore

$$\delta = \frac{E}{D}\{m(B+\beta) - n(A+a)\},$$

and if there is no observable deflexion, then we know that the quantity enclosed in brackets cannot differ from zero by more than a certain small quantity, depending on the power of the battery, the suitableness of the arrangement, the delicacy of the galvanometer, and the accuracy of the observer.

Suppose that B has been adjusted so that there is no apparent deflexion.

Now let another conductor A' be substituted for A, and let A' be adjusted till there is no apparent deflexion. Then evidently to a first approximation $A' = A$.

To ascertain the degree of accuracy of this estimate, let the altered quantities in the second observation be accented, then

$$m(B+\beta)-n(A+a)=\frac{D}{E}\delta,$$

$$m(B+\beta)-n(A'+a)=\frac{D'}{E'}\delta'.$$

Hence $$n(A'-A)=\frac{D}{E}\delta-\frac{D'}{E'}\delta'.$$

If δ and δ', instead of being both apparently zero, had been only observed to be equal, then, unless we also could assert that $E=E'$, the right-hand side of the equation might not be zero. In fact, the method would be a mere modification of that already described.

The merit of the method consists in the fact that the thing observed is the absence of any deflexion, or in other words, the method is a Null method, one in which the non-existence of a force is asserted from an observation in which the force, if it had been different from zero by more than a certain small amount, would have produced an observable effect.

Null methods are of great value where they can be employed, but they can only be employed where we can cause two equal and opposite quantities of the same kind to enter into the experiment together.

In the case before us both δ and δ' are quantities too small to be observed, and therefore any change in the value of E will not affect the accuracy of the result.

The actual degree of accuracy of this method might be ascertained by making a number of observations in each of which A' is separately adjusted, and comparing the result of each observation with the mean of the whole series.

But by putting A' out of adjustment by a known quantity, as, for instance, by inserting at A or at B an additional resistance equal to a hundredth part of A or of B, and then observing the resulting deviation of the galvanometer needle, we can estimate the number of degrees corresponding to an error of one per cent. To find the actual degree of precision we must estimate the smallest deflexion which could not escape observation, and compare it with the deflexion due to an error of one per cent.

*If the comparison is to be made between A and B, and if the positions of A and B are exchanged, then the second equation becomes

* This investigation is taken from Weber's treatise on Galvanometry. *Göttingen Transactions*, x. p. 65.

$$m(A+\beta)-n(B+a)=\frac{D'}{E}\delta',$$

whence $$(m+n)(B-A)=\frac{D}{E}\delta-\frac{D'}{E}\delta'.$$

If m and n, A and B, a and β are approximately equal, then

$$B-A=\frac{1}{2nE}(A+a)(A+a+2r)(\delta-\delta').$$

Here $\delta-\delta'$ may be taken to be the smallest observable deflexion of the galvanometer.

If the galvanometer wire be made longer and thinner, retaining the same total mass, then n will vary as the length of the wire and a as the square of the length. Hence there will be a minimum value of $\frac{(A+a)(A+a+2r)}{n}$ when

$$a=\tfrac{1}{3}(A+r)\left\{2\sqrt{1-\frac{3}{4}\frac{r^2}{(A+r)^2}}-1\right\}.$$

If we suppose r, the battery resistance, small compared with A, this gives $$a=\tfrac{1}{3}A;$$

or, *the resistance of each coil of the galvanometer should be one-third of the resistance to be measured.*

We then find $$B-A=\frac{8}{9}\frac{A^2}{nE}(\delta-\delta').$$

If we allow the current to flow through one only of the coils of the galvanometer, and if the deflexion thereby produced is Δ (supposing the deflexion strictly proportional to the deflecting force), then

$$\Delta=\frac{mE}{A+a+r}=\frac{3}{4}\frac{nE}{A}\text{ if } r=0 \text{ and } a=\frac{1}{3}A.$$

Hence $$\frac{B-A}{A}=\frac{2}{3}\frac{\delta-\delta'}{\Delta}.$$

In the differential galvanometer two currents are made to produce equal and opposite effects on the suspended needle. The force with which either current acts on the needle depends not only on the strength of the current, but on the position of the windings of the wire with respect to the needle. Hence, unless the coil is very carefully wound, the ratio of m to n may change when the position of the needle is changed, and therefore it is necessary to determine this ratio by proper methods during each course of experiments if any alteration of the position of the needle is suspected.

The other null method, in which Wheatstone's Bridge is used, requires only an ordinary galvanometer, and the observed zero deflexion of the needle is due, not to the opposing action of two currents, but to the non-existence of a current in the wire. Hence we have not merely a null deflexion, but a null current as the phenomenon observed, and no errors can arise from want of regularity or change of any kind in the coils of the galvanometer. The galvanometer is only required to be sensitive enough to detect the existence and direction of a current, without in any way determining its value or comparing its value with that of another current.

347.] Wheatstone's Bridge consists essentially of six conductors connecting four points. An electromotive force E is made to act between two of the points by means of a voltaic battery introduced between B and C. The current between the other two points O and A is measured by a galvanometer.

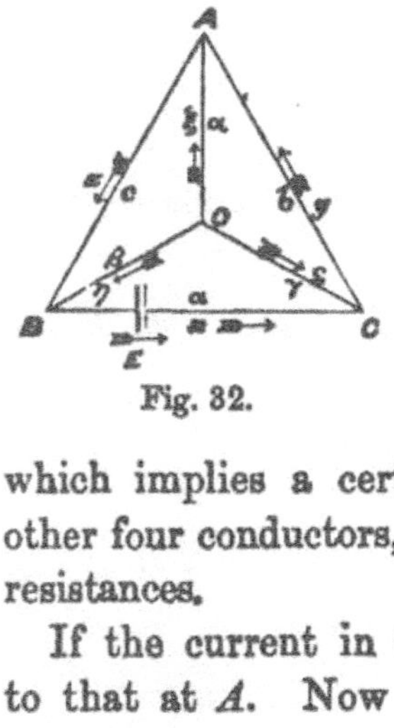

Fig. 32.

Under certain circumstances this current becomes zero. The conductors BC and OA are then said to be *conjugate* to each other, which implies a certain relation between the resistances of the other four conductors, and this relation is made use of in measuring resistances.

If the current in OA is zero, the potential at O must be equal to that at A. Now when we know the potentials at B and C we can determine those at O and A by the rule given in Art. 275, provided there is no current in OA,

$$O = \frac{B\gamma + C\beta}{\beta + \gamma}, \quad A = \frac{Bb + Cc}{b + c},$$

whence the condition is $b\beta = c\gamma$,

where b, c, β, γ are the resistances in CA, AB, BO, and OC respectively.

To determine the degree of accuracy attainable by this method we must ascertain the strength of the current in OA when this condition is not fulfilled exactly.

Let A, B, C and O be the four points. Let the currents along BC, CA and AB be x, y and z, and the resistances of these conductors a, b and c. Let the currents along OA, OB and OC be ξ, η, ζ and the resistances α, β and γ. Let an electromotive force E act along BC. Required the current ξ along OA.

Let the potentials at the points A, B, C and O be denoted by the symbols A, B, C and O. The equations of conduction are

$$\begin{aligned} ax &= B-C+E, & \alpha\xi &= O-A, \\ by &= C-A, & \beta\eta &= O-B, \\ cz &= A-B, & \gamma\zeta &= O-C; \end{aligned}$$

with the equations of continuity

$$\begin{aligned} \xi+y-z &= 0, \\ \eta+z-x &= 0, \\ \zeta+x-y &= 0. \end{aligned}$$

By considering the system as made up of three circuits OBC, OCA and OAB, in which the currents are x, y, z respectively, and applying Kirchhoff's rule to each cycle, we eliminate the values of the potentials O, A, B, C, and the currents ξ, η, ζ, and obtain the following equations for x, y and z,

$$\begin{aligned} (a+\beta+\gamma)x-\gamma y \quad\quad -\beta z &= E, \\ -\gamma x \quad +(b+\gamma+\alpha)y-\alpha z &= 0, \\ -\beta x \quad -\alpha y \quad\quad +(c+\alpha+\beta)z &= 0. \end{aligned}$$

Hence, if we put

$$D = \begin{vmatrix} a+\beta+\gamma & -\gamma & -\beta \\ -\gamma & b+\gamma+\alpha & -\alpha \\ -\beta & -\alpha & c+\alpha+\beta \end{vmatrix},$$

we find $$\xi = \frac{E}{D}(b\beta-c\gamma),$$

and $$x = \frac{E}{D}\{(b+\gamma)(c+\beta)+\alpha(b+c+\beta+\gamma)\}.$$

348.] The value of D may be expressed in the symmetrical form, $D = abc+bc(\beta+\gamma)+ca(\gamma+\alpha)+ab(\alpha+\beta)+(a+b+c)(\beta\gamma+\gamma\alpha+\alpha\beta)$ or, since we suppose the battery in the conductor a and the galvanometer in α, we may put B the battery resistance for a and G the galvanometer resistance for α. We then find

$$\begin{aligned} D = BG(b+c+\beta+\gamma)&+B(b+\gamma)(c+\beta) \\ &+G(b+c)(\beta+\gamma)+bc(\beta+\gamma)+\beta\gamma(b+c). \end{aligned}$$

If the electromotive force E were made to act along OA, the resistance of OA being still α, and if the galvanometer were placed in BC, the resistance of BC being still a, then the value of D would remain the same, and the current in BC due to the electromotive force E acting along OA would be equal to the current in OA due to the electromotive force E acting in BC.

But if we simply disconnect the battery and the galvanometer, and without altering their respective resistances connect the battery to O and A and the galvanometer to B and C, then in the value of D we must exchange the values of B and G. If D' be the value of D after this exchange, we find

$$D-D' = (G-B)\{(b+c)(\beta+\gamma)-(b+\gamma)(\beta+c)\},$$
$$= (B-G)\{(b-\beta)(c-\gamma)\}.$$

Let us suppose that the resistance of the galvanometer is greater than that of the battery.

Let us also suppose that in its original position the galvanometer connects the junction of the two conductors of least resistance β, γ with the junction of the two conductors of greatest resistance b, c, or, in other words, we shall suppose that if the quantities b, c, γ, β are arranged in order of magnitude, b and c stand together, and γ and β stand together. Hence the quantities $b-\beta$ and $c-\gamma$ are of the same sign, so that their product is positive, and therefore $D-D'$ is of the same sign as $B-G$.

If therefore the galvanometer is made to connect the junction of the two greatest resistances with that of the two least, and if the galvanometer resistance is greater than that of the battery, then the value of D will be less, and the value of the deflexion of the galvanometer greater, than if the connexions are exchanged.

The rule therefore for obtaining the greatest galvanometer deflexion in a given system is as follows:

Of the two resistances, that of the battery and that of the galvanometer, connect the greater resistance so as to join the two greatest to the two least of the four other resistances.

349.] We shall suppose that we have to determine the ratio of the resistances of the conductors AB and AC, and that this is to be done by finding a point O on the conductor BOC, such that when the points A and O are connected by a wire, in the course of which a galvanometer is inserted, no sensible deflexion of the galvanometer needle occurs when the battery is made to act between B and C.

The conductor BOC may be supposed to be a wire of uniform resistance divided into equal parts, so that the ratio of the resistances of BO and OC may be read off at once.

Instead of the whole conductor being a uniform wire, we may make the part near O of such a wire, and the parts on each side may be coils of any form, the resistance of which is accurately known.

We shall now use a different notation instead of the symmetrical notation with which we commenced.

Let the whole resistance of BAC be R.

Let $c = mR$ and $b = (1-m)R$.

Let the whole resistance of BOC be S.

Let $\beta = nS$ and $\gamma = (1-n)S$.

The value of n is read off directly, and that of m is deduced from it when there is no sensible deviation of the galvanometer.

Let the resistance of the battery and its connexions be B, and that of the galvanometer and its connexions G.

We find as before

$$D = G\{BR+BS+RS\}+m(1-m)R^2(B+S)+n(1-n)S^2(B+R) + (m+n-2mn)BRS,$$

and if ξ is the current in the galvanometer wire

$$\xi = \frac{ERS}{D}(n-m).$$

In order to obtain the most accurate results we must make the deviation of the needle as great as possible compared with the value of $(n-m)$. This may be done by properly choosing the dimensions of the galvanometer and the standard resistance wire.

It will be shewn, when we come to Galvanometry, Art. 716, that when the form of a galvanometer wire is changed while its mass remains constant, the deviation of the needle for unit current is proportional to the length, but the resistance increases as the square of the length. Hence the maximum deflexion is shewn to occur when the resistance of the galvanometer wire is equal to the constant resistance of the rest of the circuit.

In the present case, if δ is the deviation,

$$\delta = C\sqrt{G}\xi,$$

where C is some constant, and G is the galvanometer resistance which varies as the square of the length of the wire. Hence we find that in the value of D, when δ is a maximum, the part involving G must be made equal to the rest of the expression.

If we also put $m = n$, as is the case if we have made a correct observation, we find the best value of G to be

$$G = n(1-n)(R+S).$$

This result is easily obtained by considering the resistance from A to O through the system, remembering that BC, being conjugate to AO, has no effect on this resistance.

In the same way we should find that if the total area of the acting surfaces of the battery is given, the most advantageous arrangement of the battery is when

$$B = \frac{RS}{R+S}.$$

Finally, we shall determine the value of S such that a given change in the value of n may produce the greatest galvanometer deflexion. By differentiating the expression for ξ we find

$$S^2 = \frac{BR}{B+R}\left(R + \frac{G}{n(1-n)}\right).$$

If we have a great many determinations of resistance to make in which the actual resistance has nearly the same value, then it may be worth while to prepare a galvanometer and a battery for this purpose. In this case we find that the best arrangement is

$$S = R, \qquad B = \tfrac{1}{2}R, \qquad G = 2n(1-n)R,$$

and if $n = \frac{1}{2}$, $G = \frac{1}{2}R$.

On the Use of Wheatstone's Bridge.

350.] We have already explained the general theory of Wheatstone's Bridge, we shall now consider some of its applications.

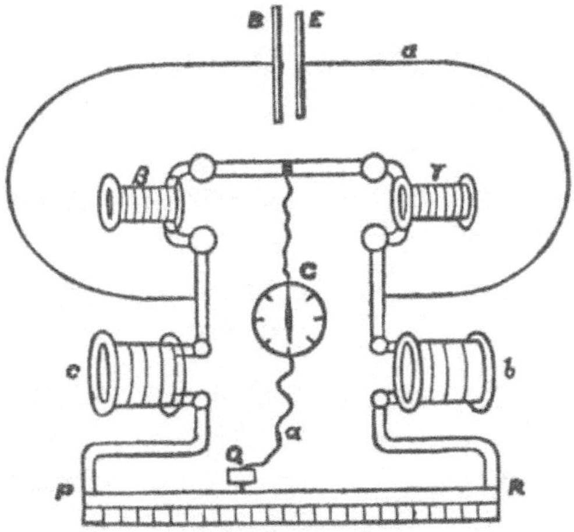

Fig. 33.

The comparison which can be effected with the greatest exactness is that of two equal resistances.

Let us suppose that β is a standard resistance coil, and that we wish to adjust γ to be equal in resistance to β.

Two other coils, b and c, are prepared which are equal or nearly equal to each other, and the four coils are placed with their electrodes in mercury cups so that the current of the battery is divided between two branches, one consisting of β and γ and the other of b and c. The coils b and c are connected by a wire PR, as uniform in its resistance as possible, and furnished with a scale of equal parts.

The galvanometer wire connects the junction of β and γ with a point Q of the wire PR, and the point of contact at Q is made to vary till on closing first the battery circuit and then the galvanometer circuit, no deflexion of the galvanometer needle is observed.

The coils β and γ are then made to change places, and a new position is found for Q. If this new position is the same as the old one, then we know that the exchange of β and γ has produced no change in the proportions of the resistances, and therefore γ is rightly adjusted. If Q has to be moved, the direction and amount of the change will indicate the nature and amount of the alteration of the length of the wire of γ, which will make its resistance equal to that of β.

If the resistances of the coils b and c, each including part of the wire PR up to its zero reading, are equal to that of b and c divisions of the wire respectively, then, if x is the scale reading of Q in the first case, and y that in the second,

$$\frac{c+x}{b-x} = \frac{\beta}{\gamma}, \qquad \frac{c+y}{b-y} = \frac{\gamma}{\beta},$$

whence
$$\frac{\gamma^2}{\beta^2} = 1 + \frac{(b+c)(y-x)}{(c+x)(b-y)}.$$

Since $b-y$ is nearly equal to $c+x$, and both are great with respect to x or y, we may write this

$$\frac{\gamma^2}{\beta^2} = 1 + 4\frac{y-x}{b+c},$$

and
$$\gamma = \beta\left(1 + 2\frac{y-x}{b+c}\right).$$

When γ is adjusted as well as we can, we substitute for b and c other coils of (say) ten times greater resistance.

The remaining difference between β and γ will now produce a ten times greater difference in the position of Q than with the original coils b and c, and in this way we can continually increase the accuracy of the comparison.

The adjustment by means of the wire with sliding contact piece is more quickly made than by means of a resistance box, and it is capable of continuous variation.

The battery must never be introduced instead of the galvanometer into the wire with a sliding contact, for the passage of a powerful current at the point of contact would injure the surface of the wire. Hence this arrangement is adapted for the case in which the resistance of the galvanometer is greater than that of the battery.

When γ, the resistance to be measured, a the resistance of the battery, and α the resistance of the galvanometer, are given, the best values of the other resistances have been shewn by Mr. Oliver Heaviside (Phil. Mag. Feb. 1873) to be

$$c = \sqrt{a\alpha},$$

$$b = \sqrt{\alpha\gamma\frac{a+\gamma}{\alpha+\gamma}},$$

$$\beta = \sqrt{a\gamma\frac{\alpha+\gamma}{a+\gamma}}.$$

On the Measurement of Small Resistances.

351.] When a short and thick conductor is introduced into a circuit its resistance is so small compared with the resistance occasioned by unavoidable faults in the connexions, such as want of contact or imperfect soldering, that no correct value of the resistance can be deduced from experiments made in the way described above.

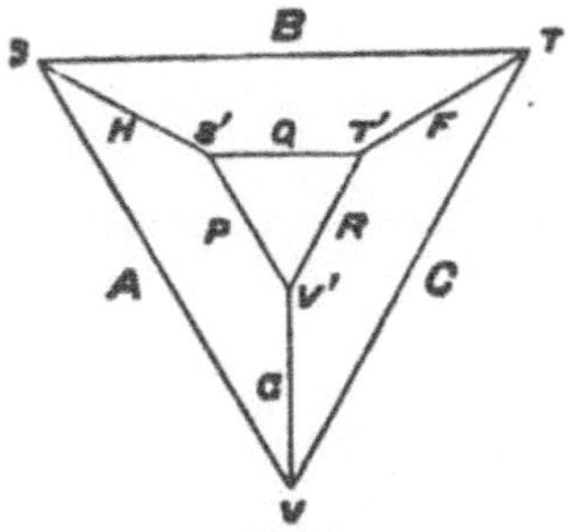

Fig. 34.

The object of such experiments is generally to determine the specific resistance of the substance, and it is resorted to in cases when the substance cannot be obtained in the form of a long thin wire, or when the resistance to transverse as well as to longitudinal conduction has to be measured.

Sir W. Thomson* has described a method applicable to such cases, which we may take as an example of a system of nine conductors.

* *Proc. R. S.*, June 6, 1861.

The most important part of the method consists in measuring the resistance, not of the whole length of the conductor, but of the part between two marks on the conductor at some little distance from its ends.

The resistance which we wish to measure is that experienced by a current whose intensity is uniform in any section of the conductor, and which flows in a direction parallel to its axis. Now close to the extremities, when the current is introduced by means of electrodes, either soldered, amalgamated, or simply pressed to the ends of the conductor, there is generally a want of uniformity in the distribution of the current in the conductor. At a short distance from the extremities the current becomes

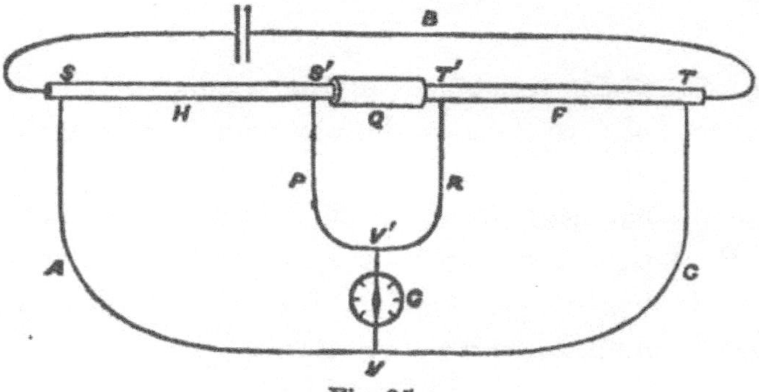

Fig. 35.

sensibly uniform. The student may examine for himself the investigation and the diagrams of Art. 193, where a current is introduced into a strip of metal with parallel sides through one of the sides, but soon becomes itself parallel to the sides.

The resistances of the conductors between certain marks S, S' and T, T' are to be compared.

The conductors are placed in series, and with connexions as perfectly conducting as possible, in a battery circuit of small resistance. A wire SVT is made to touch the conductors at S and T, and $S'V'T'$ is another wire touching them at S' and T'.

The galvanometer wire connects the points V and V' of these wires.

The wires SVT and $S'V'T'$ are of resistance so great that the resistance due to imperfect connexion at S, T, S' or T' may be neglected in comparison with the resistance of the wire, and V, V' are taken so that the resistances in the branches of either wire leading to the two conductors are nearly in the ratio of the resistances of the two conductors.

Calling H and F the resistances of the conductors SS' and $T'T$.

" A and C those of the branches SV and VT.

Calling P and R those of the branches $S'V'$ and $V'T'$.
„ Q that of the connecting piece $S'T'$.
„ B that of the battery and its connexions.
„ G that of the galvanometer and its connexions.

The symmetry of the system may be understood from the skeleton diagram. Fig. 34.

The condition that B the battery and G the galvanometer may be conjugate conductors is, in this case,

$$\frac{F}{C}-\frac{H}{A}+\left(\frac{R}{C}-\frac{P}{A}\right)\frac{Q}{P+Q+R}=0.$$

Now the resistance of the connector Q is as small as we can make it. If it were zero this equation would be reduced to

$$\frac{F}{C}=\frac{H}{A},$$

and the ratio of the resistances of the conductors to be compared would be that of C to A, as in Wheatstone's Bridge in the ordinary form.

In the present case the value of Q is small compared with P or with R, so that if we assume the points V, V' so that the ratio of R to C is nearly equal to that of P to A, the last term of the equation will vanish, and we shall have

$$F:H::C:A.$$

The success of this method depends in some degree on the perfection of the contact between the wires and the tested conductors at S, S', T' and T. In the following method, employed by Messrs. Matthiessen and Hockin*, this condition is dispensed with.

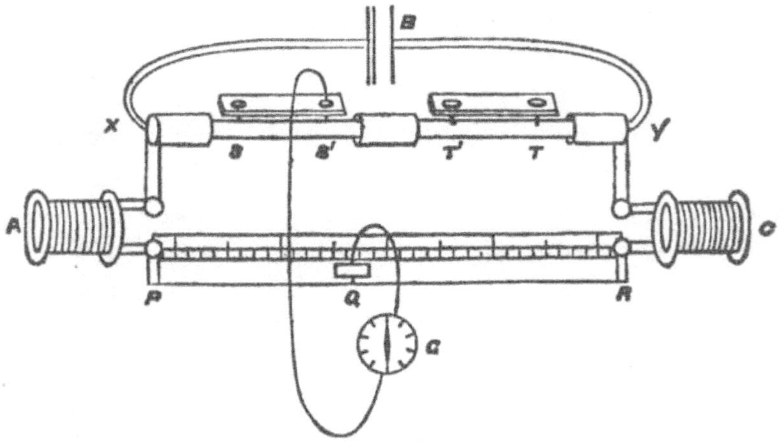

Fig. 36.

352.] The conductors to be tested are arranged in the manner

* *Laboratory*. Matthiessen and Hockin on Alloys.

already described, with the connexions as well made as possible, and it is required to compare the resistance between the marks SS' on the first conductor with the resistance between the marks $T'T$ on the second.

Two conducting points or sharp edges are fixed in a piece of insulating material so that the distance between them can be accurately measured. This apparatus is laid on the conductor to be tested, and the points of contact with the conductor are then at a known distance SS'. Each of these contact pieces is connected with a mercury cup, into which one electrode of the galvanometer may be plunged.

The rest of the apparatus is arranged, as in Wheatstone's Bridge, with resistance coils or boxes A and C, and a wire PR with a sliding contact piece Q, to which the other electrode of the galvanometer is connected.

Now let the galvanometer be connected to S and Q, and let A_1 and C_1 be so arranged, and the position of Q so determined, that there is no current in the galvanometer wire.

Then we know that
$$\frac{XS}{SY} = \frac{A_1 + PQ}{C_1 + QR}$$
where XS, PQ, &c. stand for the resistances in these conductors.

From this we get
$$\frac{XS}{XY} = \frac{A_1 + PQ_1}{A_1 + C_1 + PR}.$$

Now let the electrode of the galvanometer be connected to S', and let resistance be transferred from C to A (by carrying resistance coils from one side to the other) till electric equilibrium of the galvanometer wire can be obtained by placing Q at some point of the wire, say Q_2. Let the values of C and A be now C_2 and A_2, and let
$$A_2 + C_2 + PR = A_1 + C_1 + PR = R.$$

Then we have, as before,
$$\frac{XS'}{XY} = \frac{A_2 + PQ_2}{R}.$$

Whence
$$\frac{SS'}{XY} = \frac{A_2 - A_1 + Q_1 Q_2}{R}.$$

In the same way, placing the apparatus on the second conductor at TT' and again transferring resistance, we get, when the electrode is in T',
$$\frac{XT'}{XY} = \frac{A_3 + PQ_3}{R},$$

and when it is in T,

$$\frac{XT}{XY} = \frac{A_4 + PQ_4}{R}.$$

Whence

$$\frac{T'T}{XY} = \frac{A_4 - A_3 + Q_3 Q_4}{R}.$$

We can now deduce the ratio of the resistances SS' and $T'T$, for

$$\frac{SS'}{T'T} = \frac{A_2 - A_1 + Q_1 Q_2}{A_4 - A_3 + Q_3 Q_4}.$$

When great accuracy is not required we may dispense with the resistance coils A and C, and we then find

$$\frac{SS'}{T'T} = \frac{Q_1 Q_2}{Q_3 Q_4}.$$

The readings of the position of Q on a wire of a metre in length cannot be depended on to less than a tenth of a millimetre, and the resistance of the wire may vary considerably in different parts owing to inequality of temperature, friction, &c. Hence, when great accuracy is required, coils of considerable resistance are introduced at A and C, and the ratios of the resistances of these coils can be determined more accurately than the ratio of the resistances of the parts into which the wire is divided at Q.

It will be observed that in this method the accuracy of the determination depends in no degree on the perfection of the contacts at S, S' or T, T'.

This method may be called the differential method of using Wheatstone's Bridge, since it depends on the comparison of observations separately made.

An essential condition of accuracy in this method is that the resistance of the connexions should continue the same during the course of the four observations required to complete the determination. Hence the series of observations ought always to be repeated in order to detect any change in the resistances.

On the Comparison of Great Resistances.

353.] When the resistances to be measured are very great, the comparison of the potentials at different points of the system may be made by means of a delicate electrometer, such as the Quadrant Electrometer described in Art. 219.

If the conductors whose resistances are to be measured are placed in series, and the same current passed through them by means of a battery of great electromotive force, the difference of the potentials at the extremities of each conductor will be proportional to the resistance of that conductor. Hence, by connecting the electrodes of the electrometer with the extremities, first of one conductor and then of the other, the ratio of their resistances may be determined.

This is the most direct method of determining resistances. It involves the use of an electrometer whose readings may be depended on, and we must also have some guarantee that the current remains constant during the experiment.

Four conductors of great resistance may also be arranged as in Wheatstone's Bridge, and the bridge itself may consist of the electrodes of an electrometer instead of those of a galvanometer. The advantage of this method is that no permanent current is required to produce the deviation of the electrometer, whereas the galvanometer cannot be deflected unless a current passes through the wire.

354.] When the resistance of a conductor is so great that the current which can be sent through it by any available electromotive force is too small to be directly measured by a galvanometer, a condenser may be used in order to accumulate the electricity for a certain time, and then, by discharging the condenser through a galvanometer, the quantity accumulated may be estimated. This is Messrs. Bright and Clark's method of testing the joints of submarine cables.

355.] But the simplest method of measuring the resistance of such a conductor is to charge a condenser of great capacity and to connect its two surfaces with the electrodes of an electrometer and also with the extremities of the conductor. If E is the difference of potentials as shewn by the electrometer, S the capacity of the condenser, and Q the charge on either surface, R the resistance of the conductor and x the current in it, then, by the theory of condensers,

$$Q = SE.$$

By Ohm's Law,

$$E = Rx,$$

and by the definition of a current,

$$x = -\frac{dQ}{dt}.$$

Hence

$$-Q = RS\frac{dQ}{dt},$$

and

$$Q = Q_0 e^{-\frac{t}{RS}},$$

where Q_0 is the charge at first when $t = 0$.

Similarly $$E = E_0 e^{-\frac{t}{RS}}$$
where E_0 is the original reading of the electrometer, and E the same after a time t. From this we find

$$R = \frac{t}{S\{\log_e E_0 - \log_e E\}},$$

which gives R in absolute measure. In this expression a knowledge of the value of the unit of the electrometer scale is not required.

If S, the capacity of the condenser, is given in electrostatic measure as a certain number of metres, then R is also given in electrostatic measure as the reciprocal of a velocity.

If S is given in electromagnetic measure its dimensions are $\frac{T^2}{L}$, and R is a velocity.

Since the condenser itself is not a perfect insulator it is necessary to make two experiments. In the first we determine the resistance of the condenser itself, R_0, and in the second, that of the condenser when the conductor is made to connect its surfaces. Let this be R'. Then the resistance, R, of the conductor is given by the equation

$$\frac{1}{R} = \frac{1}{R'} - \frac{1}{R_0}.$$

This method has been employed by MM. Siemens.

Thomson's Method for the Determination of the Resistance of the Galvanometer.*

356.] An arrangement similar to Wheatstone's Bridge has been

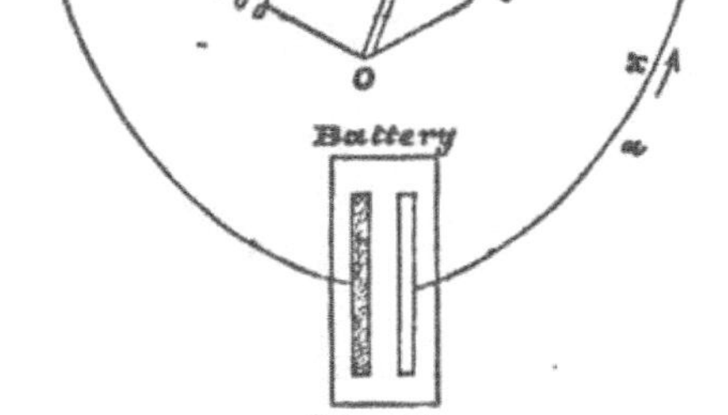

Fig. 37.

employed with advantage by Sir W. Thomson in determining the

* *Proc. R. S.*, Jan. 19, 1871.

resistance of the galvanometer when in actual use. It was suggested to Sir W. Thomson by Mance's Method. See Art. 357.

Let the battery be placed, as before, between B and C in the figure of Article 347, but let the galvanometer be placed in CA instead of in OA. If $b\beta - c\gamma$ is zero, then the conductor OA is conjugate to BC, and, as there is no current produced in OA by the battery in BC, the strength of the current in any other conductor is independent of the resistance in OA. Hence, if the galvanometer is placed in CA its deflexion will remain the same whether the resistance of OA is small or great. We therefore observe whether the deflexion of the galvanometer remains the same when O and A are joined by a conductor of small resistance, as when this connexion is broken, and if, by properly adjusting the resistances of the conductors, we obtain this result, we know that the resistance of the galvanometer is

$$b = \frac{c\gamma}{\beta},$$

where c, γ, and β are resistance coils of known resistance.

It will be observed that though this is not a null method, in the sense of there being no current in the galvanometer, it is so in the sense of the fact observed being the negative one, that the deflexion of the galvanometer is not changed when a certain contact is made. An observation of this kind is of greater value than an observation of the equality of two different deflexions of the same galvanometer, for in the latter case there is time for alteration in the strength of the battery or the sensitiveness of the galvanometer, whereas when the deflexion remains constant, in spite of certain changes which we can repeat at pleasure, we are sure that the current is quite independent of these changes.

The determination of the resistance of the coil of a galvanometer can easily be effected in the ordinary way of using Wheatstone's Bridge by placing another galvanometer in OA. By the method now described the galvanometer itself is employed to measure its own resistance.

Mance's Method of determining the Resistance of the Battery.*

357.] The measurement of the resistance of a battery when in action is of a much higher order of difficulty, since the resistance of the battery is found to change considerably for some time after

* *Proc. R. S.*, Jan. 19, 1871.

the strength of the current through it is changed. In many of the methods commonly used to measure the resistance of a battery such alterations of the strength of the current through it occur in the course of the operations, and therefore the results are rendered doubtful.

In Mance's method, which is free from this objection, the battery is placed in BC and the galvanometer in CA. The connexion between O and B is then alternately made and broken.

Now the deflexion of the galvanometer needle will remain unaltered, however the resistance in OB be changed, provided that OB and AC are conjugate. This may be regarded as a particular case of the result proved in Art. 347, or may be seen directly on the elimination of z and β from the equations of that article, viz. we then have

$$(a\alpha - c\gamma)x + (c\gamma + c\alpha + cb + b\alpha)y = E\alpha.$$

If y is independent of x, and therefore of β, we must have $a\alpha = c\gamma$. The resistance of the battery is thus obtained in terms of c, γ, α.

When the condition $a\alpha = c\gamma$ is fulfilled, the current through the galvanometer is then

$$\frac{E\alpha}{cb + \alpha(a+b+c)}, \text{ or } \frac{E\gamma}{ab + \gamma(a+b+c)}.$$

To test the sensibility of the method let us suppose that the condition $c\gamma = a\alpha$ is nearly, but not accurately, fulfilled, and that

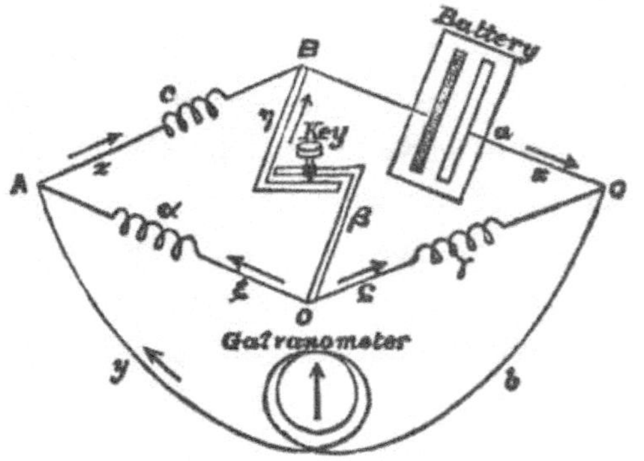

Fig. 38.

y_0 is the current through the galvanometer when O and B are connected by a conductor of no sensible resistance, and y_1 the current when O and B are completely disconnected.

To find these values we must make β equal to 0 and to ∞ in the general formula for y, and compare the results.

The general value for y is

$$\frac{c\gamma+\beta\gamma+\gamma a+a\beta}{D}E,$$

where D denotes the same expression as in Art. 348. Making use of the values of y given above we can then easily shew that the expressions for y_0 and y_1 are approximately

$$y+\frac{c(c\gamma-a\alpha)}{\gamma(c+\alpha)}\frac{y^2}{E},$$

and $$y-\frac{b(c\gamma-a\alpha)}{\gamma(\gamma+\alpha)}\frac{y^2}{E}.$$

From these values we find

$$\frac{y_0-y_1}{y}=\frac{\alpha}{\gamma}\,\frac{c\gamma-a\alpha}{(c+\alpha)(\alpha+\gamma)}.$$

The resistance, c, of the conductor AB should be equal to a, that of the battery; α and γ should be equal and as small as possible; and b should be equal to $\alpha+\gamma$.

Since a galvanometer is most sensitive when its deflexion is small, we should bring the needle nearly to zero by means of fixed magnets before making contact between O and B.

In this method of measuring the resistance of the battery, the current in the galvanometer is not in any way interfered with during the operation, so that we may ascertain the resistance of the battery for any given strength of current in the galvanometer so as to determine how the strength of the current affects the resistance *.

If y is the current in the galvanometer, the actual current through the battery is x_0 with the key down and x_1 with the key up, where

$$x_0=y\left(1+\frac{b}{\gamma}+\frac{ac}{\gamma(a+c)}\right),\qquad x_1=y\left(1+\frac{b}{\alpha+\gamma}\right),$$

the resistance of the battery is

$$a=\frac{c\gamma}{\alpha},$$

and the electromotive force of the battery is

$$E=y\left(b+c+\frac{c}{\alpha}(b+\gamma)\right).$$

* [In the *Philosophical Magazine* for 1857, vol. i. pp. 515-525, Mr. Oliver Lodge has pointed out as a defect in Mance's method that as the electromotive force of the battery depends upon the current passing through the battery, the deflexion of the galvanometer needle cannot be the same in the two cases when the key is down or up, if the equation $a\alpha=c\gamma$ is true. Mr. Lodge describes a modification of Mance's method which he has employed with success.]

The method of Art. 356 for finding the resistance of the galvanometer differs from this only in making and breaking contact between O and A instead of between O and B, and by exchanging a and β we obtain for this case

$$\frac{y_0-y_1}{y}=\frac{\beta}{\gamma}\frac{c\gamma-b\beta}{(c+\beta)(\beta+\gamma)}.$$

On the Comparison of Electromotive Forces.

358.] The following method of comparing the electromotive forces of voltaic and thermoelectric arrangements, when no current passes through them, requires only a set of resistance coils and a constant battery.

Let the electromotive force E of the battery be greater than that of either of the electromotors to be compared, then, if a sufficient

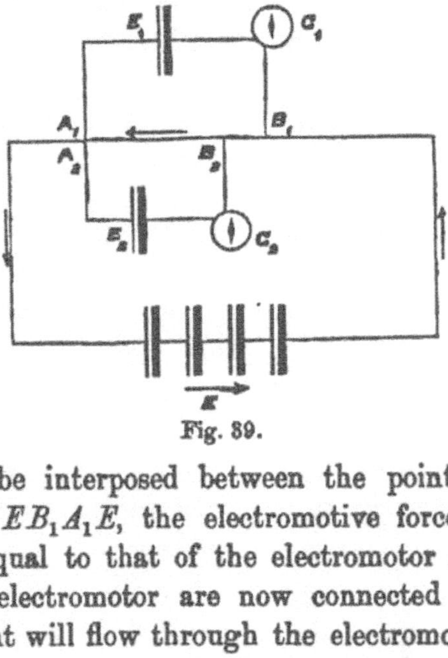

Fig. 39.

resistance, R_1, be interposed between the points A_1, B_1 of the primary circuit EB_1A_1E, the electromotive force from B_1 to A_1 may be made equal to that of the electromotor E_1. If the electrodes of this electromotor are now connected with the points A_1, B_1 no current will flow through the electromotor. By placing a galvanometer G_1 in the circuit of the electromotor E_1, and adjusting the resistance between A_1 and B_1, till the galvanometer G_1 indicates no current, we obtain the equation

$$E_1 = R_1 C,$$

where R_1 is the resistance between A_1 and B_1, and C is the strength of the current in the primary circuit.

In the same way, by taking a second electromotor E_2 and placing its electrodes at A_2 and B_2, so that no current is indicated by the galvanometer G_2,

$$E_2 = R_2 C,$$

where R_2 is the resistance between A_2 and B_2. If the observations of the galvanometers G_1 and G_2 are simultaneous, the value of C, the current in the primary circuit, is the same in both equations, and we find

$$E_1 : E_2 :: R_1 : R_2.$$

In this way the electromotive force of two electromotors may be compared. The absolute electromotive force of an electromotor may be measured either electrostatically by means of the electrometer, or electromagnetically by means of an absolute galvanometer.

This method, in which, at the time of the comparison, there is no current through either of the electromotors, is a modification of Poggendorff's method, and is due to Mr. Latimer Clark, who has deduced the following values of electromotive forces:

			Concentrated solution of		Volts.
Daniell I.	Amalgamated Zinc	$HSO_4 + 4$ aq.	$Cu\,SO_4$	Copper	= 1.079
II.	"	$HSO_4 + 12$ aq.	$Cu\,SO_4$	Copper	= 0.978
III.	"	$HSO_4 + 12$ aq.	$Cu\,NO_6$	Copper	= 1.00
Bunsen I.	"	" "	$H\,NO_6$	Carbon	= 1.964
II.	"	" "	sp. g. 1.38	Carbon	= 1.888
Grove	"	$HSO_4 + 4$ aq.	$H\,NO_6$	Platinum	= 1.956

A Volt is an electromotive force equal to 100,000,000 *units of the centimetre-gramme-second system.*

CHAPTER XII.

ON THE ELECTRIC RESISTANCE OF SUBSTANCES.

359.] THERE are three classes in which we may place different substances in relation to the passage of electricity through them.

The first class contains all the metals and their alloys, some sulphurets, and other compounds containing metals, to which we must add carbon in the form of gas-coke, and selenium in the crystalline form.

In all these substances conduction takes place without any decomposition, or alteration of the chemical nature of the substance, either in its interior or where the current enters and leaves the body. In all of them the resistance increases as the temperature rises.

The second class consists of substances which are called electrolytes, because the current is associated with a decomposition of the substance into two components which appear at the electrodes. As a rule a substance is an electrolyte only when in the liquid form, though certain colloid substances, such as glass at 100°C, which are apparently solid, are electrolytes. It would appear from the experiments of Sir B. C. Brodie that certain gases are capable of electrolysis by a powerful electromotive force.

In all substances which conduct by electrolysis the resistance diminishes as the temperature rises.

The third class consists of substances the resistance of which is so great that it is only by the most refined methods that the passage of electricity through them can be detected. These are called Dielectrics. To this class belong a considerable number of solid bodies, many of which are electrolytes when melted, some liquids, such as turpentine, naphtha, melted paraffin, &c., and all gases and vapours. Carbon in the form of diamond, and selenium in the amorphous form, belong to this class.

The resistance of this class of bodies is enormous compared with that of the metals. It diminishes as the temperature rises. It is difficult, on account of the great resistance of these substances, to determine whether the feeble current which we can force through them is or is not associated with electrolysis.

On the Electric Resistance of Metals.

360.] There is no part of electrical research in which more numerous or more accurate experiments have been made than in the determination of the resistance of metals. It is of the utmost importance in the electric telegraph that the metal of which the wires are made should have the smallest attainable resistance. Measurements of resistance must therefore be made before selecting the materials. When any fault occurs in the line, its position is at once ascertained by measurements of resistance, and these measurements, in which so many persons are now employed, require the use of resistance coils, made of metal the electrical properties of which have been carefully tested.

The electrical properties of metals and their alloys have been studied with great care by MM. Matthiessen, Vogt, and Hockin, and by MM. Siemens, who have done so much to introduce exact electrical measurements into practical work.

It appears from the researches of Dr. Matthiessen, that the effect of temperature on the resistance is nearly the same for a considerable number of the *pure* metals, the resistance at 100°C being to that at 0°C in the ratio of 1.414 to 1, or of 100 to 70.7. For pure iron the ratio is 1.645, and for pure thallium 1.458.

The resistance of metals has been observed by Dr. C. W. Siemens* through a much wider range of temperature, extending from the freezing point to 350°C, and in certain cases to 1000°C. He finds that the resistance increases as the temperature rises, but that the rate of increase diminishes as the temperature rises. The formula, which he finds to agree very closely both with the resistances observed at low temperatures by Dr. Matthiessen and with his own observations through a range of 1000°C, is

$$r = \alpha T^{\frac{1}{2}} + \beta T + \gamma,$$

where T is the absolute temperature reckoned from -273°C, and α, β, γ are constants. Thus, for

Platinum......$r = 0.039369\, T^{\frac{1}{2}} + 0.00216407\, T - 0.2413$,

Copper.........$r = 0.026577\, T^{\frac{1}{2}} + 0.0031443\, T - 0.22751$,

Iron............$r = 0.072545\, T^{\frac{1}{2}} + 0.0038133\, T - 1.23971$.

* *Proc. R. S.*, April 27, 1871.

From data of this kind the temperature of a furnace may be determined by means of an observation of the resistance of a platinum wire placed in the furnace.

Dr. Matthiessen found that when two metals are combined to form an alloy, the resistance of the alloy is in most cases greater than that calculated from the resistance of the component metals and their proportions. In the case of alloys of gold and silver, the resistance of the alloy is greater than that of either pure gold or pure silver, and, within certain limiting proportions of the constituents, it varies very little with a slight alteration of the proportions. For this reason Dr. Matthiessen recommended an alloy of two parts by weight of gold and one of silver as a material for reproducing the unit of resistance.

The effect of change of temperature on electric resistance is generally less in alloys than in pure metals.

Hence ordinary resistance coils are made of German silver, on account of its great resistance and its small variation with temperature.

An alloy of silver and platinum is also used for standard coils.

361.] The electric resistance of some metals changes when the metal is annealed; and until a wire has been tested by being repeatedly raised to a high temperature without permanently altering its resistance, it cannot be relied on as a measure of resistance. Some wires alter in resistance in course of time without having been exposed to changes of temperature. Hence it is important to ascertain the specific resistance of mercury, a metal which being fluid has always the same molecular structure, and which can be easily purified by distillation and treatment with nitric acid. Great care has been bestowed in determining the resistance of this metal by W. and C. F. Siemens, who introduced it as a standard. Their researches have been supplemented by those of Matthiessen and Hockin.

The specific resistance of mercury was deduced from the observed resistance of a tube of length l containing a weight w of mercury, in the following manner.

No glass tube is of exactly equal bore throughout, but if a small quantity of mercury is introduced into the tube and occupies a length λ of the tube, the middle point of which is distant x from one end of the tube, then the area s of the section near this point will be $s = \frac{C}{\lambda}$, where C is some constant.

The weight of mercury which fills the whole tube is

$$w = \rho \int s\,dx = \rho C \Sigma \left(\frac{1}{\lambda}\right) \frac{l}{n},$$

where n is the number of points, at equal distances along the tube, where λ has been measured, and ρ is the mass of unit of volume.

The resistance of the whole tube is

$$R = \int \frac{r}{s}\,dx = \frac{r}{C} \Sigma (\lambda) \frac{l}{n},$$

where r is the specific resistance per unit of volume.

Hence $$wR = r\rho \Sigma (\lambda) \Sigma \left(\frac{1}{\lambda}\right) \frac{l^2}{n^2},$$

and $$r = \frac{wR}{\rho l^2} \frac{n^2}{\Sigma (\lambda) \Sigma \left(\frac{1}{\lambda}\right)}$$

gives the specific resistance of unit of volume.

To find the resistance of unit of length and unit of mass we must multiply this by the density.

It appears from the experiments of Matthiessen and Hockin that the resistance of a uniform column of mercury of one metre in length, and weighing one gramme at 0°C, is 13.071 Ohms, whence it follows that if the specific gravity of mercury is 13.595, the resistance of a column of one metre in length and one square millimetre in section is 0.96146 Ohms.

362.] In the following table R is the resistance in Ohms of a column one metre long and one gramme weight at 0°C, and r is the resistance in centimetres per second of a cube of one centimetre, according to the experiments of Matthiessen *.

	Specific gravity		R	r	Percentage increment of resistance for 1°C at 20°C.
Silver	10.50	hard drawn	0.1689	1609	0.377
Copper	8.95	hard drawn	0.1469	1642	0.388
Gold	19.27	hard drawn	0.4150	2154	0.365
Lead	11.391	pressed	2.257	19847	0.387
Mercury	13.595	liquid	13.071	96146	0.072
Gold 2, Silver 1. .	15.218	hard or annealed	1.668	10988	0.065
Selenium at 100°C		Crystalline form		6×10^{13}	1.00

* *Phil. Mag.*, May, 1865.

On the Electric Resistance of Electrolytes.

363.] The measurement of the electric resistance of electrolytes is rendered difficult on account of the polarization of the electrodes, which causes the observed difference of potentials of the metallic electrodes to be greater than the electromotive force which actually produces the current.

This difficulty can be overcome in various ways. In certain cases we can get rid of polarization by using electrodes of proper material, as, for instance, zinc electrodes in a solution of sulphate of zinc. By making the surface of the electrodes very large compared with the section of the part of the electrolyte whose resistance is to be measured, and by using only currents of short duration in opposite directions alternately, we can make the measurements before any considerable intensity of polarization has been excited by the passage of the current.

Finally, by making two different experiments, in one of which the path of the current through the electrolyte is much longer than in the other, and so adjusting the electromotive force that the actual current, and the time during which it flows, are nearly the same in each case, we can eliminate the effect of polarization altogether.

364.] In the experiments of Dr. Paalzow * the electrodes were in the form of large disks placed in separate flat vessels filled with the electrolyte, and the connexion was made by means of a long siphon filled with the electrolyte and dipping into both vessels. Two such siphons of different lengths were used.

The observed resistances of the electrolyte in these siphons being R_1 and R_2, the siphons were next filled with mercury, and their resistances when filled with mercury were found to be R_1' and R_2'.

The ratio of the resistance of the electrolyte to that of a mass of mercury at 0°C of the same form was then found from the formula

$$\rho = \frac{R_1 - R_2}{R_1' - R_2'}.$$

To deduce from the values of ρ the resistance of a centimetre in length having a section of a square centimetre, we must multiply them by the value of r for mercury at 0°C. See Art. 361.

* *Berlin Monatsbericht*, July, 1868.

The results given by Paalzow are as follow :—

Mixtures of Sulphuric Acid and Water.

	Temp.	Resistance compared with mercury.
H_2SO_4	......15°C	96950
H_2SO_4 + 14 H^2O	......19°C	14157
H_2SO_4 + 13 H^2O	......22°C	13310
H_2SO_4 + 499 H^2O	......22°C	184773

Sulphate of Zinc and Water.

$ZnSO_4$ + 23 H^2O	......23°C	194400
$ZnSO_4$ + 24 H^2O	......23°C	191000
$ZnSO_4$ + 105 H^2O	......23°C	354000

Sulphate of Copper and Water.

$CuSO_4$ + 45 H^2O	......22°C	202410
$CuSO_4$ + 105 H^2O	......22°C	339341

Sulphate of Magnesium and Water.

$MgSO_4$ + 34 H^2O	......22°C	199180
$MgSO_4$ + 107 H^2O	......22°C	324600

Hydrochloric Acid and Water.

HCl + 15 H^2O	......23°C	13626
HCl + 500 H^2O	......23°C	86679

365.] MM. F. Kohlrausch and W. A. Nippoldt * have determined the resistance of mixtures of sulphuric acid and water. They used alternating magneto-electric currents, the electromotive force of which varied from $\frac{1}{2}$ to $\frac{1}{74}$ of that of a Grove's cell, and by means of a thermoelectric copper-iron pair they reduced the electromotive force to $\frac{1}{429000}$ of that of a Grove's cell. They found that Ohm's law was applicable to this electrolyte throughout the range of these electromotive forces.

The resistance is a minimum in a mixture containing about one-third of sulphuric acid.

The resistance of electrolytes diminishes as the temperature increases. The percentage increment of conductivity for a rise of 1°C is given in the following table.

* Pogg., *Ann.* cxxxviii. p. 286, Oct. 1869.

Resistance of Mixtures of Sulphuric Acid and Water at 22°C *in terms of Mercury at* 0°C. MM. Kohlrausch and Nippoldt.

Specific gravity at 18°5	Percentage of H_2SO_4	Resistance at 22°C (Hg = 1)	Percentage increment of conductivity for 1°C
0.9985	0.0	746300	0.47
1.00	0.2	465100	0.47
1.0504	8.3	34530	0.653
1.0989	14.2	18946	0.646
1.1431	20.2	14990	0.799
1.2045	28.0	13133	1.317
1.2631	35.2	13132	1.259
1.3163	41.5	14286	1.410
1.3547	46.0	15762	1.674
1.3994	50.4	17726	1.582
1.4482	55.2	20796	1.417
1.5026	60.3	25574	1.794

On the Electrical Resistance of Dielectrics.

366.] A great number of determinations of the resistance of gutta-percha, and other materials used as insulating media, in the manufacture of telegraphic cables, have been made in order to ascertain the value of these materials as insulators.

The tests are generally applied to the material after it has been used to cover the conducting wire, the wire being used as one electrode, and the water of a tank, in which the cable is plunged, as the other. Thus the current is made to pass through a cylindrical coating of the insulator of great area and small thickness.

It is found that when the electromotive force begins to act, the current, as indicated by the galvanometer, is by no means constant. The first effect is of course a transient current of considerable intensity, the total quantity of electricity being that required to charge the surfaces of the insulator with the superficial distribution of electricity corresponding to the electromotive force. This first current therefore is a measure not of the conductivity, but of the capacity of the insulating layer.

But even after this current has been allowed to subside the residual current is not constant, and does not indicate the true conductivity of the substance. It is found that the current continues to decrease for at least half an hour, so that a determination of the resistance deduced from the current will give a greater value if a certain time is allowed to elapse than if taken immediately after applying the battery.

Thus, with Hooper's insulating material the apparent resistance at the end of ten minutes was four times, and at the end of nineteen hours twenty-three times that observed at the end of one minute. When the direction of the electromotive force is reversed, the resistance falls as low or lower than at first and then gradually rises.

These phenomena seem to be due to a condition of the gutta-percha, which, for want of a better name, we may call polarization, and which we may compare on the one hand with that of a series of Leyden jars charged by cascade, and, on the other, with Ritter's secondary pile, Art. 271.

If a number of Leyden jars of great capacity are connected in series by means of conductors of great resistance (such as wet cotton threads in the experiments of M. Gaugain), then an electromotive force acting on the series will produce a current, as indicated by a galvanometer, which will gradually diminish till the jars are fully charged.

The apparent resistance of such a series will increase, and if the dielectric of the jars is a perfect insulator it will increase without limit. If the electromotive force be removed and connexion made between the ends of the series, a reverse current will be observed, the total quantity of which, in the case of perfect insulation, will be the same as that of the direct current. Similar effects are observed in the case of the secondary pile, with the difference that the final insulation is not so good, and that the capacity per unit of surface is immensely greater.

In the case of the cable covered with gutta-percha, &c , it is found that after applying the battery for half an hour, and then connecting the wire with the external electrode, a reverse current takes place, which goes on for some time, and gradually reduces the system to its original state.

These phenomena are of the same kind with those indicated by the 'residual discharge' of the Leyden jar, except that the amount of the polarization is much greater in gutta-percha, &c. than in glass.

This state of polarization seems to be a directed property of the material, which requires for its production not only electromotive force, but the passage, by displacement or otherwise, of a considerable quantity of electricity, and this passage requires a considerable time. When the polarized state has been set up, there is an internal electromotive force acting in the substance in the reverse direction, which will continue till it has either produced a reversed current equal in total quantity to the first, or till the state of polarization has quietly subsided by means of true conduction through the substance.

The whole theory of what has been called residual discharge, absorption of electricity, electrification, or polarization, deserves a careful investigation, and will probably lead to important discoveries relating to the internal structure of bodies.

367.] The resistance of the greater number of dielectrics diminishes as the temperature rises.

Thus the resistance of gutta-percha is about twenty times as great at 0°C as at 24°C. Messrs. Bright and Clark have found that the following formula gives results agreeing with their experiments. If r is the resistance of gutta-percha at temperature T centigrade, then the resistance at temperature $T+t$ will be

$$R = r \times 0.8878^t,$$

the number varies between 0.8878 and 0.9.

Mr. Hockin has verified the curious fact that it is not until some hours after the gutta-percha has taken its temperature that the resistance reaches its corresponding value.

The effect of temperature on the resistance of india-rubber is not so great as on that of gutta-percha.

The resistance of gutta-percha increases considerably on the application of pressure.

The resistance, in Ohms, of a cubic metre of various specimens of gutta-percha used in different cables is as follows *.

Name of Cable.	
Red Sea	$.267 \times 10^{12}$ to $.362 \times 10^{12}$
Malta-Alexandria	1.23×10^{12}
Persian Gulf	1.80×10^{12}
Second Atlantic	3.42×10^{12}
Hooper's Persian Gulf Core	74.7×10^{12}
Gutta-percha at 24°C	3.53×10^{12}

368.] The following table, calculated from the experiments of

* Jenkin's *Cantor Lectures*.

M. Buff, described in Art. 271, shews the resistance of a cubic metre of glass in Ohms at different temperatures.

Temperature.	Resistance.
200°C	227000
250°	13900
300°	1480
350°	1035
400°	735

369.] Mr. C. F. Varley * has recently investigated the conditions of the current through rarefied gases, and finds that the electromotive force E is equal to a constant E_0 together with a part depending on the current according to Ohm's Law, thus

$$E = E_0 + RC.$$

For instance, the electromotive force required to cause the current to begin in a certain tube was that of 323 Daniell's cells, but an electromotive force of 304 cells was just sufficient to maintain the current. The intensity of the current, as measured by the galvanometer, was proportional to the number of cells above 304. Thus for 305 cells the deflexion was 2, for 306 it was 4, for 307 it was 6, and so on up to 380, or $304+76$ for which the deflexion was 150, or 76×1.97.

From these experiments it appears that there is a kind of polarization of the electrodes, the electromotive force of which is equal to that of 304 Daniell's cells, and that up to this electromotive force the battery is occupied in establishing this state of polarization. When the maximum polarization is established, the excess of electromotive force above that of 304 cells is devoted to maintaining the current according to Ohm's Law.

The law of the current in a rarefied gas is therefore very similar to the law of the current through an electrolyte in which we have to take account of the polarization of the electrodes.

In connexion with this subject we should study Thomson's results, described in Art. 57, in which the electromotive force required to produce a spark in air was found to be proportional not to the distance, but to the distance together with a constant quantity. The electromotive force corresponding to this constant quantity may be regarded as the intensity of polarization of the electrodes.

370.] MM. Wiedemann and Rühlmann have recently † investi-

* *Proc. R. S.*, Jan. 12, 1871.

† *Berichte der Königl. Sächs. Gesellschaft*, Oct. 20, 1871.

gated the passage of electricity through gases. The electric current was produced by Holtz's machine, and the discharge took place between spherical electrodes within a metallic vessel containing rarefied gas. The discharge was in general discontinuous, and the interval of time between successive discharges was measured by means of a mirror revolving along with the axis of Holtz's machine. The images of the series of discharges were observed by means of a heliometer with a divided object-glass, which was adjusted till one image of each discharge coincided with the other image of the next discharge. By this method very consistent results were obtained. It was found that the quantity of electricity in each discharge is independent of the strength of the current and of the material of the electrodes, and that it depends on the nature and density of the gas, and on the distance and form of the electrodes.

These researches confirm the statement of Faraday * that the electric tension (see Art. 48) required to cause a disruptive discharge to begin at the electrified surface of a conductor is a little less when the electrification is negative than when it is positive, but that when a discharge does take place, much more electricity passes at each discharge when it begins at a positive surface. They also tend to support the hypothesis stated in Art. 57, that the stratum of gas condensed on the surface of the electrode plays an important part in the phenomenon, and they indicate that this condensation is greatest at the positive electrode.

* *Exp. Res.*, 1501.

PLATES.

Fig. I.

Art. 118.

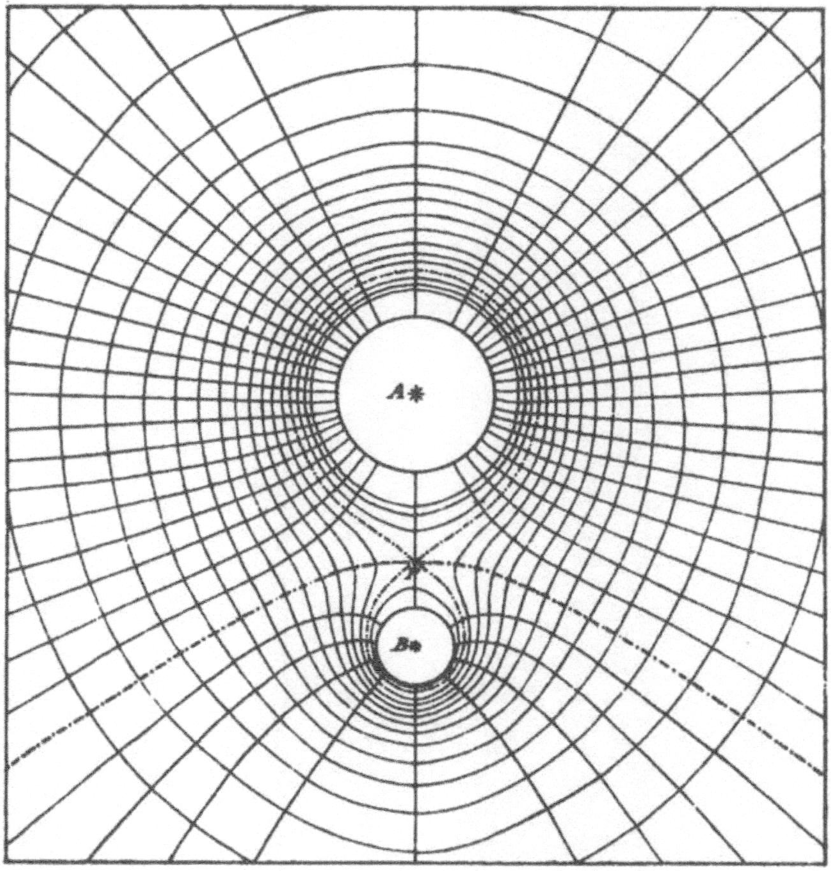

Lines of Force and Equipotential Surfaces.

$A = 20$. $B = 5$ P, Point of Equilibrium. $AP = \frac{2}{3} AB$.

FIG. II.
Art 119.

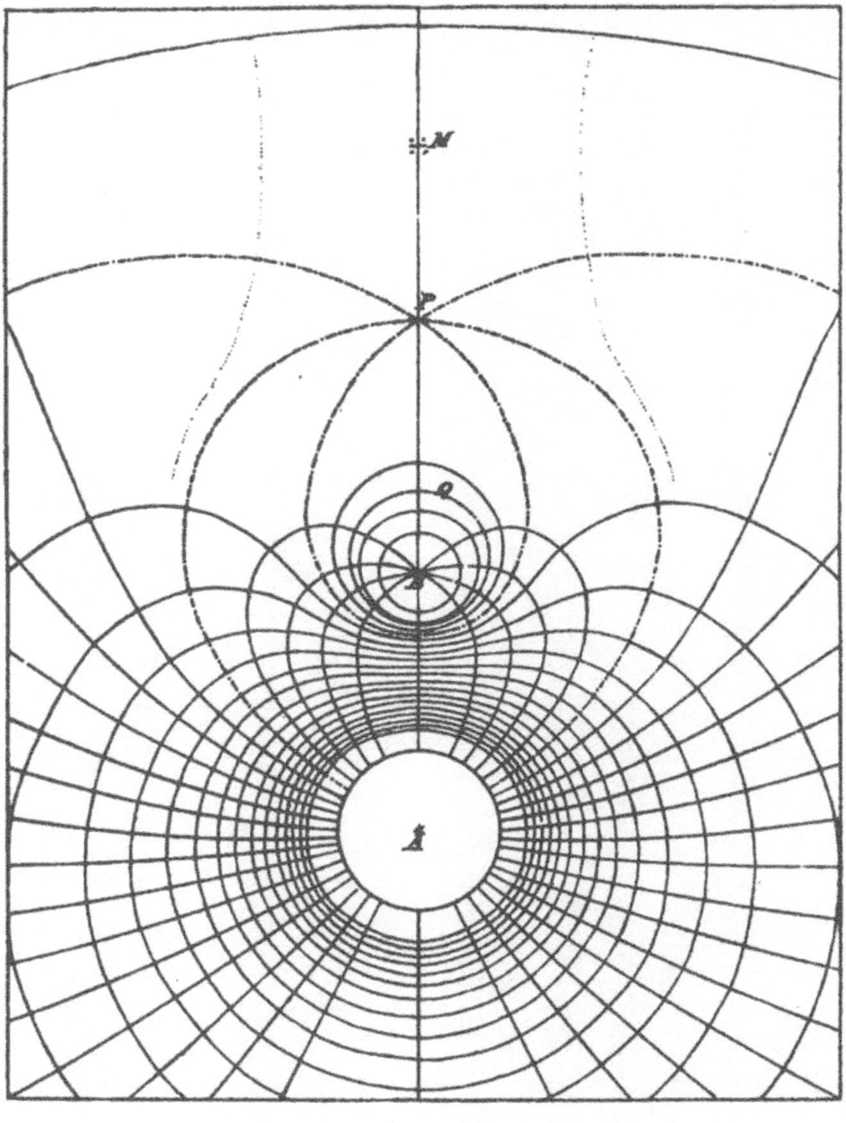

Lines of Force and Equipotential Surfaces.

A = 20 B = −5 P, Point of Equilibrium. AP = 2 AB

Q, Spherical surface of Zero potential.

M. Point of Maximum Force along the axis.

The dotted line is the Line of Force Ψ = 0.1 thus ———

FIG. III.

Art. 120.

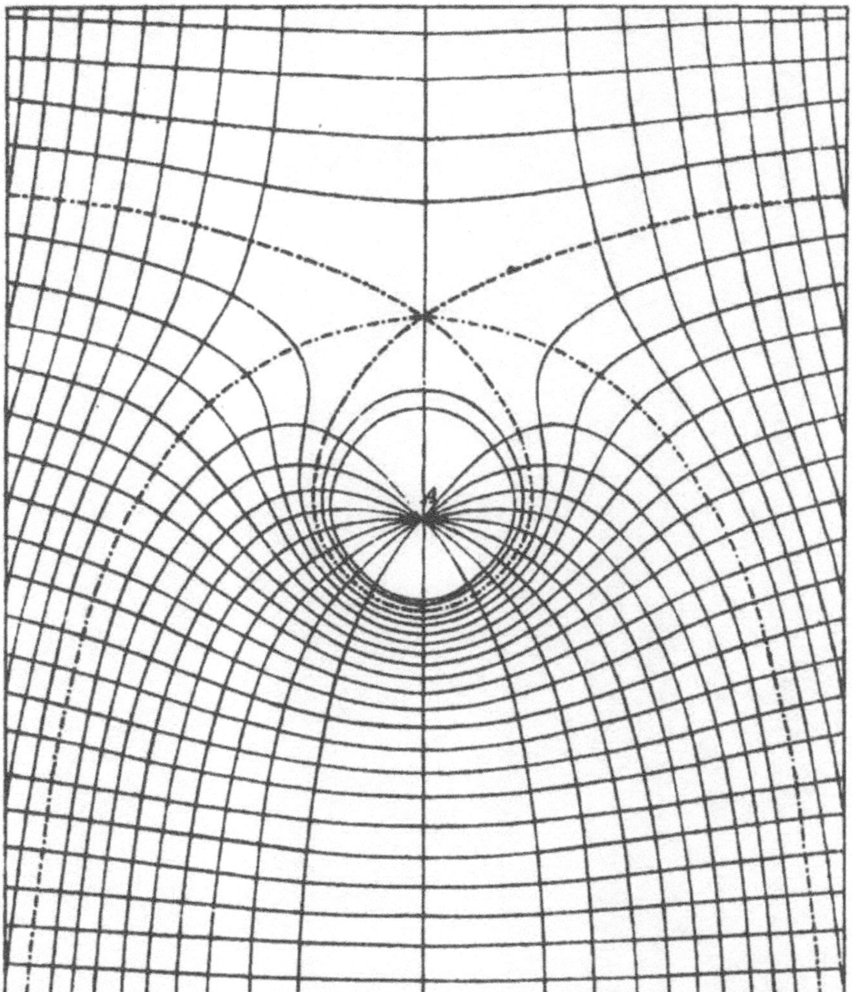

Lines of Force and Equipotential Surfaces.

$A = 10$.

FIG. IV.
Art. 121.

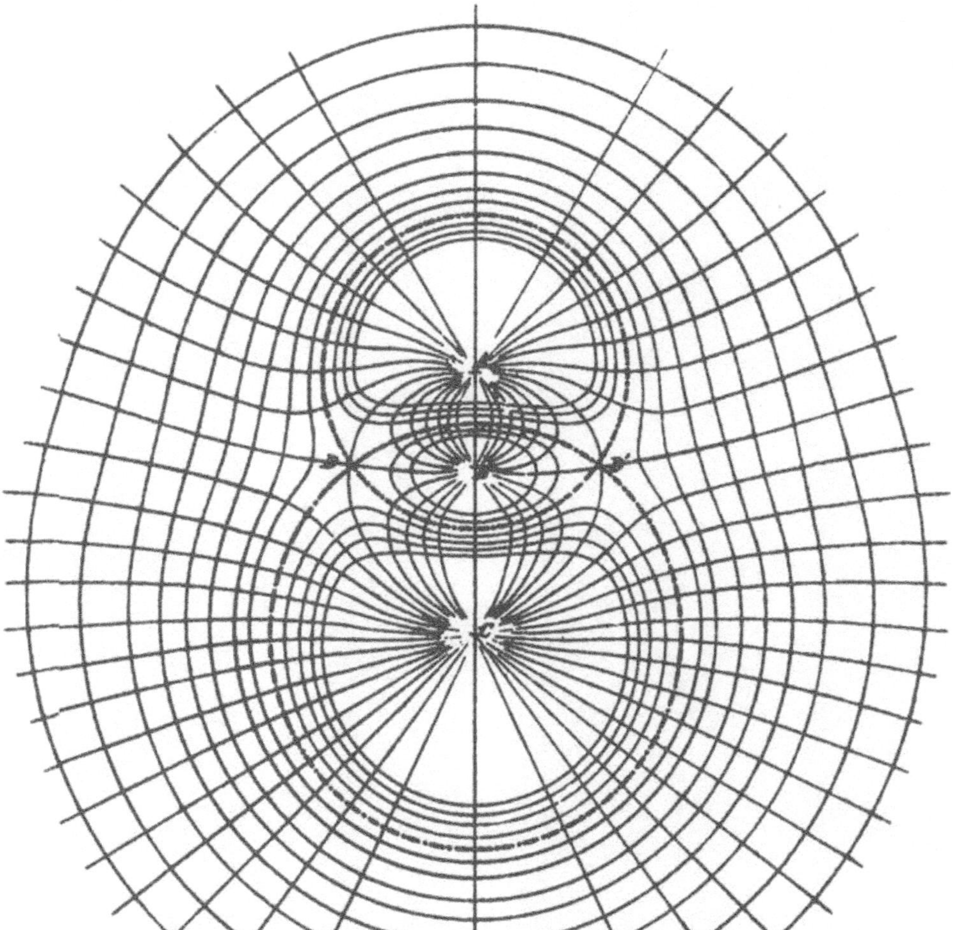

Lines of Force and Equipotential Surfaces.

A = 16. *B* = −12. *C* = 20.

FIG. V.

Art 143

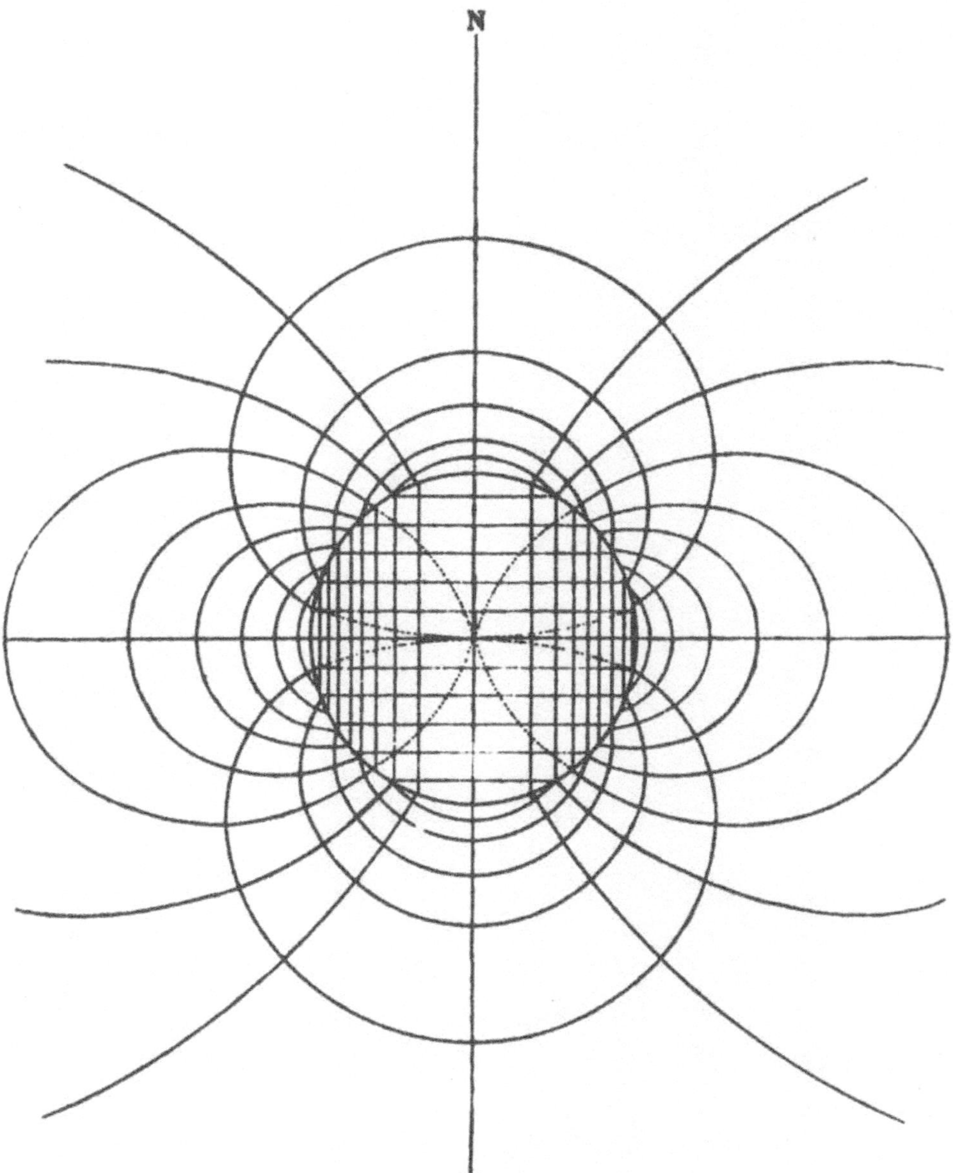

Lines of Force and Equipotential Surfaces in a diametral section of a spherical Surface in which the superficial density is a harmonic of the first degree.

FIG VI.

Art 143.

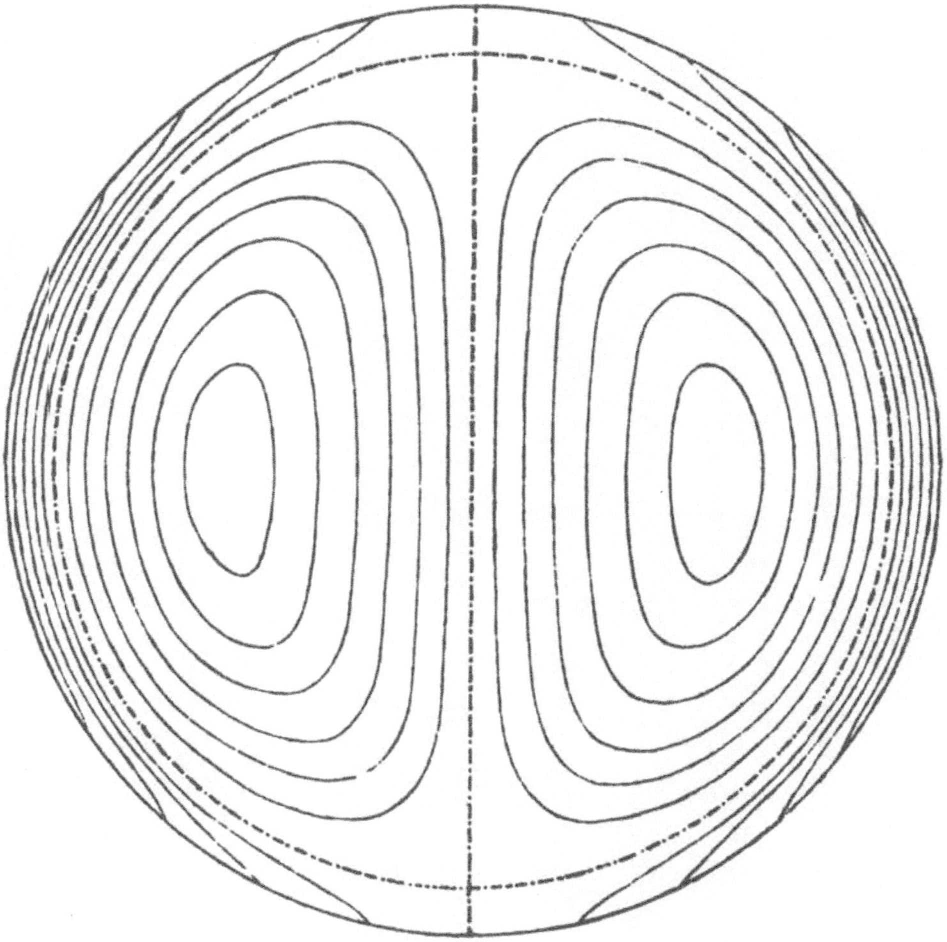

Spherical Harmonic of the third order

$n = 3$ $\sigma = 1.$

FIG. VII.

Art. 143.

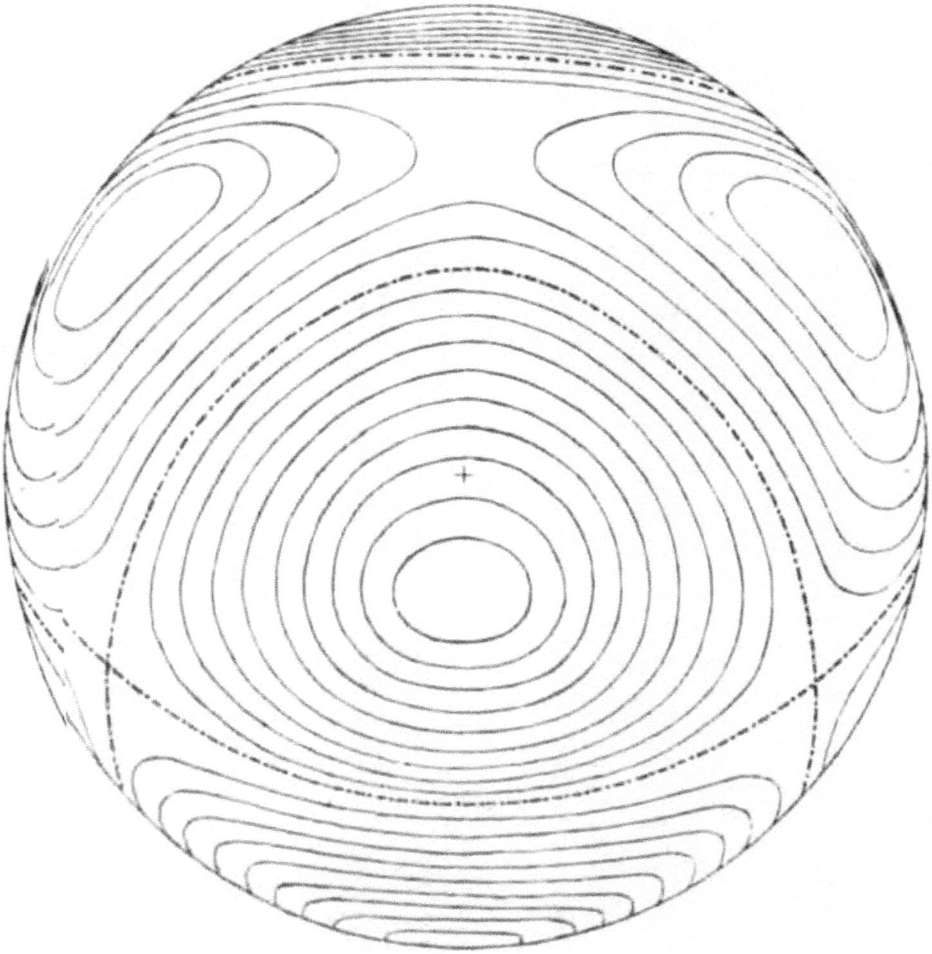

Spherical Harmonic of the third order.

$n = 3$

FIG. VIII.
Art 143

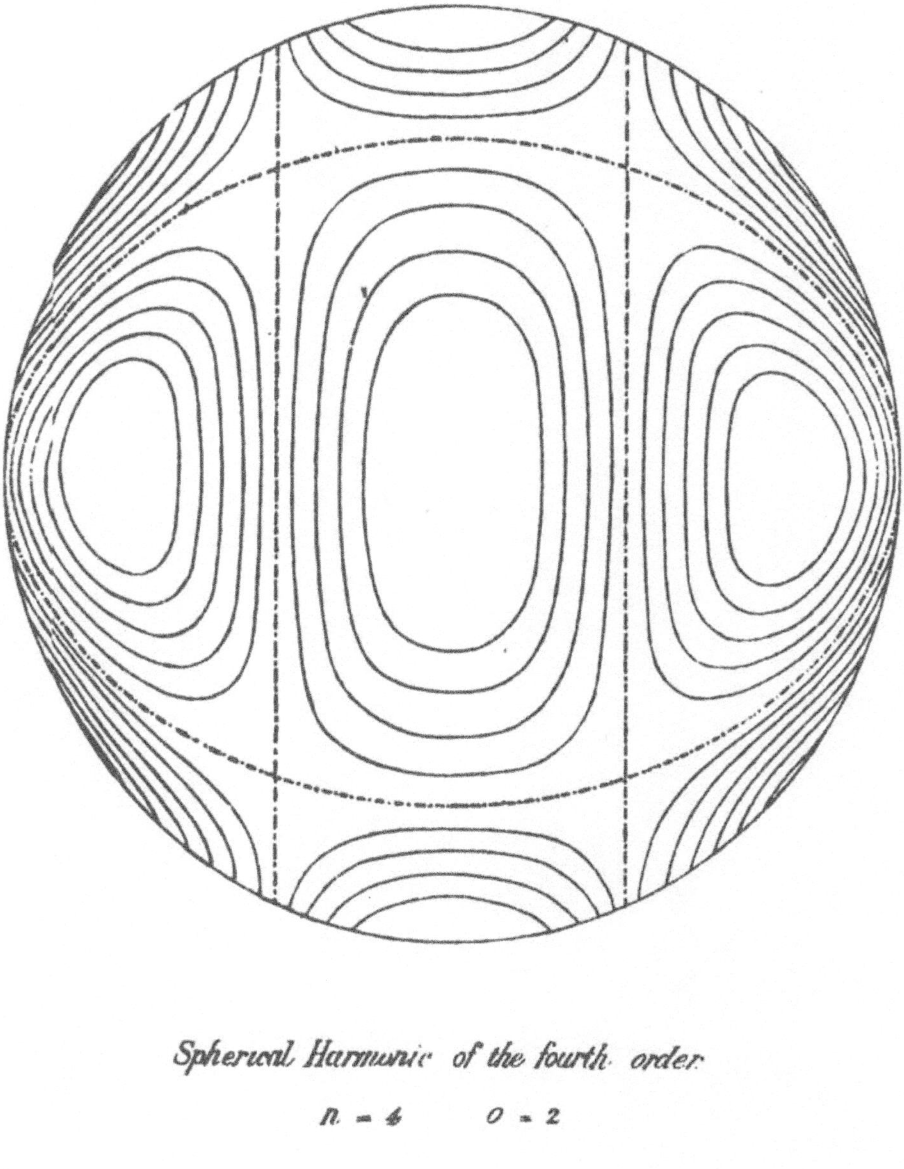

Spherical Harmonic of the fourth order.

$n = 4 \qquad \sigma = 2$

FIG IX.

Art. 143.

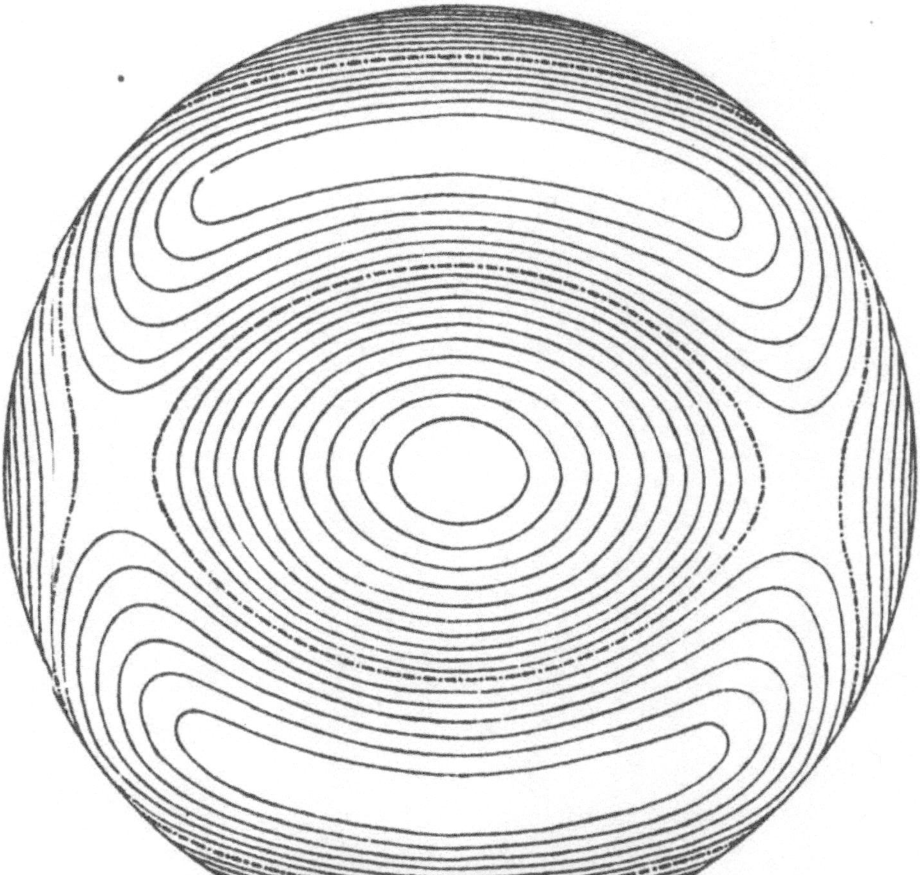

Spherical Harmonic of the fourth order.

FIG. X.

Art. 192.

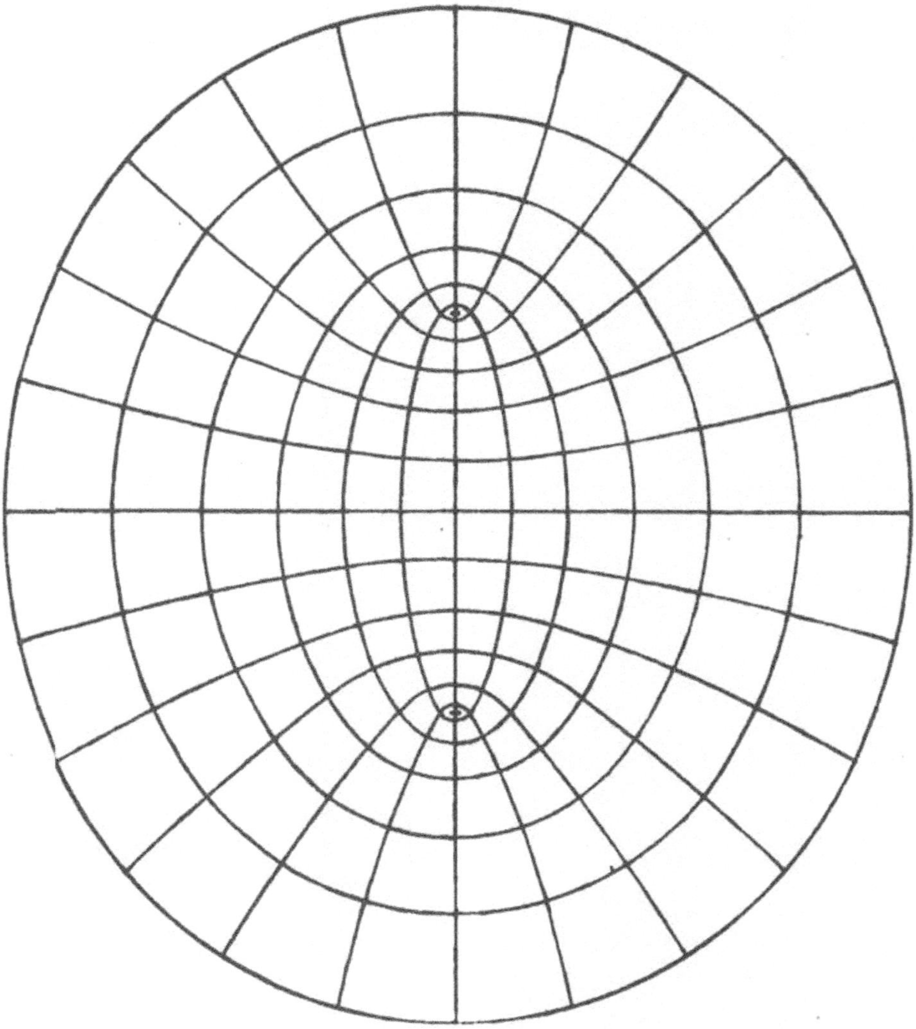

Confocal Ellipses and Hyperbolas

Fig. XI.

Art 193.

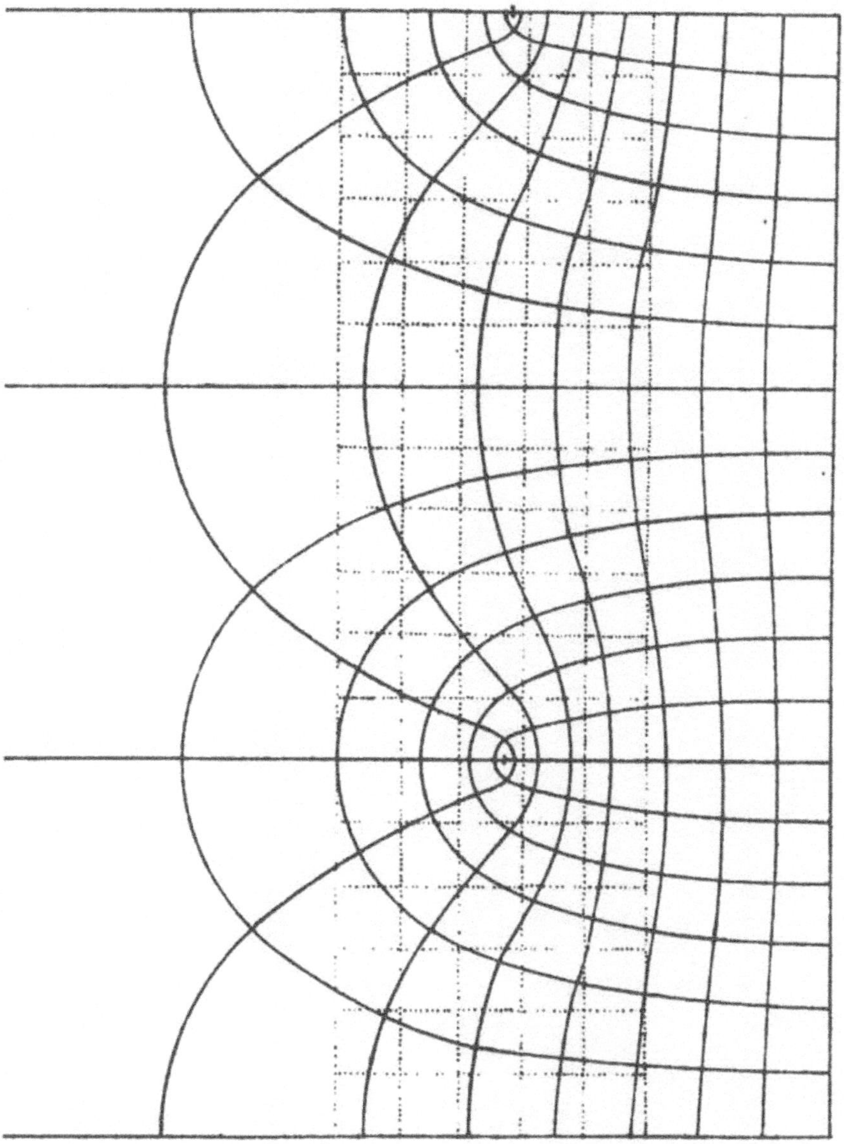

Lines of Force near the edge of a Plate.

Fig. XII.

Art. 202

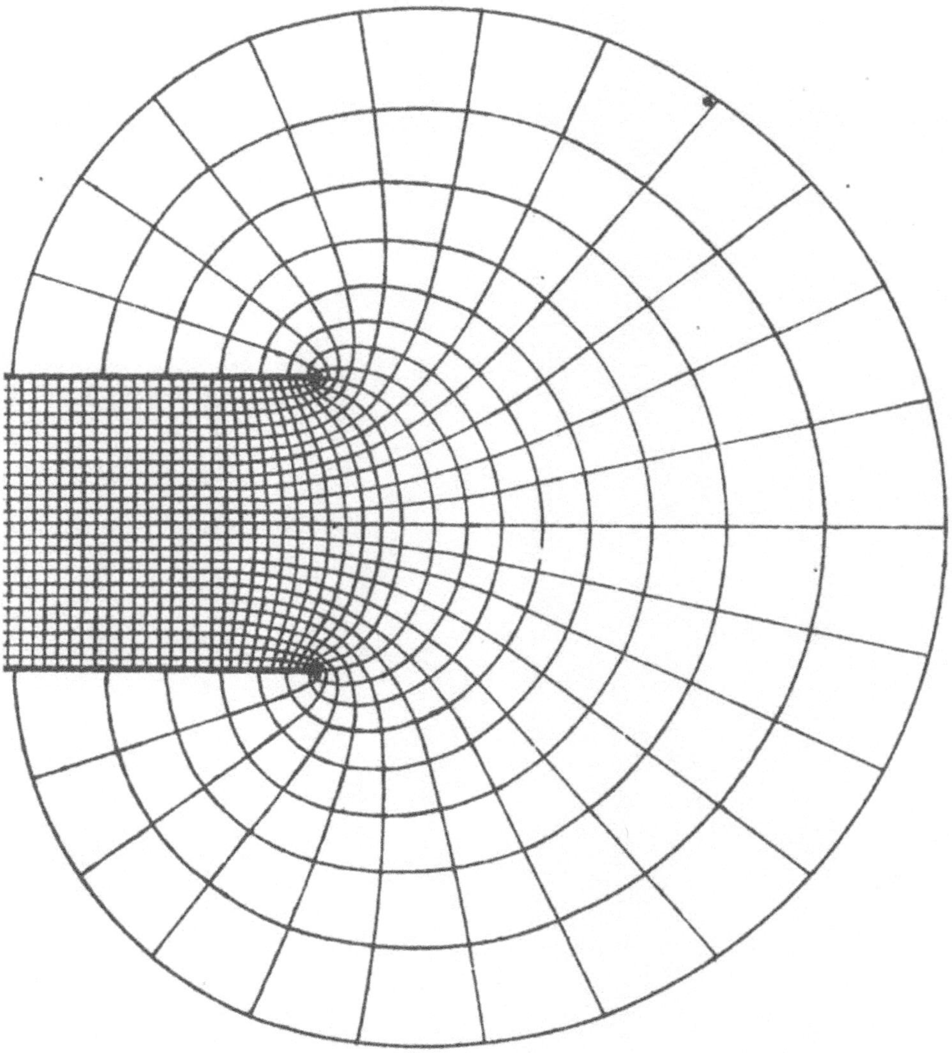

Lines of Force between two Plates

FIG. VIII.

Art 143

Spherical Harmonic of the fourth order

$n = 4 \qquad \sigma = 2$

FIG IX.

Art. 143.

Spherical Harmonic of the fourth order.

FIG X.

Art. 192.

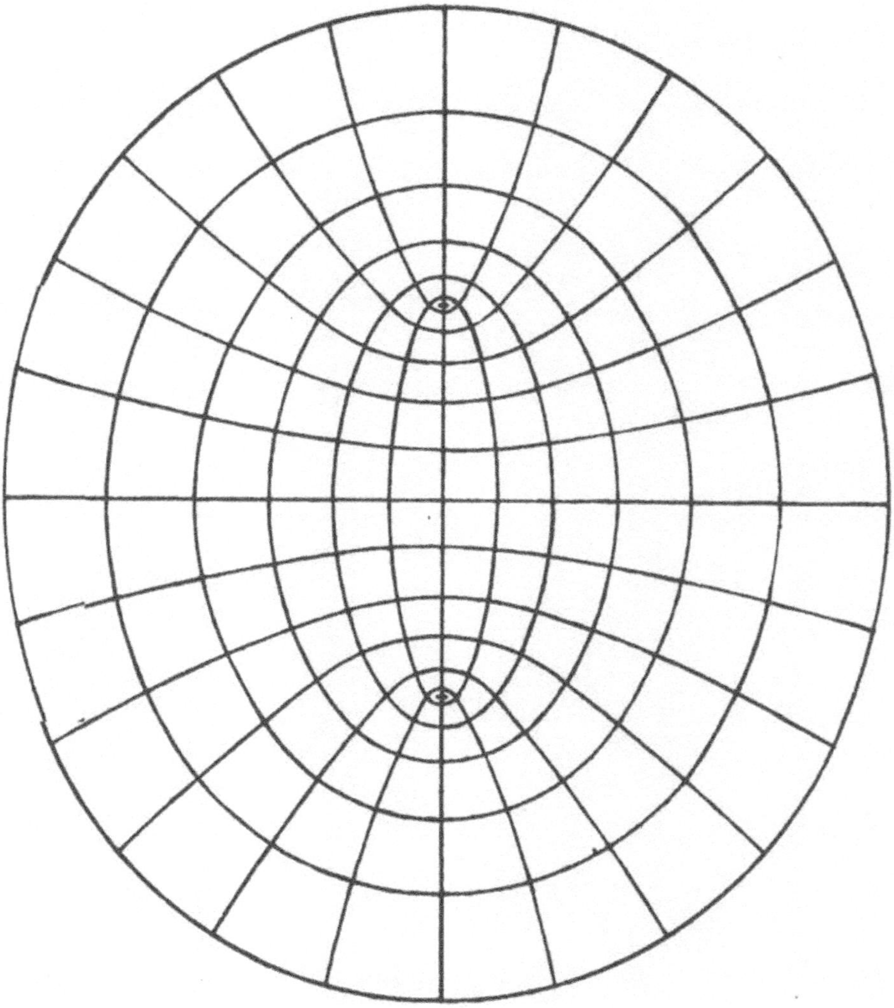

Confocal Ellipses and Hyperbolas

FIG. XI.

Art 193.

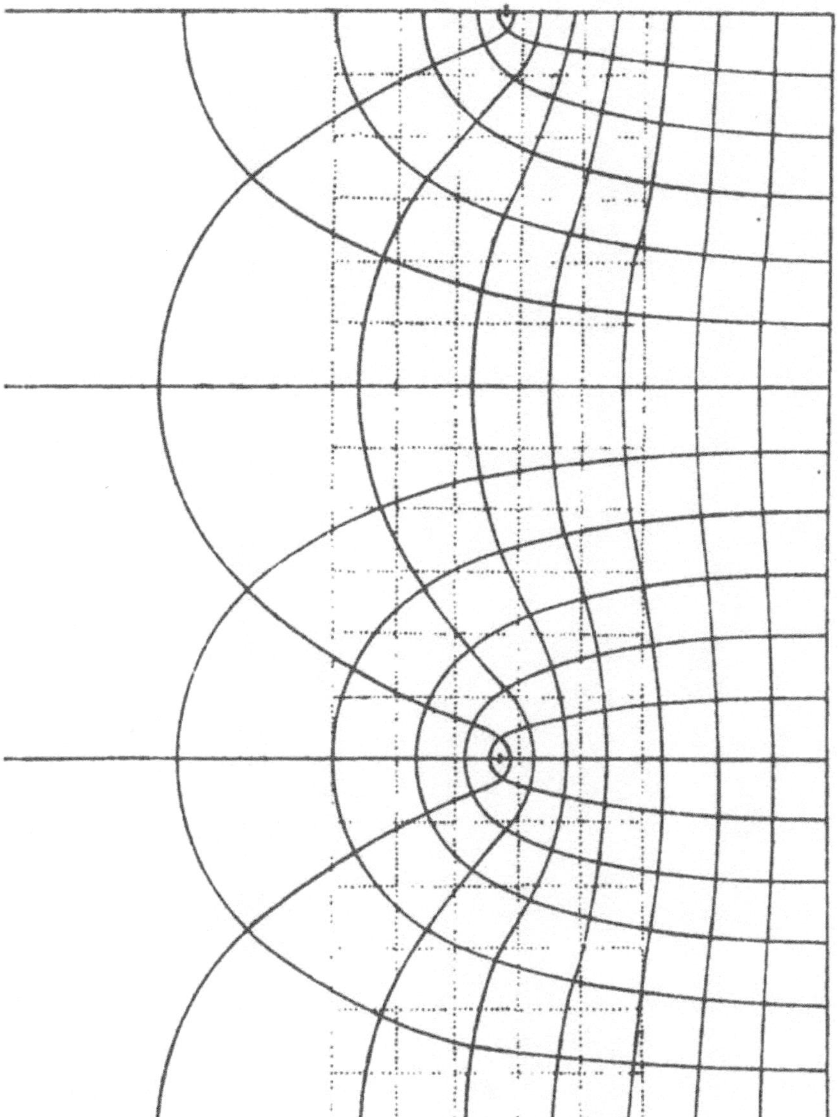

Lines of Force near the edge of a Plate.

Fig. XII.

Art. 202

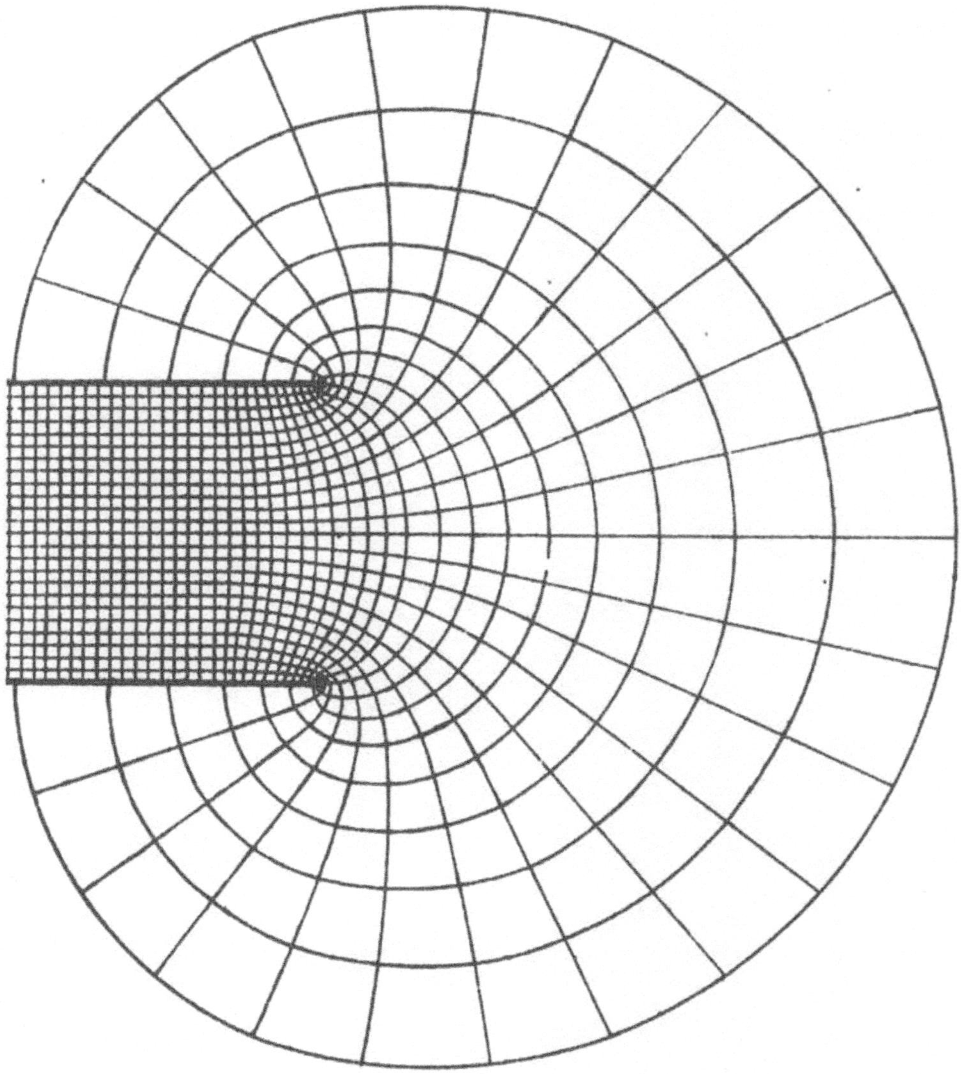

Lines of Force between two Plates

FIG. XIII.
Art. 203

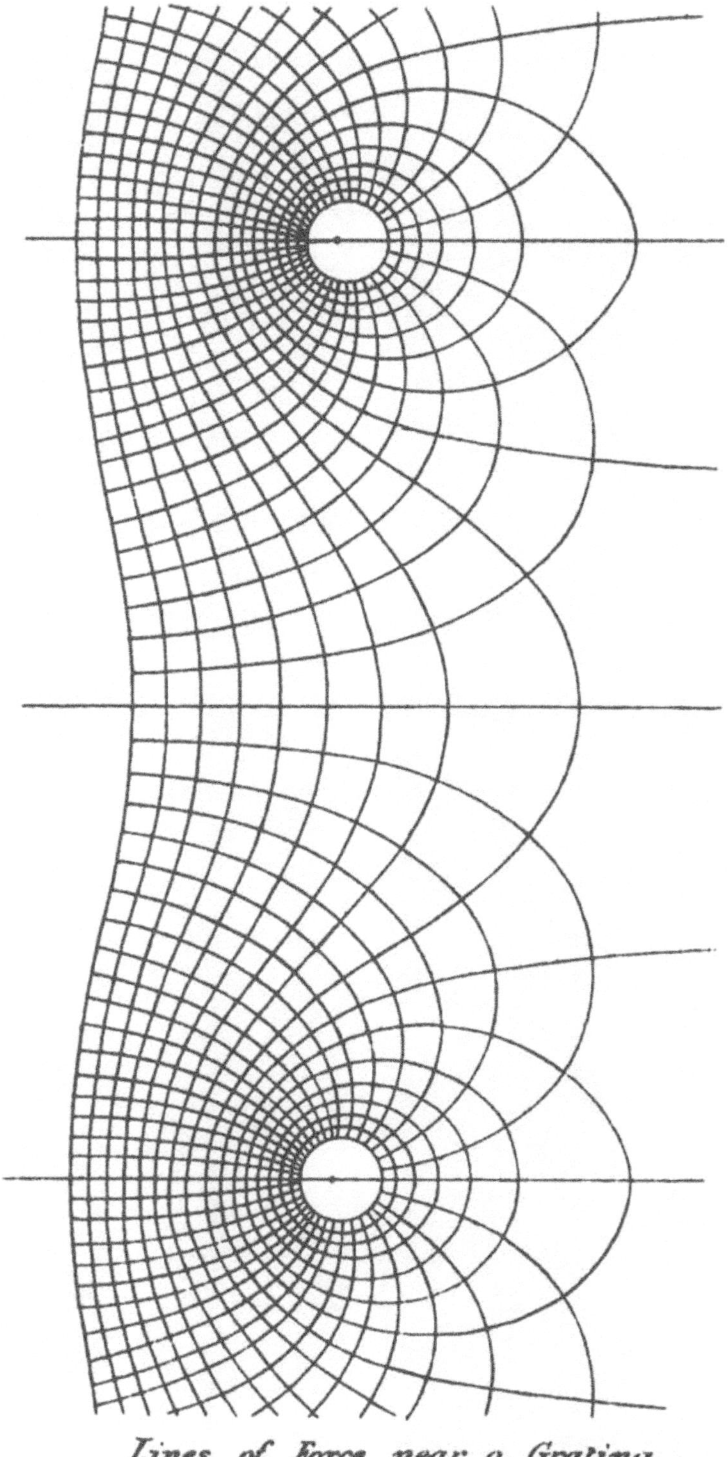

Lines of Force near a Grating.

A TREATISE ON ELECTRICITY AND MAGNETISM
IN 2 VOLUMES AND IN THREE EDITIONS

"He deserves to be remembered with affection wherever the name of James Clerk Maxwell is honoured or beloved."

- Lewis Campbell, August, 1882.

THE AUTHENTIC BIOGRAPHY OF JAMES CLERK MAXWELL

WRITTEN BY: LEWIS CAMPBELL AND WILLIAM GARNETT

"James Clerk Maxwell's biography has been written by 'Professors Lewis Campbell and William Garnett with so much skill and appreciation of their subject that nothing further remains to be told."

- Royal Naval College, Greenwich, August, 1890.

ISBN: 978-1773752341

To give an exact insight into the theory of Relativity to those readers who, from a general scientific and philosophical point of view.

NOTES

Made in United States
Cleveland, OH
05 November 2026

47670552R00297